PLANTS

Osteichthyes
(Bony Fish)

Angiosperms
(Flowering Plants)

Cycads and Kin

Filicinae (Ferns)

Ginkgoales
(Ginkgo and Kin)

Conifers
(Cone-Bearing Plants)

Lycopodiales (Club Mosses)

Bryophytes (Mosses and Liverworts)

Thallophytes (Algae and Fungi)

N

Chondrichthyes

Placoderms

Pteridosperms
(Seed Ferns)

Lepidodendrales
(Scale Trees)

Arthrophytes

Agnatha

Cordaites

TERTIARY

CRETACEOUS

JURASSIC

Psilophytes

CAMBRIAN

ORDOVICIAN

SILURIAN

DEVONIAN

MISSISSIPPIAN

PENNSYLVANIAN

PERMIAN

TRIASSIC

Essentials of Earth History

THIRD EDITION

Essentials

WILLIAM LEE STOKES

Professor of Geology, University of Utah

# of Earth History

**AN
INTRODUCTION
TO
HISTORICAL
GEOLOGY**

Prentice-Hall, Inc., Englewood Cliffs, New Jersey

To my parents, who taught me to observe nature

Illustrations by Joseph M. Sedacca, Juan Barberis, and Simon Siflinger. Typographical design by John J. Dunleavy.

Library of Congress Cataloging in Publication Data

Stokes, William Lee, 1915–
    Essentials of Earth History.

    Includes bibliographies.
    1. Historical geology. I. Title.
QE33.S76   1973   551.7   72-1749
ISBN 0-13-285932-7

10   9   8   7   6   5   4   3   2   1

Prentice-Hall International, Inc., London
Prentice-Hall of Australia, Pty. Ltd., Sydney
Prentice-Hall of Canada, Ltd., Toronto
Prentice-Hall of India Private Ltd., New Delhi
Prentice-Hall of Japan, Inc., Tokyo

PREFACE

Since the publication of the second edition of *Essentials of Earth History*, a vast amount of new and exciting information has practically revolutionized the field of historical geology. A unifying concept of the first order, one that seems to rival in its historical implications the pronouncement of Darwin's theory of organic evolution, has suddenly been revealed. The concept of global tectonics, which proposes a worldwide, dynamic, and ongoing interaction of vast plates of solid surficial rock material upon a mobile substratum, has now become a standard against which the whole body of observational and experimental geologic science may be tested. Geology will never be the same, even though there is not yet universal agreement concerning all the postulated implications of this new theory. Hundreds of relevant technical papers have been presented, but it is only fair to say that thorough and defensible analysis lags behind the massive influx of new data.

The task of putting together a geology textbook has become more demanding and challenging. A writer must be aware that his expanding subject is only one among many competing for the time and attention of the modern student. Brevity of treatment is necessary, and this edition represents a conscious effort to cover the subject of historical geology without unduly lengthening the text. The organization plan of the second edition has been retained. There are six chapters on materials and methods, ten on the historical development of the earth and its inhabitants, and four on the interactions of living things among themselves and with their physical environment. Chapter 8 deals with the core, mantle, crust, lithospheric plates, oceans, and atmosphere in a historical manner to emphasize that the physical world has had a progressive, ongoing development and is not merely a static backdrop for organic evolution. Chapter 17 contains a new and expanded treatment of organic evolution with stress on the contributions of paleontological science. An annotated and illustrated section on biological classification, which clarifies the relationships between fossil organisms and those still alive, has been appended to Chapter 5.

Other topics that have been deemed in need of improved or broader treatment include the nature of scientific inquiry as pursued in historical subjects, the early development of earth science, paleontology as it is practiced today, and paleoecology, the study of ancient organisms in relation to their environments. Needless to say, there have been numerous interpolations of up-to-date information derived from discoveries of the past half-decade, especially those having to do with continental drift and its implications. About a third of the book has been rewritten, and most of the illustrations are new or updated. The paleogeographic maps and cross sections have been redrawn, and the new two-color format has improved their clarity.

Many grateful acknowledgments are in order. The perceptive guidance of William H. Grimshaw, geology editor of Prentice-Hall, in matters of general organization has been invaluable. The meticulous and painstaking editing of Raymond Mullaney of Prentice-Hall's Project Planning Department has improved the manuscript in every way; if any errors of fact or interpretation persist, I must take full responsibility for them.

Norma Karlin of Prentice-Hall performed the tedious task of obtaining permissions, and Faye Patrick aided immensely in reorganizing and typing the manuscript at home base at the University of Utah. Special mention should be made of the Prentice-Hall artists whose work has greatly enhanced the appearance of this book, and so a word of thanks is owed to John J. Dunleavy, Rita Ginsburg, Ingeborg Schalkwyk, and Winifred Schneider.

WM. LEE STOKES

CONTENTS

19

EXTINCTION AND REPLACEMENT

20

SOME SELECTED FAMILY HISTORIES

Essentials of Earth History

1 THE HISTORY OF PREHISTORY

We often think of the past as being divided into historic and prehistoric periods. The historic period includes only the last few thousand years, that time during which man has kept written records of events that he felt were worth recording. The prehistoric period lasted at least 4.5 billion years. Although the term "prehistoric" seems to imply that there were no records until man learned to write, there is indeed an extensive and important account of prehistory. This account is not set down in words but in the materials of which the earth itself is composed.

The study of the development of our physical world and its life forms during the prehistoric period is called "historical geology" and is the subject matter of this book. Because historical geology includes both science and history, a discussion of the characteristics and methodology of science in general, and historical science in particular, will be useful here in clarifying the peculiar problems of the discipline of historical geology. After we have discussed some very basic concepts—the elements of a science, the relationship between science and truth, the nature

of, and distinction between, historical and non-historical science—the basic assertions of historical geology will be more easily grasped.

Science and Truth

James B. Conant, a noted scientist, has suggested that science is "an interconnected series of concepts and conceptual schemes that have developed as a result of experimentation and observation and are fruitful of further experimentation and observations." Science may also be said to be one kind of attempt at explaining the material universe (myths, for example, are another kind of attempt). This explanation must be based on observed phenomena and their orderly relationships and should be capable of being tested.

There are many other possible definitions, but in general they contain the following common elements: There is an awareness of a problem, a questioning attitude, and a desire for explanations; a body of information relevant to the question or problem is gathered and tested by experimentation and observation; answers derived from the information are either agreed upon or not by qualified persons, depending on the strength of the evidence.

Perhaps the clearest definition of science can be derived from a description of the scientific method, a series of six ideal operations often used by scientists to arrive at explanations of phenomena in the material universe. The procedure is as follows: (1) A problem is stated or recognized; (2) observations relevant to the problem are collected; (3) a possible solution to the problem, one that is consistent with the observations, is formulated; (4) predictions of other observable phenomena are deduced from, or various tests devised on the basis of, the proposed solution; (5) occurrences or nonoccurrences of the predicted phenomena are observed; (6) the proposed explanation is accepted, modified, or rejected in accordance with the degree of fulfillment of the prediction. Although this procedure may not always be followed strictly, it embodies the basic steps in any ideal scientific investigation.

How do preliminary explanations and proposed solutions come to be regarded as true? How do scientists obtain agreement as to the validity of their explanations? The answers to these questions have been the subject of many philosophical treatises whose complexity and profundity are due to the abstract nature of the thinking process, the nature of human perceptions, the gradational shades of certainty and uncertainty, and the diverse logical, mathematical, and legal systems that have been constructed to deal with proof and disproof. Here our discussion will be as simple and as brief as possible but will nevertheless present the basic elements of any scientific inquiry.

Scientific investigation *may begin* when a man experiences something. He may observe countless phenomena in the world around him; but the scientist not only observes but also questions. The layman may simply look at a sunrise and say "That's beautiful," whereas a scientist might view the same sunrise and then ask "How can I find out why the sun rises in the east?" The layman might eat fruit from an apple orchard; but a Newton will wonder how an apple falls from the branch to the ground.

Based on his various observations and in response to his questions, the scientist formulates a possible explanation for what he has seen. This preliminary explanation is a *hypothesis*. The hypothesis is an "educated guess" in that it has no firm evidence to support it but a lot of observations that make it plausible. The scientist now tries to substantiate his hypothesis by accumulating evidence. He makes further observations, gathers relevant facts, and uses controlled experimentation.

If all the evidence—observations, facts, and experiments—point to the validity of the proposed hypothesis, the scientist may then propose a *theory*. A good theory is one that can be tested in significant ways and that leads to the discovery of additional new facts. The theory should also predict results. One scientist observes countless apples and other objects falling to the ground and theorizes that this is caused by the great mass of the earth, which attracts the smaller mass of the other objects. Another scientist hundreds of years later uses this same theory to help put a satellite into orbit around the earth. Newton certainly did not dream of artificial satellites when he was observing apples, but the validity of his theory is proved by continual new applications and observations.

Figure 1.1 *The Great Sphinx and Pyramid of Giza. These monuments of ancient Egypt typify the beginnings of human civilization. We do not know exactly when the Great Pyramid was erected; it is estimated that it was begun about 2600 B.C. Although such ancient monuments symbolize the antiquity of human culture, they are but products of yesterday when considered in the perspective of geologic time. The Sphinx is carved partly from on-site rock, which was deposited, solidified, uplifted, and eroded during countless centuries before man gave it its present form. (Arab Information Center.)*

If a theory has been subjected to all possible tests and experiments and has predicted results with unvarying accuracy, it may be designated a natural law, a law that is based on some inherent quality of matter or energy. Unlike institutional laws, which are passed by men for the governing of society, natural laws are simply discovered or formulated. They have an existence of their own and are not affected by any actions of men.

A scientific law has been defined as a statement of invariable associations, of uniform connections between different phenomena—if event A happens, then event B will invariably also happen. This is essentially a cause-and-effect relationship. Must a scientist prove that this cause-and-effect relationship will remain stable in the future? A scientist need not be gifted with prophecy. He has no way of knowing whether the laws he proposes will remain valid in the future and so he has only to show that the asso-

ciation he posits has never been observed to vary—no one has ever witnessed event A without seeing event B as an effect. All the available evidence should indicate, however, that the law will *probably* be valid for all time and in all places.

New and more refined observations have shown that many of the so-called natural laws formulated by physical scientists in the eighteenth and nineteenth centuries are simply specific cases of more general laws or simply not laws at all. These "laws" are no longer valid in the form in which they were originally presented and have been revised or amended. Because of this, today's scientists tend to be quite cautious with regard to the formulation of new laws.

This then is how a scientist moves from observations to generalized laws. The process does not stop with the formulation or designation of a natural law. There is continual testing and experimentation. New observations are made and new applications are found. Though laws represent the ideal end of any scientific investigation, they are in a sense the beginning of new insights as knowledge of the material universe broadens and deepens.

Science is therefore a means and a method for arriving at rational explanations of the material universe. It is the unknown becoming known. It is a search for "truth."

Historical and Nonhistorical Science

Historical geology is a science in that it searches for truth by means of the scientific method. By definition, historical geology is also a historical science. Other sciences, such as physics and chemistry, are nonhistorical. What is the difference between these two types of science?

Nonhistorical sciences are not bound by the limitations of time and space. Events and phenomena with which these sciences deal are continually recurring in nature and many are of a scope small enough to permit laboratory experimentation. These experiments are repeatable at will, and hundreds of physicists or chemists can recreate and study the same process. They can arrange a set of causes and then study the results.

The historical sciences on the other hand are bound by temporal and spatial limitations. The events that historical scientists study happened once in the past and can never be exactly duplicated. These events and similar phenomena are

uncontrollable and variable and of so vast a scope that laboratory experimentation is limited (though not impossible). Because of this, the historical scientist must rely heavily on careful and precise observation and on intuitions and judgments derived from such observation. Unlike the physicist who moves from controlled and regulated causes to predictable results, the historical scientist moves backward from observed results to possible causes.

All this does not mean that there is no overlap between the methods of testing used by the two types of science. Nonhistorical scientists certainly employ observation, and historical scientists can at times use experimentation. What we have described here are the general characteristics.

The Task of the Historical Geologist

Now that we have discussed the basic characteristics and methodology of both the historical and nonhistorical sciences, we are ready to say more about the particular subject of this book, historical geology.

The Roman poet Vergil wrote, "Happy is he who has been able to learn the causes of things." Everyone tends to think in terms of cause and effect, and the historical geologist is no exception. He wants to know what caused all that he sees around him—the mountains and valleys, the lakes and rivers, and every other feature on or below the surface of the earth. He also wants to solve the complex problems of the origin and destiny of all the living things that inhabit the earth, including man himself. He desires to reconstruct and understand the causes of everything.

Before he can assign results to their proper causes, the historical geologist must determine the correct sequence of past events. It is not enough to know simply that something happened. The true significance of an event lies in its chronological relationship with other events, and a true history can be established only after the order of events is known.

Because the geologist studies prehistoric events, there are no written records to guide him in his reconstruction of the past. Where then does he find the evidence for these events? The geologist relies on the "testimony of things." Every mineral, rock, fossil, and physical feature of the earth is the result of the causes that he seeks to reconstruct. And so the geologist's evidence consists of all the materials of the earth in all their mutual interrelationships. Geologists must therefore be above all keen observers of the world around them. Whether they are in the field picking away at rocks with their hammers and other tools or in a laboratory studying small specimens with powerful and sensitive instruments, geologists must carefully analyze and evaluate all that they see.

Geologists pursue their craft everywhere—not only on land but also under the seas and oceans. Surface exposures are not enough. What lies

Figure 1.2   Universal interest in the life of past ages is typified by school children viewing the dinosaur skeletons in the Field Museum of Natural History in Chicago. Gorgosaurus, the great flesh-eating dinosaur, is shown towering over his prey, the duck-billed Lambeosaurus. (Field Museum of Natural History, Chicago.)

below, out of sight, is of equal importance. Geologists want to know what is found in every deep canyon, every cave, every hole drilled for oil, gas, or water, every mine tunnel, road cut, or canal. Geologically speaking the earth is a giant jigsaw puzzle, and the task of the geologist is to put the pieces of this puzzle together in order to reveal the history of the past.

Because of the great volume and extent of the evidence that they must study, geologists use maps and specimen collections to reduce the evidence to a scale that permits accurate observation. Mapping in the field is the only way that large areas such as counties, states, countries, and continents can be reduced to a comprehensible scale. Today, after a great deal of concentrated map-making effort, it is finally possible to compile a fairly comprehensive map of the entire earth and to begin to explain why it is the way it is.

Mapping has been accompanied by a prodigious amount of collecting and analysis of rocks, minerals, and fossils. Specimens are simply samples of the earth that can be brought indoors for study, comparison, and preservation. The search for these specimens has been worldwide and has been marked by international cooperation and sharing of data.

Evaluation of all this evidence has proceeded according to well-tested scientific methods. As a matter of course, all ideas and discoveries at all levels have been disclosed in various publications and have been subjected to the critical analysis of interested and qualified students.

The Assertions of Historical Geology

On the basis of all this evidence of the past history of the earth, geological science makes two main assertions: (1) The earth is extremely old and has a complex history, and (2) present life forms have evolved from preexisting species by means of slow, gradual changes over long periods

of time. When these assertions were first made, they met with a great deal of opposition because they seemed to contradict long-standing traditional beliefs and the Biblical explanation of the origin of the earth and man. But during the years that followed the initial uproar, evidence for the validity of these theories steadily mounted. The vast majority of men eventually realized that direct observation of nature, coupled with the methodology of scientific investigation, was superior to literal Scripture and tradition in explaining the nature of the earth and living things. Much of the evidence for the validity of these geologic assertions is discussed in the chapters that follow.

Geology as a Practical Science

Geology not only adds to the store of information concerning the past but also helps man in his search for the resources that have become economic necessities of civilization. All earth materials, including water, soils, minerals, fossil fuels, and building materials are subject to geological investigation. The study of these things appeals especially to those who prefer to work out-of-doors. Many who choose geology as a profession find employment in industries whose object is the discovery and eventual extraction of raw materials.

One particular field, the oil industry, finds geology a particularly essential tool. Here all the fundamental ideas and principles of the science are put to a test. The geologist does not invariably succeed in finding oil pools, but he knows where they cannot exist and where they are most likely to be found, thus making the search much less expensive.

Geology is also useful in connection with engineering projects such as the building of highways and dams, the digging of foundations and tunnels, and the dredging of harbors and canals. These activities are usually accomplished with greater economy and safety when geologic infor-

(a)

(b)   (c)   (d)

Figure 1.3   Painstaking and detailed laboratory analysis and the accurate preparation of maps and reports go hand in hand with outdoor exploration. Here geologists are shown (a) compiling a special report from maps, photographs, and other references, (b) preparing a geologic map, (c) examining ore-bearing rock in the laboratory, and (d) processing seismic data on sensitive instruments. (a and c, U.S. Geological Survey; b, U.S. Bureau of Reclamation; d, Geophysical Service, Inc.)

(a)

(b)

(c)

Figure 1.4 *A great deal of the geologist's time is spent in the field studying formations and samples in their natural surroundings. In this series of photographs, geologists are shown (a) studying submarine rock exposures, (b) measuring the velocity and volume of stream flow, (c) conducting a topographical survey in the western United States, and (d) exploring the frozen Antarctic. (a, official photograph, U.S. Navy; b and d, U.S. Geological Survey; c, U.S. Bureau of Reclamation.)*

(d)

mation about the underlying soil and rock has been gathered before operations commence.

Not to be overlooked in this age of travel is the contribution geologic study makes to our understanding of the natural features of the landscape. To know what has happened or is happening to produce present-day scenery adds to a person's appreciation of his natural surroundings. Perhaps in the long run this contribution to an understanding of nature is fully as important as the discovery of oil fields and the tracing of mineral deposits.

FOR ADDITIONAL READING

Adams, F. D., *The Birth and Development of the Geological Sciences.* New York: Dover Publications, 1954.

Albritton, C. C., Jr., ed., *The Fabric of Geology.* San Francisco: Freeman, Cooper, 1963.

Hempel, C., *Philosophy of Natural Science.* Englewood Cliffs, N.J.: Prentice-Hall, 1966.

Reichenbach, Hans, *The Rise of Scientific Philosophy.* Berkeley: University of California Press, 1966.

Toulmin, Stephen, and June Goodfield, *The Discovery of Time.* New York: Harper & Row, 1965.

Walker, M., *The Nature of Scientific Thought.* Englewood Cliffs, N.J.: Prentice-Hall, 1963.

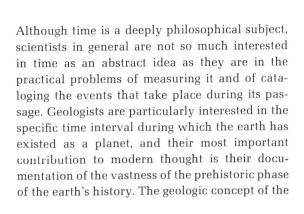

TIME AND ITS MEASUREMENT

Although time is a deeply philosophical subject, scientists in general are not so much interested in time as an abstract idea as they are in the practical problems of measuring it and of cataloging the events that take place during its passage. Geologists are particularly interested in the specific time interval during which the earth has existed as a planet, and their most important contribution to modern thought is their documentation of the vastness of the prehistoric phase of the earth's history. The geologic concept of the

Symbolic of man's attempt to measure and understand time is ancient and mysterious Stonehenge on Salisbury Plain, southern England. (British Information Service.)

immensity of time must rank alongside other modern discoveries regarding space, matter, and energy.

Geologists do not seek only to measure the duration of past time periods; they want to understand, as far as possible, the events that have brought the earth to its present condition. As an essential step toward their goal, they are working to construct an accurate calendar of past events by assigning dates or ages to rock formations and their fossil contents. In this chapter, we shall be concerned primarily with the various means that geologists have employed in attempting to determine the age of many kinds of specimens.

Absolute and Relative Dating

Ideally, any historical event should be identified by a specific date that relates it to the present. Written records and calendar systems enable historians to assign *positive,* or *absolute, dates* to

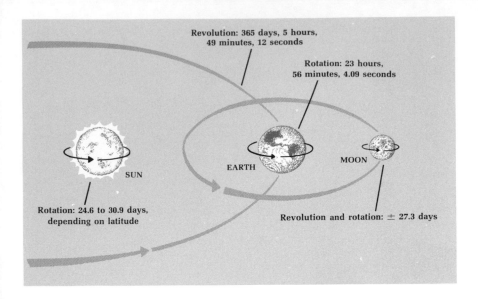

Rotation: 23 hours,
56 minutes, 4.09 seconds

Revolution: 365 days, 5 hours,
49 minutes, 12 seconds

SUN

EARTH

MOON

Rotation: 24.6 to 30.9 days,
depending on latitude

Revolution and rotation: ± 27.3 days

Figure 2.1 Astronomical "clocks," some more reliable than others. Notice the times of revolution and rotation. This diagram is not intended to portray true relative sizes of the earth, moon, or sun, or their orbits.

many events in the history of man. It is even possible to assign absolute dates to many prehistoric events by means of a number of techniques that will be described later in this chapter. This means that some ancient rocks, minerals, fossils, and geologic features can be identified as having originated a specific number of years ago.

If absolute dating methods are not possible, the geologist can still arrange events in their proper order of occurrence so that an event or object can be identified as being relatively younger or older than other events or objects. This is called *relative dating* or *placement.* Practically all geologic specimens are dated in a relative way by applying the *law of superposition,* which we shall discuss in Chapter 4. In general, fossil-bearing rocks usually can be dated by relative means, whereas nonfossiliferous rocks are more easily dated by absolute methods. The two methods tend to check and countercheck each other.

Measures of Time

Time is measured by recurrent regular events that can be counted by man. The beat of a heart, the swing of a pendulum, the cycle of vibration of an atom, the rotation of the earth, the revolution of the moon around the earth and of the earth around the sun are measures of time. Of these, the revolution and rotation of the earth provide two very important and useful measures, the *year* and the *day* respectively. Crude observations of the changing seasons and of the positions of the sun and stars were succeeded by more accurate scientific measurements, and we now know that the length of the year is 365 days, 5 hours, 49 minutes, and 12 seconds, plus or minus a few seconds. Weeks, hours, minutes, and seconds are purely artificial, earth-bound, fractional time measures invented by man.

Strangely, the earth's period of revolution appears to be more reliably determinable than the period of rotation. The position of the earth in space can be checked by reference to other astronomical bodies, and the period of revolution has been found to be invariable during the period of instrumental observation. On the other hand, studies have shown that the length of the day varies in a measurable amount because of unpredictable changes that occur in the earth's speed of rotation, which slows down or speeds up rather quickly. Disturbances in the liquid portions of the earth's interior and the shifting of air masses on the surface are thought to be responsible for the changes in speed.

In addition to the more or less erratic changes that affect the earth's rotation, there is substantial evidence that the day is gradually getting longer. This lengthening is due to the gravitational pull of the moon on the earth's tides, which are steadily slowing the rate of rotation. About 1.5 billion years ago the length of the day would have been about 11 hours, and 400 million years ago it would have been 20 hours. In addition to the good theoretical reasons for suspecting this variation, there is fossil evidence that the year had about 400 days in the Devonian Period about 400 million years ago. This is arrived at by counting minute "growth rings" in fossil corals; each ring supposedly represents a day's growth. The space of a year can be determined by noting seasonal variations in the size of the growth rings.

The second has been defined as 1/31,556,925.9 of the tropical year 1900. But this seemingly very precise measurement is not quite accurate. Geologists have for a long time suspected that the time it takes the earth to rotate on its axis has

been getting longer, and recently astronomers have shown that the earth is indeed slowing down. Although the loss is only about 1 second per year, this cumulative loss does introduce troublesome errors in any attempt at precise timekeeping. And so, by international agreement, the second was redefined in 1967 as 9,192,631,770 cycles of vibration of the cesium atom. By using this new device, scientists have been able to limit errors to 1 in 10 billion, thus restricting variation to the loss or gain of 1 second over a 300-year span. In order to keep the atomic and astronomic clocks in step, it will be necessary to make small periodic adjustments of about 1 second per year.

YEARS AND SEASONS

Has the earth always made an annual journey around the sun? So far as we know, the earth has revolved around the sun ever since it took form as a planet. Changes in the seasons give us strong evidence for this fact. Because the axis of the earth is inclined to the plane along which it travels, there are characteristic changes in light, temperature, and precipitation that mark the various seasons. These changes exercise a pro-

Figure 2.2 (a) Shell of a modern pelecypod (Pecten diegensis) and (b) a diagram showing a particular growth period. The shell measures about 4 centimeters across. Note that each growth line represents one day's growth. (Courtesy of George R. Clark II.)

(a)

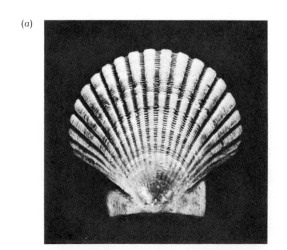

(b)

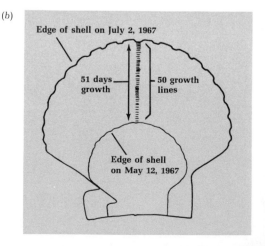

found influence on the food supply and growth pattern of plants and animals and, locally at least, on the erosion, transportation, and deposition of sediments. The effects of seasonal changes are recorded in living and nonliving materials in many ways and have become permanently recorded in the earth's crust.

GROWTH RINGS AND VARVES

Growth Rings in Plants. The best-known seasonal records that are preserved in living organisms are *tree rings*. The width and spacing of the rings depend on temperature, light, and moisture variations that are largely of a seasonal nature. Each ring consists of two parts: the so-called "summer wood," which has small cells and thick walls, and the "spring wood," which has larger cells and thinner walls. The study of tree rings, called *dendrochronology*, has helped archaeologists to date various archaeological sites, espe-

cially in the arid portions of the American Southwest, where wood is easily preserved. A continuous chronology going back to 1550 B.C. has been pieced together, and many important ruins have been dated by pieces of structural wood.

An even longer sequence of rings can be found in the trunks of some sequoia trees in California. Studies of the annual rings prove that some of these trees are over 3,000 years old. The sequoia trees have long been regarded as the oldest living things on earth, but this claim has recently been challenged by a less spectacular tree, the bristlecone pine, which is found in various drier parts of California, Nevada, and Utah. One tree of this species so far dated is more than 4,800 years old.

Growth Rings in Animals. Water-living animals, like plants, respond to seasonal variations in temperature and food supply. Seasonal effects are most evident in lakes in temperate regions where there are extreme seasonal temperature variations. Freshwater clams usually show annual rings that are much like tree rings. A dark, narrow zone marks the colder season when little or no new material is added to the shell; a wider

Figure 2.3   Growth rings in a highly magnified section of the wood of the Douglas fir. The cells of the summer wood are small with thick walls. Spring wood has large cells with thinner walls. (General Biological Supply House, Inc., Chicago.)

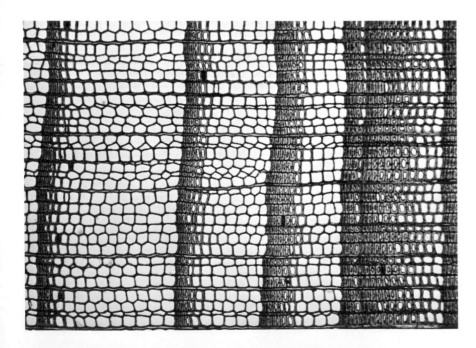

Figure 2.4 *Bristlecone pines (Pinus aristata) growing in the Inyo National Forest, California. Portions of a tree may continue to grow after other parts have died. Specimens from this area are among the oldest known living things. (U.S. Department of Agriculture, photo by Leland J. Prater.)*

band of lighter color indicates the season when food is abundant. The seasonal nature of these variations has been proved by cutting small notches in the shells of living clams and noting how and when new shell material is added. Comparisons between fossil specimens and their modern counterparts show that seasonal effects were operating on water-living animals far back in time. Recent studies of the internal skeletons of extinct types of squidlike mollusks show that the deposits of winter and summer can be distinguished on the basis of differences in the chemical makeup of the layers. Among vertebrates, the best-known evidences of annual growth are found in fish scales. The ages of members of certain species can be determined from studying the minute rings, or *annuli,* that form in the scales. Fossil scales also show similar rings. Pre-

Figure 2.5 *Freshwater clam (Legumia recta) from Lake Michigan, showing well-marked annual growth rings. The shell is 8.9 centimeters long.*

Figure 2.6 *Scale of the Alaska herring, showing growth rings, or annuli. The scale shows that the fish was in its seventh year of growth.*

liminary research has revealed rings in the bones of higher vertebrates, but the seasonal nature of these rings has not yet been established.

Varves—the Sedimentary Record of Seasonal Change. The effects of seasonal change are not confined to living things. A great deal of the earth is exposed to recurrent variations in precipitation during wet and dry seasons. Under favorable conditions these variations are reflected in the erosion and deposition of sediment. Ideally, there should be an interval of little or no deposition followed by one of rapid deposition, corresponding to seasons of low and high stream flow.

Any deposit that reflects a yearly cycle is called a *varve.* The most clearly marked and easily interpreted varves are associated with glacial activity. In and near most ice fields, the seasons of melting and freezing are sharply marked, and there are abundant lakes and ponds in which deposits may be preserved. During the warm season, when snow and ice is melting, a large quantity of sediment is deposited in bodies of water, and a relatively thick layer of coarse sediment is laid down. With the onset of winter, when water ceases to flow and the ponds and lakes are covered by ice, deposition slows down. During this quiet period, very fine clay particles and some dead organic matter slowly settle to the pond and lake bottoms, forming a thinner layer of finer and usually darker material. As a result of these conditions, each varve consists of two gradational parts. Observations of varves in the

Figure 2.7 *Varved sediments. The large cylindrical core is from the Paradox Formation in southern Utah. The dark-colored varves are salt, whereas the thin, light-colored ones are dolomite. The thick varves measure about 2.54 centimeters. The small core at the right is from the Green River Formation in eastern Utah. In this specimen the individual varves are only about 0.01 centimeter thick. The bands visible in the photograph are aggregations of many varves. The rough specimen at the left is from the Castile Formation in Texas. The varves are chiefly pure and impure gypsum. The varve-based estimate of the total period of deposition in the Castile is 300,000 years.*

process of formation and examination of ancient deposits reveal that glacial varves are usually rather thin, ranging from a few millimeters to several centimeters in thickness. The record of hundreds of years may readily accumulate in one lake or pond.

Varves, like tree rings, may be thick or thin, depending on the length and relative warmth of the seasons. As a matter of fact, the sequence of thick and thin varves permits geologists to correlate one set of varves with another, just as one tree may be correlated with another on the basis of growth rings. The most complete sequences of varves are created in lakes that are formed as large continental glaciers melt away. The oldest varve accumulation lies near the point of maximum glacial extent, and the latest may still be forming at the glacier's edge. Between the oldest and newest varves lie lake beds with overlapping varve histories.

The record of retreat of the last ice sheets in Europe and America has been traced by means of varves. Unfortunately, there are gaps in the records that can be filled in only by estimates. In Scandinavia a connection with the present has been established and the varve chronology, with certain portions estimated, extends from the present to about 10,000 to 15,000 years ago. In North America the varve chronology, again with some estimated portions, reaches back some 20,000 years. In view of the uncertainties and imperfections of the method, it is impossible to say exactly how far back continuous varve records go. Evidence seems adequate, however, to carry the count back from the present for at least 15,000 years.

Glacial varves have been found in very old deposits laid down during earlier ice ages. The antiquity of these records is indicated by the fact that the rocks containing them have been deeply buried and then hardened by heat or pressure. Undoubtedly, these older varves were formed under conditions not significantly different from conditions prevailing today in glacial areas.

Many banded deposits that are thought to

Figure 2.8 Varves formed of alternating layers of silt and clay. This site is near the mouth of Sherman Creek, Ferry County, Washington. (U.S. Geological Survey.)

Figure 2.9 *Varved sediments from the Green River Formation, Uinta Basin, Utah. The photo shows two rock cores taken from wells 29 kilometers apart. The matching of the two specimens layer by layer is strong evidence for the quiet conditions and slow deposition that prevailed in this area during the formation of these varves about 50 million years ago. (Courtesy of H. D. Curry.)*

record yearly events are found in nonglacial sediments. Lake deposits, shale formations of marine origin, and salt beds commonly contain varves or varvelike beds that probably have resulted from seasonal changes in the sediment supply. Several dozen deposits or formations of this type, each containing hundreds or even thousands of varves, have been discovered. Cores taken from

one lake deposit of Eocene age in the western United States have 71 varves per centimeter. The total thickness of the deposit indicates that the lake probably existed for about 12 million years. Needless to say, this set of varves does not connect with the present.

The evidence provided by more or less regular growth rings in fossil wood and shells and by varved sediments from all geologic ages seems to indicate that seasonal changes have been affecting life forms and the surface of the earth for a very long time.

How Old Is the Earth?

Granting that the earth has been making a yearly circuit of the sun for a very long time, we are led to ask just how many such journeys it may have made. In other words, how old is the earth? Until fairly recent times, the origin and age of the earth were not considered to be subjects for serious inquiry. Hebrew Scripture, basis of the Christian faith of the Western world, was considered to be the final and sufficient word on the subject. In 1654, Archbishop James Ussher concluded from Scriptural analysis that the earth had been created in 4004 B.C. This date was printed as an explanatory note in several editions of the Bible and was generally accepted by most Christians. A few years later a learned Biblical scholar, Dr. John Lightfoot of Cambridge, felt that he could be even more specific and wrote that "heaven and earth, center and circumference, were made in the same instance of time, and clouds full of water, and man was created by the Trinity on the 26th of October 4004 B.C. at 9 o'clock in the morning." The idea of a 6,000-year-old earth was entirely satisfactory as long as there were no reasons for believing otherwise. It is interesting to note that ancient Hindu thinkers had placed the age of the earth at almost 2 billion years.

As the spirit of scientific inquiry began to assert itself, the age of the earth became a subject for serious consideration. Facts were few, and meaningful observations were just beginning. Some men felt that every natural feature of the landscape gave evidence of great antiquity. The cutting of valleys, the advance and retreat of glaciers, the destruction of coasts by erosion and their restoration by deposition, all seemed to take place over very long periods of time. But quantitative data were needed, and in the eighteenth and nineteenth centuries a few preliminary attempts were made at actual measurement and evaluation of certain properties of the earth in order to determine its age.

Among the natural phenomena that seemed to offer clues in the search were (1) the saltiness of the ocean, (2) the internal heat of the earth, and (3) the rate of deposition of sediments.

It was assumed in the first case that the original ocean was fresh and that salt had been added at approximately the current rate ever since rivers commenced to run. If we divide the amount of sodium now in the ocean by the amount brought in annually, we theoretically could determine the age of the ocean. In 1899 an Irish physicist, John Joly, announced that by using this method he had calculated that the age of the oceans was 99.4 million years.

The second method carried with it the assumption that the earth must have cooled from an originally molten condition. Because the rate of cooling and present temperature can be approximated, the entire period of cooling may be calculated. This gives a period of 20 to 40 million years. As an incidental argument it was contended that no source of heat then known could have supported the sun's output for much longer than 20 million years and that the earth could not possibly be older than the sun.

Finally, with regard to deposition of sediments, it was reasoned that, if we determine how many meters of sediment have been laid down and how long it takes a meter to accumulate under average conditions, we may by simple division arrive at an estimate of how long erosion and deposition have been going on. Latest figures show that the cumulative maximum thickness of rock laid down since abundant fossils appeared is at least 137,195 meters. Although rates of deposition vary from time to time and place to place, an average of 0.305 meter, in 1,000 years may not be far wrong. At this rate, the fossil-bearing sedimentary rocks would take over 450 million years to accumulate.

Although these earlier hypotheses were well conceived and the supporting calculations were mathematically correct, they involved so many unknowns and gave such varied results that no one now has much confidence in any of them. It is likely that the seas have always been about as salty as they now are, and we know from the presence of thick salt beds that much salt has been returned to the lands from the seas. It is also known that the earth contains its own heat-producing radioactive elements, which would totally confuse any calculations based on gradual cooling from an original molten state. The heat of the sun is now known to be provided by nuclear reactions and not by ordinary combustion as once supposed. The rates of sediment formation range from thousands of years for 1 meter of limy ooze to several hours for a similar amount of river-laid sand, so that it seems impossible to arrive at reliable rates of deposition. Though these estimates failed to provide a reliable figure for the age of the earth, they did indicate that Ussher's contention that the earth was 6,000 years old was far off the mark.

Radioactivity

Near the close of the nineteenth century, when there seemed to be little hope of finding out the secret of the earth's age, a discovery was made in another field that was to shed unexpected light on the subject. In 1896, Antoine Becquerel, a French physicist, announced that the constituents of certain natural ores are capable of spontaneously generating a mysterious form of energy that affected photographic plates, even through

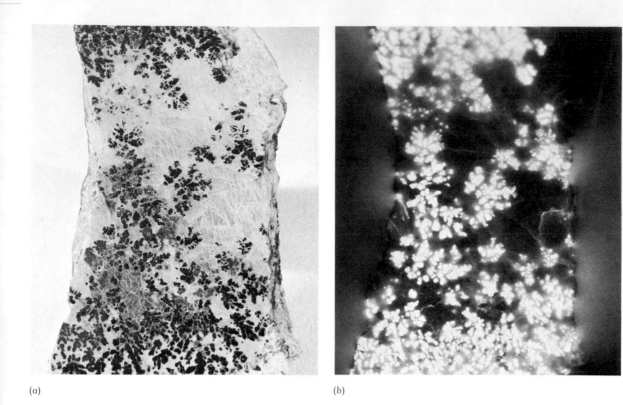

(a)

(b)

Figure 2.10 A radioactive specimen takes its own photograph. The piece of ore (a) contains dark-colored uranium minerals in a matrix of nonradio-active material. Placed against a photographic film (b), the radioactive areas affect the silver compounds in the same manner as light. (Wards Natural Science Establishment, Inc.)

materials that were opaque to light. The Curies, Marie and Pierre, isolated the energy-producing elements from uranium ores, and in 1898 they announced the discovery of *radium*. Soon physicists were demonstrating that the property of radioactivity involves a spontaneous breakup of the atomic nuclei of certain unstable substances.

Through radioactive disintegration, one element is transformed into another. During the process, three types of products are emitted: *alpha (α) rays, beta (β) rays,* and *gamma (γ) rays.* The α rays are nuclei of helium atoms, the β rays are electrons, and the γ rays are electromagnetic radiations, like light, but of very short wave-

length. The detection of these products by suitable instruments, such as the well-known Geiger counter, indicates the presence of radioactive ores in nature. The effects of radioactivity on living things and its connection with atomic fission are subjects too broad to be covered here, but their implications for geologic age determination are very significant.

ISOTOPIC DATING

All radioactive isotopes (similar forms of the same element but with different mass numbers) are subject to disintegration from the moment they come into existence. A specific atom may disintegrate immediately or it may remain intact for millions of years. The spontaneous behavior of the individual atom is unpredictable and is not governed by any known law. Nevertheless, aggregations of atoms, like aggregations of people, are subject to mathematical rules. There is no way we can predict the death of a specific person, but we can forecast quite accurately what proportion of a group will die yearly and how many will be alive at any given time. Because the potential life of a radioactive atom may be infinitely long, the total period of activity of a large group of atoms is impossible to predict. It is easier and more meaningful to measure the time interval in which half of the atoms of a large group or specimen have disintegrated. This interval is called the *half-life,* a term widely used in nuclear studies. If half of a certain population has disintegrated

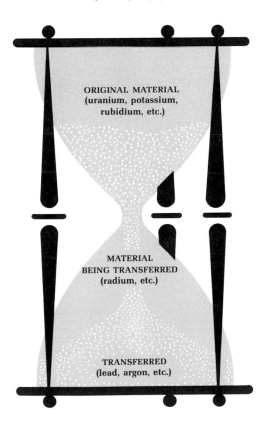

Figure 2.11 *Radioactive disintegration operates like an hourglass. In both cases a given amount of original, or parent, material is transferred or changed so that it can be measured separately. The difference lies in the rate of transfer. (After Howell, 1959.)*

ORIGINAL MATERIAL
(uranium, potassium, rubidium, etc.)

MATERIAL
BEING TRANSFERRED
(radium, etc.)

TRANSFERRED
(lead, argon, etc.)

in a million years, then half of those remaining will disintegrate in the next million years, and so on. The process is infinite. At first the decline in abundance of a radioactive material is rapid, but later on, as it approaches its end stages, the decline is progressively slower.

The half-life existence of different radioactive isotopes ranges from a fraction of a second to billions of years. Many kinds of atoms with short half-lives that were probably once common in the universe have declined to the vanishing point. Others with longer periods have traveled but a fraction of the way to extinction. In recent years, scientists have created artificially many types of radioactive substances that in general have relatively short half-lives. Also not to be overlooked are radioactive substances that are produced in the atmosphere and on the earth by powerful cosmic rays issuing from space.

Several fundamental characteristics of radioactivity are of interest to geologists. In the first place, the rate of radioactive disintegration is not significantly influenced by external terrestrial conditions. Specimens subject to extremes of temperature and pressure show no variation in activity, and we assume that disintegration has always proceeded at the same rate no matter what conditions within the earth's crust may have been. The second important property of radioactivity is its ability to produce heat energy within the earth, energy that is capable of causing earth movements, volcanic eruptions, and other manifestations of internal change.

The dating of radioactive materials depends on the correct determination of the ratios of original, or parent, materials to derived, or daughter, products. We can make a rather crude comparison between this principle and the operation of the old-fashioned hourglass, which "told time" by the passage of sand from one compartment to another.

If the *rate of transfer* (decay or disintegration) of the material is known, the ratio between the amount of material already transferred and that not transferred will tell how long the "clock" has

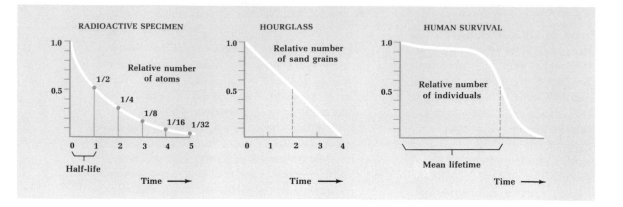

| RADIOACTIVE SPECIMEN | HOURGLASS | HUMAN SURVIVAL |

Figure 2.12 Radioactivity, hourglass, and human survival compared. Each curve indicates a decrease over time, but the rates are different. Half of the radioactive material has disintegrated quite early in the total period of existence of the specimen; the remainder disintegrates at the same rate, diminishing by half during each half-life period until it approaches the zero point. In an hourglass the sand runs steadily; half the sand has been transferred to the bottom of the glass when the time is half gone, and at the end of the specified time all the sand is gone. Half of the persons in a given segment of the human population are dead when the mean lifetime has been reached; the decline is slow at first, rapid later.

been operating. Obviously, the sample must have remained in a closed system with nothing added or subtracted since it began to operate. A cracked hourglass that allows sand grains to escape will not reliably tell time. In nature, the escape from, or addition to, the chemical systems poses the most serious problem: for example, some radioactive products are gaseous and may be driven off by heat and pressure through minute cracks or pores in rock. Geologists take all precautions to see that their specimens are as unweathered and free of fractures as possible.

The *radioactive transformations* that are most valuable in determining geologic age are uranium to lead, thorium to lead, rubidium to strontium, potassium to argon, and carbon to nitrogen.

In the next few pages, we shall discuss these transformations and a few other minor dating methods.

URANIUM-THORIUM DATING

Uranium is the most important and spectacular of the radioactive elements, and its transformations have received more attention because this element is the chief fuel of nuclear reactors, the parent of radium, and the essential ingredient of the atom bomb. Uranium occurs in natural ores as two isotopes (similar forms of the same element but with different mass numbers): uranium with the mass number 235 (U^{235}) and uranium with the mass number 238 (U^{238}). These occur together, but U^{238} is about 140 times more abundant than U^{235}. Uranium ores also contain radioactive thorium 232 (Th^{232}) which is the only natural isotope of that element. All three isotopes pass through a series of changes that end in stable lead. Table 2.1 presents the transformation series for U^{235}.

Uranium 235 (U^{235}) decays to lead 207 (Pb^{207}) with a half-life of 713 million years; uranium 238 (U^{238}) decays to lead 206 (Pb^{206}) with a half-life of 4.5 billion years; thorium 232 (Th^{232}) produces lead 208 (Pb^{208}) with a half-life of 13.9 billion years. It should be noted that helium gas is also a by-product of each of the completed reactions.

As originally applied, uranium-lead dating was designed to determine the ratio of unaltered uranium remaining in a sample to the lead produced and remaining in the same sample. This method assumes that no lead of any kind was present to begin with, an assumption that carries with it a degree of uncertainty. It has been found, however, that the mineral zircon ($ZrSiO_4$) commonly contains uranium, the atoms of which can enter into the crystal structure as it forms; lead atoms cannot do this. Thus, any lead found in a zircon crystal is derived from a radiogenic source and gives a true basis for time calculations. A clue to how much lead may have been in a rock or mineral originally is furnished by the amount of another isotope of lead, Pb^{204}, which is not produced radiogenically. The amount of this isotope in a specimen is a guide to the amount of Pb^{206} and Pb^{207} that may have been present at the time the specimen was formed. A key to this crucial relationship can be found in metallic meteorites. These bodies are usually poor in uranium, and any lead they contain must have been derived from the original mix of which the local planetary system was formed. Because the meteorites probably formed at about the same time as the earth it is assumed that the earth also had about the same ratios of lead isotopes. Analysis shows that well-mixed samples of earth lead, such as that taken from the oceanic oozes, contain almost twice as much Pb^{206} and Pb^{207} as the metallic meteorites. This fact gives some indication of how much the lead content of the earth has been enriched by radioactivity since its formation.

If we assume that specific amounts of parent material were present originally and that there was a subsequent steady rate of decay, the ratio of Pb^{206} to Pb^{207} can also be valuable in age determination. This is the lead-lead method. The problem here is similar to the classic algebra exercise of determining how long two automobiles have been traveling if we know their rates of travel and how far apart they are at present.

Table 2.1 Decay Series for U^{235}

Isotope	Mass Number	Particle Emitted	Half-life
Uranium (U)	235	α	713,000,000 years
Thorium (Th)	231	β	25.6 hours
Proactinium (Pa)	231	α	34,300 years
Actinium (Ac)	227	β	13.5 years
Thorium (Th)	227	α	18.9 days
Radium (Ra)	223	α	11.2 days
Radon (Rn)	219	α	3.917 seconds
Polonium (Po)	215	α	0.00185 second
Lead (Pb)	211	β	36.1 minutes
Bismuth (Bi)	211	α	2.16 minutes
Titanium (Ti)	207	β	4.76 minutes
Lead (Pb)	207		Stable

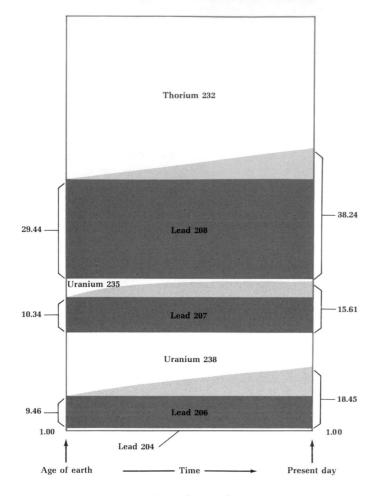

Thorium 232

Lead 208

29.44 — — 38.24

Uranium 235

10.34 — — 15.61

Lead 207

Uranium 238

9.46 — — 18.45

Lead 206

1.00 — — 1.00

Lead 204

Age of earth ———— Time ————► Present day

Figure 2.13 Evolution of isotopic ratios of common lead for the earth as a whole during geologic time. Areas in light blue represent isotopic additions. Pb²³⁴ = 1, and the amount of present-day U²³⁵ is exaggerated. (Adapted from D. L. Eicher, Geologic Time, *Prentice-Hall, Inc., 1968.)*

Obviously the longer the system has been operating the greater the divergence in the end products.

POTASSIUM-ARGON METHOD

The potassium-argon method of age dating was discovered in 1948 and is widely used to date rocks of varied origin and from many different geologic periods. Potassium 40 (K^{40}) produces either calcium 40 (Ca^{40}) or argon 40 (Ar^{40}) and has a half-life of 1.3 billion years. In using this method, it is necessary to measure both the amount of argon and the amount of potassium in a mineral by extremely delicate analytical procedures. It is assumed that no potassium or argon has been added or subtracted during the lifetime of the mineral and that no argon was present to begin with.

The great value of the potassium-argon method lies in the fact that it can be applied to many common minerals, such as biotite and muscovite (both micas), and to sanidine, hornblende, glauconite, pyroxene, and certain fine-grained volcanic rocks. Mica is almost universally present in igneous and metamorphic rocks and is also found in beds of ash thrown out by volcanoes. Because these ash beds are found in sedimentary deposits, these too may be dated. Dating by mica and hornblende is currently widely used with satisfactory results.

The accuracy of the potassium-argon method is limited because argon can be lost at high temperatures. Argon may be driven off at temperatures between 50° and 200°C, depending on the type of rock. This heat, in effect, upsets the closed chemical system essential to accurate results. If a rock has been metamorphosed, it may not yield a reliable date of formation, but the contained minerals will indicate the time when the last metamorphism occurred. Metamorphism resets the atomic clock, so to speak.

The potassium-argon method, like the uranium-lead method, places the age of the earth at about 4.5 billion years.

RUBIDIUM-STRONTIUM METHOD

About 28 percent of naturally occurring rubidium is the radioactive isotope rubidium 87 (Rb^{87}), which decays with a single beta emission to stable strontium 87 (Sr^{87}). The half-life of Rb^{87} is about 47 billion years, and so it provides a convenient scale for dating extraordinarily ancient rocks. Difficulties arise from the fact that radiogenic Sr^{87} may occur in the same rocks with nonradiogenic Sr^{86}. The nonradiogenic strontium must be detected and then eliminated from calculations if a true age is to be obtained.

Minerals that lend themselves to rubidium-strontium dating are muscovite, biotite, lepidolite, microcline, and glauconite. In addition, the method is applicable to whole-rock dating of igneous and metamorphic samples. This is a useful method for dating rocks that are so fine-grained that the individual minerals cannot be segregated. The entire specimen is ground up, thoroughly mixed, and analyzed as though it were a pure mineral. The analysis is aimed at detecting the ratio of Sr^{87} and Sr^{86}.

Figure 2.14   Diagram showing the half-life system of measuring radioactive decay of rubidium 87.

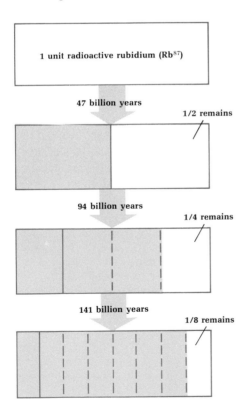

THORIUM AND PROACTINIUM DATING OF OCEANIC SEDIMENTS

Thorium 230 (Th^{230}) is produced in the U^{238} disintegration series and is useful in dating marine sediments. Whereas uranium dissolves readily in seawater, Th^{230} precipitates and becomes part of the sediment. There it decays with a half-life of 75,000 years. If steady deposition has been concurrent with thorium decay, there will be a measurable decrease of thorium downward so that little will remain in sediment deposited over several hundred thousand years ago. Owing to the slow rate of sedimentation in the deep sea, this method can be used to date the relatively short cores that are taken there.

Another method of dating deep sediments is the Th^{230} and proactinium 231 (Pa^{231}) method. The decay of U^{235} produces Pa^{231}, which, like Th^{230}, precipitates rapidly on the sea floor. Once deposited, the two isotopes decay at different rates so that the relative amount of each changes regularly with time. This method is independent of the rate of deposition and is useful for ages up to 150,000 years. It fills a gap between the useful limits of the carbon 14 (C^{14}) and potassium-argon methods but only in very restricted environments.

CARBON 14 METHOD

Probably the most spectacular success in the field of radioactive dating has involved the discovery and use of carbon 14 (C^{14}). This radioactive isotope is continually being created in the atmosphere by the impact of cosmic rays on nitrogen. The carbon thus created joins with oxygen to form carbon dioxide, which enters into organic compounds of all kinds. It is assimilated by living things, and its disintegration can be measured by radiation counters. As long as an organism is alive, a balance is maintained between the radioactive C^{14} and the ordinary variety. New C^{14} is added as fast as the old disappears. When an

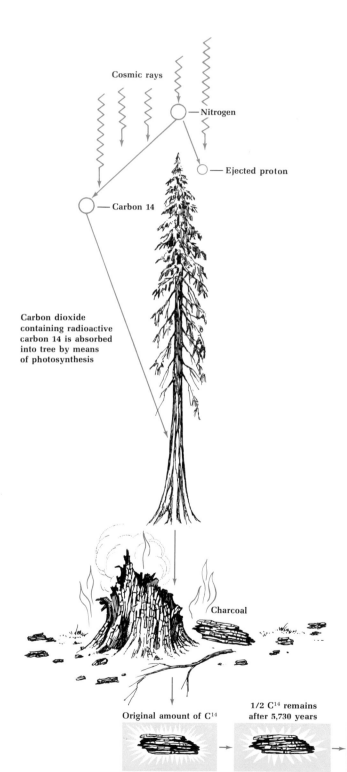

Cosmic rays

Nitrogen

Ejected proton

Carbon 14

Carbon dioxide
containing radioactive
carbon 14 is absorbed
into tree by means
of photosynthesis

Charcoal

organism dies, however, radiocarbon is no longer added, and the C^{14} "clock" gradually runs down by giving off beta particles and reverting to nitrogen. The half-life of C^{14} is about 5,730 years, a period so brief that the amount remaining after 40,000 or 50,000 years is difficult to measure.

Radiocarbon dating has been applied to a great variety of materials, including wood, peat, seeds, shells, charcoal, leaves, bone, manuscripts, rope, and cloth. The method has been checked by applying it to human artifacts and tree rings whose ages are known by other methods. The agreement is good and lends confidence to the dating of older materials. Literally thousands of dates have been obtained that are of interest not only to geologists but also to anthropologists, archaeologists, and historians. See Table 2.2 for a list of various materials that have been dated by means of the C^{14} and other radiometric methods.

FOSSIL FISSION TRACKS AS AGE INDICATORS

As early as 1900 it was observed that mica will show discolored areas around radioactive particles embedded in it. These *pleochroic halos* are caused by the bombardment of the mica by high-energy fragments set free by spontaneous fission of scattered uranium atoms. Workers at the General Electric Research Laboratory at Schenectady, New York, studied pleochroic halos under high magnification and found numbers of small tubes,

Figure 2.15 Formed from nitrogen in the atmosphere, C^{14} is incorporated in all living things. The rate of addition and disintegration is assumed to be constant and in equilibrium. By reverting to nitrogen C^{14} disappears from dead organic material; the amount of C^{14} remaining in a dead specimen is a measure of age.

Original amount of C^{14} | 1/2 C^{14} remains after 5,730 years | 1/8 C^{14} remains after 17,190 years | 1/32 C^{14} remains after 28,650 years | Insignificant C^{14} remains after 45,000 years

Carbon 14 reverts to nitrogen

Table 2.2   Materials Dated by Various Radiometric Methods*

Sample	Approximate Age in Years
Cloth wrappings from a mummified bull taken from a pyramid in Dashur, Egypt. This date agrees well with the age of the pyramid as estimated from historical records.	2,050
Charcoal recovered from bed of ash near Crater Lake, Oregon, is from a tree burned in the violent eruption of Mount Mazama that created Crater Lake. This eruption blanketed several states with ash, providing geologists with an excellent time zone.	6,640
Tamarack tree fragment recovered from glacial debris near Green Bay, Wisconsin, dates the time of the last advance of the continental ice sheet into the United States from the north.	11,640
Marine shells recovered from layer of clay at Norridgerock, Maine, at an elevation of 76.2 meters above present sea level. After the retreat of glaciers in this part of New England, the land's surface stood at least 76.2 meters lower because it had been depressed by the weight of the ice.	11,950
Volcanic ash collected at Olduvai Gorge, East Africa, from strata that sandwich the fossil remains of *Zinjanthropus* and *Homo habilis*—possible precursors of modern man.	1,750,000
Monzonite of copper-bearing rock from vast open-pit mine at Bingham Canyon, Utah.	37,500,000
Quartz monzonite collected from Half Dome, Yosemite National Park, California.	80,000,000
Conway Granite collected from Redstone Quarry in the White Mountains of New Hampshire.	180,000,000
Rhyolite collected from Mount Rogers, the highest point in Virginia.	820,000,000
Pikes Peak Granite collected on top of Pikes Peak, Colorado.	1,030,000,000
The Old Granite from outcrops in the Transvaal, South Africa. These rocks intrude even older rocks that have not been dated.	3,200,000,000
Morton Gneiss from outcrops in southwestern Minnesota are believed to represent the oldest rocks in North America.	3,600,000,000

*Adapted from "Geologic Time," U.S. Geological Survey (1970).

like minute bullet holes, that had been plowed out by the outward-moving fission fragments. These holes can be enlarged by etching with acid and then counted and studied in detail.

In order to obtain an age estimate it is necessary to know how many uranium atoms have already disintegrated in a given volume of mineral and how many have not yet disintegrated. The first figure can be determined by counting the naturally occurring tracks; to get the second figure the sample is artificially bombarded in a reactor with neutrons that cause the remaining uranium atoms to disintegrate, leaving another set of tracks that can in turn be counted.

About ten minerals and natural glasses have been found to contain fission tracks suitable for dating. It has been discovered that the tracks are "erased," or heal, when the containing material is heated above a few hundred degrees centigrade. Samples that have been near the surface of the earth, and thus have been subject to cosmic rays, also give unreliable dates. The best results are expected to come in the study of materials less than about 100 million years old. Already the method has been successfully applied to the study of many rocks and minerals and to the meteoritic fragments called *tektites*. It is fre-

Figure 2.16   Fossil fission tracks in biotite mica. The crystal was etched in hydrofluoric acid to enlarge the tracks, which radiate from a microscopic impurity containing a larger number of uranium atoms. Magnification about 3,000×. (Courtesy of P. Buford Price.)

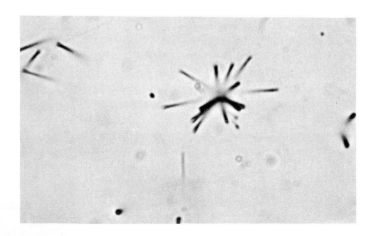

quently possible to date the same specimen both by fission tracks and by some other methods, thus testing the accuracy of both procedures.

LIMITATIONS AND CRITICISMS
OF RADIOMETRIC DATING METHODS

The theory behind dating by use of radioactive decay phenomena is relatively simple, but the practical applications are not always easy and the results are not always beyond question. Difficulties have arisen in the purely mechanical processes of measuring the fantastically small amounts of material involved, from uncertainties regarding the past history of the specimens being tested, and from contaminating agents that frequently affect buried substances. Lead-thorium, lead-uranium, and lead-lead methods are criticized because of assumptions made about the early history of the earth, specifically, whether or not radiogenic material was present to begin with. The potassium-argon method is subject to limitations because argon will escape at the high temperatures that accompany metamorphism or mountain building. We have already seen that the fission-track method can be disrupted by heat effects.

Carbon 14 dating is beset with several difficulties. It is assumed that the rate of influx of cosmic rays is constant and that the amount of carbon 14 present today in living things (about 15.3 disintegrations per minute per gram) also applied to ancient living things. If the rate of production has varied, this assumption is invalid. The problem of contamination by carbon from the surrounding water, soil, rock, or vegetation is also serious.

The chief limitation of *any* dating method arises from the fact that dates derived from small samples or restricted areas are difficult to relate to their surroundings in a meaningful way. A coin found on the street, although itself accurately dated, tells us nothing about when nearby buildings were erected. A vein of uranium ore in a mountain of igneous rocks tells us little

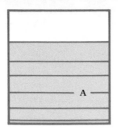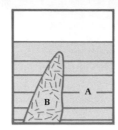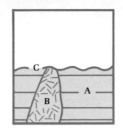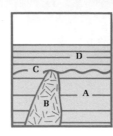

Figure 2.17 Absolute and relative dating. From left to right, cross sections represent serial geologic events that pose a problem in dating. Sediments A are deposited in regular order. Then igneous rock B, dated at 20 million years by radiometric methods, is intruded from below. Erosion produces surface C and cuts away part of the igneous intrusion. Finally, sediments D are laid down. From the order of events, we know that the relative ages are A, B, C, and D. The absolute date for the intrusion indicates that sediments A are older than 20 million years, whereas surface C and sediments D are younger. Sediments containing minerals suitable for radiometric measurement can be dated more precisely.

about the age of the sedimentary formations that surround the mountain. In order to date the sedimentary formations, a direct physical link with the *igneous* formation must exist. If eroded pieces of the vein are found embedded in the sediments, it is obvious that the vein came first and is older than the sediment. If a dated igneous body is found intruding into a sedimentary formation with fossils whose place in the standard geologic time scale has been fixed, then we know the vein is younger than the sediments. If the vein has been cut off by erosion and covered by a second sedimentary formation, then we know that the igneous rock is not only younger than the lower formation, but also older than the upper one. *The igneous material, therefore, furnishes a maximum age for the younger formation and a minimum age for the older one.* Unfortunately, such ideal situations are rare. With newer methods we are not burdened by the necessity of relating the igneous rocks to the sediments, for the sediments themselves are providing minerals that can be dated directly.

In general, we may say that radiometric dating methods are improving. There are reasonable explanations for most discrepancies and contradictions. Techniques and instruments have been improved constantly, especially since World War II, largely because of the tremendous advances that have been made in atomic research. Geologists are learning to select samples with greater care, and the search for suitable materials is now a recognized part of geologic exploration. The concerted efforts of scientists in many fields may be expected to yield still more important and interesting results.

Other Techniques of Absolute and Relative Dating

Organic materials may absorb elements from surrounding solutions in a relatively steady way so that the amount absorbed is a measure of how long the absorption has been going on. One element that is commonly taken up and retained by hard animal tissues is fluorine. Fluorine analysis has value in providing evidence as to whether or not bones from associated or neighboring deposits are contemporaneous—they should have the same amounts of fluorine if they are of the same age. This method was employed in detecting the famous Piltdown hoax in which bones of different ages had been placed together to create an "artificial" fossil that misled many anthropologists for a number of years.

Another age-dating method based on rate of accumulation is obsidian dating. This method relies on the fact that a freshly exposed surface of obsidian, a natural glasslike volcanic material, will absorb surrounding water to form a measurable altered, or weathered, layer. This layer thickens with time and thus will give a rough measure of how long the process has been operating. This procedure appears to be applicable to the period between 2,000 and 50,000 years ago. It is especially good for dating obsidian objects shaped by man.

Astronomical Considerations

MOON ROCKS

A variety of specimens of rock collected from the surface of the moon contain radioactive isotopes that have been useful in age dating. Samples include fine aggregates of "soil," solidified broken fragments called breccia, and unaltered basalt. The most significant specimen thus far is the so-called "genesis rock" collected by astronauts David R. Scott and James B. Irwin on their Apollo

Figure 2.18 The so-called "genesis rock" brought back by the Apollo 15 lunar mission. The upper limit on the age of this sample is 4.35 billion years. The specimen is about 8 centimeters long. (NASA.)

15 moon exploration. Scientists estimate that this rock is about 4.15 billion years old, with a margin of error of about 200 million years. The term "genesis rock" was applied to this specimen because scientists had hoped that it would prove to be as old as the solar system itself—an estimated 4.6 billion years. The discovery of this rock lends support to the theory that the moon was once completely molten; scientists feel that rocks such as this one would have floated to the surface of the moon because they are less dense than other lunar minerals. Another sample brought back by the Apollo 15 mission, a piece of basalt, has been estimated to be about 3.3 billion years old.

METEORITES

Astronomers believe that most meteorites originated within the solar system through the breakup of larger bodies such as asteroids, comets, or small planets. In any event, they provide actual pieces of material that must have originated when the solar system was formed and under conditions not greatly different from the conditions that accompanied the birth of our earth.

Meteorites are usually classified as stony or metallic according to their composition. Both varieties contain elements that are of value in age determination. The stony varieties can be dated by the uranium-lead system, the strontium-rubidium system, and the potassium-argon system. The results obtained have not been entirely consistent, probably because of the effects of cosmic rays or heating in space or from contamination after reaching the earth. The metallic meteorites contain lead, and the ratio of Pb^{207} to Pb^{206} has given a very consistent age of about 4.5 billion years, the point in time at which the parent body or bodies are thought to have formed. That essentially the same age has been assigned to the earth must indicate that planet formation was a well-defined event accomplished in a relatively short time.

A SUMMARY STATEMENT

A major contribution of geology is the concept that the earth is very old. From evidence of seasonal changes, such as tree rings, and from annual sedimentary deposits called *varves*, it is thought that the earth has been revolving and rotating in a relatively steady manner for at least 3 billion years. However, the day may have been getting shorter, due to the slowing of the earth's rotation by action of the moon on the earth's tides.

Attempts to answer the question "How old is the earth?" on a scientific basis began in the late eighteenth century. Crude estimates based on the amount of salt in the ocean, the rate of deposition of sediments, and the rate of cooling of the earth, did little more than indicate that the traditional idea of a 6,000-year-old earth was almost certainly in error.

The discovery of the spontaneous breakup of certain elements—known as *radioactivity*—opened up a new and more certain method of determining not only the age of the earth but also of many rocks and minerals of its crust. Because the rate of breakup of radioactive atoms is not influenced by ordinary terrestrial conditions, several natural dating methods are provided by measuring the amount of material already disintegrated (as known from the daughter products) as compared with that yet to disintegrate. The chief radioactive transformations that are useful in age determination are *uranium to lead, thorium to lead, potassium to argon, rubidium to strontium,* and *carbon 14 to nitrogen.*

Analysis of thousands of specimens has given the following "time table of creation": earth solidified about 4.5 billion years ago; first extensive rock masses formed about 3.5 billion years ago; fossil record begins about 3.3 billion years ago; abundant fossil record commences with the beginning of the Cambrian Period about 600 million years ago.

The age of most meteorites for which radiometric dating is possible is also about 4.5 billion years. This is also the maximum date of lunar samples. Widely divergent age estimates for the universe range between a low of 6 billion and a high of 20 billion years.

FOR ADDITIONAL READING

Eicher, Don L., *Geologic Time.* Englewood Cliffs, N.J.: Prentice-Hall, 1968.

Faul, Henry, *Ages of Rocks, Planets, and Stars.* New York: McGraw-Hill, 1966.

———, ed., *Nuclear Geology.* New York: Wiley, 1954.

Harbaugh, John W., *Stratigraphy and Geologic Time.* Dubuque, Iowa: Brown, 1968.

Hurley, Patrick M., *How Old Is the Earth?* Garden City, N.Y.: Doubleday, 1958.

Kulp, J. L., "Geologic Time Scale." *Science,* Vol. 133, 1961, 1105–1114.

Libby, W. F., *Radiocarbon Dating,* 2nd ed. Chicago: University of Chicago Press, 1959.

Toulmin, Stephen, and June Goodfield, *The Discovery of Time.* New York: Harper & Row, 1965.

Zeuner, F. E., *Dating the Past: an Introduction to Geochronology.* London: Methuen, 1952.

UNIFORMITARIANISM

AND THE HISTORY OF GEOLOGY

Someone has said that before we can ask a question we must already know part of the answer. So it was with the early history of geology—no worthwhile hypotheses could even be proposed before there had been some preliminary thought and observation bearing on the origin of the earth and the duration of its past history.

In ancient times men saw the natural world as a product of supernatural events or as the work of powerful gods. Many creation myths emphasize the role of the supernatural or miraculous in the beginning of things, but they usually

include man as the central figure. The Creation account in Genesis seems to imply that all living things were miraculously brought forth over a short period of time and only a few thousand years ago in ways beyond the comprehension of man. The belief in a literal 6-day creation and a 6,000-year-old earth was entirely acceptable for a time because there was no evidence against it.

Regardless of how the chronology is interpreted, the Hebrew scriptures teach that the earth has had a gradually unfolding history. Furthermore, the same Western culture and civilization that gave rise to the idea of a short and violent history of the earth also produced its opposite, the concept of a long and gradual evolutionary development. For centuries there have thus been two ways of looking at the problem of the origin of the earth and living things, one view tending toward supernatural catastrophism, the other toward slow, natural, uniform processes. As the science of historical geology developed and advanced, new evidence concerning the history of the earth caused a gradual adjustment in both views.

Catastrophism

Catastrophism is the name given to the theory that the earth has been affected by extraordinary disasters, by sudden and violent disturbances of existing things arising from natural, not man-made, causes. Catastrophism has been viewed not only as a method of divine punishment but as an ally of creation. In the theological scheme it may be the only proper method of eliminating the old and imperfect to make way for something better. Catastrophism supports the idea of a young earth and a short period of creation because it explains how processes that would normally take place over a long period of time can on the contrary be accomplished rather quickly.

Believers in catastrophism assert that the earth in past times has been affected by universal, world-shaking events that surpass in violence anything known in modern time. Man has always been fascinated in some way by violence and destruction, and this attitude may encourage a belief in catastrophism. The element of natural disaster looms large in fiction and plays an essential role in the epic tales of man. Floods, tornadoes, earthquakes, volcanic eruptions, and tidal waves are front-page items in newspapers and are many times magnified out of true proportion, especially if they occur in heavily populated districts.

The greatest of all past catastrophes described in the Bible with any detail is Noah's Flood, or the Deluge. Controversy over the Flood still exists, but there was a time not many decades past when the debate was even more heated and serious. When geology was still in its infancy, it was customary to explain practically everything geologic in terms of the Deluge. The carving of the landscape was attributed to the currents and storms that accompanied the Flood, and mountains and hills were thought to be formed by the violent stirrings and mixings of the waters. The Flood offered also a convenient explanation for the existence of fossil remains, not only on the earth's surface but also in caves or deep in the earth. In addition, it was supposed that the flood waters had lifted and dispersed huge blocks of ice and left them to melt in southern lands. Early generations of scientists, as well as theologians, turned to the Flood to explain practically all evidences of geologic change.

CUVIER, THE CATASTROPHIST

The doctrine of catastrophism was supported by many of the best minds of the eighteenth century. One of the chief supporters of the theory was Baron Georges Cuvier, who was among other things an able anatomist and student of geology. He wrote twelve large volumes on vertebrate

Figure 3.1 *Baron Georges Cuvier (1769–1832), versatile French genius whose studies of fossil remains from the environs of Paris proved that life has varied and changed in the past. He favored the idea of catastrophic change and argued against the theory of organic evolution. (Brown Brothers.)*

fossils and knew more about the subject than any of his contemporaries. He studied especially the remains of fossil animals found in the environs of Paris. He realized that these extinct and exotic forms were indeed remains of past inhabitants of the earth, but he was convinced that each different group was the result of a separate creation and had ultimately disappeared as the result of a violent catastrophe. According to Cuvier:

Life upon the earth in those times was often overtaken by these frightful occurrences. Living things without number were swept out of existence by catastrophes. Those inhabiting the dry lands were engulfed by deluges, others whose home was in the waters perished when the sea bottom suddenly became dry land; whole races were extinguished leaving mere traces of their existence, which are now difficult of recognition, even by the naturalist. The evidence of those great and terrible events are everywhere to be clearly seen by anyone who knows how to read the record of the rocks.

Cuvier believed that Noah's Flood was universal and had prepared the earth for its present inhabitants. The Church was happy to have the support of such an eminent scientist, and there is no doubt that Cuvier's great reputation delayed the acceptance of the more accurate views that ultimately prevailed.

Uniformitarianism

The idea that catastrophism might be wrong may have been a reaction against theological dogmatism, but for most men it was an outgrowth of actual observation of nature. The concept of *uniformitarianism,* meaning that the earth has been shaped by uniform, gradual processes, is opposed to catastrophism. Early thinkers, as soon as they began to doubt the time-honored traditions, were faced with some difficult questions. Did the physical world originate by means of miraculous

events over a short period of time as suggested by Genesis, or was a much longer period, even a limitless amount of time, required? The question was the following: 6 days, or forever? Because an infinite variety of things could have happened in an infinity of time, some were led to believe our world is merely a chance occurrence among all manner of other possible worlds. If, however, time is limited, chance lessens.

As we saw in the last chapter, the study of absolute measures of time has provided philosophers with a firmer basis for thought and a better starting point for their hypotheses. It now seems certain that the development of the earth and its life must be confined to 4.5 billion years, a time span that is neither infinitely long nor extremely short. But this realization merely replaces the earlier question with another no less challenging one: Are a few billion years sufficient for chance factors alone to have produced the earth and its inhabitants? This problem involves weighty and important matters that we shall consider under the general question of whether or not the present is a reliable guide to the past.

Although most educated men feel that uniformitarianism is the only dependable and rational approach to a study of the history of the earth, we need to be reminded that it is not the only approach. It is a method of thinking and not an empirical generalization. Uniformitarianism rejects supernatural (miraculous or inexplicable) causes as long as natural ones will suffice. It appeals to known laws or principles rather than to unproved or unprovable suppositions. It seeks explanations of the past that are based on processes that can be observed in action at the present, and not those based on pure imagination. Furthermore, uniformitarianism asserts that these natural, everyday, observable processes acted at approximately the same rates and on the same scales in the past as they now do, provided that these rates and scales could accomplish the observed results in the time available.

Although uniformitarianism favors certain explanations of the past as more probable than others, it would be wrong to say that it must always deny or exclude other explanations. If natural, known, provable, observable processes acting at rates and on scales that are known to be possible cannot explain an observed situation, then we must consider alternatives—even unprovable ones, if necessary.

It is with regard to rates and scales of action that uniformitarianism is most frequently questioned. In the past, there have obviously been more violent and extensive volcanic eruptions than any recorded in human history. The evidence of the great Ice Age proves that modern glaciers are puny and feeble compared with those of the past. Craters on the moon signify activity of a kind not going on every day. But the physical laws that govern volcanic eruptions, ice action, and the impact of meteorites cannot be said to have changed; only the intensity of action has.

In the sense that uniformitarianism implies the operation of timeless, changeless laws or principles, we can say that nothing in our incomplete but extensive knowledge disagrees with it. Furthermore, it is by application of the uniformitarian principle even down to the level of rates and scales of action that we have had our greatest success in unravelling the past. By uniformitarianism we achieve the most reasonable and rational explanation involving not only the least expenditures of energy but also a minimum of hypothesis. Perhaps the best expression of what geologists mean by uniformitarianism is still the time-worn saying, "the present is the key to the past."

Catastrophism or Uniformitarianism— The Evidence

Any concept that cannot be illustrated by specific examples is difficult to comprehend. Most geologic processes are not difficult to visualize as such, but their total effects are not generally appreciated. The difficulty obviously lies in comprehending the vast intervals of time over which the processes operate. Once the proper perspective of time is established, the objections to uniformitarianism largely disappear.

What examples might constitute convincing evidence for either catastrophism or uniformitarianism? Our best examples are those for which there is some type of documentary or eyewitness evidence. This imposes severe restrictions. We are confined to the last few centuries, during which reliable observations and measurements have been made, and to those areas where civilized populations capable of making reliable records have lived. Anything else is either hearsay or is inferred from indirect evidence.

A number of illustrations of the operation of ordinary geologic processes can be offered as evidence of what has been accomplished during the last small segment of time that we know best.

DIASTROPHISM

Diastrophism is a general term for processes involving faulting, folding, and warping, which deform the earth's crust. The energy manifested by these movements is derived from within the earth, and the effects contrast strongly with the purely superficial results of weathering and erosion. Faults, for example, cut upward through great thicknesses of rock below the surface dislocation. Although the original surface expression of a fault may in time be erased by erosion or be covered by sediment, the deeper subsurface continuation remains as a more or less permanent record of the disturbance that caused it.

All the major types of diastrophism have been operating during the span of human observation. Literally hundreds of active faults have been traced, many of which are associated with recorded earthquakes. Faulting, with its accompanying earthshocks and sudden surface disruptions, is more easily detected than foldings and widespread warpings, which nevertheless produce profound and lasting effects.

Regional effects of both slow and rapid diastrophism have been recorded at various places.

Careful measurements made in the Buena Vista oil field in the southern San Joaquin Valley in California show that the land is being slowly uplifted at the rate of about 1.2 meters every hundred years. This figure may seem trivial to some, but in 25,000 years the area would be some 300 meters higher than it now is and in 200,000 years it would be as high as the Cascade Range, assuming, of course, that no erosion took place. To make a Mount Everest at this rate would require about 2 million years. The great earthquake of March 27, 1964, with an epicenter near Anchorage, Alaska, caused an area of approximately 77,700 square kilometers to sink as much as 2 meters, and an adjacent area of perhaps 129,500 square kilometers to be elevated, in places as much as 11 meters.

The most severe seismic disturbance recorded in North America was a series of three major shocks that struck on December 16, 1811, and January 23 and February 7, 1812. The epicenter was near New Madrid, a Mississippi River town in southeast Missouri. The region at that time was sparsely settled and modern seismic instruments had not yet been devised; nevertheless, there can be no doubt that the New Madrid

Figure 3.2 Effects of the Alaskan earthquake of March 27, 1964. The shattered condition of the ground at this locality is due chiefly to landslides set off by the quake. (U.S. Geological Survey.)

Figure 3.3 Large rockslide and lake that resulted from an earthquake centering in southwestern Montana, August 17, 1959. A mass of rock and soil, dislodged from the scar on the right, descended suddenly into the valley damming the Madison River and causing a number of deaths among campers in the vicinity. (U.S. Department of Agriculture.)

quake was of the most severe type by any standard of reference. In addition to the three major shocks, literally hundreds of lesser quakes occurred over a period of a year and a half and were felt over hundreds of square kilometers.

The New Madrid quake was physically felt over two-thirds of the United States, an area of about 2.6 million square kilometers was severely shaken, and major topographic changes occurred over an area of 130,000 square kilometers. Many elongate tracts were uplifted, fissuring was widespread, and landslides took place along long stretches of the Mississippi River. A tract 240 kilometers long and 64 kilometers wide in Tennessee and Arkansas sank from 1 to 3 meters, and water from the Mississippi River rushed into the resulting depression. Reelfoot Lake in Tennessee and St. Francis Lake in Arkansas still remain as evidence of this subsidence. Large quantities of muddy water and gas were violently ejected from the earth, destroying much vegetation and covering fertile topsoil with sand.

The cause of the New Madrid earthquakes is not well understood, because the area is far removed from active fault systems or growing mountain ranges. Perhaps the disturbances reflect adjustments to the great weight of sediments built up by the Mississippi River on its alluvial plain, the area that was chiefly affected.

CHANGES IN SEA LEVEL

Sea level is the great gauge of geologic change—nature's plane for recording relative movements of land and sea throughout the globe. Some effects are more apparent than real; at times it is the land itself and not the ocean that changes elevation. Effects that are actually caused by worldwide changes in sea level are designated as *eustatic* to distinguish them from local uplift or subsidence of the land. Results of sea-level change are generally dramatic even though the change is not fully carried out in one lifetime or even in many lifetimes. However, the gradual appearance of islands or of broad coastal areas

is a historical fact, and the sinking and disappearance of sections of the shoreline have not gone unnoticed by those who have laid claim to them.

There are difficulties in reading the sea-level gauge because it varies for many and complex reasons. There is evidence that the earth may not always keep the same shape. At present it is slightly flattened at the poles, and the ocean is apparently flooding over polar regions because water tends to keep a more nearly spherical shape than the solid earth. If the earth does change shape, the ocean adjusts readily and may then flood some areas and leave others high and dry.

Sea level rises as sediment is dumped into the ocean or as submarine mountains rise on the sea bottom. An extensive flooding of all continents during the Cretaceous Period is attributed to the rise of a large submarine ridge on the Pacific floor. Drastic effects ensue from the growth and melting of large ice caps on land. As much as 5

Figure 3.4 Collecting samples from a submerged terrace off Mission Beach, California. This terrace, which lies at a depth of about 180 meters, contains shallow-water fossils of latest Pleistocene age. (Official photograph, U.S. Navy.)

Figure 3.5 The art and architecture of Venice is threatened by a general rise of sea level combined with subsidence of the land. In this photograph, note how the water is undermining the lower levels of the buildings. (Italian Government Travel Office.)

percent of the earth's total water may move onto land during glaciation with a consequent fall in sea level of at least a hundred meters. Conversely, as the ice melts, sea level rises.

Systematic underwater exploration of La Jolla Canyon, a submerged watercourse near La Jolla, California, shows a gradual rise in sea level thought to be due to return of seawater from glaciers of the last Ice Age. Carbon 14 dates from successive levels show that sea level has been about the same as at present for 4,000 years. Before that, about 4,200 years ago, it was 12 meters lower. At a level between 3 and 12 meters below the surface, a campsite of ancient men with over 2,000 stone bowls was found on a submerged beach. About 6,500 years ago sea level was below the campsite, about 18 meters below the present level; 13,200 years ago the level was about 38 meters lower than at present. Old beaches have been noted elsewhere at depths as great as 180–210 meters. An average rise of 15 centimeters per century is indicated, but there have been irregular spurts that accelerate the general submergence.

The Mediterranean Sea was the center of ancient civilization, and many important cities and monuments were erected on its shores and islands. Evidence of sea-level change can be found on the island of Crete. Submerged stone tombs thought to date from the fourth or fifth century of our era have been found beneath the water on the south side of the island. This tract is thought to have submerged at least 4 meters since the time of Christ. The old Roman city of Limani Chersonisos was built well inland; now the foundations are washed by the surf. There is also evidence of elevations: shell beds in their place of accumulation have been found in areas that are nearly 10 meters above the present beach.

The great city of Venice is a victim of both sinking land and rising sea—truly a geologic disaster area. On the one hand the city is affected by the general rise of sea level due to the melting of Antarctic ice, a rise that amounts to about 0.14 centimeter per year. On the other hand the land itself is sinking 0.27 centimeter per year due mainly to the withdrawal of underground water by thousands of wells on the nearby mainland. (As the water leaves the porous sand beneath the area, there is a small but measurable settling.) The combined effect is that Venice is sinking slowly but visibly into the sea. As a geologic sidelight, the deplorable decay of the statuary and ornamental marble is due to the reactions of calcium carbonate in the stone with industrial pollutants in the air. Sulfur is most detrimental because it combines with the calcium to produce gypsum, which is soft and expandable, so that affected surfaces spall away or literally explode in unsightly patches.

A well-known example of changes in relative level of land and sea is revealed at the famous ruins of the so-called Temple of Jupiter Serapis erected about 200 B.C. near Pozzuoli, a small town on the Bay of Naples. This building, constructed near sea level as a port facility, originally had forty-six stone columns about 12 meters high. Only three of these remain standing, but they record a series of very significant changes. Up to about 4 meters above their pedestals, the columns are smooth and relatively unmarred because they were buried and protected in mud. The next 3 meters or so have been perforated in numerous places by the marine, stone-boring mollusk *Lithodomus lithophagus,* whose shells still occupy some of the pear-shaped borings. The top of the bored zone is about 7 meters above the present highwater mark of the Bay of Naples.

The general area around Pozzuoli seems to be subsiding at a fairly uniform rate of about 11 millimeters per year, but on several occasions the downsinking has been rather suddenly reversed by strong upward movements. Such upheavals seem to have taken place in 63 and 79 B.C., and again in A.D. 1197, 1302, 1416, 1488, and 1536–38.

The Temple of Jupiter Serapis apparently sank into the bay during the quiet period preceding the twelfth century and then rose above sea level near the end of the fifteenth century. The building was cleared of mud and volcanic ash in 1749. Again sinking took over and in 1950 the floor was

Figure 3.6 Ruins of the so-called Temple of Jupiter Serapis on the shores of the Bay of Naples. The three intact columns clearly show a zone of borings made by marine mollusks when the site was under at least 6 meters of water. (Italian Government Travel Office.)

Figure 3.7 A sandstone remnant being destroyed by coastal wave erosion about 16 kilometers north of Antofagasta in Chile. (U.S. Geological Survey.)

about at sea level. The latest movements took place in March 1970, when the general area rose gradually about 90 centimeters with relation to sea level. No earthquakes accompanied this movement, but there was nevertheless great concern because a number of old stone walls were cracked.

The importance of this record to our present topic is that the movements were generally so gradual and unspectacular that no accounts were kept of the appearance or disappearance of the original floor, and because the action was never violent enough to overthrow the three telltale columns.

EROSION

Erosion is a dominant process over much of the earth's surface and strongly affects the activities of man. Sedimentation, the redepositing of the products of erosion, goes on in invisible realms beneath seas or lakes or on limited areas of floodplains and deltas that are occasionally inundated by water. Many books and articles have been written about the erosion of the superficial layers of soil on which the well-being of men and nations depend, and most of us are more or less familiar with certain aspects of the subject. Although soil erosion is important, particularly from an economic standpoint, the geologist is concerned mainly with the more extensive effects of erosion throughout geologic time. Observation teaches us that the forces of erosion need only time to accomplish tremendous transformations of the landscape. The brief span of an individual human life encompasses only an infinitesimal part of the process, but through intervals of several hundred or a few thousand years the results are measurable and unmistakable. Early proponents of uniformitarianism taught that rivers cut their own valleys; their opponents believed that valleys, gorges, and canyons were created by the splitting of the earth or by the sudden and violent flow of Noah's Flood. Unbiased observations and careful measurements confirm the uniformitarian viewpoint.

Coastal Erosion. The tides, currents, and waves of the ocean are obviously powerful erosive agents. Their attack on all shorelines is continuous and persistent. Where relatively soft rocks meet the full force of waves from the open ocean, the effects during an ordinary lifetime are easily measurable. Careful studies of the coast of

Figure 3.8 Horseshoe Falls at the head of the Niagara gorge. Over 90 percent of the river's water passes over these falls. The dashed line shows the approximate position of the falls when they were first sketched in 1678. (N.Y. State Department of Commerce.)

East Anglia in Great Britain show an average rate of retreat of about 3 meters per year for the last century. At one locality some 24 meters of retreat was caused by a single 1953 storm. This illustrates the importance of the "rare" event in creating an average rate.

The destruction of land by the sea has also been studied near Santa Barbara, California. Effects vary from place to place, but the average rate amounts to about 15 meters per century. Investors in shoreline property may be paying for something that will cease to exist in their own lifetime.

River Erosion—Niagara Falls. Spectacular evidence of river erosion is provided by the retreat of Niagara Falls. Here the concentrated drainage of the Great Lakes area plunges over a 50-meter fall on the Niagara River between Lake Erie and Lake Ontario. This tremendous spectacle is more than a waterfall; it is a demonstration of the cutting of a deep gorge under rather special geologic conditions. The lip, or upper rim, of the falls is very resistant rock over which the water passes with practically no erosive effect. Below the hard layer is a thicker formation of soft rock. The powerful swirling action of the plunging water cuts into the softer rock, undermining the hard ledge above and leaving it unsupported. As piece after piece of the lip tumbles and is carried away, the falls move backward. They have moved upstream toward Lake Erie at least 150 meters since they were first described in the seventeenth century. Exact surveys have been kept since 1841. At times, the rate has been about 1.5 meters per year, but now that large quantities of water are being diverted to produce hydroelectric power, the process has been slowed down. The gorge below the falls is about 11 kilometers long and provides a clue to the rate and duration of past erosion.

The falls probably began their retreat at the face of a long, clifflike bluff that crosses the general area, sometime between or since advances of glaciers during the Ice Age. Obviously the river could not exist when the area was under ice. The retreat of the falls from the bluff to their present position must have required about 10,000 years, which would correspond to an average rate of about 1 meter per year. But regardless of the exact time required, the evidence seems conclusive that the Niagara River has cut its own gorge by a mechanism that can still be seen operating today.

The problem of how rapidly unprotected rock can be removed by ordinary precipitation has been investigated in the Painted Desert area of Arizona. Measurements taken on stakes driven deep into shale indicate that the surface is being lowered at an average rate of 2 centimeters every

Figure 3.9 *A measure of erosion in soft sediments. In an experiment to determine the rate of erosion in the soft Chinle Shale of northern Arizona, marked stakes were driven into the earth and allowed to remain for specific periods of time. In this photo the distance between the groove around the stake and the surface of the ground indicates the erosion that took place over a 2-year period. (Courtesy of Edwin H. Colbert.)*

Figure 3.10 *A near view of the sloping face of the Great Pyramid of Cheops, near Cairo, Egypt. The disintegrated material that has fallen from the stone blocks represents the erosion of 1,200 years. The pyramid was originally covered by flat slabs of marble, which were removed in the ninth century of our era, at which time the erosion began. In spite of this disintegration, it has been calculated that the pyramid will endure for another 100,000 years. (Courtesy of K. O. Emery.)*

4 years, which would be equal to 50 centimeters per century. The material in this case is relatively soft, but the hardest rocks eventually wear away.

SEDIMENTATION

All material removed from any part of the earth's surface must somewhere come to rest. The laying down, or deposition, of eroded fragments and chemical precipitates, such as salt, is called *sedimentation*. The most obvious effects of sedimentation can be seen at the mouths of large rivers where silt and other suspended materials scoured by the river from extensive areas upstream is deposited. Because the mouths of many rivers are important sites of ports and may support large populations, the effects of sedimentation are matters of great concern.

The Mississippi Delta. The delta of the Mississippi River is growing forward into the Gulf of Mexico at a rate of some 1.5 kilometers every 16

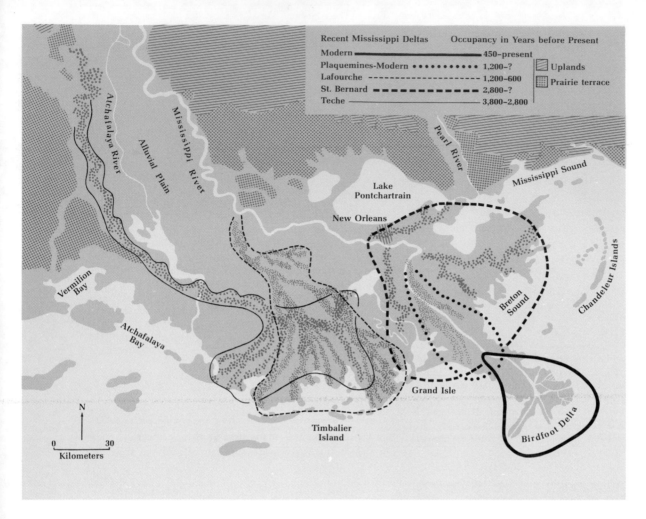

Figure 3.11 Sequence and age of deltas and abandoned Mississippi River courses on the Mississippi deltaic plain. (*Adapted from H. R. Gould, "The Mississippi Delta Complex,"* in Deltaic Sedimentation, Modern and Ancient, *Society of Economic Paleontologists and Mineralogists, 1970.*)

years, or a little less than 30 centimeters per year. It not only grows forward but is also built upward by silt deposited along the banks and in the channels. When one channel becomes sufficiently high, there is a tendency for the river to break through its banks to take up a new course at a lower position on adjacent ground. The general area of the Mississippi delta is made up of at least five subdeltas that mark former courses and deposits of the river. According to detailed studies, the river has occupied its present course for only 450 years. There is imminent danger that the river will again break through its banks above New Orleans to find a shorter route to the gulf by way of the Atchafalaya Channel unless the waters are restrained by levees.

The Mississippi emerges on a wide expanse of coast, but other rivers, such as the Po in Italy,

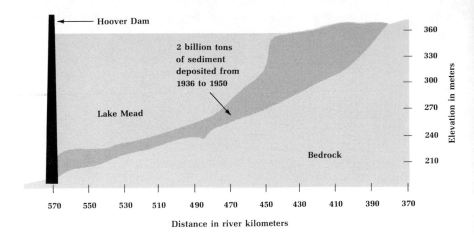

Figure 3.12 The filling of Lake Mead by the Colorado River. The cross section shows in diagrammatic form the dam, the lake, and the amount of sediment added during the first 14 years of the lake's history. The vertical scale is greatly exaggerated; the lake is over 320 kilometers long and less than 150 meters deep.

empty into narrow embayments where the sediment is confined to a restricted area. Estimates indicate that the delta of the Po has advanced about 22 kilometers since the time of Augustus; the town of Adria, once a seaport, is now situated far inland.

All large deltas are constantly changing. Those of the Nile, the Tigris-Euphrates, the Hwang Ho, and the Ganges are heavily populated, and the shifts in the channels have been of great historical interest. The shifting of channels from place to place on a broad surface creates the typical fanlike delta shape. Diversions and shifts must be considered as normal incidents in the development of river plains and deltas.

The Colorado River. In some instances the future effects of sedimentation must be carefully calculated because of their bearing on important engineering projects. Until 1964, the entire flow of the Colorado River was ponded in Lake Mead, behind Hoover Dam. It is estimated that some 129 million metric tons of sediment came to rest in the lake every year. At this rate the dam would have become useless in 100 to 200 years. To solve this problem, other dams have been built upstream. The older deposits of the Colorado occupy a deep trench that is a continuation of the Gulf of California. Here lie the sediments that

have been eroded from 630,000 square kilometers over a span of many thousands of years. The amount of material eroded during the formation of the Grand Canyon is but a small fraction of the total that has been carried through the canyon from upstream sources.

Deep-sea Sedimentation. Although the rates of deposition in the deep oceans cannot be directly observed, it is now possible, because of extensive drilling, to make reliable estimates. Average rate over many broad tracts is about 6 meters per million years or about 2.5 centimeters in 4,500 years.

VOLCANIC ACTION

Volcanoes have always attracted a great deal of attention. Ancient peoples called them "burning mountains" and "lakes of fire and brimstone." The old belief that the earth has a molten interior and a thin external crust traces directly to volcanic observation. Leaving aside the many geologic effects of volcanic action, we shall direct

Figure 3.13 Looking into the crater of Mount Vesuvius from an airplane. This view is intended to emphasize the curving cliffs of Mount Somma (above and to the right of Vesuvius), which are the remnants of the much larger and older cone destroyed by the exceptionally violent explosion of A.D. 79. (Courtesy of Cecil B. Jacobson.)

our attention here to the relationship between *volcanism* and uniformitarianism.

The sudden, dramatic appearance of smoking volcanic islands in the seas and oceans and of volcanic vents in the land has been observed for thousands of years. On the basis of these eruptions, people naturally argued that volcanoes, regardless of size, are created in sudden violent cataclysms. Some observers thought that volcanic mountains were pushed up from beneath the earth's surface like huge bubbles and were underlain by extensive caverns from which the lava had escaped or that they were ordinary mountains that were somehow connected with a source of subterranean lava. Always present in older theories was the idea of sudden, catas-

trophic growth. The possibility that volcanoes build their own cones arose after this idea was swept away.

Vesuvius. The history of Vesuvius, one of the world's greatest volcanoes, is instructive in respect to the emergence of our modern theory of volcanic action. The early Romans thought the volcano was totally inactive. Trees and other vegetation grew on the crater slopes, and in 72 B.C. the mountain was the site of the battle between Roman forces and the gladiators of Spartacus. About A.D. 63, earthquakes began to be felt in the area, and in August, 79, the volcano we now know as Vesuvius broke through the crater of its prehistoric ancestor, Mount Somma, and destroyed the populous cities of Pompeii and Herculaneum, the former under a rain of ashes and dust, the latter under a flood of mud derived from the heavy volcanic rains. From 79 to 1036 there were seven eruptions, and lava began coursing down the slopes for the first time. During the period from 1139 to 1631 little activity took place.

Trees and other foliage appeared once again, and the slopes of the mountain supported vineyards and towns. Then, in 1631, there was a tremendous explosion that emptied the crater and took thousands of lives. Since then, Vesuvius has been in a continuous state of unrest. The most violent action of the twentieth century took place during the Allied invasion of Italy in 1944.

Looking at Vesuvius in the light of geologic perspective, we see that the volcano's known history is merely a late phase of a long-continued process. On the northern side of the present cone is a lofty, semicircular cliff that is clearly the wall of a vastly greater prehistoric crater. The various beds of ash and lava that can be seen in this wall resemble the beds that were produced over the last 2,000 years and cannot be reasonably attributed to any process other than ordinary volcanic action.

Volcanic Islands. Amid clouds of steam and violent submarine eruptions, a new volcanic island appeared off the coast of Iceland on November 14, 1963. In one month it was 90 meters high, with a diameter ten times its height. This dramatic incident illustrates how the whole of Iceland has been created. About 150 to 200 volcanoes exist on this island, 30 of which have been active within the last 1,100 years.

Among the greatest volcanic groups on earth are the Hawaiian Islands. Mauna Loa on Hawaii extends about 4,170 meters above sea level, and the combined island group rests on the bed of the ocean 5,200 meters below. This huge mass consists chiefly of solidified lava and minor amounts of organic material. The presence of lava flow upon lava flow, each having a streamlike or threadlike form, gives an impressive picture of an extremely long past history. Even if the accumulation had been more rapid in the past, there is no evidence that it took place by processes other than those now in operation. Not all lava piles up in mountainous form. Vast, level plains, with fairly uniform sheetlike flows, occur in many places. There is no reason to believe that these sheets all erupted in one great cataclysmic upheaval, because the tops of many individual beds show the effects of weathering, and between many of the layers are old fossil soils, lake beds, peat bogs, and abundant fossil remains.

Figure 3.14 *A newly emerged volcano in eruption off the south coast of Iceland, February 1964. (Courtesy of James Crowe, NAVOCEANO.)*

Lava is one type of rock that can easily be observed in the process of formation. Its emergence in fluid form and its immediate consolidation into solid rock leave no doubt about its origin. Furthermore, lava is distinguished by various types of surfaces that form as the lava cools and by the inclusion of bubble-formed cavities. When it flows into water, lava congeals in characteristic pillowlike forms. Identical forms and structures are found in lava beds at all levels within the earth's outer crust and stand as convincing evidence of uniform action.

Development of Historical Geology

In a book of this type, there is neither the space nor the time for a complete survey of the development of historical geology. Nevertheless we should at least mention some of the key characters who played an important role in the development of geology as a science. In the fifth century B.C. the Greek historian Herodotus visited Egypt and reported:

. . . Egypt . . . is an acquired country, the gift of the river. . . . The following is the general character of the region. In the first place, on approaching it by sea, when you are still a day's sail from the land, if you let down a sounding-line you will bring up mud, and find yourself in eleven fathoms of water, which shows that the soil washed by the stream extends to that distance.

This is good thinking along uniformitarian lines. But Greek, Roman, and early Christian thinkers in general were obsessed with celestial influences and neglected to study the evidence that the earth itself provided regarding its origin.

With the Renaissance came pioneers of science. Leonardo da Vinci (1452–1519) was, among other things, an engineer who paid close attention to the materials of his roads and canals. His notes contain numerous passages that indicate that he understood many geologic processes, including the formation of beds of fossil shells in the same ways that shells originate and accumulate today.

A truly outstanding geologic thinker of the seventeenth century is Niels Stensen, usually known as Steno (1638–1687), a Dane who was trained in France and Italy. Among other accomplishments he wrote a small volume titled, *Prodomus, a Dissertation on Bodies Naturally Contained in Other Bodies* (1669). Steno wrote:

. . . That the beds of the earth, for the place and manner of their production, agree with those which turbid waters let fall. . . . That those bodies, which being digged out of the earth are altogether like the parts of plants and animals, were produced in the same manner and place in which the very parts of plants and animals are produced.

At what time these were formed any Bed, the Matter incumbent upon it was all fluid, and by consequence, when the lowest Bed was laid, none of the upper beds was extant.

When any bed was formed, its inferior surface, and that of its sides did answer to the surfaces of the inferior body and of the bodies lateral; but the superior surface was, as far as possible, parallel to the Horizon. . . . Hence it follows, that Beds either perpendicular to the Horizon, or inclined to it, have been at another time parallel to the same.

The changed situation of beds is the chief original of mountains.

All mountains to this day have not existed from the beginning of things.

In these few paragraphs one can detect a fundamental understanding of sedimentation, origin of fossils, order of superposition, original horizontality of strata, deformation of rocks, and the origin of folded or uplifted mountains.

The writings of da Vinci and Steno, perceptive as they were, appear to have had little effect on their contemporaries who were, in the main, catastrophists. But confrontations were inevitable as more and more people from practical necessity or intellectual curiosity began to take an interest in the earth.

Figure 3.15 Diagrams from Steno's Prodomus, published in Florence in 1669. These stages, starting with sketch number 25, illustrate Steno's concept of the evolution of Tuscany. To account for the creation of the valley in 23 Steno postulated the collapse of a large cavern shown by the blank space in 24. Subsequent filling of the valley and creation of another cavern are shown in 22 and 21. This method of valley formation is not possible on the scale Steno imagined, but he does try to come to grips with the problem of making tilted beds out of level ones.

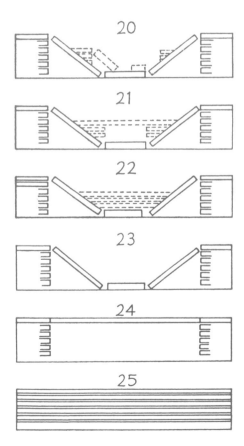

One of the battles of the formative period centered about the origin of the rocks of the earth's crust. This particular problem has little or no theological implication and, unlike the question of the origin of fossils or uniformitarianism, was argued without much reference to Scripture. Important and influential persons became involved, significant lessons about scientific thinking were learned, and the outcome was a turning point in the history of geologic science. One school, following the teachings of the famous German mineralogist Abraham Gottlob Werner, held that the crustal rocks were all precipitated or deposited from a universal ocean that covered the entire earth. This vast sea was not Noah's Flood but an earlier, primitive ocean. The opposing group, led by an equally important figure, the Scottish geologist James Hutton, contended that fire as well as water played a part in producing successive geologic "formations." The opposing factions were dubbed *Neptunists* and *Plutonists*, in allusion to the respective roles of water and fire that they emphasized in their hypotheses.

Werner, the Neptunist. According to the Neptunists, the earth was once submerged in an ocean of thick, muddy water that contained all the elements necessary to form the rocks of the crust. Arguing from the materials that he saw exposed in his native Germany, Werner supposed that the first deposits laid down by the seas consisted of granite, which he thought existed everywhere as a foundation for succeeding deposits. On top of the granite was deposited a mixture of more or less crystalline rocks, including gneisses, schists, and porphyries, all of which were considered to be chemical precipitates. As time passed and as the mother ocean began to subside, a third series of rocks formed, partly chemical precipitates and partly broken or eroded fragments. These rocks, which Werner called the "transitional series," include slate, schist, and similar rocks containing a few fossils, mainly of sea creatures. Next followed a series of chiefly sedimentary or fragmental rocks, including such common types as sandstone, conglomerate, chalk, gypsum, salt, and coal, generally with numerous fossils. A final veneer of beds, deposited after the ocean had subsided and exposed some of the higher lands, Werner called the "alluvial series." The "alluvial series" was made up of sand, clay, gravel, and other more or less unconsolidated materials that Werner thought were derived in part by erosion of the older formations.

The Neptunists thought that all the beds below the "alluvial series" were of universal extent and encased the nucleus of the earth like the shells of an onion. If Werner had traveled or been willing to accept the words of others, he would have realized that the arrangement of rock types differs greatly from place to place and that "universal formations" do not exist. Werner explained the presence of folded, tilted, and disturbed beds by saying that the material had settled in irregular ways and in places had slipped down from higher to lower places of deposition. He believed that deep valleys resulted from powerful currents and storms that had swept the primitive ocean. He offered no logical explanation for the final disappearance of the waters that once covered the earth except to say that they may have escaped into space. The contest between the Neptunists and the Plutonists eventually hinged largely on the question of the origin of *basalt,* a dark, volcanic rock that is common throughout the earth and on the ocean floor. Werner felt that it was formed as a precipitate from his primeval ocean. Hutton considered it to be an extrusion of once-melted rock from the depths of the earth. Observation of streams of basalt issuing from active volcanoes finally convinced most geologists that Werner could not possibly be right. The application of uniformitarianism led to the downfall of the Neptunists.

More about Hutton. James Hutton (1726–1797), leader of the Plutonists, was a man of unusually wide attainments. He was qualified as a doctor of medicine and practiced in Edinburgh and Paris. Later he turned to practical agriculture, and it was not until his middle age that he came to devote his entire time to geology. He traveled and observed for 17 years and in 1785 presented his views on geology in a paper entitled *Theory of the Earth, or an Investigation of the Laws Observable in the Composition, Dissolution, and Restoration of Land Upon the Globe.* Underlying this paper was the idea that the basis for understanding the earth lay in a study of its materials. Hutton stressed his belief that present-day rocks are composed mainly of the waste of older rocks carried down by rivers and deposited in water. He reasoned that loose sediment can be converted to solid rock by great heat and pressure. The subsequent reappearance of rocks at the surface, often contorted and disrupted, stemmed from the expansive power of subterranean heat. Veins and masses of molten rocks that were forced into overlying rocks or that burst through the earth's surface in the form of lava or ash were further manifestations of the power of heat. Hutton noticed that when any mass of rock was exposed to the atmosphere, it became subject to weathering and to the effects of running water. To him the cycle seemed interconnected, coherent, and continuous. Sediment worn from the mountains is carried to the sea where it becomes distributed in even layers, or beds. Under pressure from overlying material and heated by subterranean sources, the originally loose sediment hardens into solid rock.

Through the intervention of uplifting forces accompanied by intrusions of molten rock, the beds of sediment are thrust upward, broken, buckled, and folded to become land. The weather immediately attacks the land to produce new sediments and launch another cycle. Little wonder that Hutton should say that in this process he saw "no trace of a beginning and no prospect of an end."

As long as he dealt with observable processes, Hutton made few mistakes. In trying to explain what went on beneath the surface to convert sediment to rock, he was less successful. He overemphasized the effects of heat and understood little about pressure and the cementing effects of mineral matter from underground solutions. For example, he conceived that puddingstone, a variety of conglomerate, was formed by a mass of "melted flint" being forced into a bed of loose gravel, whereas actually the "flint" is silica deposited from underground water.

Hutton's success lay in field observation. He paid particular attention to matters that his rivals considered trivial. When he found thin "veins" of igneous rock penetrating upward into sedimentary rock, he realized that here was evidence not only that certain rocks originated below the earth's surface but also that they were younger than the rocks they transected. When he found level undisturbed beds over crumpled and disturbed layers, he recognized that the interval between their deposition must have been long enough not only for the crumpling to have taken place but also for much of the older rocks to have been leveled by erosion. In interpreting these unconformities between rocks, Hutton was far ahead of Werner, who attributed the unconformities vaguely to sliding or crumpling.

Unfortunately, Hutton's style of writing was heavy and involved, and his original works held little general appeal. In 1802, 5 years after Hutton's death, his friend John Playfair wrote a book giving a more popular and precise account of Hutton's idea. This well-written work, called *Illustrations of the Huttonian Theory of the Earth*, exercised a wide influence in spreading the doctrine of uniformitarianism. The central theme of uniformitarianism as stated by Playfair still stands as a guide to scientific thinking:

Amid all the revolutions of the globe the economy of Nature has been uniform, and her laws are the only things that have resisted the general movement. The rivers and the rocks, the seas and the continents, have been changed in all their parts; but the laws which direct those changes

Figure 3.16 *Sir Charles Lyell (1797–1875). His writings established historical geology on a firm basis and greatly influenced later workers, including Charles Darwin. (Crown copyright, Geological Survey photograph. Reproduced by permission of the Controller of Her Britannic Majesty's Stationery Office.)*

and the rules to which they are subject, have remained invariably the same.

Lyell—Observer and Writer. Hutton had laid the groundwork for historical geology by setting forth the idea of uniformitarianism. It remained for others to gather the facts that would support or condemn it. The main burden of proof was furnished by a man born in Scotland in 1797, the year of Hutton's death. This man was Charles Lyell, son of a naturalist father and trained in law. He had an intense interest in geology and was motivated by one great urge: to go and see. He was able to travel extensively in many European countries and in America, where he was constantly on the lookout for signs of change and alteration of natural features. Everywhere he found evidences of uplift and subsidence, of erosion and deposition, of volcanic activity and long-continued glacial action. These evidences were too strong to be ignored, and Lyell proclaimed them boldly. His books were very influential and many editions were published.

Lyell states his case for uniformitarianism in the following manner:

. . . in attempting to explain geological phenomena, the bias has always been on the wrong side; there has been a disposition to reason a priori on the extraordinary violence and suddenness of changes, both in the inorganic crust of the earth, and in organic types, instead of attempting strenuously to frame theories in accordance with the ordinary operations of nature.

THE MODERN VIEWPOINT— A COMPROMISE

It is unfortunate that the debate concerning the validity of catastrophism or uniformitarianism has been carried out on a "one or the other" basis. Certainly no well-informed geologist would today deny the importance of catastrophes in creating the geologic record. However to escape some of the connotations of the word "catastrophe" we might prefer the term "unusual event." It now appears that in the development of the earth there has been, and continues to be, a complete gradation from ordinary everyday processes to earth-shattering events of cosmic proportions. Where in this gradational series should we begin to apply the term "catastrophic"? Where would we feel safe in defining any process as nonuniform?

Mankind has recorded some unusually intense natural disturbances, and these were without doubt catastrophes to those immediately involved. The Alaskan earthquake of March 27,

Figure 3.17 *Mount Katmai, an Alaskan volcano, after the massive eruption of 1912. Trees in the foreground were killed by the heavy fall of ash that covered an area of hundreds of square kilometers. The eruption completely destroyed the upper part of the mountain and produced an estimated 20 cubic kilometers of pulverized rock. (Photograph by R. F. Griggs, copyright National Geographic Society.)*

1964, which was 8.5 on the Richter scale of earthquake magnitude (the most intense earthquake ever recorded was 8.9 on the same scale), lasted only 3 or 4 minutes. Its damage zone covered some 129,000 square kilometers, and shock waves were felt over an area ten times as large. The maximum uplift of land was about 11 meters, and the area of crustal deformation was 259,000 square kilometers.

The greatest recorded volcanic eruption is that of Krakatoa in 1883. An island with an area of 47 square kilometers and whose highest point was some 427 meters above sea level was destroyed so completely that all that was left of it was a submarine depression. A vast blanket of pulverized dust and ash covered sea and land over hundreds of square kilometers and obscured the sun up to 160 kilometers away. Noise of the explosion was heard 5,000 kilometers from the island, and giant sea waves rolled outward for 12,500 kilometers. At least 36,000 persons were killed.

Other even more devastating and more widespread events have probably been witnessed by man but are known only from vague and exaggerated legends. One of these, now verified by modern research, is the eruption of the volcano Santorini, on the Greek island Thira, about 1400 B.C. It is believed that the explosion was twice as powerful as Krakatoa. Most of eastern Crete would have been under a thick layer of ash, and the Minoan civilization that flourished there was abruptly snuffed out. Vast tidal waves may have devastated many Mediterranean coasts and perhaps caused a widespread cultural decline. Some scholars feel this event may be the cause of the plagues mentioned in the Bible. The widespread prevalence of flood myths is not to be taken lightly, and such tales certainly record an impressive natural event of unusual magnitude. However, scientific verification is lacking in most cases.

Evidence of any event that occurred before the coming of man can be found only in the geologic record. Look at a map of Hudson Bay

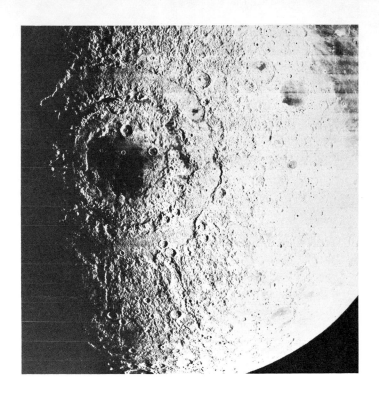

Figure 3.18 Large impact feature on the moon, the Mare Orientale, as shown in a photograph taken by Orbiter IV. Only the edge is visible from earth. The outer circle is about 1,050 kilometers in diameter. Sometimes called the "Bullseye," this feature can be nothing less than the impact scar of an unusually large asteroid or comet. (NASA.)

and note the circular configuration of the southeastern shore just north of James Bay. This arcuate configuration is thought to be a scar left by a gigantic meteorite. Other effects of this collision can be imagined. Geologists cannot deny that such extraterrestrial intrusions probably have occurred and are responsible for some features of the geologic record. The millions of impact craters on the moon tell us something—could the earth have totally escaped the bombardments that have pockmarked the moon? Bear in mind that erosive and tectonic effects can erase marks on earth that would be permanently preserved on the dead moon. Space contains a fearsome array of missiles. More than 1,650 large asteroids have been observed traveling between the orbits of Mars and Jupiter, and thousands of smaller ones are also suspected. The random orbits of many of these pass dangerously near the earth. Eros, about 19 kilometers long and 6 kilometers thick passed within some 25 million kilometers of the earth, and there have also been other near misses. It has been calculated that a relatively large asteroid might be expected to strike the earth every 500,000 years.

The moon has been a dangerous neighbor. It was apparently very near the earth sometime before 1.5 billion years ago. If this is true, the mildest effect would be the gigantic ocean tides that could sweep over practically all the emergent land. Internal friction would simultaneously generate volcanic action, earthquakes, and even massive internal melting. Such an unquiet world is obviously ill-suited to advanced forms of life.

This topic should not be expanded into the realm of fantasy. A scientist is guided best by what he actually observes. Some good questions might be the following: How much of what the rocks reveal can be explained by ordinary, gradual, everyday effects; how much by the rare but known events such as a large earthquake or hurricane; how much by the very rare and powerful events such as an asteroidal impact; and how much is unexplainable by any known natural agent? Although a person will be guided most by what he knows from personal experience, it is nevertheless true that practically all details of the geologic record can be explained by pure uniformitarianism—that is by known, well-understood principles. A residue of large-scale effects, perhaps reflecting powerful, short-lived events, must probably be termed cataclysmic.

Now that we know that the large-scale or first order features of the earth almost certainly originate through slow gradual processes, we are left with only a few situations for which we can offer no clear explanation. Most of these concern the fossil record. Why did life apparently expand greatly at the beginning of the Cambrian, and what caused the great exterminations at the end of the Paleozoic and Mesozoic? We await acceptable explanations for these events.

On one thing geologists stand firm: The history of the earth must probably be confined within a period of about 4.5 billion years. By contrast most catastrophists must maintain their position because of their belief in a 6,000-year-old earth. Catastrophes are essential if all known events of earth history and the succession of life forms are to be compressed into such a short period.

A SUMMARY STATEMENT

There are two opposing viewpoints regarding the origin and past history of the earth. *Catastrophism,* arising from a supposed need to compress earth history into 6,000 years to satisfy a literal view of Scriptures, was the predominant view well into the seventeenth century. *Uniformitarianism,* the belief that the earth has developed slowly and gradually by the same processes we observe operating today, has displaced catastrophism and gained almost universal acceptance. There have been adjustments in thinking however, and today's view is somewhat of a compromise. No proof for a succession of earth-shaping violent catastrophes has been discovered, though some catastrophes probably occurred. Likewise the history of the earth is known not to be an endless succession of similar cycles as Hutton appeared to believe; changes have been progressive and forward-moving, not repetitious and self-destructing.

Examples of uniformitarianism are seen in the historic effects of marine and non-marine erosion, the retreat of shorelines, the rise and fall of sea level, and in earthquakes and volcanic eruptions. Some of these are violent and locally very upsetting but none is truly catastrophic in the sense of being universal and permanent in effect.

Application of the established laws of physical science over an ample span of time would seem to explain in a satisfactory way what is seen in the stratified rock record. Beyond this—during the early history of the earth for which there is no sedimentary record—there were undoubtedly catastrophic, or at least very widespread and powerful, events. The older and larger craters of the moon are reminders of cosmic violence that must have affected the earth.

FOR ADDITIONAL READING

Adams, Frank Dawson, *The Birth and Development of the Geological Sciences*. New York: Dover, 1954.

Albritton, C. C., Jr., ed., *The Fabric of Geology*. San Francisco: Freeman, Cooper, 1964.

Fenton, C. L., and M. A. Fenton, *Giants of Geology*. Garden City, N.Y.: Doubleday, 1952.

Geikie, Archibald, *The Founders of Geology*. London: Macmillan, 1905.

Gillespie, C. C., *Genesis and Geology*. New York: Harper & Row, 1959.

Lyell, Sir Charles, *The Principles of Geology*, 2 vols., 12th ed. London: Murray, 1875.

Moore, Ruth, *The Earth We Live On*. New York: Knopf, 1956.

Pearl, Richard M., *Geology*. New York: Barnes & Noble, 1960.

Playfair, John, *Illustrations of the Huttonian Theory of the Earth*. Edinburgh, 1802. (Reprinted by University of Illinois Press, Urbana, 1956.)

Toulmin, Stephen, and June Goodfield, *The Discovery of Time*. New York: Harper & Row, 1965.

LAYERED ROCKS
AND THE LAW OF SUPERPOSITION

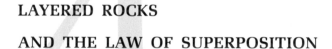

The chief characteristic of sedimentary rocks is that they are formed in layers or beds. This is because the materials of which they are composed have accumulated in an intermittent way under the influence of gravity. Most sediments are made up of small particles of preexisting rocks that have been picked up, transported, spread out, and deposited by currents of air or water. Other sediments, such as salt, are precipitated directly from water; a few peculiar types, such as coal, consist of organic material buried in the place of growth.

Horizontal, well-exposed strata in the upper part of the Grand Canyon, Arizona. (U.S. Bureau of Reclamation.)

Although sedimentary beds vary greatly in thickness, composition, color, internal structure, fossil content, and reaction to erosion, they have the common characteristic of having been formed in layers that are (or were originally) level or nearly so. In the language of geology, a single bed is called a *stratum,* a number of beds are called *strata.* The study of strata is called *stratigraphy,* and a person who specializes in the subject is a stratigrapher. Stratigraphy is a very important branch of geology as we shall see in later chapters. It concerns the definition and inter-pretation of stratified rocks, the conditions of their formation, their characteristics, arrangements, sequence, age, distribution, and correlation by the use of fossils and other means.

Beds and Bedding Planes

Successive beds of sedimentary rocks are separated by bedding planes, or surfaces of weakness and discontinuity. These surfaces may be many feet apart or so closely spaced that paper-thin

Figure 4.1 Stratification planes permit easy splitting of flagstones. The very evenly bedded rock from which the slabs have been taken is seen in the background. Quarry at Saõ Paulo, Brazil. (Courtesy of Darry Closs.)

layers are produced. Bedding surfaces are mostly smooth and flat because of a number of actions that take place during the interruptions in deposition. Currents of air or water may have leveled a surface by removing high spots and filling low ones; cohesive forces may draw the particles of one bed together before the next episode of deposition; or actual differences in composition may

Figure 4.2 Variety in stratified rocks. (a) Sandstone alternating with shale in thin, even beds. (b) Extremely thin shale with a high content of organic material (notice that the beds are so flexible that they bend without breaking). (c) Sandstone in beds of uneven thickness with thin partings of shale marking the bedding planes. (d) Limestone beds with thin shale breaks between. (U.S. Geological Survey.)

(a)

(b)

(c)

(d)

Figure 4.3 *Cross section of a buried stream channel exposed by road building, central Utah. This records the cutting and gradual filling with sand of a small stream bed in a floodplain environment. The concentration of coarse sand in the stream bed and the deposits of clay and silt along the sides and over the banks are characteristic. Because only a short time-period is indicated and because there is no basic change in the local geologic activity, the bottom of this channel would be classed as a diastem.*

lead to planes of weakness. Whatever the cause may be, the individual beds do not merge with, or solidly join, those above and below. It is the presence of bedding surfaces that makes possible the splitting of flagstones and building stones into thin slabs and also allows weathering and erosion to produce the alternating protruding and receding surfaces seen on cliffs and hillsides.

It seems probable that in most geologic formations the total time represented by the bedding planes equals or exceeds that represented by the beds themselves. If this is true, much of the time involved in laying down a geologic formation is not represented by actual material. For the significant but relatively short interruptions between beds the term *diastem* is used. Because diastems usually separate rocks that are similar in composition and origin, they do not signify drastic changes in environment or type of material.

UNCONFORMITIES

Between many rock layers there is an abrupt, striking change in composition or material. Most of these changes are indicative of major lapses of deposition and are of great importance in interpreting the local or regional history. Many such interruptions represent very long periods, even many millions of years, during which much erosion, extensive earth movements, or other important events may have taken place. We may think of these major interruptions in the record as ancient erosion surfaces cut into older rocks and buried by younger ones. They are, in fact, buried landscapes or seascapes. Major discontinuities in the record, during which deposition ceased for a considerable span of time, are called *unconformities*, of which several types are recognized. If the strata below a break had been tilted or deformed before being buried beneath those

Figure 4.4 Tilted strata. Aerial view of a thick succession of thin beds of limestone, southern Oklahoma. The limestone has been tilted at a moderately high angle, eroded to rolling hills and laid bare so that the structure is revealed. Height of camera about 305 meters. (Courtesy of Frank Melton.)

Figure 4.5 Angular unconformity between tilted beds of Mesozoic age and horizontal beds of Cenozoic age, southern Nevada. Truck in foreground gives scale. (Courtesy of Stephen D. Olmore.)

above an *angular unconformity* results. If the rock below is igneous or has lost its bedding by metamorphism the plane of juncture is a *nonconformity*. If the older and younger sets of beds are parallel but separated by a well-marked surface of erosion, the interruption is a *disconformity*. Many unconformities are difficult to locate because the rocks above and below are similar, there are few evidences of erosion, and the break to all appearances resembles an ordinary bedding plane.

CONTACTS

Any fairly sharp or well-marked surface that marks a juncture between two bodies of rocks is called a *contact*. Unconformities and bedding planes are two types of contacts that occur in sedimentary rocks, and there are special types involving igneous and metamorphic rocks and even individual mineral or rock fragments. Some contacts may be selected as convenient places to divide rocks into formations for purposes of mapping or description. These contacts will appear on maps as lines between the units we wish to differentiate. If two rock masses grade into

Figure 4.6 Well-exposed contact between two sedimentary formations in the San Rafael Swell, Emery County, Utah. Below is the light-colored Navajo Sandstone, a cross-bedded, massive deposit of windblown sand; above is the darker, flat-bedded Carmel Formation, a deposit of limestone and shale containing marine fossils. The contact represents a passage of time during which the environment changed from a dry desert to a shallow sea bottom. This contact is an important one in the western United States and can be traced for hundreds of kilometers. (Courtesy of Harry D. Goode.)

each other and do not exhibit sharp, abrupt changes, it is customary to draw the lines on some convenient but arbitrary basis.

ROCK EXPOSURES AND OUTCROPS

We scarcely need to be reminded that solid rock is not visible everywhere on the earth's surface. Vegetation, soil, water, and the works of man cover extensive tracts, and the solid *bedrock* may lie far below the surface. If the rock does appear at the surface, we say that it is *exposed,* or "crops out." Areas that are thus made visible are called *exposures,* or *outcrops.* The geologist is frequently greatly hampered by not being able to find unobscured bedrock and has to search diligently along stream banks or in road cuts and quarries. He may even have to dig or drill through the surface material to get samples of the rocks beneath.

Rock exposures are best seen in arid regions where there is little soil or vegetation. The Grand Canyon of the Colorado is cut in an area where the rock layers are clearly exposed. The same rocks in a humid area would be covered by soil and vegetation, and the steplike profile of hard and soft layers would be replaced by more rounded contours.

The Making of Rocks

Historically, the application of uniformitarianism to an understanding of the formation of sedimentary rocks was a slow process. Translating everyday sights and experiences into a concise theory of the meaning of rock layers was a difficult process because men knew very little about conditions within the earth. It is common enough for someone to observe firsthand the upper surfaces of newly deposited sediments. Simply by walking across sand dunes, beaches, floodplains, or tidal flats we can see the beginning of geologic formations and leave our footprints on potential bedding planes. An observant person can readily identify the natural agents that have carried the sediments to the spot where they now rest, and with a little detective work he can determine from what direction and from what source the sediments were moved. He may observe ripple-

(a)

(b)

Figure 4.7 A study in ripple marks. (a) A large patch of ripple-marked sand on Slim Island in the Yukon. Only the light patch is dry. (b) An exposure of solidified ripple-marked sandstone turned on end. The surface was exposed by mining operations near Golden, Colorado. Ripple marks are most common in sedimentary rocks and are useful in determining the mode of origin of beds that display them. (U.S. Geological Survey.)

like markings or smooth surfaces being formed by water or air, and he may see dead organisms being buried and may recognize that these are potential fossils.

Here, unfortunately, everyday experience ends. If the sediments we see in the making are eventually buried and become solid rocks, they will be hidden from view for a very long time and will certainly be greatly changed in the meantime. More important, when they are exposed again, probably only their edges will be showing (as in the Grand Canyon) and not the original upper surfaces. A geologist must mentally reconstruct three-dimensional rock masses from the two-dimensional views he sees on canyon walls or mountain slopes.

The sedimentary rock layers with which the geologist deals are like a book. When the book is closed we see only the edges of numerous pages, but we know that each page has certain measurable dimensions and bears information in the form of writing or illustrations. The book of

the stratified rocks is so vast that the geologist must be content with a study of the edges of the pages, though he may be permitted a more extensive view of their surfaces here and there. Because of these limitations the geologist seldom has all the information he would like and must reconstruct invisible subsurface conditions in his imagination, aided as far as possible by mathematical calculations, diagrams, and projected cross sections.

The ability to visualize underground conditions on the basis of limited surface views requires not only experience but also extensive training and a high degree of native imagination. An understanding of the hidden regions of the earth's crust is obviously important in finding oil pools and deposits of valuable materials of any kind. Although haphazard mining or drilling might uncover these deposits, it is helpful to be able to pinpoint as closely as possible the best places to explore before expensive operations are begun.

(a)

(b)

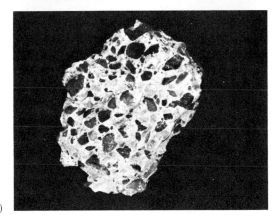

(c)

Figure 4.8 *The consolidation of rock illustrated by stages in the cementation of gravel. (a) A pile of loose, stream-worn pebbles. (b) A specimen to which only enough material has been added to hold the fragments together. (c) A well-cemented mass of pebbles in which the process of cementation has been completed and has resulted in a very solid lump of material called "conglomerate."*

Rock to Sediment and Sediment to Rock

Everyone has noticed at one time or another that weathering and erosion of rocks produce particles that can be carried away by the wind or water. That rocks may be broken down into sediment needs no special explanation, but the manner in which loose sediment may again become solid rock is not so obvious. The conversion of loose sediments to solid rock is called *consolidation*, a process that includes any action that increases the solidity, firmness, or hardness of sediments. Geologists are thoroughly familiar with the process in spite of the fact that it is mainly a subsurface phenomenon. The many thousands of drill holes that are put down in search of oil, water, and minerals have yielded a tremendous variety of sediments in all stages of consolidation. These samples reveal most of the actual changes by which, for example, sand becomes sandstone and mud becomes shale. Laboratory experiments in which heat and pressure are applied to loose material have also added a great deal to our understanding of the rock-making processes.

CEMENTATION AND COMPACTION

The two most important actions that convert loose sediment to solid rock are *cementation* and *compaction*. Cementation is the process by which certain material, usually silica (SiO_2) or calcite ($CaCO_3$), is deposited between sand grains or other fine particles by underground water. The cementing agent then hardens or crystallizes and binds the particles together. Sediments may be firmly or loosely cemented according to the kind of cement and the amount that is deposited, and we may speak of hard or tough rocks or soft and friable ones depending on the strength of the cement. On rare occasions the cementing agent may be a valuable material such as copper, silver, or uranium minerals.

Compaction is the process by which small fragments are brought closer together with or without pressure, so that they occupy less space and form a more dense type of material. Cohesion brought about by the attractive forces be-tween atoms and molecules is important. Observations and experiments indicate that a bed of loose sand may be compacted to 80 percent of its original volume, whereas a clay that has been buried to a depth of about 900 meters may occupy only 60 percent of its original volume. As compaction proceeds, any gas or liquid that a

Figure 4.9 Cross-bedding, a common feature of sandstones. (a) Wind, or aeolian, cross-bedding in the Navajo Sandstone of central Utah. Here erosion has cut into solidified sand dunes. Aeolian cross-bedding shows numerous curving laminations and the individual sets are not uniform in thickness or in direction of inclination. (b) Stream, or fluvial, cross-bedding in Lower Cretaceous beds near Grand Junction, Colorado. In this type of bedding the individual groups, or sets, of bedding are quite uniform in appearance and each cross-bed is inclined in the direction of stream flow. (c) Small-scale cross-bedding in coarse-grained sediment. This type is formed in shallow but rapidly flowing water. Specimen is about 10 centimeters long. (a, U.S. Bureau of Reclamation; c, courtesy of Eliot Morris.)

(a)

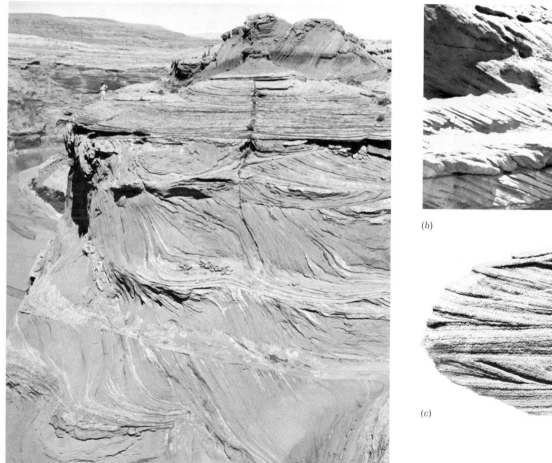

(b)

(c)

Figure 4.10 Rounded and frosted sand grains such as these bear evidence of a long period of transportation by wind or water. Highly magnified. (U.S. Geological Survey, photograph by K. B. Ketner.)

rock contains may be forced out and either flow elsewhere underground or rise to the surface of the earth.

Certain materials, such as ordinary salt, that crystallize or precipitate out of solution may be firm and somewhat hard from the moment they are formed, but burial and pressure usually makes them denser and more compact.

Historical Meaning of Rock Features

The realization that loose material can become rock just as surely as rock can crumble to loose material was an important advance but was only the beginning insofar as our understanding of the meaning of rocks is concerned. What of the large and small structures within formations and the relationships between entire beds or formations? To the modern geologist, rock formations are much more than mere beds of shale or limestone or conglomerate that were once mud or ooze or gravel. They are manifestations of events, environments, and conditions of past geologic ages.

It is not enough, for example, to say that sandstone was once sand, for there are many types of sand and many conditions under which it accumulates. Sand is the common material that makes up beaches, stream beds, and desert dunes, where we can see it being deposited at the present time. By noticing how sand bodies are forming under these differing conditions, and by applying our observations to ancient solidified deposits, we can reconstruct something of the past. Figure 4.9 shows the common types of bedding that occur in sand laid down by wind and water. Even the study of individual sand grains is revealing. Grains that have been shifted about by wind for long periods tend to become rounded and smoothed. Grains that have traveled only a short distance may be angular and may retain some of the original crystal form. Although most sand is quartz, there may be here and there a grain of other material. The minor constituents serve to tell us where a sand originated and how far it has traveled.

Pebbles making up gravel or conglomerate also tell many interesting stories. By their size we

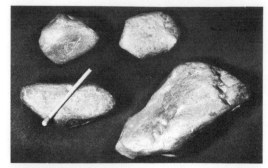

(a)

(b)

(c)

Figure 4.11 Every stone has a history. Certain characteristic signs enable the geologist to distinguish pebbles shaped by different agents and to reconstruct something of the environment in which they were formed. (a) Glacial pebbles have been picked up and carried along by glaciers. They are characterized by blunted edges and flattened faces with numerous striations and scratches. (b) Gastroliths are pebbles that are said to have been picked up and swallowed by dinosaurs to aid in the digestive process. The pebbles are usually brightly colored, extremely hard, and highly polished. (c) Wind-etched pebbles have been subjected to the impact of wind-driven particles and may be pitted, etched, and faceted. They occur in the vicinity of glaciers and in desert regions.

can guess the strength of the currents that moved them; by their degree of rounding we judge how far they may have traveled; by their composition we learn what formation supplied them; and by marks on their surface we can in some cases detect evidences of wear by ice or running water or by wind-driven sand.

Paleogeographic Reconstructions

Such minor matters as the shapes and surface markings of pebbles, the wear displayed by sand grains, and the type of bedding that a rock stratum contains are all understandable clues to local past conditions. But more fundamental interpretations are possible: *When all available clues are assembled and evaluated in their proper relationships, it is possible to reconstruct the environment in which specific sedimentary deposits originated.* This information is usually presented in the form of *paleogeographic maps,* which show the geography of specific intervals of the past. A number of these are presented in later chapters to illustrate generalized conditions of the successive geologic periods.

A most important line drawn by map makers is that between land and sea. This is also a boundary of great significance to paleogeographers. The tracing of ancient shorelines is, in fact, one of the major tasks of geologists. When former shorelines are found and followed they serve to outline the continents, islands, seas, lakes, and oceans of the past.

A commonsense approach goes far. Marine conditions are indicated chiefly by the saltwater organisms that abound on coral reefs or on favorable areas of the sea bottom. Distance from shore and depth of water are indicated not only by different types of organisms but also by the kind of sediment in which they are buried. Judged by modern conditions, certain organisms, such as oysters, prefer shallow, near-shore environments; others, such as barnacles, must have solid places of attachment; other life forms, including most types of clams, thrive best on sandy bottoms; still others, such as sea cucumbers, prefer mud or ooze. Reefs are almost universally indicative of shallow, near-shore conditions. Considering type of sediment as such, we know that gravel and sand are most commonly found along the actual beaches; farther out, in calmer water, medium-to-fine sand and silt come to rest; in deeper tracts, along the continental shelves, still finer mud or clay is found; and in the open ocean, hundreds or even thousands of kilometers from land, only the very finest sediment, called ooze, is produced.

Gradations in sediment are indicative chiefly of shifting energy conditions; as a rule, the greater the local display of energy, the coarser the sediment. Cobbles and pebbles indicate strong waves and currents; sand suggests moderate current action; fine clay or ooze implies very little if any movement of the overlying water. Energy conditions are also made manifest by internal or primary structures. The example of cross-bedding has been given. Sand moved consistently along the surface by water or by wind currents tends to show cross-bedding; on the other hand, finer material, settling directly downward, will accumulate in even, very extensive layers.

Although nonmarine rocks are far less common in the geologic record than are marine varieties, those that are present are of unusually great significance. From these we learn about land life and the environments with which man is more familiar. From land-laid sediment we gather fossils of land animals—bones of mammals, reptiles, and birds (together with their tracks). Here we also find rich stores of plant material—leaves, trunks, seeds, and pollen grains. As we know from experience, remains of land organisms may float out to sea to become buried in near-shore deposits; naturally we find many of these in fossil form and must judge their importance less in terms of where they are buried than in terms of where they actually lived.

Land environments in general are less stable, more transitory, and have more energy than marine ones. Thus we find evidence of constant movement, strong current action, erosion, reworking, and redeposition in most land-laid deposits. Coarser grades of sediment, including much conglomerate and coarse sand, are common. Finer deposits are rare. A little thinking and observation will reveal even to a casual observer a great number of characteristics that might be expected of land-laid sedimentary deposits. The important, but not so obvious, point is that few land-laid sediments will become long-lasting parts of the geologic record. Anything that is in the open, and thus subject to erosion and weathering, is transitory. Only when rock material reaches a position either in a permanent body of water or under deep cover of sediment is it likely to be preserved for a very long period.

The foregoing paragraphs present in a brief way the correlations of everyday experience, observation, and synthesis that go into the location of ancient shorelines and related features. In an ideal case we would be able to find the transition from marine to nonmarine deposits by observing the fossil contents of the rocks, the type of sediment represented, the energy requirements of deposition, and the environments that can be reconstructed. For each of these there will be gradation indicating progression either toward or away from the shoreline. To grasp the meaning of the gradations it is necessary to study many samples or exposures. Understandably, much work, time, and expense is involved, especially if the information must come from holes drilled for water, oil, or gas.

After the broad outlines of lands and seas have been established it is frequently possible to fill in lesser geographic features. Of these a few may be mentioned together with the geologic evidence for their existence.

River valleys: Characterized by irregular, interbedded deposits of somewhat coarse material representing channel fillings and finer, more evenly stratified deposits representing floodplains. Deposits contain characteristic plant and animal fossils.

Deltas: Intergrading marine and nonmarine deposits as judged by their contained fossils and types of sediment are characteristic. Fragmentary or broken fossils of land organisms are found among less-disturbed marine types. Gradation from coarse materials near the river mouth (apex of delta) outward toward finer deposits around the margin. Coal may be found locally near the seaward edge.

Bays, lagoons, and estuaries: Mixed deposits containing varied plant and animal fossils. Evidence of obstacles such as reefs or ridges that have restricted circulation. The degree of restriction that prevented a "normal," or open-sea, type of environment can be judged by the presence of unusual sediments, such as salt, black organic-rich shale, or coal.

Lakes: Fossils are of paramount importance in deciding whether an extensive water-laid deposit originated in fresh water or salt water. Here the evidence provided by fish, arthropods, and mollusks is critical. If no fossils are found it may be difficult to distinguish lake deposits purely on the basis of the type of sediment or energy levels. The presence and configuration of the barriers between the ocean and lake may sometimes be determined.

Deserts: Thick, cross-bedded deposits of well-rounded sand grains are usually of desert origin. Other evidence in the form of abraded or grooved pebbles is helpful. Although life is rare in deserts, the presence of fossils of land organisms, especially those known to favor drier environments, is diagnostic. Evidence of temporary water bodies (including the presence of salt beds) helps in the reconstruction. Although red-colored sediments are thought to reflect arid or semiarid conditions, this criterion must be used with caution.

Mountain ranges: Few mountain ranges are preserved as such in the geologic record because the burial of such massive features is rarely possible. Nevertheless, the following are some of the many indirect, but satisfactory, clues that may be found: (1) the preservation of the lower levels ("stumps" or "roots") of ranges whose upper levels have been eroded away (the Appalachian Range is obviously a worn-down remnant of a formerly higher range); (2) the presence of eroded and transported debris, especially coarse conglomerate or sand appropriate in size and composition to a nearby elevated tract; (3) evidence that a barrier once existed as shown by different life forms on each side (Isthmus of Panama, Andes Range). Most great mountain ranges are marked by volcanoes and deep-seated igneous intrusions. Remnants of igneous rocks or of identifiable eroded fragments, no matter how small, help localize volcanic belts and occasionally even individual volcanoes.

Glaciers and ice caps: Glacial deposits have definite characteristics; included are largely unstratified mixtures of rock fragments of all sizes called till (tillite if converted to rock) and local lenses of layered or banded fine material representing marginal lakes. Especially significant are smoothed, polished, or grooved surfaces worn down by passage of the ice.

Other examples, especially those representing subenvironments of the seas or oceans could be given, but space is limited. Many sedimentary environments are known to favor the accumulation of natural mineral products such as coal, salt, placer gold, sulfur, uranium, diamonds, iron ore, and most types of building stone. To locate or outline the characteristic environment is obviously most helpful in the search for a particular mineral product.

The Law of Superposition

From observations and reasoning about the field relations of sedimentary rock layers came one of the fundamental guides of historical geology, the *law of superposition*. Briefly, this means that *in* any undeformed sequence of sedimentary rocks (or, other surface-deposited material such as lava) each bed is younger than the one below it and older than the one above it. This statement expresses such a simple and self-evident fact that it seems scarcely worth emphasizing. Nevertheless, it remains the most important generalization in the whole realm of earth history. If this principle is not reliable, geologists will have to abandon any hope of unraveling the past history of the earth.

Although the idea of superposition is easily understood, we may have difficulty in grasping the scale of the masses of rock involved and in determining the original arrangement of layers in a local area so that the past history of the area can be reconstructed.

Superposition in ideal form is illustrated by

a complete file of newspapers neatly stacked with the first issue at the bottom and the latest at the top. Even with the dates removed, the newspapers still furnish a reliable history of events in their correct order of occurrence. But nature provides few well-arranged piles of rock; there are many gaps and omissions in the geologic story. Some events are well recorded; others are only hinted at. Many individual pages are disordered and scattered, and some pages are altered and obliterated by heat and weather. In spite of these imperfections, we must conclude that any event, any condition, or any fossil found recorded in a lower stratum must precede any condition, event, or fossil recorded in a higher one.

Figure 4.12 may help you understand the ideas we have just presented. The arrangement of the playing cards in the diagram corresponds very closely to a typical geologic map of a section of the earth's crust with flat-lying sedimentary formations. The analogy is accurate because we are dealing in both cases with relatively thin but widespread sheetlike bodies that are piled one upon another in an overlapping manner. Note in Figure 4.12 the following:

1. No one spot on the "map" has all the cards (or formations) present in the area. By applying

Figure 4.12 A stack of playing cards illustrates the law of superposition. The arrangement from bottom to top shows the order in which the cards were laid down. The boxes on the right give the order of superposition for all the cards except the one that does not touch the others. The "age" of this card cannot be determined by superposition alone. Compare this with the simplified map in Figure 4.13, which shows how superposition is used to determine the relative ages of stratified rocks.

Explanation

5 ♥

9 ♠

2 ♣

4 ♥

10 ♦   8 ♦

7 ♠

2 ♦

6 ♣

8 ♠

Figure 4.13 *A geologic map of part of north-central Texas. Here the rocks are all of sedimentary origin and are almost flat-lying. The correct order of deposition can be determined by superposition of the various formations. Problem: How do we know that the rocks on the left side are older than those on the right side, and which way do the rocks of the older and younger sets of formations dip, or incline? Hint: If you were standing on the banks of one of the rivers where it passes from one formation to another, would you see younger or older beds in the hills on either side? Study the key for the correct order of superposition and see if you can determine how it was arrived at.*

the law of superposition, however, we can determine the correct sequence of all cards except the one isolated by itself. The correct order of cards is given by the key, or legend, on the right.

2. Once we understand the arrangement of the different cards, we can predict with fair accuracy the ones that will be found under any given spot. This procedure is comparable to the methods a geologist uses in predicting what formations will be penetrated by an oil well.

3. A card (or formation) that does not touch or interfinger with other units cannot be placed in its correct position by superposition alone. Such areas must be "dated" by other means, mainly by the use of fossils, as we shall see in the next chapter.

Other Aids in Determining Order of Events

Strictly speaking, superposition refers to the placing of one thing over or upon something else. The accumulation of sedimentary layers is a perfect

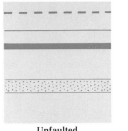  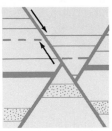

| Unfaulted sedimentary beds | Normal fault lowers left side | Second normal fault lowers right side |

Figure 4.14 Diagrammatic cross sections showing the effect of two faults on a section of horizontal beds. Notice how the second fault cuts and offsets the first one so that their relative ages can be determined.

geologic example of superposition and provides undisputable proof of the order in which the layers were formed. But the geologist does not depend entirely on strict superposition in working out geologic history. Any relationship that indicates which of two events occurred first is useful and significant. Thus, we not only look for masses of rock lying one upon another, but we observe the relationships of any earth material that may lie within, against, around, or upon other materials. In the next few paragraphs, various types of relationships are discussed.

CROSS-CUTTING RELATIONSHIPS

Although sedimentary beds do not intersect or cut across one another, faults and intrusive igneous bodies commonly show *cross-cutting relationships*. Faults may cut faults, igneous bodies may cross igneous bodies, faults may cut igneous bodies, and, of course, both may also transect sedimentary layers. Conditions can understandably become very complex, but careful observation and mapping usually leave little doubt about the correct order of events.

When two igneous bodies intersect or cross, it is obvious that the older one must have been opened up to provide space for the younger one to intrude. In other words, the younger one is

continuous, the older one discontinuous at the place of juncture, regardless of the size of the bodies involved.

INTRUSIVE RELATIONSHIPS

Igneous material arises from the depths of the earth and may solidify either below or on the surface. Lava flows and beds of volcanic ash are commonly deposited on the surface in distinct layers, and the correct order of formation can readily be determined by superposition. But superposition does not apply to subterranean

Figure 4.15 Cross-cutting relationships shown by younger light-colored vein in dark-colored older rock, Bear Creek Canyon, Jefferson County, Colorado. (U.S. Geological Survey.)

Figure 4.16 *Gravel—the winnowed debris of vanished formations. A typical beach or stream bed contains rock fragments derived from all the older formations in the vicinity. A study of the pebbles in gravel or conglomerate frequently yields valuable clues concerning the order of geologic events. (U.S. Geological Survey.)*

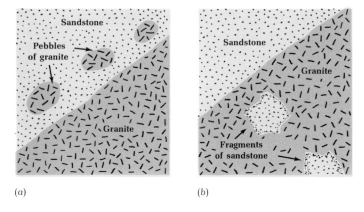

(a)   (b)

Figure 4.17 *Age relationships of intrusive and sedimentary rocks. (a) Pebbles of the granite rock are found in the overlying sandstone, proving that the igneous rock is older. (b) Unmelted pieces of sandstone are shown within the granite, indicating that the sandstone is older.*

igneous bodies that have been forcibly injected between or among preexisting sedimentary beds. The igneous bodies known as *sills* originate from molten material that has been forced along bedding planes so as to be parallel with the enclosing layers. After a sill has solidified, it may be folded and faulted in the same manner as the enclosing

sediments. Although the rock types composing sills are easily distinguished from sediments, it is difficult to distinguish a lava flow, deposited on the surface and then buried, from a sill intruded between older sediments. In making this distinction, we note that the heat of a sill will affect adjacent rock both below and above, whereas the heat of a flow affects only the rock below. Then, too, the surface of a lava flow may show bubblelike cavities and cracks, indicating exposure to the air, but these effects are absent in sills.

Faults are merely planes of movement and have no mass. When one fault crosses another, their relative ages must be determined by considering offsetting relationships. The effects of earlier faults are disrupted by later ones. Fault relationships are determined by using key beds, which in sedimentary rocks are usually thin, distinctive units that can be traced and recognized over wide areas. In general, the youngest fault follows a straight course; others are offset or disrupted.

With regard to faults and intrusions an important corollary to the law of superposition may now be stated: *Intrusive rocks are younger than the rocks they intrude, and rocks cut by faults are older than the faults.*

Relatively large rock fragments are frequently found embedded in other rocks in a manner that indicates that they must have come from some preexisting source and are therefore older than the matrix in which they are embedded. *Conglomerate,* which is consolidated gravel, is composed mostly of these so-called derived fragments. A typical conglomerate has pieces of all the harder rocks in its source area, picked up and brought together by stream or wave action. These fragments are usually large enough to contain diagnostic minerals, rocks, or even fossils that can be related specifically to older deposits.

A geologist may have to examine certain fossils very closely before he can say whether or not they came from an earlier deposit or originated with the formation in which they now occur. Derived or reworked fossils are usually broken and waterworn. Of course, entire skeletons or large plants could not be exhumed, transported, and reburied intact.

Another type of derived material is that dislodged and engulfed by igneous material as it moves upward through the earth. Obviously such pieces must be older than the surrounding matrix. Unmelted pieces of sedimentary rock in a large granite mass prove the sediment to be older. If, on the other hand, we find rounded pebbles of granite in a sedimentary bed, we know the granite is older.

A second corollary to the law of superposition is obvious: *Any rock is younger than its externally derived component materials.* In other words, a sedimentary bed is younger than the intact grains or pebbles of which it is composed.

ORDER OF GROWTH OF CRYSTALS AND ORGANISMS

Order of formation is an important consideration in the study of mineral deposits. Geologists are interested in finding out the proper succession of events in the formation of a particular deposit,

Figure 4.18 Successive buried fossil forests of Yellowstone Park. In the cliffs along the south side of the Lamar River in the northeastern part of Yellowstone Park, Wyoming, geologists have counted the remains of twenty-seven superimposed fossil forests. As shown by the photograph, the broken trunks stand erect in their places of growth. Because the enclosing rocks consist of volcanic fragments, it is evident that each successive forest took root and grew during a period of quiet and was then overwhelmed and partly buried by a volcanic eruption. The sketch, prepared by Dr. Erling Dorf, depicts the successive superimposed forests and the nature of the enclosing rocks. The age of the forests is middle Eocene. (Courtesy of Erling Dorf.)

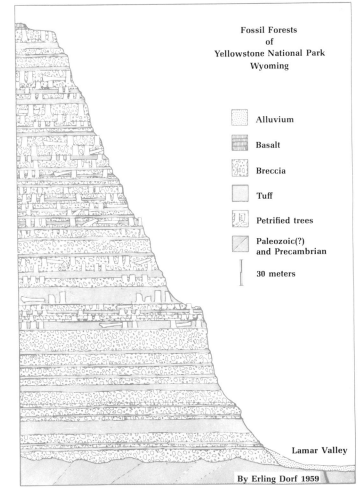

Fossil Forests
of
Yellowstone National Park
Wyoming

Alluvium

Basalt

Breccia

Tuff

Petrified trees

Paleozoic(?)
and Precambrian

30 meters

Lamar Valley

By Erling Dorf 1959

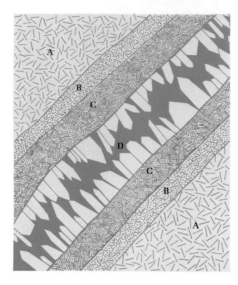

Figure 4.19 Succession of mineral deposits within an open fracture to produce a vein. A indicates original wall rock; B and C are successive coatings deposited in the fracture; D is a growth of crystalline material; E is unfilled space. (Adapted from W. L. Stokes and S. Judson, Introduction to Geology: Physical and Historical, Prentice-Hall, Inc., 1968.)

for such information may yield clues to the location of other deposits. Many minerals grow or crystallize in open spaces that provide ample opportunity for later generations of crystals to grow upon or over preceding ones. Crusts and bands are produced as recently deposited material surrounds earlier deposits. Many different periods of deposition may be recorded in veins and crevices. The study of ore deposits is complicated by the fact that minerals may grow *within* other minerals as well as *upon* them and that heat and pressure may completely obliterate a great deal of the evidence. The order of events eventually may have to be decided from microscopic study of minute specimens.

Organisms, like crystals, also grow over and upon each other in various ways. The great coral reefs are impressive examples of superimposed

growth. The shells of dead animals become overgrown by living organisms and lime-secreting plants, and layers of many generations may accumulate on even a small object. And, of course, superposition indicates the order of growth.

SUCCESSION
IN LANDSCAPE DEVELOPMENT

What we have already said about methods of determining the order of geologic events applies chiefly to solid rocks. Our discussion would be incomplete without some mention of the methods used in unraveling the history of the surface features of the earth's crust. Even the most monotonous landscapes show the effects of successive events and changing climates, and we know that the features were not all produced at the same time. The geologist has the interesting problem of reconstructing the history of landscapes from the various forms that compose them. For example, the peaks of many high mountain ranges were sharpened and furrowed by glaciers that have melted away, and the areas once occupied by ice are now being reshaped by running water. In many areas there are steplike terraces or beaches that were cut by the waves of lakes that have since dried up. These old lake beds may be invaded by sand dunes or cut by streams, depending on subsequent climatic changes.

A study of the changes in rivers, lakes, deserts, and glaciers involves the fundamental idea of superposition, but there is the added complication of explaining the surface forms. River terraces, especially, are highly instructive but difficult to interpret, for streams are very sensitive to climatic changes, and the same river may run slowly or swiftly, swell in volume or dwindle away, and remove or deposit sediment, all within the same valley or *floodplain*. Thus a terrace may have been cut on bedrock or on previous river deposits; it may have been made by a small river meandering from side to side or by a large river

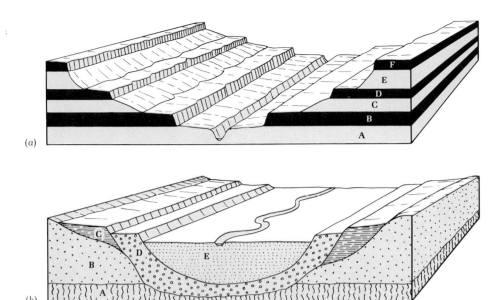

(a)

(b)

Figure 4.20 Terrace formation. In (a) the valley profile is steplike, as in the Grand Canyon. The different ledges and slopes are arranged according to superposition: oldest at the bottom, youngest at the top. In (b) the effects of cut-and-fill show that the successive benches are older above and younger below. The true order of age is determined by superposition in all cases. Thus, in (b) the valley was once filled with material C, which was cut into and mostly removed. Then, D was laid down on and against C, and the process was repeated. Terrace formation and cut-and-fill can be very complex processes.

occupying the entire valley. The river valley may have been filled and partly emptied of sediment several times, leaving a variety of deposits to record its changes.

Lakes in arid regions expand and shrink as climates change. The rise and fall of water levels are recorded by terraces and beaches, but it is surprisingly difficult to determine just what the true order of events has been. Likewise, as glaciers expand and retreat they leave surface evidences and thin layers of debris to record their fluctuations.

In dealing with the history of landscapes, we look chiefly for evidence indicating which of two events has effaced or disturbed the other. The sequence of cutting and filling of river valleys can be determined by noticing which deposits cut into or lie against or on other deposits. Contrary to the situation in canyons where bedrock is exposed, we may find that older deposits make up higher terraces and younger ones lower terraces. Some examples of cut-and-fill relationships are shown in Figure 4.20.

Superposition in Other Sciences

Geology is not the only science that relies heavily on the law of superposition. Archaeologists are concerned with remains or evidences that are dug up in caves or from sites of ancient villages or cities. Because many sites have been occupied more or less continuously for hundreds or even thousands of years, it is only natural that extensive piles of refuse and debris should have accu-

Figure 4.21 *An unusual example of superposition in landscape development near Searls Lake, California. Three different processes have left their marks. Terraces A were cut on the lower foothills by waters of Searls Lake when it was deeper and higher. Alluvial fan material B has been derived from the mountain and has obliterated the lake terraces where they have not been protected. A fairly recent fault scarp C cuts, and is consequently younger than, the alluvium. (U.S. Geological Survey, photograph by G. I. Smith.)*

Figure 4.22 *Superposition of stratified layers in Danger Cave, Utah. The correct order of deposition of human artifacts, animal bones, and plant remains is being determined by students making a careful study of the order of superposition at this important archaeological site. (Courtesy of Jesse D. Jennings.)*

Figure 4.23 *Superposition of lunar craters. Many examples of young craters superimposed upon older ones are evident. Note particularly the arrangement just to the right of the center of the photograph. (NASA.)*

that had accumulated naturally at an earlier time. So-called intruded burials of artifacts have caused many perplexing problems and must be constantly watched for.

The law of superposition has recently been applied to the study of the surface of the moon. Here thousands of successive craters have been formed and it is possible to determine the order of formation by noting that older craters have been partly obliterated by younger ones. Without ever setting foot on its surface, lunar geologists were able to compile a history of the moon's major external features.

Problems of Applying the Law of Superposition

We have already inferred that the law of superposition does not apply to rock layers that have been greatly disturbed after their deposition. Beds that were originally perfectly level may be broken, tilted, and even turned completely over during mountain-building movements. It is com-

mulated. Trenches dug into these ancient mounds or cave deposits reveal the successive cultural levels in order of decreasing age from bottom to top. Bones and artifacts in the successive layers tell the history of the human inhabitants. Interpretations must be made with care, for men are prone to bury many things in the earth; and important items of evidence may come to rest out of their proper place, next to or among things

Figure 4.24 *The overturning of sedimentary layers illustrated by ripple marks and cross-bedding. The distinctive appearance of certain original features enables us to know whether they are still in their original position or if they have been tipped up and overturned.*

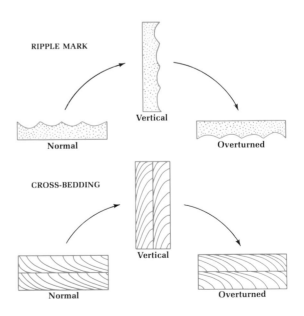

RIPPLE MARK

Normal Vertical Overturned

CROSS-BEDDING

Normal Vertical Overturned

Figure 4.25 *Folded beds of the Franciscan Chert, a deep-sea deposit of western California. The folding moves younger beds to a position beneath older beds and also turns all beds upside down. This illustrates why the law of superposition must be applied with due regard for disturbances that have affected strata since their deposition. (Courtesy of Mary Hill.)*

Figure 4.26 *Near view of a fault plane. In this unusual photograph the actual plane of movement of a great thrust fault is seen. View is looking into a shallow excavation along the fault. Above, forming the roof, is solid quartzite of Precambrian age; below is softer, much contorted shale of Cretaceous age. The quartzite has moved across the shale for an estimated 48 kilometers. By the purely mechanical action of faulting very old rock is brought above much younger rock. Compare with the normal sedimentary contact in Figure 4.6. Pencil gives scale. (Courtesy of R. C. Wilson.)*

mon in mountainous areas to see beds that have been tipped up at steeper and steeper angles until they are standing on end. If they go beyond this angle, they are said to be *overturned,* and the bedding planes that originally faced upward now face downward. Superposition cannot apply to rocks in this disturbed position, and they must be restored, in theory at least, to their original position before they can be correctly interpreted. The problem of determining which side of a bed of rock was originally up and which down presents interesting possibilities for geologic detective work. Thus, the sharp crests of ripple marks point upward, and their rounded troughs curve downward. The reverse appearance indicates overturning. Small channels that were cut and filled before burial are convex downward if undisturbed. If turned over, they are convex upward and appear unnatural.

Folding is not the only action by which order of rock layers is reversed. Actual breaking or fracturing of strata may also force older rocks upon younger ones. If vast sections of the earth are powerfully compressed, faults will appear along which great masses are pushed upward and forward. These faults are known as *overthrust faults,* or simply *thrusts.* They are areas in which dislocated masses have moved forward many miles to occupy positions on top of rocks of entirely different age and origin. If, during these movements, rocks that were originally older come to rest on younger ones, the law of superposition obviously cannot apply. All such instances must be carefully examined. If the rocks are well exposed, fault planes will be discovered with evidences of intense mechanical disruption, such as crushed and broken rock material and gouged or grooved surfaces that indicate movement.

Closely allied to the blocks moved by thrust faults are large landslide masses. The force of gravity acting on elevated rock masses, especially those that are saturated with water or those in submarine environments, may cause them to slip

Figure 4.27 *Chief Mountain, Montana. At the base of this imposing mountain is a great horizontal thrust fault along which the overlying older rocks have been carried forward upon the younger rocks of the foothills. This fault plane is much like that shown in Figure 4.26. The movement along the fault is about 24 kilometers. (U.S. Geological Survey.)*

downward over lower beds. The process of sliding may be slow or rapid, and the masses involved may cover many square kilometers. If older rocks slide over or fall on younger rocks, we have another instance of deceptive superposition. Landslides and similar materials usually

Figure 4.28 *Determining the sequence of past events. In (a) we see the present landscape and subsurface conditions. Removing the evidence of successively older events reveals the past history of the area. In (b) the effects of erosion (as shown by the canyon) are removed, showing that the lava flow was continuous across the area. In (c) the volcano and its subterranean conduit and surface flows are removed, exposing what was once a gravel-covered plain. In (d) the original height of the fault is restored together with some of the debris that has weathered from the scarp. In (e) the fault is removed by sliding the beds back to their original positions so that corresponding beds match. This reveals the folded condition of the beds as they were before the faulting. Finally, in (f) the beds are unfolded and the original condition of the area is revealed.*

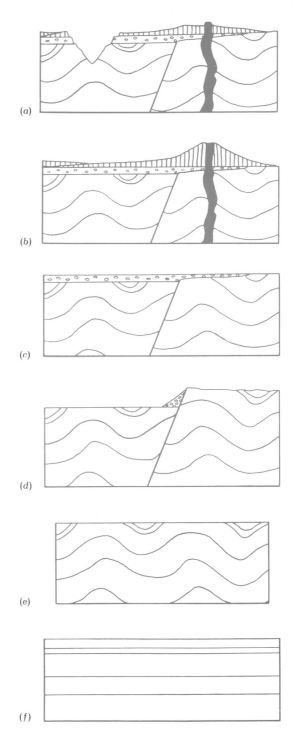

Figure 4.29   Flows from a volcano on the brink of the Grand Canyon in Arizona have descended over the cliffs to the river bottom. Obviously the canyon was already in existence in essentially its present form when the eruption occurred. (U.S. Geological Survey.)

show many indications of breakage and contortion and may even occur in the form of broken fragments called *breccias*. Careful study may reveal the source of the disrupted rocks and provide a logical reason why they have moved from their original positions.

Reconstructing Past Events

Even though a long series of events may have disturbed the sedimentary rocks of a region, it is still possible to reconstruct the major features of the past and to apply the law of superposition. The reconstruction must be done diagrammatically or with models and always commences with the present and proceeds backward in time. In effect, the geologist mentally removes the evidences of each significant event, thus restoring the condition of things before the event took place. Therefore, if the last event in a particular area was a lava flow, the geologist "removes" the igneous rock and reconstructs a picture of conditions as they were before the eruption occurred. If faulting had preceded the lava flow, he moves the faulted strata to their original unbroken position. If there is evidence of folding, he graphically straightens out the beds to a horizontal position. He "replaces" eroded material as far as possible to recreate mountains or larger land areas. Unconformities and loss of the record will leave many gaps, but additional evidence may be recovered in nearby areas. Thus, through painstaking mapping and study, the broad features of the history of any area of sedimentary rocks may be reconstructed.

A SUMMARY STATEMENT

All sedimentary rocks occur in *beds,* or *layers,* separated by *bedding planes.* Major planes of discontinuity indicating erosion and, hence, lapses in the record, are known as *unconformities.* Certain conspicuous planes representing major changes in composition or mode of origin are designated on geologic maps to divide the geologic formations.

The *principle of uniformitarianism* is helpful in understanding the formation of rocks, for even today we can find material at all stages of the transformation from loose particles to solid rock. Structures that are in process of formation today are also found in sedimentary rocks of all ages, indicating the operation of the same forces through time.

The *law of superposition,* which states that in any pile of undisturbed sediments any bed is older than the overlying bed and younger than the underlying one, is of fundamental importance in reconstructing the past. Other evidences indicating the correct order of events include cross-cutting relationships, presence of included or derived fragments, intrusive effects, and growth sequences in minerals or fossils.

The original level condition of sediments may be disturbed by *faulting, folding,* and *sliding,* and in extreme cases superposition may not apply. Through diagrammatic elimination of the successive superimposed effects, we can "restore" past conditions and interpret something of the history of the formations involved. The events that have shaped a landscape are also revealed to a considerable extent by superposition.

FOR ADDITIONAL READING

Cloud, Preston, ed., *Adventures in Earth History.* San Francisco: Freeman, 1970.

Donovan, D. T. *Stratigraphy.* Chicago: Rand McNally, 1966.

Dunbar, C. O., and John Rodgers, *Principles of Stratigraphy.* New York: Wiley, 1957.

Kay, Marshall, *North American Geosynclines.* New York: Geological Society of America, Memoir 48, 1951.

Krumbein, W. C., and L. L. Sloss, *Stratigraphy and Sedimentation.* San Francisco: Freeman, 1951.

Kunen, P. H., *Marine Geology.* New York: Wiley, 1950.

Laporte, Leo F., *Ancient Environments.* Englewood Cliffs, N.J.: Prentice-Hall, 1968.

Pettijohn, F. J., *Sedimentary Rocks,* 2nd ed. New York: Harper, 1957.

Poldervaart, Arie, ed., *Crust of the Earth.* New York: Geological Society of America, Special Paper 62, 1955.

Selley, R. C., *Ancient Sedimentary Environments.* Ithaca, N.Y.: Cornell University Press, 1970.

Termier, Henri, and Genevieve Termier, *Erosion and Sedimentation.* New York: Van Nostrand, 1963.

FOSSILS AND THE SUCCESSION OF LIFE ON EARTH

The term "fossil" originally referred to any curious object obtained by digging in the earth—mineral specimens, crystals, old artifacts of man, odd stones of any kind, and the remains of ancient plants and animals. Gradually the term became restricted to the last category—remains or evidences of ancient organisms that have been preserved by natural means in the earth's crust. Usually fossils are preserved in solid rock, but many specimens have been found in loose sand, in bogs, and in tar pools, and some have even been found frozen in ice. Most fossils cannot be

Excavation at the La Brea tar pits, near Los Angeles, California. Bones of many species of animals are mingled together in the tarry soil. Note tar oozing from the sides of the pit. (Natural History Museum of Los Angeles County.)

precisely dated in terms of years; the majority are prehistoric and incredibly old.

The number of fossils preserved in the earth is beyond calculation. Many common rock types, such as limestone and chalk, are composed almost entirely of countless small fossils. A few cubic centimeters of diatomaceous earth contain millions of skeletons of a small floating type of alga called a diatom. Literally cubic kilometers of sediment may be made up of animal shells. The Great Barrier Reef of Australia is composed of the shells and skeletons of billions of marine organisms. Similar reefs are found in sedimentary formations throughout the earth. Coal beds are accumulations of fossil plants, and petroleum is primarily altered material that once constituted various kinds of living organisms.

The Origin of Fossils

As a rule, dead things disintegrate and disappear. A fossil is the remains of a dead organism whose disintegration has been halted. Thus the orga-

(a)

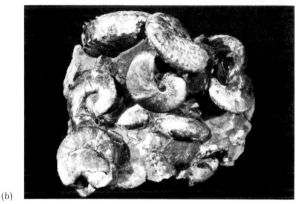

(b)

(c)

Figure 5.1 Masses of fossils. (a) A tangle of leaves from an ancient coal forest of Australia. (b) A closely packed group of extinct Cretaceous ammonites (Scaphites) from South Dakota. (c) A Late Devonian fish (Holoptychius) from the Old Red Sandstone in Scotland. (b, courtesy of Karl Wagge; c, Department of Geology, Royal Scottish Museum.)

nism leaves a record of existence as a living thing. Normal processes of decay may be retarded or prevented in a great number of ways, and the type of preservation, or fossilization, depends largely on the nature of the animal or plant, its life habits, and the conditions under which it died and was buried. Two common, but not absolutely essential, conditions for fossilization are the possession of hard parts and immediate burial of the organism. An organism with a hard shell, such as an oyster, is much more likely to be fossilized than a jellyfish, and an animal that drowns in water-saturated mud will be fossilized much more readily than a creature that dies on a grassy plain or forest floor where scavengers and bacterial action may destroy it before it is covered over. Soft and delicate organisms are sometimes fossilized, however. Many specimens of preserved feathers, fruits, flowers, insects, bacteria, fish eggs, leaves, and filmy sea animals have been recovered.

Everyday experience provides us with very little opportunity to observe and understand the process of fossilization. Weathering, erosion, decay, and disintegration are characteristic of the surface of the earth. Fossilization takes place out of sight beneath the surface, where destructive action is inhibited or halted. In fact, most fossils are formed beneath shallow water, where sediments are being deposited rapidly and continually over long periods of time. In submarine environments near the continents, and especially near the mouths of large rivers, the constant precipitation of sediment slowly and effectively buries the remains of creatures that are washed down from the land or that die on the spot. Whole colonies or beds of marine animals may be buried by sudden shifts of currents, and a fairly complete sample of the life of the area will be preserved. Once covered, such sites may not be exposed to erosion for countless geologic ages, during which time the remains are petrified and preserved. Generally, large deltas are sinking gradually, which means that layer upon layer of fossil-bearing beds will be piled up and will pro-

Figure 5.2 *Hard parts favor fossilization. The large gastropod shell and the molar tooth of an elephant are both old enough to be regarded as fossils. In both, the original material has been bleached and softened by removal of organic constituents. Fossilization of such objects could proceed through stages of petrifaction if they are buried in proper surroundings.*

vide a record of life over very long periods. The remains of nonmarine plants and animals are best preserved in deposits of river plains and freshwater lakes. Surrounding vegetation supplies leaves and fruits, insects fall into the waters, and land animals inhabiting the shores leave their bones to be covered and preserved. The periodic overflowing of rivers and the shifting of channels catches animals unaware and entombs their bones in more or less perfect condition.

Methods of Fossilization

CASTS AND MOLDS

Shells of invertebrate animals provide the best-known fossils and illustrate a common method of preservation. A shell buried in sediment may undergo a variety of transformations while the surrounding matrix solidifies to stone. Slowly moving underground water may dissolve the shell entirely, leaving only a hollow space, or *mold*, that faithfully reflects the shape and surface markings but gives no hint of the internal

Figure 5.3 *Molds of snail shells. The cavity in the rock shows the faint markings of the outer surface of the shell and is a mold of the exterior; the upper specimen, freed from the rock, is a filling or mold of the interior.*

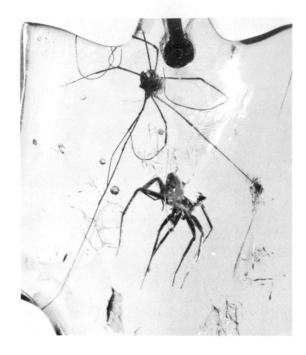

Figure 5.4 Two arachnids (Phalangium and Oxy-opes) are preserved in this single piece of amber from the Tertiary of East Prussia. (By permission of the Trustees of the British Museum, Natural History.)

structure. If the mold shows the outside, it is an *external mold*; if it shows the inside it is an *internal mold*. If later on these hollow spaces are filled with mineral matter, *casts* of the original objects may be created, but again, no cellular structure is preserved. The material that fills the hollow spaces may be somewhat different from the material of the matrix, or surrounding rock,

Figure 5.5 A near-recent "fossil" from the ruins of Pompeii. This is a cast of the body of a dog killed by ash from the eruption of Vesuvius in A.D. 79. It was made artificially by filling with plaster the hollow space once occupied by the body. This illustrates the process by which casts of many natural fossils are formed. (Alinari-Art Reference Bureau, Inc.)

because it has been brought in by solutions at a later time. Natural casts and molds make up a very large proportion of the fossils that are found in sandstones and limestones. The famous fossils of insects preserved in amber are in fact natural molds. The outlines are hollow, all that remains of the insects are films of the body covering or wings and other appendages (Figure 5.4).

A fossil that faithfully preserves the original shape and markings but not the internal structure may be termed a *natural replica*. *Artificial replicas* may be made by filling a natural mold with wax, clay, or plaster (Figure 5.5).

PETRIFACTION—
REPLACEMENT AND IMPREGNATION

Under favorable conditions organic materials may be replaced by mineral matter that preserves the form and structure of the original specimen by means of a process called *petrifaction* (converting to stone). The process is most effective when it acts on hard substances such as bone, tooth, wood, and shell. Soft tissues are rarely preserved by this method. Commonly only the small internal cavities of cells or other open spaces are filled with precipitated mineral matter. Organic material that is only partly replaced is said to be *permineralized*.

Figure 5.6 *A silicified tree trunk in position of growth in Yellowstone National Park. (Courtesy of Erling Dorf.)*

Preservation by rapidly deposited, noncrystalline *silica* (SiO_2) is a special case. Under conditions not yet well understood, silica gel may be precipitated suddenly within and around tissues or cells of any kind in a sort of embalming or impregnating process before bacterial decomposition has time to destroy the fine details of the specimen. Such rapidly deposited siliceous material seems to harden immediately and, if it occurs in large masses, it may become the tough, incompressible rock known as chert. Occasionally, delicate fossils, such as the spider shown in Figure 5.7, may result. The chief advantage of this type of rapid preservation is that much, if not all, of the actual organic material is preserved. Individual algal cells showing internal structures have been discovered in chert over 2 billion years old.

Circulating solutions of mineral-bearing water are essential for petrifaction. In addition to silica, the most common material carried by ground water is *calcite* ($CaCO_3$). Fossils preserved by either of these substances are accordingly said to be *silicified* or *calcified*. But because animals commonly use these same materials in constructing their shells and skeletons, it is sometimes impossible to tell just how much of the final product is original and how much has been introduced. Scientists recently have demonstrated that some of the complex amino acids that are built up only by living animals may be preserved for millions of years in fossil remains. Comparisons prove that the chemical makeup of the preserved compounds is identical to the compounds found in living relatives of the fossil forms.

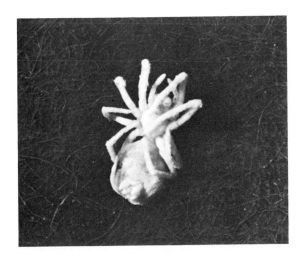

Figure 5.7 *A fossil spider (Argenna fossilis) from Miocene lake deposits of the Mojave Desert, California. The specimen has been impregnated by a combination of minerals. (U.S. Geological Survey.)*

Figure 5.8 *A fossil leaf (Aralia) preserved as carbonaceous residue in fine-grained shale. Specimen measures about 9 centimeters. (Utah Museum of Natural History.)*

A certain amount of bleaching and destruction of the softer organic portions of shells and bones almost always precedes petrifaction. Shells that have been exposed on the sea bottom or to the waves for only a short time are usually bleached and whitened, and the color patterns, confined chiefly to a thin outer organic layer, are soon worn away. In older sediments, the fossils gradually become darker and take on the color of the surrounding rocks. Their final colors are determined by the petrifying agent and the containing rock and usually bear no relation to the living material. Bone that originally was white may become black or brown; wood may take on brilliant shades of red, yellow, or blue; opalized fossils may assume iridescent colors of great beauty.

In addition to calcite and silica, a great variety of other minerals act as petrifying agents. Some of these are pyrite, marcasite, dolomite, barite, fluorite, gypsum, hematite, galena, sulfur, and talc. In sandstone formations in the western United States, fossil vegetation has been replaced by rich uranium-bearing minerals or even by pure native silver.

CARBONACEOUS RESIDUES AND IMPRESSIONS

A type of fossilization called *carbonization,* or *distillation,* is especially effective in preserving leaves and delicate animal forms. Carbonization occurs when the liquid and gaseous constituents of an organism are squeezed out by high pressure and are absorbed by the surrounding matrix. All that remains of the organism is a thin, filmy residue that consists chiefly of carbon. So delicate is this kind of fossilization that fine veins of leaves and even individual cell walls are preserved. Where preservation has been exceptionally favorable, the film of carbon can be stripped from the rock and studied as a mounted specimen. If little or no carbon is retained, but there is an indentation or imprint of the original object, the fossil may best be called an *impression.*

FROZEN AND MUMMIFIED REMAINS

We should mention here some of the rare but spectacular specimens of frozen mammals discovered in the frigid parts of the globe. These specimens consist of more or less entire carcasses of animals that have been frozen in ice or sediment and preserved for periods of as long as 20,000 years. All parts, including blood, hair, fat, and even undigested food, may be well preserved. Several complete woolly mammoths and less complete portions of this and other animals with hair, skin, and other tissues intact have been uncovered by workmen dredging for gold or have been exposed by earth slides and by the present thawing of the Northlands. These remains were buried during final stages of the last glacial period and are thus comparatively recent in the scale of geologic time.

Remains of the woolly rhinoceros, a contemporary of the woolly mammoth, have been found preserved in oil seeps in Poland. The oil inhibited bacterial action and decay, and so skin, hair, and soft parts survived the vagaries of time and the attacks of the elements. Excellently preserved human bodies of pre-Christian date have been discovered in peat deposits in Denmark and Holland. In these specimens, the skin and other tissues, including internal organs, have escaped destruction and are easily studied. Antiseptic or anaerobic swamp water, which prohibits the growth of decay-inducing organisms, will preserve anything that has been immersed in it and subsequently buried in its bottom sediments.

Other Evidence of Past Life

TRACKS, TRAILS, AND BURROWS

Practically every type of organism that moves has left somewhere a record of its passage. These range from "worm" trails that are among the oldest of fossils to the footprints of man in hardened volcanic ash. Many impressions have been made by animals that have left us no other clues to their existence on earth; a few have defied all attempts at classification and remind us only of the incompleteness of the geologic record. All fossils that do not reveal the body form are called "trace fossils."

Burrows and borings are made by organisms as they work their way through subsurface layers of mud or other material in search of food or protection. The openings thus made are usually filled with sediment similar to that which the organisms penetrate.

Tracks and trails are surface impressions made primarily in wet sediment near bodies of water where periods of submergence and emergence alternate rather rapidly but where sedimentation prevails over erosion. Thus, tracks are common in muddy environments where wet sand and mud receive impressions of the feet of passing animals and where succeeding layers of sedi-

Figure 5.9 Fossil insect (Palaeobittacus) *from the Green River Formation of Eocene age near Rio Blanco, Colorado. Length of body is about 1.3 centimeters. (Courtesy of Frank M. Carpenter.)*

Figure 5.10 *Fossil burrows. The soft sediment through which the organism burrowed has eroded away leaving in relief the hardened sand that filled the burrows.*

ment may cover the tracks before they are disturbed or eroded.

The typical track consists of two complementary parts—the initial depression made by the foot, which is a mold, and the filling of the depression, which forms a cast, or reproduction, of the foot. Probably more casts than molds are preserved, because molds are more likely to occur in soft material, which is easily eroded.

INDIRECT EVIDENCE

A few objects that have not themselves been alive and are not organic are also called fossils. The so-called "stomach stones," or *gastroliths*, that were swallowed by various types of animals and deposited with their bones provide interesting evidence about the food habits of early creatures. Petrified stomach contents and excrement (*coprolites*) provide clues to the diet and physiology of ancient animals. Artifacts of man, if sufficiently old, may also be considered as indirect evidences of life. An arrow point or scraper proves the existence of a human being just as surely as a piece of human bone or a footprint.

PSEUDOFOSSILS

Many inorganic objects are found that bear a deceptive or superficial resemblance to fossils. Such things are called *pseudofossils* ("false fos-

Figure 5.11 *Tracks in the Coconino Sandstone of the Grand Canyon, Arizona. Length of slab is about 38 centimeters. (Smithsonian Institution.)*

Figure 5.12 *Pseudofossils—objects that resemble, but are not, true fossils. On the right is a pebble that strong stream action has shaped into the form of an egg. On the left is a dendrite, a fernlike, but entirely inorganic, growth of manganese minerals that has spread along a narrow fracture in a rock.*

sils"). Thus pebbles may be shaped like eggs by erosion, and nodules or concretions may resemble turtle shells, bones, fish, or even more complicated forms. Perhaps the most common pseudofossil is the *dendrite,* a leaflike or fernlike crystal growth found in narrow crevices or fractures in rock. When such forms are present in transparent or semitransparent rock they are called "mossagates" and may be cut into semiprecious ornaments.

The Study of Fossils— Paleontology

The scientific study of fossils is called *paleontology.* Some of the most important subdivisions of the subject are invertebrate paleontology, vertebrate paleontology, paleobotany, and micropaleontology. Each of these fields has its own specialists and embraces such a vast amount of information that no one person can satisfactorily master more than a small fraction of the subject matter of any one of the subdivisions. Likewise, as explained in the following paragraphs, the techniques and aims of the various fields are significantly different.

Invertebrate Paleontology. Invertebrate paleontology is perhaps the oldest and the best-established subdivision of the subject of paleontology. Dealing as it does with the most abundant types of fossils, it has attracted correspondingly more attention than its sister studies. Most professional paleontologists begin their studies with basic courses in this field.

Vertebrate Paleontology. Vertebrate paleontology deals with the highly organized backboned animals. Fossils of the vertebrates are relatively less common than invertebrate fossils and are restricted mainly to land deposits of the later geologic periods. Vertebrates offer scientists the most convincing and understandable evidence for organic evolution, and a good part of the

Figure 5.13 The petrified bones of the hind leg and foot of the Jurassic carnivorous dinosaur Allosaurus. This is part of an in-place exhibit of dinosaur bones to be seen at the Dinosaur National Monument, near Vernal, Utah. (U.S. National Park Service.)

study of vertebrate specimens involves establishing lines of descent and tracing adaptations to changing environments.

The preparation and display of vertebrate specimens is a highly specialized field in itself. Massive skeletons and lifelike models of spectacular creatures such as the extinct dinosaurs occupy prominent places in museum displays. Because of their characteristic teeth and rapid evolution, the higher vertebrates are most useful in the study of land-laid sediments in which other fossils may be rare.

Figure 5.14 *The minute fossils studied by micropaleontologists. The inked squares on the rock specimens indicate the location of fossils. Some of these have been picked out of the rock and are mounted on a special microscope slide for study. (Standard Oil Company of California.)*

Paleobotany. Paleobotany is the study of plant fossils, and a great deal of its subject material has been gathered in the great coal fields of the world. *Palynology,* the study of fossil spores and pollen, is a rapidly growing subdivision of paleobotany. Students of fossil plants have made important contributions to our understanding of the formation of coal and also have aided in the discovery of numerous coal beds. Plant fossils provide the paleontologist with evidence of past geographic conditions because plants are especially sensitive to climatic influences. The plant world supports the animal world, and evolutionary advances in plants have been followed by corresponding changes in the animals that feed on them.

Micropaleontology. Micropaleontology, as the term implies, is the study of all fossils that are so small that they must be analyzed with a microscope. Practically all groups of animals and plants are represented in this field, either as specimens or as parts of complete specimens. Plant spores, pollen grains, and seeds are included, as are minute teeth, scales, and bones of vertebrates. Most of the work of the micropaleontologist, however, centers about certain shell-bearing protozoans belonging to the order Foraminifera. These small forms are found in tremendous numbers in marine rocks and are commonly associated with oil-bearing formations. More than half of all professional paleontologists are engaged in the field of micropaleontology.

A thorough and complete paleontological investigation demands great skill and wide training. The field collections must be carefully made; the discovery sites must be accurately located both geographically and geologically to enable future investigations to be carried out. Delicate specimens require special treatment; it may take weeks, for example, to remove a skeleton of a vertebrate from the earth because the bones may be fragile and easily destroyed by careless handling. Cleaning, labeling, and storing of specimens calls for painstaking care. Misplaced or incorrectly labeled specimens are worthless for detailed study and may lead to erroneous and costly mistakes.

The proper evaluation of specimens in terms of classification and evolutionary significance requires a broad knowledge of zoology. The paleontologist must take into account living as well as extinct relatives before he can determine the exact significance of a particular fossil. Selecting a proper, distinctive name is not the least of the problems to be solved because each name must be unique—used only once, for only one organism.

The History of Paleontology

Long before fossils came to be studied systematically, they were collected as curiosities, charms, and medicines. They have been found among the ornaments, grave offerings, and temple relics of many primitive and prehistoric peoples who probably ascribed mystical or magical properties to them without clearly understanding their real origins and meanings.

Fossils intrigued many classical and medieval thinkers. Their explanations of the so-called "figured stones" are strange, fantastic, or ridiculous in the light of our present knowledge, but must be considered in terms of the superstitious setting out of which they came. In the opinion of many ancients, the stony fragments were not remains of living things at all. They were created by obscure "plastic" forces, or were ornaments for the interior of the earth, "jokes" or "sports" of nature, the result of vapors, emanations, or even of spontaneous generations. Some thinkers, however, connected fossils with living things. Aristotle seems to have believed that the fossils of fish represented individuals who had wandered into crevices in search of food and had there been hardened into stone. Others believed that seeds, spores, and eggs may have filtered downward into the rocks to germinate and achieve a sort of cramped subterranean existence.

During the Middle Ages, when theological matters occupied so much of men's thoughts, fossils were interpreted in terms of the Creation and other religious concepts. Active opposition to the scientific study of fossils was based on strict interpretations of Scripture, which seemed to indicate a recent origin of the earth, the creation of life out of nothing, and, more important, the fixity of species. Theologians could find no ready explanations for the strange extinct forms of life or the evidences of shifting seas and lands that geology brought to light. These evidences of changes were regarded as adverse reflections upon the perfection of Creation or the omnipotence of God.

Leonardo da Vinci, great Renaissance artist and precursor of modern science, was unimpressed by the prevailing arguments and preferred direct observation and reasoned interpretation. His notes cite many valid instances that convinced him that fossils are products of normal events and have been distributed by slow changes of the ocean and land. He was well ahead of his time, and it was not until the eighteenth century that the validity of his ideas was recognized.

Arguments concerning fossils arose on all sides, public interest was aroused, and the so-called "fossil controversy" became intense. Some men theorized that the Creator had made several preliminary but unsatisfactory models of living things, which were cast aside to become fossils. Others were convinced that fossils were outright works of the Devil created specifically to deceive mankind. The most popular idea to emerge was that fossils are products of Noah's Flood. This catastrophe, it was said, could not only have killed anything and everything but also could have carried the dead remains up the highest mountains or into the deepest caves. Thus, the remains of a large amphibian put on display in 1726 were considered to be "the bony skeleton of one of those infamous men whose sins brought upon the world the dire misfortune of the Deluge."

Early students of natural history tried accordingly to reconcile their findings with Scripture. John Ray (1627–1705), the first to classify plants by their flowers and fruits, insisted that species were unchanging. Carl von Linné (1707–1778), the father of modern classification, said that all species had been made at the beginning and none had been added since. He was able to catalog about 4,500 species.

An important naturalist of the eighteenth century was Georges Louis Leclerc, Comte de Buffon (1707–1788). He took a reasonable view of fossils and realized that Noah's Flood could not possibly be responsible for all of them. He maintained that normal processes of erosion and deposition must be taken into account.

By the nineteenth century a great deal of information about the earth had been accumulated. Miners and engineers were dealing directly with fossil-bearing rocks, universities were becoming concerned with natural history, and many collections of rock, fossil, and mineral specimens had been assembled. In the history of this period three names stand out: de Lamarck, Cuvier, and Smith.

Jean Baptiste de Lamarck (1744–1829) is remembered chiefly for his comprehensive theory of evolution. He realized that species are not unchanging and asserted that the gradual evolution of species is caused by changes in the needs or wants of organisms. This caused changes in the form, structure, or habits of organisms, and the resulting modifications were somehow passed on or inherited. Although his theory of inheritance of acquired characteristics is no longer tenable, Lamarck's contributions in zoology are many, and his writings on fossil invertebrates mark him as the founder of invertebrate paleontology.

Georges Cuvier (1769–1832) was a man of unusual talents and wide interests. Among his many attainments was his authorship of twelve large volumes on fossil vertebrates. His knowledge of comparative anatomy was unsurpassed, and he realized from his studies that many fossil forms are no longer found on earth. His conclusions were somewhat reactionary, however, because he attributed extinctions to successive divinely ordered catastrophes, the last of which was Noah's Flood. His opposition to the idea of evolution retarded the development of paleontological science, but his basic contributions clearly entitle him to be known as "the father of vertebrate paleontology."

William Smith (1769–1839) was an English engineer who practiced his profession between the years 1787 and 1839. His works brought him into direct contact with rock freshly excavated from canal beds and quarries. The fossils of the formations in which Smith worked were exceedingly large and striking in appearance. Smith observed that different rock layers could be identified by the fossils they contained. From a practical viewpoint, he was thus able to predict the location and properties of rocks below the surface by means of the fossils he found in exposed rocks in his canals and quarries. He regarded his discovery as being chiefly of economic value, and his writings and maps were designed to help his colleagues discover and trace valuable mineral deposits. His more erudite and academic contemporaries, however, paid little immediate heed to his findings, and several decades passed before

Figure 5.15 William Smith (1769–1839). His observations opened the way to the systematic study of stratified rocks and their correlation by the use of fossils. (Crown copyright, Geological Survey photograph. Reproduced by permission of the Controller of Her Britannic Majesty's Stationery Office.)

he achieved his rightful title as "the father of stratigraphy." The practical success of Smith's observations proved them to be far superior to purely academic theorizing in piecing together the fragmentary geologic record.

Space permits mention of only a few other important figures in the history of paleontology. Adolphe Théodore Brongniart (1801–1876) is usually regarded as a founder of paleobotany. Sir Charles Lyell (1797–1875) published the first comprehensive geology textbook, but his contributions to historical geology were mainly along physical, not biological, lines. Charles Darwin (1809–1882) successfully welded together the study of ancient and modern life and conceived the theory of descent with modification. (We will discuss Darwin in more detail in Chapter 17.) Alfred Russel Wallace (1823–1913) began the study of animal distribution and migration and attempted to apply it to past ages. Karl Alfred von Zittel (1839–1904) compiled an exhaustive four-volume work in which known fossils were arranged in a systematic manner. This became a widely used standard reference work for students of fossils.

Some of the excellent books that deal with the history of plaeontology and related subjects are listed at the end of this chapter.

Paleontology Today

BIOSTRATIGRAPHY— A MODERN SYNTHESIS

No living thing can be fully understood apart from the surroundings in which it lives. This is the essence of ecology, a science that has gained wide public interest relative to man and his changing environment. It is becoming increasingly evident that fossils also cannot be understood without reference to their ecological relationships. A great many, perhaps a large majority, of studies dealing with ancient organisms make no reference at all to the environment in which these organisms lived, because up to now paleontologists were concerned mainly with describing their specimens in biological terms in an attempt to fit them into the framework of evolutionary classification. This method of study and presentation is gradually being expanded into the new science of *biostratigraphy* which embraces the interrelated fields of stratigraphy and paleontology.

The biostratigrapher attempts to discover not only what fossils themselves may reveal but also what the enveloping and surrounding sediments can add to the story. Because of this he should be a well-trained paleontologist and stratigrapher. Excellent results can be achieved when individuals with different skills combine to make joint or team studies on biostratigraphic problems.

Basic to all biostratigraphic studies is careful field work and collecting. The aim of this phase of investigation is to sample and record as much as possible of the field evidence before the fossils are separated from their natural surroundings. Geologic maps are made in order to determine the thickness, areal extent, and variations of the fossil-bearing units. This technique is described in Chapter 6. Fully as important is the measurement and description of *sections,* a section being the total aggregation of strata found in any specific area. As mentioned earlier, the analogy with the pages of a book is frequently cited. The geologist sees mainly the edges of strata made visible by natural outcrops or has only small samples of the strata derived from vertical drill holes. From these samples a written and graphic record, which contains all available information that may become important in a biogeologic reconstruction, is composed. The following is a typical entry describing one unit of a geologic section:

30. *Sandstone, medium-grained; light yellow, weathers brown (sample R7); individual beds, from 20 to 38 centimeters thick, are made up of finer parallel laminations; many vertical burrows; fossils (sample F10) are mainly scattered*

and buried in the upper few centimeters of each major bed. Forms a continuous cliff with distinct topographic form.

In measuring a section under natural conditions in the field, a biostratigrapher proceeds upward from bed to bed, taking detailed notes and collecting samples as he goes. Depending on how thorough he wishes to be he may sample from a few meters to several hundred meters each day. To facilitate relocation of the line of traverse by himself or by others, the route may be marked by painted index figures or by permanent stakes driven into the rock or soil.

Later in the laboratory the rock specimens may be studied in detail. The standard procedure is to cut thin sections from each different type of rock so that the specimen can be studied under high magnification. The fossils will be cleaned, classified, and recorded in relation to their parent rock types. Not only is it important to know the species that are present but also the relative numbers of each—some species are common, some rare. Careful measurement of the fossils is also important, because dimensions of various parts of specimens and overall sizes may indicate, among other things, the age distribution of the population. As with modern organisms, many statistical studies are possible with fossils. These may be aimed at detecting the degrees of variability of specimens, the presence of subspecies or varieties within the main groups, and differences in body shape that relate to environment and sediment type.

During the process of classifying his specimens, the biostratigrapher must make judgments about a matter that is not usually faced by modern biologists. This is the decision as to which organisms are near or at their place of growth and which have been transported to the place of burial and fossilization. The nature of the organisms will tell much. A large oyster bed cannot be moved intact by currents but individual lighter shells may be. Floating animals may die or settle on almost any type of bottom sediment or among any contemporary bottom-dwelling organisms. Conversely those organisms that are fixed in place, corals for example, must be preserved primarily in the environment in which they lived and died.

Among the questions to be answered are those pertaining to the overall general environment: Do the sediments represent land or water? If water was the medium of deposition, was it fresh, brackish, or marine, and cold, warm, or cool? If marine, what realm is represented—deep, moderate, or shallow? What clues can be found concerning light conditions, currents, chemistry of the water, nature of dissolved gases or suspended matter, seasonal variations, and other items of an oceanographic nature? If the sediments are land deposits, in contrast to marine ones, the biostratigrapher will want to know which of many possible environments are responsible: lake, glacial, desert, forest, grassland, swamp, river, or delta. Climate is of paramount importance in all land environments and may be hinted at by many clues in the sediments.

Some environmental factors may be identified with ease; others may never be known. The biostratigrapher will find himself turning constantly to the fossils for answers and will thus be drawn into an integrated study not unlike that carried out by a present-day ecologist. He may look upon his fossils as individual species or may consider them as operating, interrelated communities. He will be able to decide rather easily whether an assemblage is land, freshwater, or marine and can quickly identify certain forms as fixed in place (*sessile*) or freely moving (*vagrant*). If he is studying marine forms, he will try to segregate the floating (*planktonic*), swimming (*nektonic*), and bottom-living (*benthonic*) forms.

When all the data are assembled a biostratigraphic analysis may be undertaken. The aim is to achieve a complete reconstruction of each successive situation or environment and to detect possible or probable reactions of the organisms to one another and to their surroundings. Needless to say a full reconstruction is almost impossible, but this should not be unexpected in view of the fact that few living communities, with all elements intact and measurable, have been completely analyzed. The study of the past must progress by bits and pieces with the certainty that there will be increasing understanding with each successive step.

BIOMETRICS AND OTHER MATTERS

Most of the techniques of study applied to living organisms can be used also with fossils. One of these techniques is *biometrics,* the statistical study of biological observations and phenomena. The most serious hindrance to this type of study of fossils is the generally poor preservation of soft parts of organisms, but this may be compensated for by intense concentration on shells, skeletons, and teeth. One advantage that paleontology has over modern biology is the availability of fossils of successive populations spanning long intervals of time. Under ideal conditions—where successive generations of animals have lived, died, and left their remains in one area—a great many specimens can be collected, analyzed, and measured, thus making possible the careful study of those changes in organisms that reveal the trend of evolution in an unmistakable way.

Basic to any comprehensive study of fossils is a collection of many well-preserved specimens. Needless to say these are not available for all species, but many collections have been discovered and analyzed. When a large collection is segregated, measured, and charted graphically according to specific characteristics a typical bell-shaped distribution curve results. Figure 5.16 illustrates what is found when a collection of clam shells is arranged according to a specific feature, in this case the overall size of adult shells. This drawing indicates the size distribution at a specific time—preceding and succeeding distribution curves must be obtained from appropriate earlier and later collections. When such additional collections are analyzed certain changes become evident. Figure 5.17 shows a

Figure 5.16   When a collection of organisms is arranged according to a specific characteristic, a bell-shaped distribution curve results. Here, a collection of clam shells is arranged according to size.

trend to larger size, which, by the way, is a most common one in the organic world. The information in Figure 5.17 could be simplified into a straight-line diagram showing only that there has been an increase in size. This is the simplest graphic illustration of phylogenetic lineage (lineage based on natural evolutionary relationships). The question posed by these data is whether or not new species have been created by the progressive change of size; that is, is the individual

depicted at the base of the diagram a different species from the one in the center or at the top? Is one species or several species represented? This case is oversimplified. In actual practice there will be characteristics other than size that help paleontologists answer the species question. Certainly gradual change is one method by which new species have been created.

Imagine now that our basic data in the form of collections show that there is a tendency to form two lineages—one larger, the other smaller. Figure 5.18 illustrates these two trends. This information shows more than the simple lineage of the first illustration. Because it shows branching, it may be referred to as a simplified family tree. Looking at these data we again ask, how many species are represented? If the changes are well marked and have produced populations with no overlap, the answer will usually be that there are three species—one ancestor and two descendants. This illustration is oversimplified, but it does draw attention to a number of important questions that must be answered by those who study fossils.

What is a paleontological species? The chief test to determine whether a group of organisms is a distinct species concerns reproduction—with minor exceptions, members of a species can interbreed with one another but not with members of other species. However, paleontologists cannot use this test on dead organisms and so they fall

Figure 5.17   This drawing shows what happens when several collections of the same organism from different time periods are arranged according to a specific characteristic. We have again chosen clam shells and size as our example.

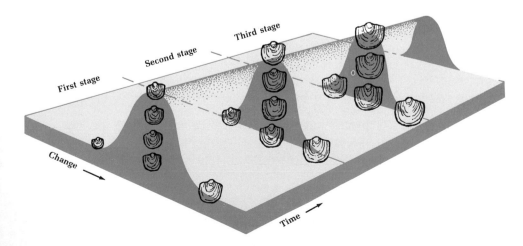

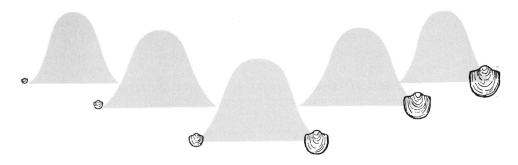

Figure 5.18 In this illustration, two separate lineages are portrayed. One branch shows a trend to larger size, the other to smaller size.

back on *morphological traits* to define their species. Thus differences in shape, size, and proportion are weighed and judged in sorting out probable subdivisions. Not only must the paleontologist make a decision regarding the species affiliations of a collection of organisms from the same geologic period; he must also relate this group to similar groups from other periods. He must decide, in effect, how long a species can exist as such. Will one species mutate to something else in ten thousand years, or a million years, or never? So many subjective factors are involved in classifying extinct organisms that differences of opinion are inevitable. At one extreme are the so-called "splitters" who tend to create species on the basis of minute differences; contrasted to the "splitters" are the "lumpers" who tolerate wide differences in the same species. In one case a "lumper" compressed into two species a total of forty-seven species that a "splitter" had previously described.

What features are important in classifying extinct organisms? We note that many diverse characteristics are employed to distinguish living organisms—floral parts in many plant groups, wing structure in certain insects, plumage patterns in birds, scale counts in reptiles, sensory hairs in spiders, gill structure in clams. Such features are of little use in dealing with extinct organisms, because these features are seldom preserved in fossils. Therefore, paleontologists classify clams, for example, on the basis of the hinge structure of the shell, not on the basis of gills. Certain measurable features are considered

to be more important than others; thus, details of tooth structure in mammals, skull bones in reptiles, details of head anatomy in trilobites, and leaf shapes in plants are most useful. Modern studies, conducted with the aid of computers, are taking into consideration multiple combinations of characteristics and are bringing to light those that have true genetic significance.

What causes species to change or vary? We know from modern studies in the realms of genetics and microbiology that changes in genetic material are spontaneous and largely unpredictable. The environmental factor of greatest influence is natural radioactivity. There is no reason to doubt that this same factor has been responsible for mutational effects in the past. We note also that certain organisms have undergone mutations that make them more fit than their contemporaries for existence in a given environment. This adaptation to environment looms large in studies of past organisms. It is obvious that certain shell shapes are adapted to sandy conditions, others to muddy conditions, still others to strong currents or floating environments, and so forth.

Study of the past reveals many striking effects of migration and isolation. Certain forms have a wide distribution, whereas others have never "left home." Investigation of fossil remains reveals a constant flux as new organisms are introduced and old ones leave by extinction or enforced migration. Because emphasis on the nature of fossil populations is growing, large well-documented collections are required. When a site is found in which fossils are abundant and well preserved, a paleontologist may spend many seasons carefully analyzing each layer of the formation.

The foregoing paragraphs demonstrate the close relationships between modern biology and paleontology. Both fields of study are essential to a full understanding of why living things are as they are.

The Principle of Faunal Succession

Based upon Smith's idea that fossils differ from rock layer to rock layer, and taking into account the worldwide experience of subsequent generations of geologists, we can now formulate a general principle of great fundamental importance to historical geology: *Groups of fossil organisms succeed one another in a definite and determinable order, and any period of time can be recognized by its respective fossils.* This is the principle of faunal succession.

It gradually became evident that fossils are more than mere tags by which various formations may be traced and recognized. Geologists became aware of the significant fact that the organic contents of the various formations do not occur in a random, haphazard, or repetitious manner. When correctly arranged in order of age, fossils show progressive changes from simple to complex. In other words, they reveal the unfolding or advancement of life that has come to be called *organic evolution.*

The proper arrangement of strata in order of age has required much labor and thought and has been guided by the law of superposition. Every action described in the preceding chapter has been applied over and over in all parts of the globe, so that geologists now feel confident that they have arranged the rocks of each major continent in the order of their formation. Notice that superposition applies to *all* sedimentary rocks, regardless of whether or not they contain fossils. This is an important fact of geology.

It cannot be said that the proof of evolutionary progression had to await the unraveling of all the complicated rock structures of the earth, for the idea developed soon after William Smith's time. After the rocks of several large regions had been mapped and classified as to relative ages, it became apparent that the various evidences of progressive change were consistently the same from continent to continent—that a coherent and ever-improving picture of the development of life was emerging. It was only after the clear demonstration of evolutionary change that fossils began to be used to "date" rocks and to place them in their proper places in time.

The proof of the principle of faunal succession does not rest entirely on superposition. It is demonstrated also by what might be called *biotic association,* or the fact that specific groups of fossils occur together consistently. Associated groups of ancient animals are called *fossil faunas,* corresponding plant groups are called *fossil floras,* and total life groups are called *fossil biotas* or *assemblages.* These life groups merge and overlap just as human generations do, but any one time period has its characteristic members.

The principle of biotic association may be clarified by the following discussion. In Figures 5.19 through 5.22 a number of large inclusive life groups have been placed in groups indicating the specific geologic interval for the specimens represented by diagrammatic figures. Thus, the generalized dinosaur skull in Figure 5.19 represents all dinosaurs, both primitive and advanced and indicates that this specimen is part of the Jurassic collection. Of course, any fossil form may be found by itself, but it is more common to find several or many kinds of fossils associated in one deposit, as indicated by the diagram. As-

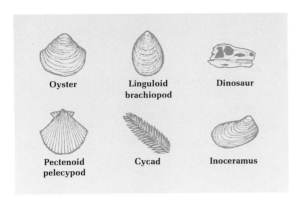

Figure 5.19   Jurassic collection.

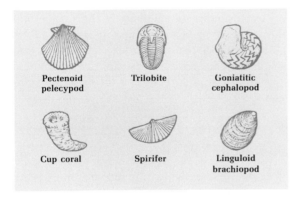

Figure 5.20   Devonian collection.

sume for purposes of illustration that the forms indicated in the Jurassic collection have been collected from a single formation. This collection, by itself, would merely prove that dinosaurs, cycads, linguloid brachiopods, *Pectens,* oysters, and *Inoceramus* pelecypods lived simultaneously, but as yet we could deduce nothing about the proper position of the Jurassic collection in time or of the meaning of its fossils in terms of evolution. Now let us turn to a second collection, gathered from rocks beneath the

Jurassic, which we classify as Devonian (Figure 5.20). As we compare the two collections, we notice many obvious differences in the specimens, but find that pectens and linguloid brachiopods are present in both. We could assume that trilobites, cup corals, and goniatitic cephalopods preceded the dinosaurs, cycads, and oysters, but details about times of first appearance or extinction would still be missing and the chart would be very incomplete.

Now assume that a third collection, representing the Triassic, has been found at some distant locality and we wish to relate it to the other collections. If the Triassic collection was found in its proper sequence between the Jurassic and Devonian, we could place it in an intermediate position without further study. For the sake of a more realistic picture, we will assume we do not know where it was found and that we must determine its relative position by studying the fossils it contains. In other words, we must decide whether to place it before the Devonian, between the Devonian and Jurassic, or after the Jurassic (Figure 5.21). If we place it before the Devonian, we are assuming that dinosaurs, oysters, and cycads have either been overlooked in Devonian collections or have appeared more than once in geologic time. If we choose to place the new collection after the Jurassic, we assume that cup corals and goniatitic cephalopods were overlooked in the Jurassic or that they originated more than once.

Our assumptions of an incomplete record or the recurrence of identical forms may conceivably be correct, but they are obviously unnecessary. By placing the Triassic collection between the Jurassic and Devonian, we are able to maintain a continuous existence for all the fossil forms and can safely add another step to the chart. Also, because the Triassic resembles the Jurassic more than it does the Devonian, we may tentatively place it closer to the Jurassic in time. As subsequent collections come to light, they will fill in the remainder of the chart and will show

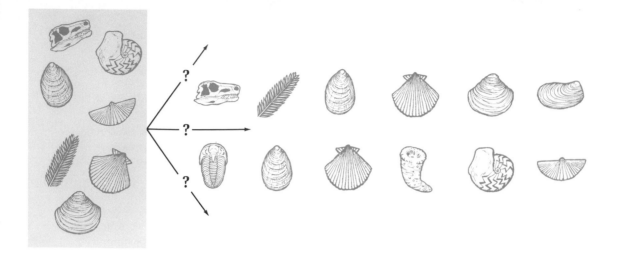

Figure 5.21 How a new collection (the Triassic) is correlated with Jurassic and Devonian collections. Why is one position more logical than either of the others?

eventually the times of arrival, periods of abundance, and final extinction of the various life forms. (See Figure 5.22.)

This discussion illustrates the type of information on which the idea of the succession of life was established. No one thought it up on a theoretical basis; it came to light only after patient, systematic searching by many diligent collectors all over the world.

The progressive evolutionary changes of life on earth can thus be demonstrated by two methods: the study of superposition and the study of associations of organisms.

Understand clearly that the principle of life succession does not mean that simple or unspecialized animals are not found in young rocks or that complex and highly specialized organisms are not present in old rocks. The principle applies only to comparisons made between the total life of successive geologic intervals and between members of the same general group. Thus, a simple sponge from a late period of time cannot be compared with a complex crustacean from an early period. Sponges must be compared with sponges and crustaceans with crustaceans. When we trace successive members of the same group through time or compare the total life of a later period with that of an earlier one, we see that significant changes have occurred. These changes constitute the main evidence for organic evolution.

Reconstructing Past Life

Because the soft tissues of plants and animals are seldom preserved, the paleontologist must base his reconstructions of extinct life forms on the hard fossil remains of the organism. Shells and skeletons are the frameworks that support and contain the other systems of the body, and a fairly accurate reconstruction can be made from hard parts alone. When a skeleton is properly assembled, the general size and shape of the living animal become clearly evident. Bones also reveal muscle attachments, courses of nerves and blood vessels, location of horns and claws, and other important facts about the soft tissues. Teeth are especially significant, for they show by their intricate and unique adaptations the type of food ingested by an organism, which in turn tells us something about the vegetation and climate of the area the animal roamed.

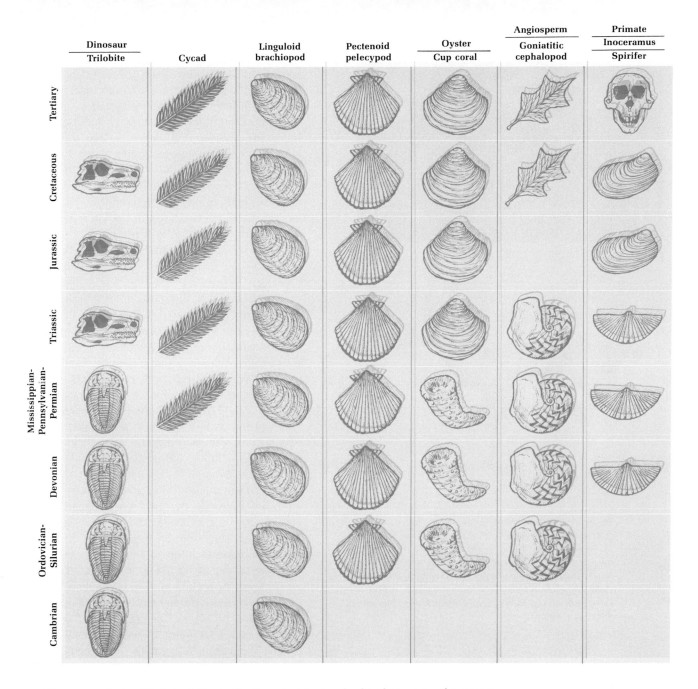

Figure 5.22   A simplified chart illustrating the principles involved in determining the correct succession of life forms on earth and how fossils are used to date rocks or periods of time. Notice that this diagram is a continuation of the three preceding illustrations. Although the drawing is diagrammatic, the individual fossil groups represented are correctly placed according to their existence in time. Example: Dinosaurs are known only from the Triassic, Jurassic, and Cretaceous, and have never been found in association with higher primates.

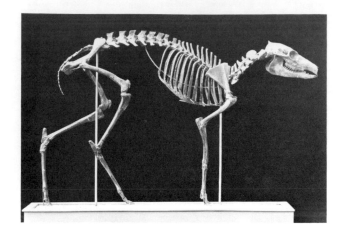

Figure 5.23   Shall these bones live? The art of restoring a living appearance to extinct animals is illustrated by the skeleton and reconstruction of this ancestral deer (Leptomeryx) from the Oligocene of South Dakota. The restoration is based on a careful study of bone and muscle structure and on comparisons with living forms. (Field Museum of Natural History, Chicago.)

One indispensable clue to reconstruction is the occasional discovery of perfect or relatively complete specimens with which fragmentary findings may be compared. Scattered and isolated bones of uncertain origin can be immediately classified if a single perfect skeleton with all parts in proper position is found.

The paleontologist also relies on comparative anatomy to help him identify fossil organisms. Fortunately, most extinct animals and plants have left descendants or relatives that we assume resemble their dead ancestors in habits, functions, and anatomy. If we know that a certain kind of tooth belongs to a particular kind of living animal, a similar fossil tooth probably belonged to an animal similar to the living one. Because feathers are characteristic of all living birds, we assume that a fossil feather is positive proof of ancient bird life. If present-day oysters feed by straining food from the ocean through their gills, we suppose that their ancestors did likewise.

The principle of anatomical correlation must be used cautiously however. For example, someone might argue that because all living elephants inhabit warm countries, an elephant fossil proves the prior existence of a warm climate. Such arguments were indeed put forth in connection with early discoveries of elephant bones in now frigid lands. But the recovery of complete, frozen specimens of the woolly, or hairy, mammoth, a kind of elephant, showed that the animal was provided with a thick coat of hair and a layer of fat that insulated him against the Arctic cold. Other clues to the appearance of the hairy mammoth are found in the cave paintings of early man (Figure 5.24). These drawings show a large, cushionlike elevation on top of the animal's head—probably a lump of fat—a characteristic that would not be apparent from the skeleton alone. Thus, by a combination of clues, we have a fairly good idea of what the mammoth actually looked like. As we go farther back to animals whose remains were not frozen or to specimens that were not depicted by man, we know less and less about the superficial details. But general conclusions, if properly balanced and weighed, may still be accepted as correct.

The imperfection of individual fossils is not the only difficulty encountered by paleontologists. There are obvious gaps in what can be learned of the total life (fauna and flora) of any

one time and place. Even though the organisms that are susceptible to fossilization may be well represented, there may have been others equally as important and abundant that have left no fossil remains. Even in an oyster bed community it has been estimated that 75 percent of the original species will not be preserved. The tendency for preferential fossilization of hard parts can be overcome by careful collecting, by the discovery of unusually complete deposits, and by appealing to established biological principles.

Under favorable conditions, the fossilized record of life in a certain area may be exceptionally complete. A thin layer of black shale, the Burgess Shale, discovered high on the slopes of Mount Wapta, in British Columbia, contains what must be a fairly representative sample of the sea life of the Middle Cambrian of that area. In addition to the usual hard-shelled specimens of the period, there are over a hundred fossil species of delicate, soft-bodied animals. From the evidence of this one deposit, the record of many forms of life has been extended far back in geologic time.

Figure 5.24 *The hairy mammoth as an art subject. The drawing was discovered on the wall of a cave at Cambarelles, southern France, and was sketched by an unknown Paleolithic artist. He saw the mammoth in the flesh. The painting is by Charles R. Knight, a famous American painter of prehistoric life whose murals adorn many of our greatest museums. He based his painting not only on the fossil bones and frozen carcasses of the mammoth but also on the sketches left by ancient man. (Smithsonian Institution.)*

Figure 5.25 The Burgess Shale. Workmen are seen quarrying and splitting the thin-bedded shale in a search for slabs containing the impressions of fossils. (Smithsonian Institution.)

Even more famous are the Solenhofen deposits in southern Germany, where limestone was quarried for many years for use in lithographic printing. This fine-grained material, thought to have been deposited in lagoons or coral atolls in Middle Jurassic time, has yielded some five hundred species of Jurassic plants and animals, including floating crinoids, jellyfish, fish, dinosaurs, flying reptiles, nearly a hundred kinds of insects, and the earliest known birds. It seems probable that the Solenhofen fossils present an almost complete picture of what actually lived at that time and place. The yield of this deposit stems partly from the fact that great quantities of rock were split open and examined for fossils during the quarrying and commercial preparation of the stone.

Another remarkable locality is the La Brea tar pits near Los Angeles, California. Here a highly efficient natural trap captured specimens of many types of late Pleistocene land life, including mammals, birds, reptiles, amphibians, land snails, insects, and plants.

Paleontologists and biologists are greatly interested in obtaining unbroken racial histories (*phylogenies*) of various life groups. Such histories, if fully complete, would clarify many of the problems of evolution. But because all group histories must be pieced together from fossil evidence, it is clear that many of these can never be complete and perfect. Although the evidence implies an unbroken continuity of life forms, there are not enough representative fossils to illustrate every step in the evolutionary process.

Gaps in the fossil record occur because a given

organism may die without being fossilized or, if fossilized, it may subsequently be destroyed by erosion or metamorphism. Many types of animals have left no fossils simply because they lived where sediment did not accumulate to bury them. Also, with the passage of time a population may shift from place to place following favorable living conditions. It may leave fossils in one place, but not in another.

The best fossil records result when a race of organisms originates, evolves, and becomes extinct in one place, leaving many representative individuals to become fossilized and buried in an uninterrupted succession of sedimentary beds. The horse, for example, had representative species in the western United States during almost its entire racial history, and numerous specimens were left as fossils to show the evolutionary progression. In Great Britain, Jurassic cephalopods lived in great numbers and left almost uninterrupted records of their progressive changes.

Fossils and Organic Evolution

Evolution in the broad general sense is the process by which anything living or nonliving progresses from a lower to a higher state, from simple to complex, or from worse to better. Organic evolution, one phase of the universal developmental process, is the progressive change of living things from simple to complex. Fossils prove that life has indeed changed, because almost all fossils are of organisms that no longer exist. That the change has been progressive is shown by the fact that successively later life forms occupy more of the available living space and have more sophisticated means of keeping alive. Keeping alive involves chiefly the procurement, protection, and utilization of energy (food in most cases). Specifically, the successive members of any group of organisms display increasing efficiency in their physiological systems. The skeletons of land animals, for example, show ever-improving mechanical efficiency, and at the same time their teeth become more closely adapted to their food sources. What is more important perhaps is the increase in volume and complexity of nervous tissue, especially that of the brain. Many other illustrations of evolutionary advances will be cited later.

The ruling theory of how life has progressed was developed by Charles Darwin. The key concept of so-called Darwinism is natural selection. Basic to this theory is the assumption that all species have arisen by normal reproductive processes from previous species by means of slow, gradual changes over long periods of time. Under this concept each species, living or extinct, is a small but integral part of an everchanging, constantly proliferating system in which there have been no discontinuities since life appeared on earth.

Darwin observed a vast, impersonal waste of living things. Perhaps a better word than "waste" is "elimination"—the apparently meaningless death of excess individuals is really of great significance in the overall evolutionary scheme. That countless numbers of organisms are doomed to failure in the struggle to survive is a fact of nature that anyone can observe, and the study of fossils reminds us that it is not only *individuals* who are highly expendable; vast numbers of *entire species* have also disappeared. According to Darwin's theory these extinct forms have been weighed in the balance of nature and been found unfit to survive.

Darwin could see that the elimination of most living things came as an inevitable consequence of their having to compete with one another for limited living space and energy resources. The question of what determines the success, or fitness, of one individual or group and the failure, or unfitness, of another must now be answered. Darwin's one-word explanation is "variation." Because there are individual differences even among offspring of the same parents, the contest to survive is not among equals. Some beings possess characteristics that give them, in one way or another, an advantage over their competitors.

The study of fossils supports and confirms not only the general concept of evolution but also Darwin's specific theory of organic evolution by natural selection. The postulates of overproduction, competition, and variation that can be visualized by observing present-day conditions explain in a unique way what is found in the fossil record. More species have been produced than have survived; these species show infinite variations that are chiefly expressions of ways to succeed in competitive situations; current inhabitants of the world are descendants of only a relatively few ancestral types. The world as we know it appears to be the type of world that natural selection would inevitably produce.

Evidence from the study of fossils does not constitute positive proof of organic evolution, in spite of the fact that many students regard it as such. Absolute proof of organic evolution would mean the positive demonstration that all life forms descended naturally from preceding ones, back to a simple beginning. This would require an unbroken succession of individuals, each with "birth certificates" that can never be supplied.

Much of the factual supporting evidence from the fossil record will be recounted in Chapters 11 through 16, and a summary of evolutionary patterns is reserved for Chapter 17.

Biological Classification

Through the science of paleontology, the life of the past is linked to life of the present. Regardless of whether an organism is represented in living form or by a fossil, it must be considered as an essential part of an integrated tree of life. Because of this fundamental unity, the same methods of naming and classifying apply both to living and to fossil organisms. Although relatively few fossils were known when the intensive study of existing plants and animals began, the number has steadily increased, and in many groups the number of fossil species exceeds the living forms. Undoubtedly, there are many more extinct forms than living ones, and future students will profit greatly by studying the fossil and living forms together.

A most important event in the classification of organisms was the publication in 1758 of *Systema Naturae* by Carl von Linné, a Swedish botanist. In this work von Linné arranged the then known plants and animals into groups according to structural similarities and gave each specimen a two-word name. The scientific names were extracted from the classical languages, chiefly Latin, and were intended to be used by scientists everywhere in place of local or common names. The first name labels the *genus* (plural, *genera*) to which a particular living thing belongs. The second word is an adjective that modifies and hence subdivides the genus. Although this second word is loosely referred to as the *species* name, it is more properly the combination of the two terms, the noun and the adjective, that indicates the species of any organism. Genus is the broader term and refers to a group of closely related species. Species on the other hand indicates a group of living things that are so similar in life habits and functions that they can freely interbreed. The generic name may be used only once, but the species name may be used any number of times, but not more than once in the same genus. In print the scientific name is usually italicized, with the generic name capitalized—for example, *Canis familiaris,* the dog.

In actual practice, scientific names are selected to tell something about the organisms to which they apply. A knowledge of Latin is useful in understanding what the names mean. The obvious impossibility of creating short, easily pronounceable names for a million or more organisms explains why many scientific names are strange and difficult to comprehend.

CATEGORIES

The key to proper classification is an understanding of degrees of relationship and of lines of descent. We have already defined genus and species. Still larger groups include more distantly related organisms of increasing diversity. In order of increasing scope, these higher groupings are the *family*, the *order*, the *class*, the *phylum*, and the *kingdom*. The following examples illustrate the application of this scheme to two familiar forms: the white pine, representing the plants, and man, representing the animals.

Kingdom *Plantae*
  Phylum *Tracheophyta* (vascular plants)
    Class *Gymnospermae* (naked seeded)
      Order *Coniferales* (cone-bearing)
        Family *Pinaceae* (the pines)
          Genus *Pinus*
            Species *Pinus strobus* (white pine)

Kingdom *Animalia*
  Phylum *Chordata* (mostly vertebrates)
    Class *Mammalia* (mammals)
      Order *Primates* (primates)
        Family *Hominidae* (men)
          Genus *Homo*
            Species *Homo sapiens*

The system of arranging organisms in successively larger groups is quite similar to a tree in which the main trunk divides successively into smaller and smaller branches that ultimately end in myriads of individual leaves. The concept of a tree is deceptively simple, however, because each group of organisms has its own structure, some with few branches, others with many.

CLASSIFICATION AND BASIC CHARACTERISTICS OF ORGANISMS

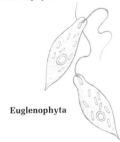

Schizophyta

Euglenophyta

Chrysophyta (diatoms)

KINGDOM MONERA

Unicellular organisms with procaryotic cell structure (cells lack nuclear membrane, plastids, and mitochondria); nutrition predominantly by absorption, but some groups are photosynthetic (energy from solar radiation) or chemosynthetic (energy from inorganic compounds); reproduction by fission or budding; locomotion by simple flagella (whiplike organelles) or by gliding.

Phylum Schizophyta: Bacteria; extremely small, with spherical, rodlike, or spiral form; fossils up to 3.2 billion years old have been discovered.
Phylum Cyanophyta: Blue-green algae, photosynthetic, numerous; stromatolites and other fossils up to 3.2 billion years old.

KINGDOM PROTISTA

Unicellular organisms with eucaryotic cell structure (cells have nuclear membrane, mitochondria and, in many forms, plastids); nutrition by photosynthesis, absorption, ingestion, or a combination of these; sexual and asexual reproduction; locomotion by flagella or other means, but some are nonmotile; plantlike or animal-like.

Phylum Euglenophyta: Motile, photosynthetic; no known fossils.
Phylum Chrysophyta: Diverse group that includes the following: (1) Diatoms, most abundant of the algal groups; secrete siliceous skeletons and inhabit salt and fresh water; fossils from Cretaceous onward. (2) Silico-flagellates; marine, Cretaceous onward. (3) Coccoliths; calcareous, marine, Cambrian onward. (4) Dinoflagellates; marine, Late Jurassic onward. Diatoms and coccoliths are important rock-making organisms.

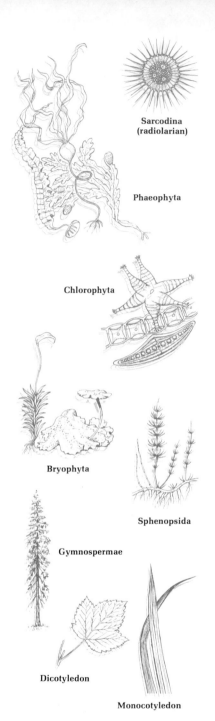

Sarcodina
(radiolarian)

Phaeophyta

Chlorophyta

Bryophyta

Sphenopsida

Gymnospermae

Dicotyledon

Monocotyledon

Eumycophyta

Angiospermae

Phylum Sporozoa: Chiefly parasitic; no known fossils.

Phylum Zoomastigina: Animal flagellates; similar to euglenoids but lack chlorophyll; no known fossils.

Phylum Sarcodina: Amoeboid forms with animal-like metabolism; includes the shell- or skeleton-bearing foraminifera (calcareous, Cambrian(?) onward) and radiolarians (siliceous, Precambrian(?) onward); both are of great value in geologic correlation.

Phylum Ciliophora: The ciliated protozoans; one small family, the Tintinnidae, has left a fossil record.

KINGDOM PLANTAE

Multicellular organisms with walled eucaryotic cells; almost all are photosynthetic and nonmotile; reproduction chiefly sexual; many fossils throughout the geologic record.

Phylum Rhodophyta: Red algae; may occur fossilized as indeterminate "sea-weed."

Phylum Phaeophyta: Brown algae; may occur fossilized as indeterminate "sea-weed."

Phylum Chlorophyta: Green algae; fossils about 1 billion years old are known.

Phylum Charophyta: Stoneworts; some (charophytes) secrete calcium carbonate and are represented as fossils from Devonian onward.

Phylum Bryophyta: Mosses and kin; soft-tissued; very rare as fossils.

Phylum Tracheophyta: Vascular plants; many are hard-tissued; abundant as fossils.

SUBPHYLUM PSILOPSIDA: Psilophytes; dominant in Devonian, very rare at present.

SUBPHYLUM LYCOPSIDA: Club mosses; dominant in the Carboniferous, rare at present.

SUBPHYLUM SPHENOPSIDA: Horsetails; dominant in the Carboniferous; one genus, *Equisetum,* survives.

SUBPHYLUM PTEROPSIDA: Ferns and seed plants.

Class Filicae: Ferns; fossils since the Carboniferous, common in the Mesozoic, subordinate to seed plants at present.

Class Gymnospermae: Conifers, cycads, ginkgoes; mostly wind-pollinated, nonflowering plants.

Class Angiospermae: Mainly insect-pollinated flowering plants; known from the Jurassic onward, become very common in the Late Cretaceous. *Dicotyledons:* Over fifty orders; net-veined leaves (oak, rose, cactus); includes almost all the flower-bearing plants; abundant fossils. *Monocotyledons:* Fourteen orders: parallel-veined leaves (grasses, palms, orchids); abundant fossils.

KINGDOM FUNGI

Mostly multinucleate organisms with eucaryotic nuclei; nutrition exclusively by absorption, hence essentially parasitic; mostly nonmotile, may be embedded in the food supply; reproduction both sexual and asexual; very few fossils.

Phylum Myxomycophyta: No fossils.

Phylum Schizomycophyta: No fossils.

Phylum Eumycophyta: Fossils up to 2 billion years old.

Porifera

Archaeocyatha

Scyphozoa

Anthozoa

Bryozoa

Inarticulata

Articulata

KINGDOM ANIMALIA

Multicellular organisms with wall-less eucaryotic cells lacking plastids and mitochondria; feeding is mainly by ingestion or, rarely, by absorption; level of organization complex; locomotion based on fibrils of many forms; reproduction predominantly sexual.

Phylum Mesozoa: Obscure wormlike, parasitic forms; no known fossils.

Phylum Porifera: Primitive multicellular animals; no true tissues or organs; skeletons of calcite, silica, or spongin; freshwater and marine habitats; fossils from Precambrian(?) onward.

Class Demospongiae: Horny sponges; skeletons of organic fibers.

Class Hexactinellida: Glass sponges; siliceous skeletons.

Class Calcaria: Chalky sponges; calcium carbonate skeletons.

Phylum Archaeocyatha: Extinct; soft tissues unknown, skeleton double-walled, conical, calcareous; Cambrian only.

Phylum Coelenterata (Cnidaria): Polyps and medusae; tissues well developed; tentacles with stinging cells; large central "gut," but no true body cavity (coelom); mostly radial symmetry; many are colonial, mostly marine; fossils from Late Precambrian onward.

Class Hydrozoa: Soft-bodied; mostly colonial and stationary; few fossils.

Class Scyphozoa: Jellyfish; few fossils.

Class Anthozoa: Sea anemones and corals; essentially nonmotile as adults; the corals generally have calcareous skeletons, and many are reef builders; abundant fossils.

Phylum Ctenopora: Comb jellies; fragile; no fossils.

NOTE: The following wormlike phyla are unrecognized or extremely rare in the fossil record: Phoronida, phoronid worms; Sipunculoidea, peanut worms; Echiuroidea, spoon worms; Acanthocephala, spiny-headed worms; Platyhelminthes, flatworms; Nemertinea, ribbon worms; Aschelminthes, roundworms, threadworms, and rotifers; Chaetognatha, arrowworms; Tardigrada, "water bears"; Pentastomida, bloodsucking parasites; Pogonophora, beard worms. Many of these live buried in sediment, and although they may have left fossil burrows, borings, or trails, these are mostly unidentifiable as to origin.

Phylum Entoprocta: Very small sedentary animals, closely related to bryozoans; no known fossils.

Phylum Bryozoa: Small, stationary, colonial, coelomate animals; ciliated tentacles around mouth; skeleton calcareous, chitinous, horny, or membranous; fossils from Late Cambrian(?) onward; freshwater and marine; important as rock builders and useful in geologic correlation.

Phylum Brachiopoda: Exclusively marine bivalves; one shell (dorsal) is slightly larger; stalklike pedicle in most species attaches the animal to the bottom; feeds by ciliated tentacles supported by a specialized, semirigid lophophore enclosed in the shell; fossils from Precambrian(?) onward, especially abundant in the Paleozoic.

Class Inarticulata: Chitinophosphatic shells and no tooth-and-socket hinge between the valves; fossils from Precambrian(?) onward.

Class Articulata: Calcareous shell and well-developed hinge structure; fossils much more abundant than Inarticulata and range from Cambrian onward.

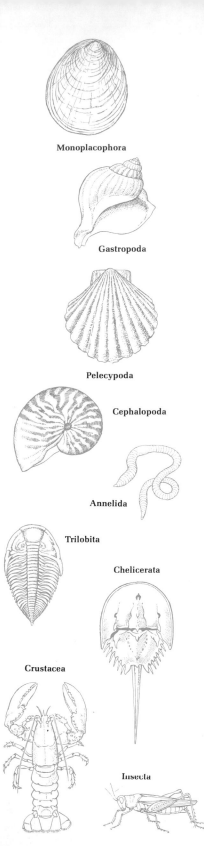

Phylum Mollusca: Extremely diverse, mostly shell-bearing animals having an alimentary canal with two openings and a true body cavity; the shell may be in one, two, or several pieces and is secreted by an underlying soft tissue, the mantle; marine, freshwater, and terrestrial habitats; fossils from Cambrian onward.

Class Monoplacophora: Primitive mollusks with cap-shaped single shells; known as rare fossils from Devonian onward.

Class Amphineura: Elongate, with eight-piece shell; marine only; rare as fossils, Ordovician to Recent.

Class Gastropoda: The most abundant class of mollusks (100,000 species); a few are shell-less, others have cap-shaped or coiled shells; locomotion by a broad creeping foot; distinct head, sense organs; marine, freshwater, and terrestrial habitats; abundant as fossils from Cambrian onward.

Class Pelecypoda: Bivalved mollusks (30,000 species); most have a laterally flattened body between two hinged mirror-image shells; mostly marine, a few are freshwater; locomotion primarily by a muscular, creeping, hatchet-shaped, foot; some (for example, oysters) are stationary; common as fossils from Early Ordovician to Recent.

Class Scaphopoda: "Tusk-shelled" mollusks; shell hollow, tapering, open at both ends; marine only; uncommon as fossils except locally, Ordovician to Recent.

Class Cephalopoda: A diverse group of shelled (pearly nautilus) or shell-less mollusks (octopus); fossil forms generally have a flat, spiral shell with internal chambering and complex folded partitions; mouth surrounded by tentacles; locomotion by jet-propulsion system; well-developed brain and sense organs; marine only; abundant as fossils from Early Ordovician onward.

Phylum Annelida: Segmented worms (for example, earthworm); six classes; marine and nonmarine; fossils consist of burrows, trails, tubes, impressions, and carbonized remains, also abundant teeth (scolecodonts); range from Precambrian(?) to Recent.

Phylum Onychophora: Onychophorians; related to arthropods and annelids; known as very rare fossils from the Cambrian onward.

Phylum Arthropoda: The most abundant phylum (estimated at a million species making up 80 percent of all known animals); diverse body form; practically all have chitinous external skeleton (exoskeleton), jointed appendages, segmented body; locomotion by burrowing, crawling, swimming, and flying; marine, freshwater, and terrestrial environments; fossils of marine forms abundant from Late Precambrian onward, land forms, such as insects, are less well represented.

Class Trilobita: Extinct segmented arthropods with a longitudinally trilobed exoskeleton consisting of head (cephalon), thorax, and tail (pygidium); simple and/or compound eyes; diverse marine habitats; many fossils from Early Cambrian to Late Permian.

Class Chelicerata: Segmented and unsegmented arthropods with pincherlike claws; terrestrial forms (spiders, scorpions, and mites) and marine forms (king crab and the extinct eurypterids); fair fossil record from Cambrian onward.

Class Crustacea: Thin- to thick-shelled arthropods (ostracods, barnacles, lobsters, and crabs), mostly with segmented bodies; also with antennae and other head appendages, compound eyes, and varied means of locomotion; marine, freshwater, and terrestrial; fossils from Cambrian onward.

Class Myriapoda: Centipedes and millipedes; terrestrial; the centipedes are represented by rare fossils from Carboniferous onward, millipedes from Silurian onward.

Class Insecta: Extremely diverse (twenty-nine orders) and numerous (over 850,000 species); wing-bearing arthropods, typically with three pairs of walking legs; food habits diverse; locomotion by burrowing, walking, swimming, and flying; primarily terrestrial; rare as fossils except locally since Devonian.

Crinoidea

Stelleroidea

Holothuroidea

Graptolithina

Conodont

Phylum Echinodermata: Echinoderms; internal skeleton (endoskeleton) of calcareous plates, some completely rigid (sea urchin), some flexible (starfish, sea cucumber); commonly with five-sided symmetry; food habits diverse; about half are sessile, the others move by burrowing or crawling; exclusively marine; common as fossils from Cambrian onward.

Class Cystoidea: Cystoids; stemmed, boxlike, rigid echinoderms with characteristic pore patterns; Early Ordovician to Late Devonian.

Class Blastoidea: Blastoids; boxlike stalked echinoderms with strong pentamerous symmetry, and eighteen to twenty-one plates in the enveloping calyx (theca); Silurian to Permian.

Class Crinoidea: Crinoids; stalked (some secondarily stalkless), pentamerous echinoderms; complex structure of calcite plates; feed by filtering seawater; Cambrian(?) onward.

Class Stelleroidea: Starfish and brittle stars, typically with five radiating symmetrical arms; calcareous plates joined by flexible connective tissue allow free movement; scavengers and predators; starfish live near the shoreline, and brittle stars mainly in deep water; Ordovician onward.

Class Echinoidea: Sea urchins and kin; boxlike, many-plated endoskeleton, covered with movable spines for protection and burrowing; five-rayed symmetry expressed by pore and spine systems; Ordovician onward.

Class Holothuroidea: Sea cucumbers, free living, flexible, with rudimentary calcareous plates; Cambrian(?) onward.

NOTE: The following classes of echinoderms are known from rare fossil representatives and are not described in detail here: Eocrinoidea, Cambrian–Middle Ordovician; Paracrinoidea, Middle Ordovician; Parablastoidea, Middle Ordovician; Edrioblastoidea, Middle Ordovician; Helycoplacoidea, Lower Cambrian; Edrioasteroidea, Lower Cambrian–Mississippian; Ophiocistoidea, confined to Paleozoic.

Phylum Hemichordata: A diverse group of mostly colonial marine organisms showing some affinities with chordates. *Order Enteropneusta:* Wormlike, no fossils. *Order Pterobranchia:* Primitive hemichordates; very rare fossils beginning in the Early Ordovician.

Class Graptolithina: Colonial marine organisms with chitinous exoskeleton taking many forms; reproduction by budding, which yields branched, simple, or, rarely, encrusting colonies; both fixed and floating forms are known; Middle Cambrian to Late Mississippian.

NOTE: One of the most puzzling of organisms is the so-called conodont animal. This creature has left innumerable toothlike objects of varied form from Middle Cambrian to Middle Triassic. It was apparently bilaterally symmetrical, free living, and marine—beyond this, little is known. The general opinion is that it represents a distinct phylum; it is mentioned here because it is also thought to be related to the chordates.

Phylum Chordata: Highly-organized bony animals and their primitive kin; extremely varied in most characteristics; flexible rod of cells (notochord) or segmented vertebral column; gill slits in embryonic and/or adult stages; food habits diverse; marine, freshwater, and terrestrial forms; fossils abundant from Ordovician onward.

SUBPHYLUM UROCHORDATA: Small, soft-bodied, exclusively marine animals; no known fossils.

SUBPHYLUM CEPHALOCHORDATA: Soft-bodied, elongate, primitive chordates with notochord; only two living genera; exclusively marine; no positive fossils.

SUBPHYLUM VERTEBRATA: Backboned animals with highly organized nervous system,

Agnatha

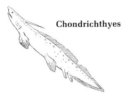

Chondrichthyes

Osteichthyes

diverse food habits, and varied means of locomotion; marine, freshwater, and terrestrial; many fossils from Ordovician onward.

Class Pisces: Fish; thirty-five orders; primarily egg-laying, cold-blooded vertebrates occupying diverse habitats in marine, brackish, and freshwater environments; food habits varied; locomotion mainly by swimming; many fossil forms from Ordovician onward.

NOTE: In the extremely diverse and ancient group known broadly as fish, the following groups or classes are recognized: Agnatha, the jawless fish, Ordovician onward (represented today by lamprey and hagfish); Placodermi, early jawed fishes, mostly armored, Silurian to Pennsylvanian(?); Chondrichthyes, the cartilaginous fish, or sharks broadly considered, Devonian onward; Osteichthyes, the bony fish including most living types, Devonian onward.

Class Amphibia: Amphibians; thirteen orders, water-dependent vertebrates with aquatic and terrestrial life stages; no marine forms positively known; rare fossils beginning with the Devonian.

Class Reptilia: Reptiles; sixteen orders, most of which are extinct; aquatic, terrestrial, and aerial forms; mostly egg laying and cold blooded; many fossils beginning with the Carboniferous.

Class Aves: Birds; thirty-three orders, mostly living; feathered, egg laying, warm blooded; most can fly; rare fossils beginning with the Jurassic.

Class Mammalia: Mammals; thirty-four orders, about half of which are extinct; warm blooded, primarily live-bearing animals with a hairy covering and milk-producing organs; highly developed, with diverse food habits, varied modes of locomotion, including flight, and great range in size; gradually increasing fossil record beginning with the Triassic.

Amphibia

Aves

Reptilia

Mammalia

A SUMMARY STATEMENT

Fossils are the remains of living organisms of the past. Preserved by a variety of processes, they occur in many types of sedimentary rocks. Those with hard parts, such as teeth, bones, or shells, are more commonly preserved than soft-bodied forms. The study of fossils, called *paleontology,* is carried on by specialists and is closely allied with biology. The chief subdivisions of the subject are *invertebrate paleontology, vertebrate paleontology, paleobotany,* and *micropaleontology.*

At first, fossils were regarded as mere curiosities, but careful collection and study eventually convinced students that they record the actual succession of life on earth. In fact, they prove that life has changed and that the change has been from simple to complex. Stages in the gradual process of change enable geologists to recognize the geologic periods by the fossils contained in the rocks. Fossils constitute the strongest evidence that *evolution* has taken place.

Although the fossil record is admittedly incomplete, the accumulation of new evidence is progressing rapidly, and the broad outlines of past biologic history are firmly established.

FOR ADDITIONAL READING

Ager, D. B., *Principles of Paleoecology.* New York: McGraw-Hill, 1963.

Andrews, H. N., Jr., *Studies of Paleobotany.* New York: Wiley, 1961.

Augusta, J., and Z. Burian, *Prehistoric Animals,* trans. G. Horn. London: Spring Books, 1956.

Barnett, Lincoln, *The World We Live In.* New York: Time Inc., 1955.

Beerbower, J. R., *Search for the Past,* 2nd ed. Englewood Cliffs, N.J.: Prentice-Hall, 1968.

Cloud, Preston, ed., *Adventures in Earth History.* San Francisco: Freeman, 1970.

Colbert, Edwin H., *Evolution of the Vertebrates,* 2nd ed. New York: Wiley, 1969.

Fenton, C. L., and M. A. Fenton, *The Fossil Book.* Garden City, N.Y.: Doubleday, 1958.

Jones, D. J., *Introduction to Microfossils.* New York: Harper, 1956.

McAlester, A. Lee, *The History of Life.* Englewood Cliffs, N.J.: Prentice-Hall, 1968.

Matthews, W. H., III, *Fossils—An Introduction to Prehistoric Life.* New York: Barnes & Noble, 1962.

Moore, R. C., C. G. Lalicker, and A. G. Fischer, *Invertebrate Fossils.* New York: McGraw-Hill, 1952.

Romer, A. S., *Vertebrate Paleontology.* Chicago: University of Chicago Press, 1966.

Shimer, H. W., and R. R. Shrock, *Index Fossils of North America.* New York: Wiley, 1944.

Stirton, R. A., *Time, Life and Man—The Fossil Record.* New York: Wiley, 1959.

6
ORGANIZING AND CORRELATING
THE RECORD

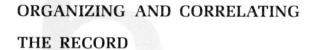

Imagine a microscopic but intelligent creature whose activities are confined to the surface of a newspaper page. To this hypothetical organism the print and pictures would appear only as a varied but apparently meaningless aggregation of light and dark patches. It might wander endlessly over these patches and, indeed, become quite familiar with certain areas or "landmarks" on the page, but it probably would not perceive anything meaningful in them. But if our imaginary organism carefully charted its world on a reduced and comprehensive scale, it would begin

to see an overall design in its surroundings and from it would perhaps learn something important about its own existence. Man's relationship to the earth is similar. The features of his natural environment are such that he cannot comprehend them without first reducing them and putting them in simplified form.

In geology we must reduce the relatively large or complex to a scale that permits study and analysis. The preparation of accurately scaled maps, charts, and diagrams quite naturally constitutes one of the chief activities of geologists.

This chapter deals with the means and methods by which information is classified, correlated, and recorded.

Discovery, Classification, and Explanation

Geologic information accumulates in three successive stages—*discovery, classification,* and *explanation.* The first stage involves the exploration of new areas and the gathering of preliminary collections of minerals, rocks, and fossil specimens. Geologic exploration carries a romantic appeal for many people because it frequently leads to remote and primitive areas and involves all the traditional elements of adventure. At times, however, geologists carry on their work of discovery in the midst of populated and highly developed areas, and amazing finds have been made during the rather unglamorous work of laying roads and streets and excavating foundations for large buildings. Much of the geologist's information comes from mines and borings. The depths of the ocean and the beds of seas and lakes are also being charted and sampled. Insofar as the geologist may be the first to examine these samples with a practiced and discerning eye, he is truly a pioneer observer.

The second stage, classification, calls for more detailed mapping, accompanied by proper identification of individual formations and structures in the field and by the cataloging of fossil and rock specimens from the area. This work may be tedious and time consuming and calls for patience and attention to detail.

If not properly classified, essential facts, measurements, and specimens are about as meaningful as a child's collection of oddities. Early efforts to devise scientific methods of geologic classification were hampered by a legacy from the nonscientific past, when fossil and rock specimens were kept in collection cabinets merely as curiosities. Haphazard classification schemes grew up, often without clearly defined objectives. No one knew exactly what or how much was being classified.

By the time formal schemes of classification were proposed and accepted, a great deal of work, both good and bad, had already been accomplished. Present methods, therefore, are mixtures of ancient and modern concepts, terms, and procedures. Unity, if it exists at all, has been slowly achieved, and the revisions and modifications that must constantly be made are a vexing but necessary phase of scientific progress. Students who are confused by the complexities of organizing a mass of detail according to man-made schemes should realize something of this early background.

The final phase of geological investigation involves synthesizing information that has been already collected and classified. Persons trained in this field search for fundamental relations and explanations. In geology the aim may be to explain the localization of ore deposits or oil fields, to learn why certain natural features are where they are, or to understand the nature and distribution of present life forms. Scientific interests vary, and it is evident that problems that fascinate one investigator will seem dull to another. Some will be led to examine the internal relationships of small units such as crystals, molecules, and atoms. Others will be attracted to problems that concern large features such as mountains, oceans, and continents. Relatively few men will be able to correlate and synthesize information about both the large and the small.

Historical geology is a subject of exceptionally wide scope, and those who study it must constantly create and test new hypotheses. Formulating successful theories is one of the chief purposes of science. Theory is not a permanent substitute for fact, but neither are theories the

Figure 6.1 Examining rock outcrops in Brazil. (UNESCO, photograph by Eric Schwab.)

products of loose or careless thinking. Theories are dangerous only when they are presented as facts or when they become so firmly established in the minds of men that they stand in the way of new knowledge.

The Geologic Column and Time Scale

The essential facts of any historical subject are generally more easily comprehended if they are presented in tabular, chronological form. Therefore, geologists organize their data concerning the earth around a fairly simple, arbitrary outline called the *geologic column*. The most difficult problem has been how and where to subdivide the record. Early observers based their subdivisions on what they considered to be natural interruptions in the rock record, interruptions that were evident, at least locally, in the form of actual physical discontinuities (unconformities) or breaks or gaps in the orderly evolution of fossil forms.

In many cases, geologists had difficulty in adapting these local classifications to the surface of the entire earth, but adjustments were made, and the present geologic column is, with few exceptions, a product of investigations carried on in Europe during the nineteenth century.

When the duration of the various subdivisions of the geologic column is given in terms of years the arrangement is called the *geologic time scale*. In this way, the same terms designate not only certain groupings of strata but also the time periods during which these strata originated. The same names thus apply both to time units and to rock units—the Cambrian *System* includes the rocks that were deposited during the Cambrian *Period*.

Efforts to create systems of classification based on superposition began about the middle of the eighteenth century. Giovanni Arduino proposed in 1760 that the rocks of the earth be divided into Primary (first), Secondary (second), and Tertiary (third) groups. Later on, the Quaternary (fourth) was introduced as a companion term to the others and included the very youngest soils and alluviums that are not solidified to rock. Primary and Secondary have been dropped by most geologists, but Tertiary and Quaternary are still used.

In 1766, Johann Gottlob Lehmann proposed three main classes of mountains: those formed since the Flood, those formed at the time of the Flood, and those formed at the creation of the earth.

As presently constituted, the time scale is divided into a number of grand divisions called *eras;* these in turn are subdivided into *periods*, which themselves are subdivided into still shorter intervals called *epochs*. Most of the eras and periods are now recognized and accepted throughout the world, but there is still a great deal of disagreement among geologists about the epochs. Only the Cenozoic epochs are generally accepted.

THE ERAS

The eras were named to indicate the characteristic stage of development of the fossils they contain. Thus, the *Cenozoic* (Greek: *kainos,* "recent," and *zoe,* "life") refers to the era of recent life; the *Mesozoic* (Greek: *mesos,* "middle") is the era of middle life; and the *Paleozoic* (Greek: *palaios,* "old") is the time of ancient life. Two additional terms, the *Proterozoic* (time of *earliest life*) and the *Archaeozoic* (time of *initial life*), are less commonly used.

THE PERIODS OR SYSTEMS

As presently organized, the geologic time scale includes twelve periods, seven of which are in the Paleozoic Era, three in the Mesozoic, and two in the Cenozoic. The arrangement of these periods is shown in Figure 6.2. A brief historical summary of the designation of the periods follows.

Figure 6.2 The geologic time scale with important biological and physical events.

ERA	PERIOD	EPOCH	MILLIONS OF YEARS AGO (approx.)	DURATION IN MILLIONS OF YEARS (approx.)
		Recent began 10,000 years ago		
CENOZOIC	Quaternary	Recent		
		Pleistocene	2.5–3	2.5–3
	Tertiary	Pliocene		13–15
		Miocene		12
		Oligocene		11
		Eocene		22
		Paleocene	68	5–7
MESOZOIC	Cretaceous			72
			140	
	Jurassic			65
			205	
	Triassic		230	25
PALEOZOIC	Permian			55
			285	
	Pennsylvanian			40
			325	
	Mississippian		350	25
	Devonian			60
			410	
	Silurian		430	20
	Ordovician			70
			500	
	Cambrian			100
			600	
PRECAMBRIAN	Upper			
	Middle			
	Lower			

Although many local subdivisions are recognized, no worldwide system of naming has been evolved. The Precambrian lasted for at least 3.5 billion years. Until more is learned, this lengthy interval may be divided into Lower, Middle, and Upper without formal names.

IMPORTANT BIOLOGICAL EVENTS

- First men
- First manlike primates
- First apes
- Grass spreads widely
- First elephants
- First horses
- Extinction of dinosaurs, giant marine reptiles, flying reptiles, and ammonites
- First primates
- Angiosperms spread widely
- First snakes
- First sequoias
- First birds
- First turtles and lizards
- First dinosaurs and mammals
- Last giant amphibians
- Extinction of trilobites, fusulinids, many corals, crinoids, and other invertebrates
- First mammal-like reptiles
- First conifers, ferns, and ginkgoes; first reptiles
- First flying insects
- First fusulinids
- Extinction of graptolites
- First seed plants
- First land-living vertebrates
- First sharks
- First forests and insects
- First ammonites
- First land vegetation and air-breathing animals
- First bone-bearing animals
- First corals and bryozoans
- First cephalopods
- First pelecypods
- First conodonts
- First graptolites
- First gastropods
- Appearance of many invertebrate phyla: arthropods, mollusks, sponges, and echinoderms

IMPORTANT PHYSICAL EVENTS OF NORTH AMERICA

West ⟵⟶ East

CASCADIAN OROGENY
Pacific border

ROCKY MOUNTAIN (LARAMIDE) OROGENY
Cordilleran region of Mexico, United States, and Canada

NEVADAN OROGENY
Western Great Basin, Sierra Nevada
COLUMBIAN OROGENY
Western Canada

LATE PALEOZOIC (APPALACHIAN) OROGENIES

ALLEGHENY OROGENY
(Mississippian-Permian)
Middle and southern Appalachians

SONOMA OROGENY
Great Basin

ANCESTRAL ROCKIES
(Mid-Pennsylvanian)
Colorado, New Mexico, Utah

OUACHITA-MARATHON OROGENY
(Mississippian-Pennsylvanian)
Texas, Oklahoma, Arkansas

ANTLER OROGENY
Great Basin

ACADIAN OROGENY
Southeastern Canada, New England, Piedmont
ELLESMERIAN OROGENY
Northern Canada

TACONIC OROGENY
Northern Appalachians, Piedmont

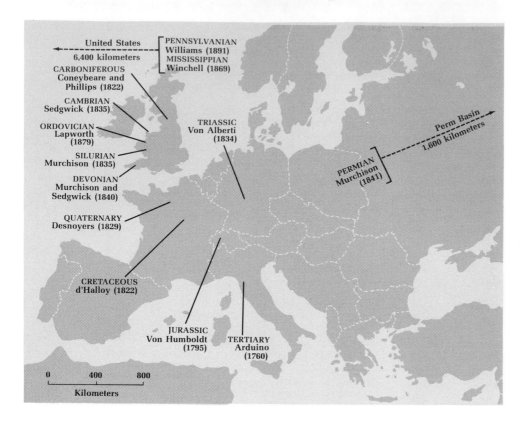

Figure 6.3 The naming of the geologic systems. (Adapted from D. L. Eicher, Geologic Time, Prentice-Hall, Inc., 1968.)

Cambrian. The name "Cambrian" was proposed by Adam Sedgwick, a British geologist, in 1835, and is taken from *Cambria,* the Latin name for Wales. As originally conceived, the Cambrian included the Lower Silurian that had been identified by Sedgwick's co-worker, Roderick I. Murchison, but this section is now recognized as a separate system, the Ordovician. At the original, or *type,* area, the Cambrian is poor in fossils; much better sections have been found elsewhere in northwestern Europe.

Ordovician. The name "Ordovician" was proposed in 1879 by Charles Lapworth, another English geologist, and commemorates the Ordovices, an ancient tribe that formerly inhab-

ited Wales. Originally, the Ordovician was both the Upper Cambrian proposed by Sedgwick and the Lower Silurian described by Murchison and was created to resolve a dispute between the two geologists, both of whom had applied names to the rocks involved. The name "Ordovician" is still not universally accepted; in Germany, for example, equivalent rocks are called Lower Silurian.

Silurian. Murchison proposed the term "Silurian" in 1835 to designate rocks that are well exposed on the borders of Wales and England, a territory formerly inhabited by the Silures, to which the name refers. As originally proposed, the term included rocks now known as Ordovician. The upper and lower boundaries are difficult to place in the type areas. The most complete example of Silurian rocks in North America is in New York State.

Devonian. The Devonian, named for Devonshire, England, was proposed by Sedgwick and Murchison in 1839. The Devonian is particularly well represented in New York State, where the standard reference sections for North America are located.

Mississippian (Lower Carboniferous). The name "Mississippian," one of the youngest period designations in use, was proposed by Alexander Winchell, an American geologist, in 1869. European geologists, however, prefer the term "Lower Carboniferous" for equivalent rocks. The term "Carboniferous" once included what is now the Permian, Pennsylvanian, and Mississippian periods. The type area for the Mississippian is in the Mississippi Basin, where excellent outcrops are exposed.

Pennsylvanian (Late or Upper Carboniferous). The term "Pennsylvanian," first used in 1891 by Henry S. Williams, an American geologist, refers to the state of Pennsylvania, where the rocks are well exposed and contain abundant coal. In Europe this portion of geologic time is designated Late Carboniferous.

Permian. The Permian was named by Murchison in 1841 and refers to widespread exposures of corresponding rocks in the Russian province of Perm. The best examples of this system in the United States are in Texas and in the Great Basin.

Triassic. The geologic period name "Triassic" dates from 1834, when it was used by Friedrich von Alberti, a German geologist. The name refers not to a particular area but rather to a striking threefold division of contrasting rocks that were first studied in Germany. The German section, however, contains few marine fossils, and better sections exist elsewhere, particularly in the western United States, Canada, and Timor.

Jurassic. The Jurassic was the first period or system to be named. Although the first intensive study of the system was carried out in England, the name originates from the Jura Alps, which lie between France and Switzerland. The name was first used in 1799 by Alexander von Humboldt, but the rocks he had in mind were later supplemented by others to make up the present system. Jurassic rocks are found on all continents.

Cretaceous. The name "Cretaceous" (Latin: *creta,* "chalk") was first applied in 1822 by J. J. d'Omalius d'Halloy, a Belgian geologist, to formations that had long been known as the "chalk formation" in England and France. The Cretaceous is represented on all continents and is a very important period from the viewpoint of volume of sediments and number of exposures.

Tertiary. As originally conceived by Arduino, the name "Tertiary" signified all relatively recent and more or less unconsolidated material that contained fossils resembling animals and plants still in existence. Geologists have thoroughly studied and subdivided the Tertiary because its rocks are young and almost everywhere are deposited over older rocks. The standard subdivisions, or epochs, of the period include, from oldest to youngest, the *Paleocene, Eocene, Oligocene, Miocene,* and *Pliocene.* These names follow the "presence-absence" method of classifying proposed by Lyell in 1833, by which the rocks are subdivided according to the degree of relationship of their fossil mollusks to those still alive. If we include the first epoch of the Quaternary (Pleistocene) in the breakdown, the scheme of classification and naming of the Tertiary is as follows:

Pleistocene Epoch (Greek: *pleistos,* "most," and *kainos,* "recent"), 90–100 percent of mollusk fossils represent modern species.
Pliocene Epoch (Greek: *pleion,* "more"), 50–90 percent modern species.
Miocene Epoch (Greek: *meion,* "less"), 20–40 percent modern species.

Oligocene Epoch (Greek: *oligos,* "little"), 10–15 percent modern species.

Eocene Epoch (Greek: *eos,* "dawn"), 1–5 percent modern species.

Paleocene Epoch (Greek: *palaios,* "ancient"), no modern species.

Lyell's original classification included only the Eocene, Miocene, and Pliocene. Later, in 1833 he applied the name Pleistocene to what had previously been Newer Pliocene. The Oligocene was added in 1854 by Heinrich Ernst von Beyrich, and the Paleocene in 1874 by Wilhelm Philipp Schimper. Lyell's idea of classifying on the basis of faunal comparison was excellent in theory, but later workers have had to rely on fossils other than those known to Lyell.

Quaternary and Its Subdivisions. The Cenozoic Era is vastly shorter than the Mesozoic or Paleozoic, including as it does only one period and part of another. The Tertiary Period, which covers about 65 million years, is on a par with earlier periods; but the Quaternary, which spans perhaps 2 or 3 million years, is just beginning.

The most significant event of the Quaternary Period was the great Ice Age. Researchers have tended to regard the onset of glacial conditions as marking the beginning of the period, but, because of the extremely diverse effects of glaciation in different areas, this time plane is difficult to establish. There is even greater doubt and disagreement among geologists over the Pleistocene Epoch itself. Until a few years ago, most investigators believed that the Ice Age had come to a close, and subsequent time was included in a later epoch, the Recent or Holocene. In other words, the Ice Age constituted the Pleistocene Epoch; all postglacial time fell within the Holocene Epoch. Many geologists today hold that the Ice Age is not yet over, however, and that we are merely in an interglacial stage between two ice advances. In view of this new development, the basis for a Holocene Epoch seems to be weakened, and geologists are more or less divided over whether or not it should still be regarded as a full-fledged epoch. The terms "Holocene" or "Recent" are used by some investigators and not by others purely as a matter of personal preference.

Problems in Applying the Traditional Geologic Column

Many of the designated divisions of the traditional geologic column were established on the basis of studies in limited areas of western Europe and do not necessarily fit conditions found elsewhere. The boundaries between most of the classical systems do not coincide with natural interruptions or discontinuities even on individual continents, and boundaries must frequently be drawn in the midst of unbroken sedimentary piles where deposition of sediment and evolution of enclosed fossils show no distinct *discontinuities* whatsoever. To overcome some of the deficiencies of the classical geologic column a great number of adjustments have been proposed, mostly without much success.

One very commendable scheme, which entails an almost complete revision of the geologic column of North America, has been proposed by the American geologist Laurence L. Sloss. The major divisions of this substitute arrangement are called *sequences,* which are analogous to, but somewhat longer than, the traditional systems. The basis for delimiting the sequences is much the same as that which hopefully, but somewhat erroneously, was used to justify the traditional, time-honored geologic column. Sequences are deposits laid down between successive widespread retreats of the ocean. In other words, a sequence begins when the ocean commences a major transgression and ends when it regresses or withdraws.

A useful bench mark for delimiting the sequences is the so-called transcontinental arch, an

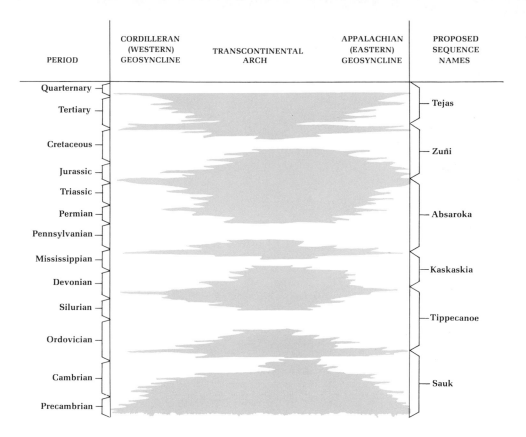

PERIOD	CORDILLERAN (WESTERN) GEOSYNCLINE	TRANSCONTINENTAL ARCH	APPALACHIAN (EASTERN) GEOSYNCLINE	PROPOSED SEQUENCE NAMES
Quarternary				
Tertiary				Tejas
Cretaceous				Zuñi
Jurassic				
Triassic				
Permian				Absaroka
Pennsylvanian				
Mississippian				Kaskaskia
Devonian				
Silurian				Tippecanoe
Ordovician				
Cambrian				Sauk
Precambrian				

Figure 6.4 *The division of geologic time into sequences based on the transgression or regression of the oceans across North America. The colored area in the diagram represents dry land. When the colored area is crossed by white, the submergence of the transcontinental arch is indicated. Note that the point of maximum emergence of land prior to the beginning of a new transgression of the waters marks the beginning of each sequence.*

elongate "backbone" of the North American continent extending from southern California to Lake Superior and beyond. Away from this ridge-like tract the ancient oceans deepened to the northwest and the southeast. With flooding seas, the water level and sediments lapped higher onto the arch to positions of maximum flooding. Occasionally the spreading seas crossed over low passes or even inundated the entire ridge but this does not lessen its usefulness as a standard of reference. As a practical matter, the divisions between sequences are well marked on the flanks of the arch, but away from it, in the deeper depositional tracts where the oceans were more or less permanent, the interruptions are fewer and shorter. Figure 6.4 shows the essential features of the Sloss scale of sequences and its relation to the traditional systems.

The Geologic Formation

Although the word "formation" as used in connection with rocks is older than the science of geology, it has come to have a specific technical meaning not generally understood by nongeologists. In simplest terms, *a formation is a mappable natural rock unit that has definite edges or contacts and certain obvious characteristics*

Figure 6.5 Formations in the walls of the Grand Canyon. Because of the scanty soil and vegetation, the formations themselves and the contacts between them are well exposed. Differences in resistance to erosion leave the harder formations standing as cliffs, whereas the softer ones form slopes. The Kaibab Limestone is especially resistant, and its top forms the level plain above the canyon rim. (Union Pacific Railroad.)

by which it may be traced from place to place and distinguished from other formations. There are formations of igneous, metamorphic, and sedimentary origin, and the methods of defining and recognizing each class are somewhat different. Sedimentary rocks are divided into formations more easily than are igneous or metamorphic rocks, and most of the named formations are of the sedimentary variety.

A sedimentary formation may be thick or thin, and it may be made up of one or several kinds of rock. It may have been deposited over a few square kilometers or over several hundred thousand square kilometers. It may have accumulated during a short period or over millions of years, or the entire formation in one area may be older or younger than the entire formation elsewhere. In spite of these wide variations, a formation should display obvious unifying characteristics that differentiate it from other formations with which it is associated. Although the formation may seem to be mainly a somewhat arbitrary convenience used by map makers, it is also a manifestation of some past event or episode that is different from succeeding and preceding ones.

In the United States and most other areas, each sedimentary formation is designated by a distinctive name. The name consists of two parts, the first taken from some geographic feature, the second indicating the rock type, if one type is dominant. As an example, the term "Madison Limestone" refers to a widespread limestone unit found, among other places, in the Madison Range of Montana. The term "Morrison Formation" refers to a unit in which no particular type of rock dominates and which is found well exposed at Morrison, Colorado.

Ideally the place from which the name is taken is an area where the formation is exposed clearly and completely, and this place becomes the type section. The formation may carry the name wherever it happens to occur, even hundreds of kilometers from the type locality. At the time the name is proposed, the newly recognized formation is described in detail, its upper and lower boundaries are clearly designated, its fossil contents are listed and dated, and all physical characteristics by which it may be recognized are described. If this procedure is followed, the work of subsequent investigators is simplified. Needless to say, in the earlier stages of exploration many units were named in a careless or incomplete manner and had to be restudied later.

Although a geologic map showing distribution of rock formations is a common and most useful product of field work, there are other types of

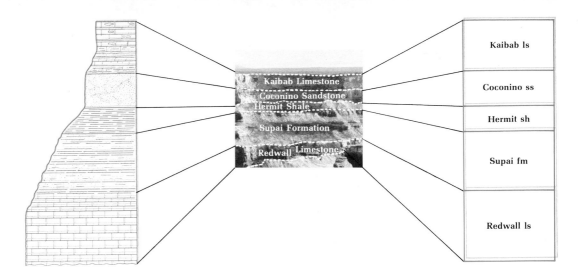

Figure 6.6 *This three-part illustration shows how a portion of the formations illustrated in Figure 6.5 can be represented by a semidiagrammatic cross section (left) and a simple columnar section (right).*

visual or graphic representation that may be made. From geologic and topographic maps it is possible to make diagrammatic cross sections to show conditions under the surface. Drill holes and mines aid in these reconstructions. Columnar sections depict what is found or expected at one particular point—as, for example, in an oil well.

Sedimentary Facies

The term *facies* is coming into use in dealing with characteristics of rocks that have to do with their environments of formation. The term is difficult to define, for it apparently means different things to different workers. As generally accepted, however, *a sedimentary facies is the sum total of the primary or original characteristics of a rock by which it can be recognized and from which its environment of deposition can be determined.* The distinguishing characteristics may be in the inorganic rock material (*lithofacies*) and/or in the fossil contents (*biofacies*). In general the term is best applied to parts of single rock units that were deposited contemporaneously. Thus we might recognize shallow-water facies and deep-water facies of the same formation.

The concept of facies avoids some of the short-comings of formation mapping and also reveals less obvious features than those that characterize formations. In other words, formations are selected on the basis of certain gross features that geologists can readily follow without being too concerned with less obvious characteristics or lateral changes within the formation; facies may be selected on the basis of any significant feature or features, large or small, subtle or obvious, and may be combined in any way to yield the desired information.

Formations and facies are not mutually exclusive or antagonistic. They may merge in some instances and serve identical purposes. The facies concept perhaps permits greater flexibility of study; no formal names need be applied, and no set rules of procedure have been adopted. Formation mapping permits rapid classification of surface outcrops according to standardized rules and procedures that yield rather uniform results. Both methods will continue to be used as needs dictate.

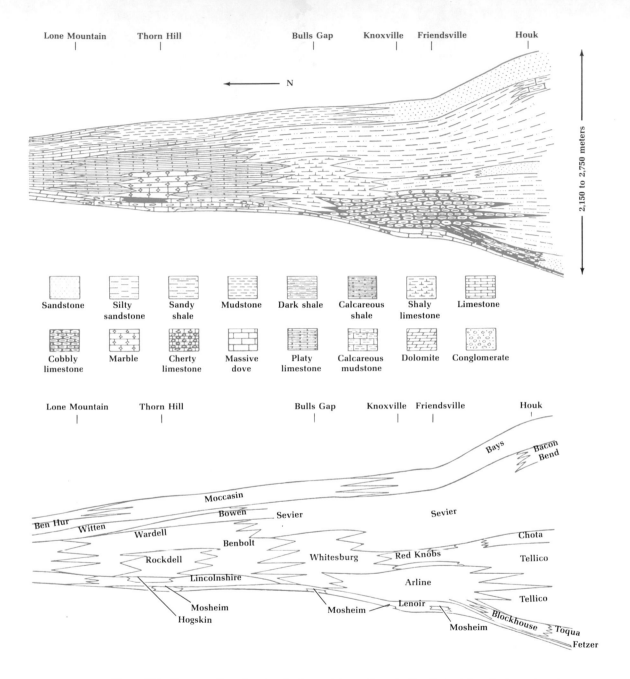

Figure 6.7 Cross section showing facies and formations in the Ordovician System of east Tennessee from north to south. In the upper cross section the individual facies are shown and designated according to the various rock types given in the boxes below the cross section. The lower section shows the same facies pattern with the formational names added. A number of rock types and formations were accumulating at any given time upon the sea bottom, and such a sequence can be mapped as either facies or formations. (Reproduced from G. Arthur Cooper, Smithsonian Miscellaneous Collections, Vol. 127.)

Time-rock Units

As geologists attempted to assemble maps of larger and larger areas and to compile a world-wide record of their work, they began encountering problematical situations that called for additional classification schemes. Single formations or facies cover only relatively small patches of the earth, not large areas such as the United States, Europe, or Australia. And on maps of large areas it was almost impossible to draw outcrops of individual formations to a convenient scale. Consequently, regional or continental maps often were neither comprehensive nor understandable. How could these difficulties be overcome? A natural solution presented itself—combine formations or facies into larger aggregations that would be suitable for depicting large areas and lump the rocks together on the basis of time of formation rather than according to physical characteristics.

The aggregations of rock identified by time of formation are called *time-rock units*. They differ from formations, which are defined on the basis of observable physical characteristics. A time-rock unit, while also a body of rock, may be thick or thin and may include any or all types of sediment and a variety of facies. Its boundaries, however, are imaginary time planes that mark the arbitrary limits of the period during which it was deposited. Consequently, the boundaries of time-rock units do not necessarily coincide with physical breaks in the rocks. Obviously, the drawing of boundaries is difficult where fossils are scarce or absent, because the recognition of time intervals has been based on fossils, and will be until other methods of dating are successfully applied.

Time-rock units bridge the conceptual gap between the tangible but incomplete rock record of the earth's crust and the intangible but unbroken flow of time, which cannot be mapped. Geologic time has been divided and subdivided so that geologic events and their physical manifestations can be grouped and arranged. The following chart summarizes what we have already said about the relation between various schemes of classification.

Time Units	Time-Rock Units	Rock Units
Era	———	Group
Period	System	Formation
Epoch	Series	Member
Age	Stage	Bed, etc.
———	Zone	

The Paleontologic Zone

Geologists place a special meaning on the word "zone" when dealing with fossils. Generally, they consider the zone to be a body of rock that is identified strictly on the basis of its contained fossils. A minority regard it merely as an interval of time during which certain organisms existed. Without entering into the pros and cons of the argument, we may cast our lot with the majority of geologists who consider the zone to be defined on the basis of certain selected fossils. The zone may consist of different kinds of rock, may be thick or thin, and may be confined to one area or have a worldwide scope. There are three basic types of paleontologic zones: A *range zone* is marked by the total life-span of one designated biological group, for example, a species. An *assemblage zone* is characterized by the coexistence of several organisms, one of which is chosen as an *index species*. But this species need not be confined to this one particular zone nor need it be found throughout every part of the zone. A third paleontologic zone, the *concurrent-range zone,* is marked by the overlapping existence of two or more species having different life-spans. Although there are several kinds of zones, we may summarize by saying that a zone, if clearly

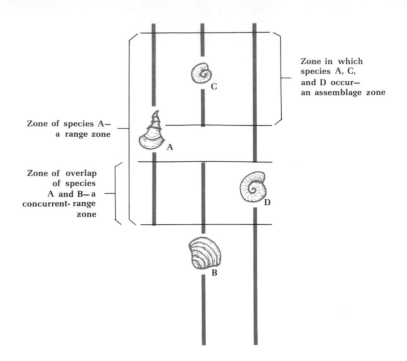

Zone in which
species A, C,
and D occur—
an assemblage zone

Zone of species A—
a range zone

Zone of overlap
of species
A and B—a
concurrent- range
zone

Figure 6.8   Paleontological zones. A zone may be defined on the basis of one fossil form or on the joint existence of two or more forms. The zone is regarded by some geologists merely as an interval of time, by others as a mass of rock.

defined, is of more value for indentification purposes than a formation, because a zone can be recognized from one area to another over very great distances without regard to the type of sediments in which the fossils lie.

Geologic Maps

A geologic map is the product of much field, office, and laboratory work and embodies all the principles of identification, correlation, and organization that geologists employ. These maps are compiled according to standards that are fairly uniform everywhere—a map made by a geologist in one country is meaningful and intelligible to a geologist in any other country.

The usual printed map depicts geologic formations upon a base or background of topography and culture (meaning the works of man). Each formation, as identified and correlated by the geologist, is represented by a particular color or pattern and designated by abbreviated symbols.

The boundaries of the formations are located and drawn with maximum possible accuracy as are faults and other structures that affect the rocks; all these are designated by appropriate symbols. A concise legend with short notations usually appears on the margins of a map, so that all features may be thoroughly understood.

· Geologists also illustrate their works with cross sections and columnar sections. A cross section is like a cut in a layer cake—it reveals what is below the surface along a particular line. A columnar section is a representation, usually in the form of a slender column, of what is known or inferred about geologic conditions under a particular point. It is a graphic depiction of what might be found in drilling a well for oil or water. A row of columnar sections with lines drawn between correlating items is a sort of skeleton cross section. Columnar sections are usually printed on long strips of paper so that they can be moved apart or brought together to reveal the possible correlations and space relations that may exist.

Geologic maps, together with cross sections, are invaluable whenever and wherever tunneling, mining, drilling, excavating, and similar activities are necessary. They serve as guides for those who search for mineral deposits and aid in the development of highway systems, river and harbor controls, and recreational areas. But the geologist sees something more on a map than a solution to practical problems. From a carefully prepared map, he can deduce much about the geologic history of the area that is depicted. The formations shown represent ancient developmental episodes that are important, if not essential, in reconstructing the past.

Paleogeographic Maps

The study and reconstruction of the geography of past times is called *paleogeography*. Just as the most significant fact of present-day geography is the distribution of natural physical features, so the arrangement of land, seas, mountains, and rivers in any past time period is important in understanding life and events during that time. Paleogeographic maps are based on all available evidence and are only as accurate as this basic

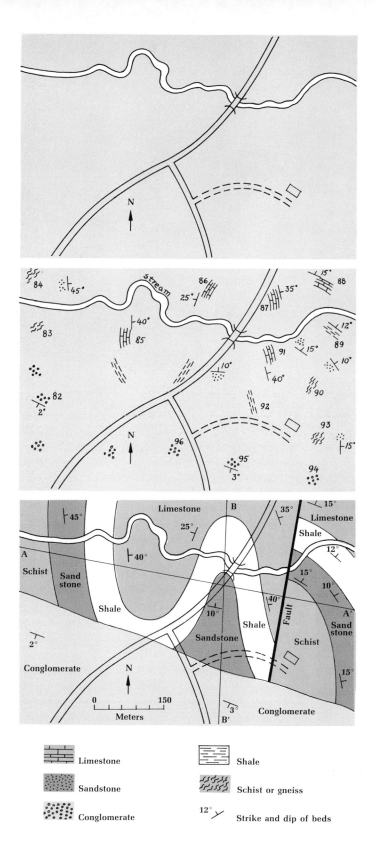

Figure 6.9 Three stages in constructing a geologic map. (a) A map containing man-made features (roads, bridges) and large natural landmarks (rivers). (b) Exposures of various types of rock have been indicated, and the strike and dip of beds has been noted in degrees. The other numbers indicate entries in the geologist's notebook. (c) Based on his study of outcrops, the geologist has added boundary lines between the different types of rock and has colored the patterns to show how the area would look if all soil were stripped away.

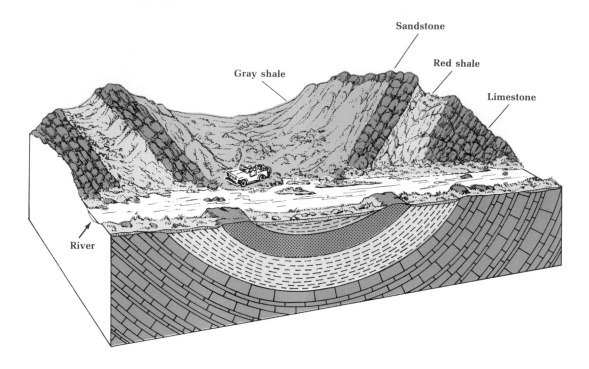

Figure 6.10   A cross-sectional view that shows how the bending of rocks forms a trough.

information permits. Some maps are more accurate than others because some regions have been studied more intensely. In most cases the information permits little more than generalized outlines of large landmasses and bodies of water, although in thoroughly studied areas, minor features such as mountain ranges, islands, deltas, and river systems can be localized. The differentiation of land and sea areas depends primarily on whether or not marine deposits are present, the shoreline being roughly located where the marine deposits disappear.

Although paleogeographic maps tend to be somewhat lacking in detail, they will continually improve as investigators collect more and more information. They contribute significantly to our understanding of earth history and facilitate the location of important commercial mineral deposits. In addition, they summarize large amounts of detailed information. As you read along, you will encounter many of them throughout the book.

Geologic Correlation

Geology, like other sciences, is studied in bits and pieces, a little at a time. The average geologic project involves a relatively small area, a few rock types, or a single group of fossils. Such local or limited studies are important and interesting, but their full value comes only when the various studies are integrated or tied together both in space and time. Here the process of *correlation* assumes importance. The word in its general usage signifies the mutual relation of things. In geology it has several specific meanings, but unfortunately usage has not always been uniform.

Geologic correlation means the degree of equivalency of rock units or the events they represent. Rocks may be "equivalent" in various ways: (1) They may be equivalent in time of formation; in other words they would be *synchronous,* or identical in age. (2) They may have the same or similar composition and thus be *lithologic* or *lithostratigraphic* equivalents. (3) They may have the same or similar fossils and thus be *biological* or *biostratigraphic* equivalents. Ideally, a formation might correlate in all three of these ways, and the task of tracing or recognizing it would be easy. But usually a formation is of slightly different age from place to place, and sometimes rocks of the same age differ in composition or have different fossils from place to place. The term *homotaxis* is used to describe situations in which we find similar fossils or a similar succession of fossils in formations that are not necessarily of the same age. Homotaxis may occur when a group of organisms migrate from one place to another; the group will be older at the place of origin than they are in the place to which they moved. Homotaxis thus signifies similarity, but not necessarily contemporaneity.

The concept of correlation must also be clarified further by considering the distance factor. Certain techniques are applied to determine whether or not specific rock samples were once part of a single, physically continuous unit; other techniques are used with rocks that were never physically continuous and are equivalent only in fossil content or in time of formation. We should remember that even though a rock formation may be continuous, not all parts of it were formed at the same time. The correlation of continuous or once-continuous beds is carried on over the relatively short distances covered by individual formations, whereas the determination of time equivalence involves the entire earth.

Most geologic work requires the correlation of outcrops or samples of rock from areas separated by faults, by eroded tracts, or merely by geologically unexplored or inaccessible country. Anyone

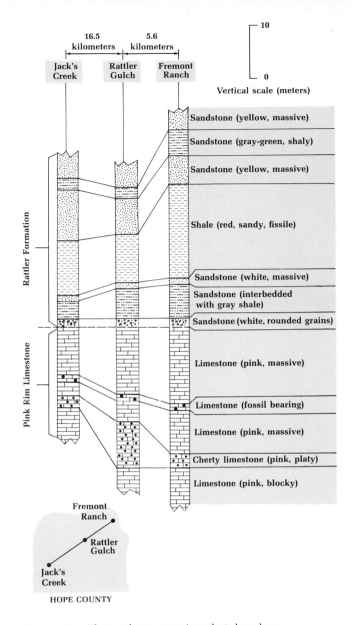

Figure 6.11   Three columnar sections that show how rock units are correlated over a given area.

who has stood on the brink of the Grand Canyon and traced the various formations from point to point with his eye has performed a simple kind of geologic correlation. Early geologic explorers concentrated their main efforts on preparing maps showing the surface distribution of rock formations. They tried to identify correctly each formation wherever it appeared and to represent it on their maps so the later investigators might find it exactly as described.

The increased importance of underground mining and drilling has caused more attention to be given to the hidden, or *subsurface*, portions of formations. Whereas early geologists placed great emphasis on the surface appearances of formations and on properties that are evident in eroded outcrops, the modern investigator relies on fine and more subtle features that are revealed by small unweathered chips or cores brought up by drilling. Because it is difficult and expensive to obtain actual specimens of buried formations, instruments have been devised to measure the properties of rocks that lie out of sight beneath the surface. Sensitive instruments are lowered into drill holes, and the results are recorded at the surface. The correlation of rock properties by instruments is a borderline field that combines techniques and applications from both geology and physics.

The broader correlations between widely separated portions of the earth's crust are absolutely essential in building up an integrated picture of past history. By combining numerous local correlations, more accurate regional or worldwide correlations can be achieved.

Correlation by Physical Criteria

LATERAL TRACING
OF SURFACE UNITS

The simplest and most direct method of correlating continuous surface exposures of rocks is to walk along their outcropping edges. This procedure involves strenuous outdoor work because the investigator must follow the outcrops over all sorts of terrain without regard for orthodox routes of travel. Where soil and vegetation are thick, the process is slow and sometimes uncertain, for only a few exposures of rock may be visible. Under these conditions, road cuts, wells, and stream banks must be carefully searched for evidence. In areas where cover is scanty or absent, the beds may be followed by eye, but vague impressions gained from a distance are never as reliable as positive identification that is made at close range.

Photogeology. Photogeology is a modern development that has partly replaced the actual "walking out" of beds. It is the technique of studying geologic formations by means of aerial photographs. A great deal of the earth has been photographed from the air, and map makers in many fields are using the results as a substitute for actual ground work. If the interpreter is skilled enough to recognize geologic features from photographs, he can compile remarkably accurate maps, but in areas of complex geologic relationships there is still need for careful and detailed inspection on the ground.

ESTABLISHING
PHYSICAL IDENTITY

If the continuity of a bed is interrupted and it can no longer be followed on the surface, the interrupted parts must be correlated by indirect means. The most common method is to establish as many points of similarity as possible. If the two separated samples have identical or sufficiently similar characteristics, they are assumed to correlate. Of course the investigator must be cautious, because similar rocks may form under similar environments at many times and places. The fact that a formation may not be entirely homogeneous throughout makes it difficult to say just which physical features are best for correlation purposes. Color, for example, is one of the most obvious characteristics of a rock, but color may be altered by weathering or by the action of heat or underground solutions.

CORRELATION BY SEQUENCE

In some instances, it is easier to correlate a sequence of beds rather than an individual member of the sequence. A red shale overlain by a thin gray limestone and underlain by a thick yellow sandstone, for example, would constitute a three-part sequence that probably would not be repeated very often. If the same sequences were found in two nearby but separate localities, there would be little doubt of their correlation. If the sequence consisted of four members, the probability of correlation would be even higher. This is illustrated by the correlation of varves as shown in Figure 6.12. Varves occur in characteristic sequences that can be correlated by the arrangement of the thick and thin units. Each

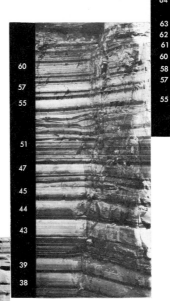

Figure 6.12 Correlation by sequence as seen in the matching of glacial varves. The series of four photographs illustrates how varves are matched and correlated from one exposure to another. These varves, in the Puget Lowland, Washington, were studied and photographed by J. Hoover Mackin of the University of Washington. (Courtesy of J. Hoover Mackin.)

Figure 6.13 Core samples cut from over 2,450 meters of rock penetrated by an oil well in the Uinta Basin, Utah. The cores have been laid out in order of recovery for the inspection of geologists. (Christensen Diamond Products.)

cess, samples of rocks are brought up in the form of broken fragments, chips, or continuous, solid, cylindrical pieces called *cores*. The physical and biological characteristics of these samples may be studied by the same methods that are used in studying samples taken on the surface, but the overall features cannot be determined until many samples are recovered. Cores and chips from newly drilled areas are usually carefully preserved for reference purposes because the rocks they represent may never be sampled again if a worthwhile oil deposit is not struck.

Geophysical Methods. After a hole is drilled, and even if no samples have been saved, it may still yield valuable scientific information. This information is obtained by lowering into the hole instruments that measure and record certain properties of the rocks, such as radioactivity or reaction to electrical currents. The electrical conductivity of the walls of the hole and of the fluids it may contain varies from formation to formation, and when properly recorded can be used to interpret the type of rock penetrated, the thickness of each bed, the contacts of formations, and other information. After the curves recorded on the *electric logs* have been related to certain properties of the rocks, the correlation of closely spaced wells may be carried on entirely from a study of these electric logs. It must be emphasized that these methods are useful only over the relatively short distances covered by individual formations.

Other geophysical instruments have been devised to detect geologic features that do not outcrop and are not penetrated by drill holes. Some of these instruments detect slight differences in magnetism or gravity; others react to vibrations caused by man-made explosions—the vibrations travel through the rocks and are reflected or refracted in various ways depending on the properties of the materials they traverse. Geophysical investigation based on these methods is useful mainly in detecting contacts or unconformities between rocks of distinctly different character.

individual layer is so much like many other layers that it could not be correlated by its physical characteristics alone.

Information from Drill-holes. The surface exposure of a rock formation is a very small sample of a much larger mass that lies out of sight beneath the earth. Many formations do not crop out at all and are known only from information discovered in wells drilled for oil. So important are these hidden formations and other conditions beneath the surface that most of the geologic work connected with drilling for oil is designated as *subsurface geology*. During the drilling pro-

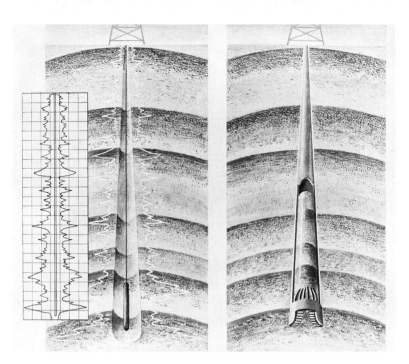

Figure 6.14 Methods of obtaining geologic information from drill holes. The figure on the right shows,
in diagrammatic form, the cutting of a core from a
subsurface formation. The rotating teeth at the end
of the drill cut a cylindrical sample of all the rocks
penetrated. These pieces may be lifted to the surface
and examined for fossils or for signs of oil. At left,
an electrode is shown being lowered into the hole.
The reactions of the surrounding rocks are recorded
in the form of a visual curve (insert) by suitable
instruments at the surface. By studying electric logs,
a geologist is able not only to correlate between
closely spaced wells but also to detect water or oil.
(Standard Oil Company of California.)

Figure 6.15 This cutaway diagram shows what
happens when a dynamite charge is set off in a drill
hole. Shock waves caused by the explosion pass
downward through the earth and are reflected by the
beds at A and B. Small microphones, called seis-
mometers, pick up the reflected shock waves and
transmit them through cables to the recording truck.
The insert at right shows a seismogram that has
been recorded by instruments in the truck. The reflecting beds at A and B are indicated by wiggles on
the seismogram. (Geophysical Service, Inc.)

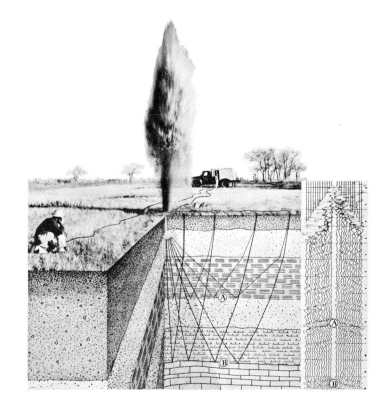

Geophysics is extremely useful in directing attention to localities that are worthy of exploration, or perhaps even drilling.

KEY BEDS

Occasionally a rock layer will be found that is easily recognized over a relatively wide area. Such a layer is called a *key bed*. It may be within a single formation or it may pass from one formation into another, or even from one basin into another. Its distinctive characteristics may be either physical properties or fossil contents. Ideally, a key bed should be relatively thin and should record a short episode or single event that

Figure 6.16 Distribution of volcanic ash from the eruptions of Mount Mazama (now Crater Lake) in southwestern Oregon and Glacier Peak in northern Washington.

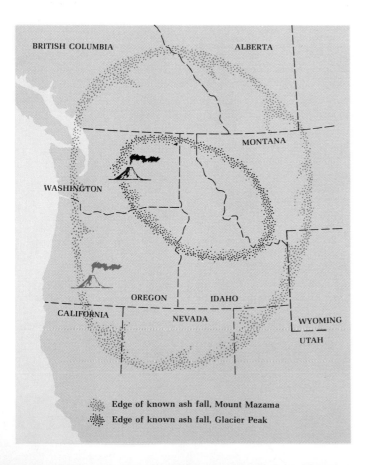

affected a wide area. Among the best-known and most widely used key beds are layers of volcanic dust. When a large eruption occurs, much pulverized rock material enters the atmosphere and is dispersed by the winds over a very wide area. On settling, the dust blankets everything beneath it. If it falls into a body of water, it may sink and be preserved from erosion indefinitely. Many, if not most, volcanic ash deposits are significantly different in composition, and so they can be accurately correlated.

A single distinctive layer of volcanic ash has been identified in two hundred postglacial peat bogs in Washington, Idaho, Montana, Alberta, and British Columbia. The eruption, which according to radioactive dating took place 6,700 years ago, covered at least 470,000 square kilometers of territory with a single layer of dust. A somewhat older eruption, which took place during the last Ice Age, showered dust over most of Nebraska and parts of nearby states. This deposit, known as the Pearlette Ash, furnishes an excellent plane of reference for dating events and life history in the Great Plains and parts of the Rocky Mountains.

Study of deep-sea cores has revealed beds of volcanic dust as widespread as those on land. One single ash fall has been traced across the Atlantic Ocean from Newfoundland to Iceland. This dust may have originated from volcanic areas in Iceland, but its exact source is unknown.

With the passage of time, a buried layer of volcanic ash gradually changes to the material called *bentonite*. Beds of bentonite are found in rocks of practically all ages and have been used by geologists to solve correlation problems in many places.

Thin beds of conglomerate, representing rapid erosion of nearby landmasses or perhaps the debris from single violent storms, have also served as key beds in continental areas. Organisms such as algae, oysters, or corals, which form solid masses, may occasionally spread out and leave a continuous thin deposit over many hundreds of square kilometers. The Aspen For-

mation of the Rocky Mountains contains countless fish scales that record the death of many fish
during a rather short period. This fish-scale bed
is particularly important in correlating rocks of
oil fields in that area. Except for undisturbed
beds of volcanic ash, there are few types of key
beds that are known to form simultaneously over
extensive areas.

Correlation by Fossils

The physical characteristics of a rock formation
are usually distinctive enough to identify it over
relatively short distances. But physical characteristics do not remain the same across extensive
areas, and if correlation between continents or
even widely separated parts of the same continent is necessary, the investigator must increasingly rely on fossils.

It may be worthwhile here to recall that the
principle of faunal succession states that each
period of time may be recognized by its respective fossil life. Thus, everywhere on the globe, on
all continents there was an ancient Age of Trilobites, followed successively by the Age of Fishes,
the Age of Coal Forests, the Age of Reptiles, and
the Age of Mammals. This is a broad generalization, however, and is a great oversimplification
of the situation. Although it is true that the major
periods and the rocks representing them are recognized rather easily by their fossils, we are interested in much finer distinctions—we want to
correlate as closely as possible. We want to know
not only that a rock belongs to a certain period,
but also which part of that period is represented.
Here the difficulties begin, for the same species
of trilobites, fish, reptiles, and mammals did not
live everywhere. Very few species are adapted to
exist over the entire earth. This is also obvious
from studies of recent animals; for example, we
find oyster beds along the coasts of all the major
continents, but each stretch of coast has a different species adapted to local conditions. Similarly,
in the past there were certain related groups of

Figure 6.17 Light-colored bentonite bed, 15 to 20
centimeters thick, in black marine shales of the
Bearpaw Formation, Oldman River, Lethbridge,
Alberta. Radiometric dating indicates that this key
bed is about 73 million years old (Late Cretaceous).
(Courtesy of R. E. Folinsbee.)

species that existed over wide areas and left
fossils that are useful in establishing correlation
between continents and across ocean basins.

How can fossils be of aid in correlating across
continents and oceans if species differ greatly
from place to place? What is known about living
organisms is helpful; for example, some organisms arc widespread, some very restricted. For
correlation we will therefore look for groups that
tend to "get around." Another obvious generality
is that similar organisms live in similar environments. We will therefore try to compare shore
dwellers with shore dwellers, lake dwellers with
lake dwellers, and so forth. Perhaps the most
important fact is that organisms of all kinds are
adapted to move or migrate from one place to
another in order to stay with the conditions most
favorable to their existence. It is the problem of
migration or dispersal that is of greatest concern
in correlating by fossils. Specifically, we may ask:
Just how fast can organisms spread from their
sources?

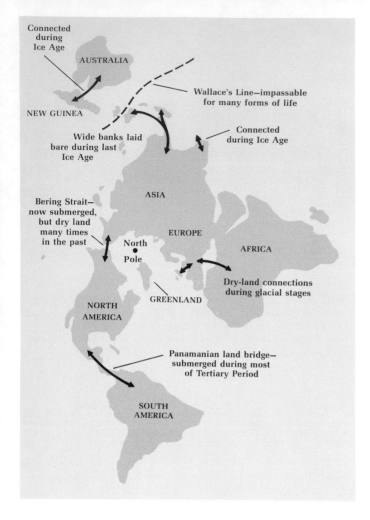

Figure 6.18 Migration routes make it possible for most forms of life to disperse widely. Land bridges and shallow shelves connect all the great landmasses, and plants and animals can spread widely if climatic conditions are favorable.

In thinking about this problem, we accept the principle that each species is unique and had its beginning at one time and in one place only. If we find the same fossil form in two widely separate places, we must assume that a certain amount of *dispersal* or travel was involved. Although we may never know the time consumed or the exact routes followed by ancient animals in their migrations, we can learn a great deal from what is happening today. Many instances are known of plants and animals spreading rapidly once they have found access to new territory. The history of the rabbit in Australia and the English sparrow in America, although influenced by man, shows what can happen under favorable conditions. It is evident that if an organism is suitably adapted and routes of migration are open, it should reach all places favorable to its existence in a few thousand years. This is a relatively short time, geologically speaking, and is of little importance in worldwide correlations.

That migration has always been possible is shown by the fact, already mentioned, that the succession of the great groups of fossils is the same on each of the large landmasses, including Australia and Greenland. Each continent tells the same story of progress from simple to complex through the ages. This means that adaptable forms of plants and animals have always had ways and means of spreading from their places of origin into all areas where their capabilities and the environment permitted. Local areas may have been temporarily separated from the rest of the world, but sooner or later, through the operation of geologic processes, all major areas have received travelers from all other areas with a resultant high degree of uniformity over the entire earth at any given time. It is by careful analysis of similarities and differences that correlation is accomplished.

GUIDE FOSSILS

The geologist pays special attention to fossils that are particularly abundant and widespread during certain periods of time; these are called *guide*, or *index*, *fossils*. Ideally, a guide fossil should be easy to identify, specimens should be relatively common, and they should be restricted to specific formations or to short periods of time. Guide fossils should be selected only after they have met these requirements.

An investigator must take dozens of collections from a specific formation before he can make a final decision about the relative abundance of its different fossil forms. Certain kinds occur too rarely; others may be abundant, but may occur also in beds above and below the one he is trying to correlate. Only those that seem to be restricted to, or abundant in, the interval in question should be designated as guide fossils. Experience has shown that infallible guide fossils are very rare, but there are a great many forms that are quite helpful if used carefully and cautiously.

Guide fossils must obviously be the remains of rather unusual organisms. By their abundance, they indicate a high degree of success in meeting competition and in finding food; by their widespread occurrence, they show that they had efficient means of getting around. Their abrupt appearance and disappearance suggests either rapid evolution or special conditions governing their survival and extinction.

Paleontologists are convinced that the most reliable marine guide fossils are left by floating organisms. An organism that floats on the surface may be widely dispersed by currents of water and it may, on dying, fall to the bottom among the creatures living there. Because it may be buried in almost any type of bottom sediment, it will be found in correspondingly diverse types of rock.

Land-living organisms that are most likely to leave good guide fossils are wide travelers that are adapted to a variety of conditions and that have relatively short life-spans. Many mammals meet these requirements. Perhaps the best of all guide fossils are furnished by spores and pollen grains that are carried by the wind over long distances; these may be buried in almost any type of sediment, in shallow water as well as on land. The study of these small fossils is still in its preliminary stages, but additional investigation and research will furnish much valuable information.

CORRELATION BY FOSSIL ASSEMBLAGES

Because many species are usually present in any fossiliferous bed, it is usually safer to correlate on the basis of a number of types rather than a single one. If species A and B overlap or have lived together for a short time, then the presence of A and B together identifies that period more precisely than either A or B alone. The record of life is a great series of overlapping family lines, and by properly selecting and localizing the overlapping portions, a reliable scheme of correlation sooner or later can be worked out.

CORRELATION BY STAGE OF EVOLUTION

Although we might debate whether or not life evolves along predestined lines, it is nevertheless true that each group passes through a series of modifications which, so far as we know, are

Figure 6.19   A useful guide fossil. At left is a section of core taken from an oil well in southern Utah at a depth of over 2,100 meters. The smaller specimen was collected on the surface in the Uinta Mountains over 160 kilometers away. Both show well-preserved specimens of Spirifer grimesi, a brachiopod, which proves correlation between the surface rocks and the deeply buried equivalent beds.

Figure 6.20   A collection of fossils—cleaned, identified, labeled, and properly stored. Collections such as this form the basis for correlation of sedimentary formations throughout the world. (California Division of Mines and Geology, photograph by Richard B. Saul.)

gradational in nature and are seldom so drastic that the offspring cannot be assigned to their proper parents. If a paleontologist uncovers fossil specimens that represent the complete history of a particular group, he can study the progressive modifications in detail and relate these changes

Figure 6.21   Succession of skulls showing evolutionary changes in a lineage of fossil mammals known as oreodonts. All the successive species are from Miocene and early Pliocene deposits of western Nebraska. The dashed line indicates the position of the break between the premolars and molars and aids in evaluating changes in proportions of the skull through time. (American Museum of Natural History.)

to particular time periods. The fossil record of the horse is an example of a well-known, fairly complete family history with all essential stages represented. If an investigator uncovers an isolated horse tooth in an undated formation, he can assign the formation to a definite age on the theory that the stage represented by the tooth in question was characteristic of only one phase of the history of the horse. Correlations of this type must be preceded by very detailed studies of long series of well-preserved fossils. As information accumulates, this method will become increasingly important, especially as applied to world-wide correlations.

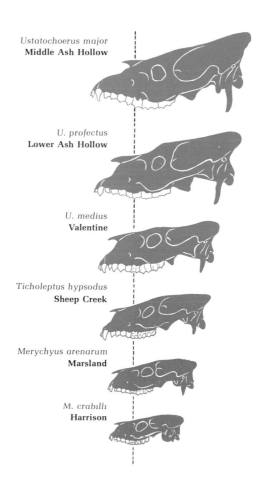

Ustatochoerus major
Middle Ash Hollow

U. profectus
Lower Ash Hollow

U. medius
Valentine

Ticholeptus hypsodus
Sheep Creek

Merychyus arenarum
Marsland

M. crabilli
Harrison

Recent Developments—
Composite Methods

Correlation between deep-sea deposits and dry-land deposits has been a most difficult problem because the types of fossils and rocks of the two environments are so greatly different. Fortunately, recent developments in several distantly related fields have combined to permit correlation of vast areas of the solid ocean bottom, sediments on the ocean bottom, many generally unfossiliferous continental sediments, lava flows, and other igneous bodies. By combining techniques of dating by sequence, by fossils, and by radiometric determinations it is now possible for the first time to achieve almost worldwide conformity to the standard geologic time scale.

The key discovery is that practically all of the ocean basins are underlain by weakly magnetized strips of basalt. These strips generally have widths of tens or hundreds of kilometers and lengths of hundreds or thousands of kilometers. How these originated is described in Chapter 8, and the process need only be very briefly discussed here. The strips differ in only one important property—magnetic polarity; that is, they have either a north- or a south-seeking magnetic response. The essential coincidences and facts are these: Basalt contains an abundance of iron, an element particularly easy to magnetize. When they emerge in liquid lava, the iron minerals become magnetized under the influence of the earth's magnetic field, and as the lava congeals or solidifies a record of the field of that moment is locked in the rocks. With proper techniques any sample of basalt, whether extruded on land or under water, will reveal the direction and something of the intensity of the magnetic field that prevailed when it became a rock.

Throughout recorded geologic time there has been a tendency of the earth's magnetic field to reverse polarity. There is no known obvious physical change; the earth does not turn over, the core of the earth remains stable—only the invisible magnetic field is reversed. The North Pole has had its present negative polarity for about 690,000 years; before that it was positive for about 100,000 years, a period that was preceded by about 60,000 years of negative polarity. Beyond this an unbroken series of reversals going back to the middle Tertiary has now been recognized. Investigation on oceanic basalts cannot be continued much further back simply because the ocean bottoms are no older than this. There are, however, many thousands of continental lava flows, dikes, and sills whose polarity can be determined and for which lengthy series of alternations have been constructed. So far, these older segments have not been connected with the present. It is known that the polarity epics are far from equal in duration and as a consequence the bands of ocean lava that are based on magnetic properties are also of unequal widths.

An analogy with tree-ring or varve-count dating is apparent. As long as we can start the count with the present and go back without breaks in the record we can give a definite absolute date for any tree ring or varve in the sequence. There are many excellent varve series of great length that cannot be correlated with the present because of gaps of unknown length that have not been filled. The same is true of the recent sequence of polarity reversals and the earlier ones—they cannot be related to each other in terms of absolute age.

Here radiometric dating is helpful. Basalt contains enough potassium-bearing minerals to permit application of the potassium-argon method. A good sample, if carefully studied, will therefore reveal two facts: an absolute date in years and a magnetic record. The important fact is that a simple radiometric date now suffices for the entire strip in which it is found. One satisfactory date will place literally thousands of square kilometers of area in its proper age relation. Furthermore, if this strip belongs say, to the tenth polarity epoch back from the present, it can be correlated by sequence with a strip somewhere else. As a result, that strip also can receive the same absolute date. Thus the tenth polarity epoch can be dated wherever it can be found in its proper sequence no matter whether it is represented by oceanic or continental rocks. It should be pointed out that most dry-land, iron-rich sediments also possess magnetic records. These are locked in as small mineral crystals settle in quiet waters and become magnetically oriented before the rock solidifies. Many iron-rich formations have formed in arid regions where life, and consequently fossils, are sparse. Such formations are difficult to date by other means so the new method is being widely applied to them. Volcanic ash also contains magnetic minerals and bridges a gap between sediments and actual lava flows.

How fossils are integrated into the scheme just described is easily understood, but the situation with regard to oceanic sediments is somewhat different from that on dry land. Fossils of land-living organisms may be found not only above

or below but also actually within material from which paleomagnetic determinations can be made. Fossils do not occur in oceanic basalt but remains of organisms will commence to accumulate on it immediately after it is formed. Thus the age of a basalt strip gives a maximum age for the fossils lying on it. This is not to say that the fossils lying on each strip are different; but neither are they identical. Gradual evolutionary changes, new arrivals, and exterminations are recorded without interruption once a seabottom is created. What paleomagnetism provides is a valuable series of punctuation marks that tell just when important organic events took place. When properly understood, the appearance or nonappearance of any fossil can in turn be given a chronological position that is as useful as a radiometric date. To obtain and thus date a fossil is much easier and less expensive than to determine a radiometric date or polarity reading from a volcanic rock. It should be understood that deep-sea fossils consist almost entirely of minute shells and skeletons that pertain to small protozoans or plants. Some of these are siliceous, some calcareous, but all are so numerous that they make up large volumes of sediments and so cannot easily be missed in any dredging or drilling operations.

Because the composite method of dating is just being integrated, it is not yet possible to know what findings may come to light as it is applied. Already it has been possible to correlate lake sediments containing evidences of ancient man in Africa with marine oozes in the Pacific and Atlantic oceans and with lava flows in Japan.

Practical Applications of Geologic Correlations

MINING PROBLEMS

The discovery and mining of valuable materials in the earth is expensive and notoriously risky. One of the traditional disappointments of the miner is to have his lode, or vein, terminated by

Figure 6.22   Round fossil coccoliths and star-shaped discoasters from a core taken in the western Pacific Ocean. The specimens are estimated to be about 4,000 years old (Pliocene). Fossils such as these are becoming increasingly important in working out the history of sedimentation in the ocean basins. Magnification about 3,000×. (U.S. Geological Survey, photograph by J. David Bukry.)

a fault. The very meaning of the word "fault" carries a certain connotation of imperfection and disruption that reflects the miner's difficulties. But there is usually some degree of hope for him, because the lost portion of the lode may have been displaced to a nearby locality and may not be lost entirely. To find the severed portion a geologist is called in.

The expert's first task is to identify the foreign material that has moved against the edge of the vein. By carefully studying the thickness and distribution of all the rock units in the vicinity, he can determine how the rocks have moved to create the fault. The miner can plan accordingly, but the results are not always favorable—the vein may have moved to another person's property, it may have sunk to inaccessible depths, or it may have been shifted to the surface and eroded away. Fault problems can be exceedingly complex, especially in areas where the appearance of the rocks has been altered by igneous or metamorphic action. Fortunes have been made or lost on the interpretation of faults. One young geologist in the West became a millionaire by locating the severed portion of an ore body in a complex fault pattern after all mining operations had been suspended as hopeless.

The principles of correlation are useful not only in finding mineral deposits in broken and disturbed areas, but also in locating areas that may contain hidden deposits. Thus coal, clay, salt, uranium, and other valuable materials occur in certain geologic settings whose characteristics are familiar and have been observed many times in the past by investigators. When geologists encounter favorable leads or indications, they logically assume that corresponding valuable materials lie nearby.

THE SEARCH FOR OIL

The intense search for oil that is carried on in widely scattered parts of the world is providing a critical test of all the older classical theories and methods of surface geology and is stimulating also the development of new techniques of all types. Oil, more than any other resource, must be sought beneath the surface of the earth in hidden deposits that give little or no surface indication of their presence. Almost without exception, oil is associated with sedimentary rocks, under a variety of structural conditions that are best interpreted by trained geologists. It is not surprising, therefore, that about 60 percent of all geologists are engaged in the search for petroleum

Figure 6.23 Geological cross section through the Tintic mining district, Utah. Note especially the effects of faults on the structure. Length of section is about 3 kilometers; maximum vertical distance about 1,000 meters.

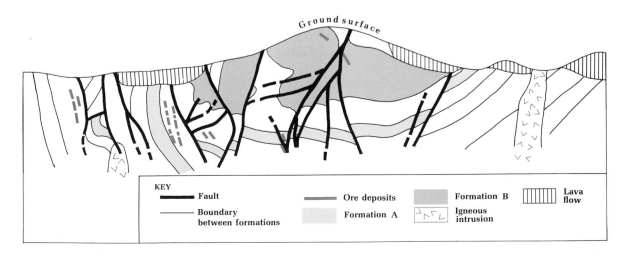

Ground surface

KEY

— Fault

— Boundary between formations

— Ore deposits

Formation A

Formation B

Igneous intrusion

Lava flow

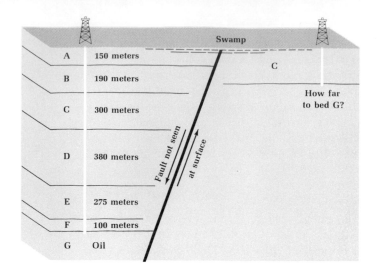

A	150 meters	
B	190 meters	C
C	300 meters	How far to bed G?
D	380 meters	
E	275 meters	
F	100 meters	
G	Oil	

Swamp

Fault not seen at surface

Figure 6.24 Correlation pays. After one well has been drilled and the correct order of formations is known, it is possible to begin to predict what will be found in nearby wells even though faults and other disturbances are present. In this illustration the well on the left has reached oil-bearing bed G and has penetrated all the overlying formations. This information enables the drillers of the well on the right to predict conditions in their well once they know the formation in which the well commences.

and related products. Their success in this field is attested to by the fact that few successful oil companies operate without geologic advice.

Modern oil prospecting requires the application of techniques, principles, and procedures from many different fields of science, but a great deal of the work involves geologic correlation. Oil is associated with certain types of rock, with certain types of structure, and with certain ages or periods of time. This association implies a correlation that is a challenge to the geologist to interpret. An ideal oil field includes a *source bed* from which the oil arises and which usually contains evidence of considerable organic matter, a *reservoir bed* with open spaces in which the oil may accumulate, and a *trap,* or local condition, where the oil concentrates to form a pool.

Any important oil-producing formation be-

comes a drilling *objective* and is thoroughly studied to determine its thickness, distribution, and depth. Figure 6.24 indicates how correlation is employed in predicting the depth of a particular bed. A deep exploratory oil well may cost as much as a million dollars, and a drilling company needs to know as much as possible about conditions at different levels. Samples of the well must constantly be inspected to detect changes in the formations, the presence of faults, and evidence of oil and gas.

A particularly productive type of rock may be found to underlie a certain tract of land. Geologists must try to predict how far this favorable area may be expected to extend, whether it will *trend* one way or the other, and what the drilling conditions may be within it.

Most large oil companies maintain staffs of scientists who work in the field and in the laboratory to solve the myriad of problems associated with oil prospecting. Geology has had to prove its value in this highly competitive industry, in the face of a certain amount of opposition. The geologist cannot guarantee to find oil; he merely removes some of the risks and uncertainties of the search by pointing out favorable sites. Many localities highly recommended by geologists have proved unproductive, and other sites drilled without geologic advice have proved successful. In the long run, however, statistics have demonstrated that sound geologic work is superior to the intuition of the "wildcatter" and to the "doodlebug" of pseudoscience.

The search for oil has not only called into use a great number of correlation techniques, but has also verified fundamental geologic ideas regarding superposition and faunal succession. The validity of any theory is measured by its success as a basis for prediction. The search for oil is a matter of prediction, each exploratory well being a test of predictions about subsurface conditions. Without the application of correlation based on superposition and faunal succession, there could be no predictions, and the search for oil would lack a scientific basis.

A SUMMARY STATEMENT

In order to be comprehensible, the earth's surface must be portrayed on a reduced scale. To do this, geologists map and classify rock formations chiefly on the basis of their relative ages. The geologic time scale is a worldwide standard of reference. The basic unit of time is the geologic *period,* the corresponding rock unit (all material deposited during the period) is the *system.* Periods may be lumped into *eras* and subdivided into *epochs.*

Names of periods (systems) were devised mainly by European geologists and were taken primarily from place names, although some came from rock types or sequential positions. Recognition of the various systems away from their original or standard sections is chiefly by means of fossils.

Other useful categories of rock masses recognized by geologists are: (1) the *geologic formation*—a natural rock unit that can be mapped because it has certain observable characteristics that distinguish it from other such units; (2) the *sedimentary facies*—a mass of rock, usually part of a formation, that has characteristics that distinguish it from other material deposited at the same time, characteristics that tell us something about its origin; (3) *time-rock units*—rock material, irrespective of its composition or thickness, that was deposited during a specified time interval; (4) the *paleontologic zone*—a mass of rock deposited during the time of existence of a particular fossil or combination of fossils. Maps may be based on any of these units. A *paleogeographic map* depicts the reconstructed geography of a particular area.

Geologic correlation is any process that seeks to establish the mutual relations of geologic events or rock formations. Usually the aim is to discover time equivalence of separated samples or exposures. Correlation is accomplished by walking along the formations, by study of photographs, by comparisons of physical features, by study of sequences, by peculiar key beds, by presence of unique fossils or assemblages of fossils, or by stage of evolution of well-known organisms.

Absolute dates can now be obtained for many types of rock. These serve as guideposts against which the other methods can be tested and corrected. Correlation is of practical importance in directing attention to sites that are favorable for exploration for mineral products and in guiding any activities having to do with moving or using rocks.

FOR ADDITIONAL READING

Berry, William B. N., *Growth of a Prehistoric Time Scale.* San Francisco: Freeman, 1968.

Donovan, D. T., *Stratigraphy, Introduction to Principles.* Chicago: Rand McNally, 1966.

Dunbar, C. O., and John Rodgers, *Principles of Stratigraphy.* New York: Wiley, 1957.

Harland, W. B., A. G. Smith, and B. Wilcox, eds., *The Phanerozoic Time Scale: Part 1, Introduction.* Geological Society of London, 1964.

Krumbein, W. C., and L. L. Sloss, *Stratigraphy and Sedimentation.* San Francisco: Freeman, 1951.

Kulp, J. L., "Geologic Time Scale." Science, Vol. 133, 1961, 1105–1114.

Lahee, Frederic H., *Field Geology,* 6th ed. New York: McGraw-Hill, 1961.

Wilmarth, M. G., *The Geologic Time Classification of the United States Geological Survey Compared with Other Classifications.* Washington, D.C.: U.S. Geological Survey, Bulletin 769, 1925.

COSMIC BEGINNINGS

7

Every curious person has probably wondered whether the earth is the only place in the universe where intelligent life exists. Is the earth a unique sort of cosmic accident? Or is it merely a peculiar aggregation of a type that must inevitably appear and evolve at many times and in many places?

In this chapter we shall concentrate rather specifically on certain aspects of astronomy that bear directly on the history of the earth. Of paramount importance is the fact that the earth is

Spiral galaxy in *Canes Venatici* with outlying satellite galaxy. (Hale Observatories.)

composed predominantly of heavy elements such as iron and has relatively little helium and hydrogen. This composition contrasts strongly with that of the universe at large and the sun in particular, both of which are rich in helium and hydrogen but poor in metals and other heavy elements. What succession of processes is capable of creating and segregating the heavier elements from lighter ones? Here we shall summarize what is known or theorized about specific developments that could lead to heavy end prod-ucts such as the earth. As a beginning, we shall consider the various theories that have been used to explain the origin of the universe.

Modern Theories of the Universe

The Evolutionary, or Big-bang, Theory. Many recent discoveries indicate that the universe is evolving, that it had a specific beginning at some point in the past and has progressed through definite stages to the present. The origin of this concept may be traced to the Belgian cosmologist Georges Lemaître, who published a comprehensive theory in 1931. The theory was expanded and modified by George Gamow, a Russian-born American astrophysicist, in 1946. According to both Lemaître and Gamow, all the matter of the universe was at one time tightly packed in a very dense region of space with a radius about equal to the distance from the earth to the sun. This state did not last for long. The initial mass was shattered by a cataclysmic explosion that hurled matter and radiant energy into space and initiated the formation of the chemical elements. The observed rate of expansion of the universe and other methods of calculation indicate that this event happened some 10 to 13 billion years ago.

Before the explosion, the initial mass consisted of elementary particles not yet organized into elements. The elements resulted from the combination of various numbers of neutrons with protons and the addition of appropriate numbers of electrons over a very short period of time during and following the explosion, when conditions for the uniting of the elementary particles were suitable. Between 5 and 30 minutes after the "bang," according to Gamow, the basic chemical elements —helium and hydrogen—were born. The rate of expansion was crucial. If it had been more rapid, only hydrogen could have formed; if less rapid, more complex elements would have had time to associate, and the universe today would consist mostly of heavier elements.

In addition to explaining the composition of primitive matter, the big-bang theory also accounts for the apparent expansion of the universe. The headlong flight of matter and energy away from the original fireball is still going on and is a basic fact that any successful theory must explain. Before carrying this theory out to the formation of galaxies, we should mention two other theories.

The Steady-state, or Continuous-creation, Theory. A second explanation of the universe has

Figure 7.1 Three theories of the universe.

been proposed by the British astronomer Fred Hoyle. Known as the steady-state, or continuous-creation, theory, it makes no mention of a specific initiating event, posits no potential end, and simply accepts the fact that the universe is expanding. As aggregations of matter move farther away from one another due to the universal expansion, new matter is created, so that the overall density of the universe remains about the same.

The steady-state theory is particularly successful in explaining the fact that a great variety of astronomical events and materials may exist simultaneously. Many, if not all, of the essential stages in the creation, growth, and decay of heavenly bodies are on display at the present time.

The Pulsation, or Expansion-contraction, Theory. The big-bang theory has given rise to a third theory—the pulsation theory—which implies the possibility of a cyclic rejuvenation of the universe. Is it conceivable that all the matter and energy in the universe might return to the condition it was in prior to the "big bang"? In other words, might expansion eventually cease and contraction take over? A few observations suggest that objects at the outermost rim of the universe may have slowed down their outward expansion or may even be moving inward. It is at least conceivable that all matter may again rush together at a central place and thus initiate another cycle of creation.

The Formation of Galaxies

Impelled by the force of the "big bang," matter and its attendant energy sped outward for millions of years. During this interval the initial radiant energy was so powerful that aggregations of matter were broken up and dispersed as soon as they were formed. Eventually, several million years after the initial explosion, gravity began to draw thinly distributed helium and hydrogen gas into separate regions of space. These clouds of matter were the primordial galaxies (*protogalaxies*). As more and more material reached the central regions stars came into being with consequent generation of light and other forms of radiant energy.

The steady-state theory asserts that galaxies are produced in essentially the same way. The only difference lies in the source of the original basic materials. As we mentioned earlier, big-bang theorists assert that hydrogen and helium were formed shortly after the explosion that marked the beginning of the universe. Steady staters on the other hand say that extremely rarified hydrogen gas appears between existing masses of matter. The pressure of light or other forms of radiation pushes the gas from all sides until it reaches a position of equilibrium, thus forming the clouds that eventually become galaxies. Whether the hydrogen comes literally from nothing or is merely the reappearance of radiation lost from stars in other regions in space need not be discussed here. The important thing is that when the hydrogen appears, the building of galaxies can proceed as in the big-bang theory.

Through the wide distribution of many striking photographs (see Figure 7.2), the swirling, spiral form of a typical galaxy is well known. Although the spiral galaxy is the most common type (80 percent of all galaxies), there are two other classes. *Ellipticals* (17 percent) range from nearly spherical to flattened lens-shaped groupings that lack concentrated bands or arms. *Irregulars* (3 percent) are shapeless and lack signs of organized motion. More than a trillion galaxies are estimated to be within the range of modern telescopes.

The problem of why the galaxies rotate and why most of them eventually develop spiral arms has not been completely solved. Even in spiral galaxies not all matter gathers into the flattened spiral form—in many cases much galactic material takes the form of globular clusters, globe-shaped groups of stars that are usually found in the halo surrounding the nucleus of the galaxy.

Our local aggregation of stars, the Milky Way galaxy, cannot be viewed in a comprehensive way from the earth, but it is known to have spiral arms, a compact central region, and a population of at least 100 billion stars. In addition, it contains other matter in the form of dust and gas equivalent to millions of additional stars. The

(a)   (b)

Figure 7.2   (a) The great spiral galaxy in Andromeda. It is similar in shape to our galaxy.
(b) A photograph of the lower left region of (a), taken with a more powerful telescope, shows
that the spiral arms consist largely of individual stars. (Hale Observatories.)

Milky Way galaxy is in a state of slow rotation; the area that contains our sun takes about 200 million years for a complete circuit around the nucleus of the galaxy. Although our galaxy is considered to be larger than average, it is not otherwise unusual. The spiral galaxy in Andromeda is the nearest spiral galaxy (750,000 light years away) and is in many ways a twin of the Milky Way.

The Milky Way galaxy has about 200 globular

Figure 7.3   The globular cluster Omega Centauri.
This cluster, containing as many as 100,000 stars, is
about 22,000 light-years away from the earth and is
probably the cluster nearest to us. (Harvard Univer-
sity Boyden Station, Bloemfontein, South Africa.)

clusters, each consisting of 100,000 to 1,000,000 stars. These clusters travel in elliptical orbits about the galactic center and appear to be accompanied by very little dust or gas. Most of the matter of our galaxy is concentrated in the nucleus and spiral arms. As a matter of fact the center is so obscured by dust and gas that it cannot be directly observed.

The Origin of Stars

Stars exist in abundant variety and may be classified according to size, color, composition, temperature, mutual relationships, or stage of evolutionary development. A description of all these classifications is beyond the scope of this book, and so our discussion shall concern primarily matters of origin and composition.

All stars form from clouds of gas and dust. Although these clouds consist mainly of helium and hydrogen, there are various amounts of other elements. As the clouds are condensed and the stellar material is squeezed into a smaller and smaller area by the force of gravity, heat is generated by collision of the closely packed particles. When the temperature reaches about 15,000,000°C, nuclear reactions begin, and the body of gas and dust becomes a self-luminous star.

Depending on the size of the mass of primary material from which it formed, a star can have a life-span ranging from a few hundred thousand years to billions of years. During the "normal" life of the star, its energy and heat come from the conversion of hydrogen atoms into helium atoms. Because the star is continually expending its supply of hydrogen, the time eventually comes when there is little or no hydrogen to convert into helium. At this point the mass of the star begins to contract under the force of gravity, thereby causing a drastic rise in the temperature at the core. The outer layers of the star are also heated. These layers then expand and cool, and the star becomes what is called a *red giant.*

Astronomers theorize that it is at this stage

Figure 7.4 Nebulosity in Monoceros. Extensive clouds of shining gas and masses of dark material appear in this striking view. New stars have been observed in process of formation from such materials. (Hale Observatories.)

in the evolution of a star that increasing temperatures cause nuclear reactions that lead to the production of heavy elements. At about 110,000,000°C, helium is converted to carbon and oxygen; at about 150,000,000°C, the burning of oxygen and carbon produces such elements as aluminum, silicon, phosphorus, and sulfur. At even higher temperatures (up to 220,000,000°C) other heavier elements up to iron may be formed. Elements that are heavier than iron are formed by neutron capture rather than by increasing temperature.

All stars do not follow this pattern of generation of elements. We know that some stars are actually forming out of clouds of gas and dust that are already rich in metals. Where does this material come from? As we said, when a star has depleted its hydrogen supply it will begin to convert helium into heavier elements and eventually will use these heavier elements as a source of energy. The time will come however when the star has very little fuel to convert into energy, and the contracting force of gravity will assume primary importance. The star will collapse and shrink, with a consequent increase in the speed of rotation. At the same time the core of the star will become hotter and hotter, so hot that certain stars of high mass may literally explode and become what is called a *supernova*. As a result of this explosion, heavy elements are formed by neutron capture and are immediately dispersed into space along with a lot of the other material of which the star was composed.

The importance of supernovae to evolutionary astronomy is that they provide an environment for the building of heavy elements out of lighter ones. Also, the dispersal of these elements into space makes them available for incorporation into new stars and planets.

The Sun and the Solar System

Like all stars, the sun began as a dense cloud of dust and gases. As stars go, the sun is of medium size and so can be expected to have a life-span of about 12 or 13 billion years. Because the sun has been in existence for about 4 or 5 billion years already, it has 8 or 9 billion years yet to go. The sun is composed chiefly of helium and hydrogen, with a small proportion of heavier elements. Its outer temperature is about 6000°C, while at its core it is about 15,000,000°C. The sun's density is 1.4 (water = 1), and its mass is over 300,000 times the mass of the earth. The stupendous flood of energy that arises and presses outward from the core keeps the sun from collapsing in on itself. This energy is currently supplied by the conversion of matter into energy through the nuclear fusion of hydrogen atoms and helium atoms. The solar furnace converts about 596 million metric tons of hydrogen into 592 million metric tons of helium every second. The difference between these two figures represents matter that has been converted into energy.

In addition to the sun, the solar family consists of nine planets of various sizes, six of which have one or more satellites. There are also more than 1,600 named asteroids and countless comets and meteors in the system. The planets are all moving in the same direction and in essentially the same plane. The sun rotates in the same direction, with its equator inclined only slightly to the planetary

Figure 7.5 The "Crab" nebula in Taurus. This nebula is composed of the remains of a supernova that appeared in A.D. 1054. (Hale Observatories.)

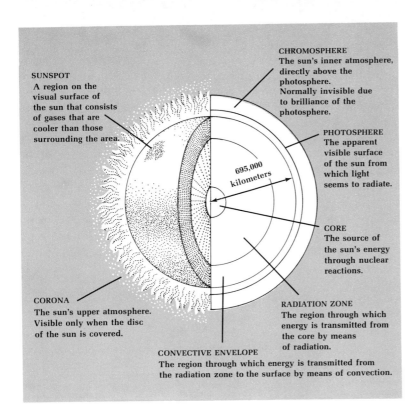

SUNSPOT
A region on the visual surface of the sun that consists of gases that are cooler than those surrounding the area.

CHROMOSPHERE
The sun's inner atmosphere, directly above the photosphere. Normally invisible due to brilliance of the photosphere.

PHOTOSPHERE
The apparent visible surface of the sun from which light seems to radiate.

695,000 kilometers

CORE
The source of the sun's energy through nuclear reactions.

CORONA
The sun's upper atmosphere. Visible only when the disc of the sun is covered.

RADIATION ZONE
The region through which energy is transmitted from the core by means of radiation.

CONVECTIVE ENVELOPE
The region through which energy is transmitted from the radiation zone to the surface by means of convection.

Figure 7.6   The structure of the sun.

Figure 7.7   The relative sizes of the planets in relation to a portion of the sun.

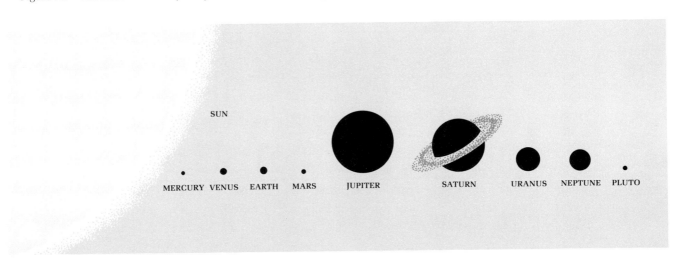

SUN

MERCURY VENUS EARTH MARS JUPITER SATURN URANUS NEPTUNE PLUTO

plane. The system is obviously highly organized. Essential statistics are given in Table 7.1.

The distribution and spacing of the individual bodies in the planetary system are surprisingly regular. In general, the distance of each member from the sun is approximately twice the distance of the preceding one (considering the asteroids to represent a planet). The mathematical expression of this relationship is called the Titius-Bode law. Some astronomers regard this regular spacing as strictly accidental; others have developed elaborate hypotheses to explain it. Another feature that demands explanation is the fact that the inner, earthlike planets (Mercury, Venus, Earth, and Mars) are physically different from the outer, or major, planets (Jupiter, Saturn, Uranus, and Neptune). The inner planets are comparatively small, with high specific densities, low rotational velocities, and few satellites. The outer planets are large, their specific densities are low, they rotate rapidly, and they have extensive moon systems. (Incidentally, Pluto, the most distant planet, may be an escaped satellite of Neptune; it is certainly not like the giant outer planets.)

One of the most difficult features of the solar system to explain is the distribution of *angular momentum*. This is *the force with which a body moves, or the product of its mass and its velocity.* The sun possesses about 99.9 percent of the total mass of the solar system but only about 2 percent of the total angular momentum because of its slow rotation. Jupiter possesses 0.1 percent of the mass but has about 59 percent of the total angular momentum. It is interesting that planets should have such high angular momenta and that the sun should rotate so slowly. Just how the planetary material could have acquired so much momentum is a problem that any acceptable theory must explain.

ORIGIN OF THE SOLAR SYSTEM

As long ago as 1644 the great French philosopher and mathematician René Descartes proposed that the solar system began as a cloud of unorganized primordial matter. The sun and planets, he believed, were accumulations caused by eddies or vortices within this cloud. In 1755 Immanuel

Table 7.1 The Solar System

	Mean Distance from the Sun, Astronomical Units	Length of the Year, Earth Units	Rotation, Earth Units		Equatorial Diameter, Kilometers	Mass, Earth = 1	Density, Water = 1	Number of Satellites
Sun	—	—	25–35	days	1,390,000	332,000	1.42	
Moon	1.00	365.26 days	27.3	days	3,475	0.012	3.36	
Mercury	0.39	87.97 days	58.64	days	4,830	0.05	5.5	0
Venus	0.72	224.70 days	243	days	12,108	0.82	5.27	0
Earth	1.00	365.26 days	23.9	hours	12,750	1.00	5.52	1
Mars	1.52	686.98 days	24.6	hours	6,800	0.11	3.56	2
Jupiter	5.20	11.86 years	9.93	hours	143,000	317.8	1.34	12
Saturn	9.54	29.46 years	10.23	hours	121,000	95.1	0.69	10
Uranus	19.18	84.02 years	10.8	hours	47,000	14.5	1.7	5
Neptune	30.06	164.79 years	15.8	hours	45,000	17.2	2.3	2
Pluto	39.5	248.4 years	6.39	days	6,000 (?)	0.8 (?)	3.0	0

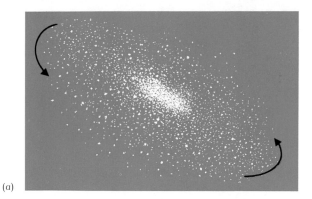

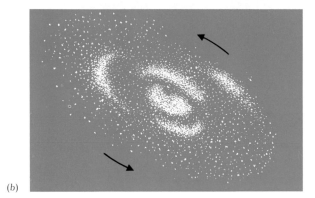

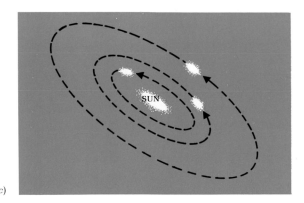

Figure 7.8 *The nebular theory of the origin of the solar system. (a) The original cloud of rotating gases and dust. (b) The formation of the protoplanets as matter condenses into globs around the sun. (c) The contraction and solidification of the protoplanets into solid bodies orbiting the sun.*

Kant published a more detailed theory that took into account the laws of gravity that Newton had described in 1687. Kant considered the combined effects of spiral motions and gravity within the primeval mass and came up with a surprisingly good explanation of the solar system.

Although the subsequent history of thought about the origin of the earth and solar system is interesting we cannot elaborate on the many diverse ideas that have been proposed. Suffice it to say that the most acceptable modern theories commence with a nebulous aggregation of diffuse matter such as Descartes envisioned centuries ago. Although the basic concept of the nebular theory is philosophically pleasing and conceptually simple, a step-by-step reconstruction of the process of sun and planet formation has turned out to be immensely complex. The chief difficulties arise in accounting for the very peculiar family of planets that accompany the sun. The sun itself probably originated by the same process of star formation going on today, the contraction of a part of the ordinary mixture of elements that prevails in the spiral arms of our galaxy.

The solar system occupies a vast, flattened, lens-shaped region of space with the planets and most of the smaller components moving in an almost perfect plane around the sun. This structure has naturally been compared to the spiral galaxies and to the planet Saturn with its satellites and puzzling rings. Although the solar system is unlike either of these aggregations in detail, the concept of a large central body with flattened encircling rings or spiral arms remains as the starting point for modern cosmological theories about the solar system.

If we assume that the total matter of the solar system was once contained in a gaseous, spherical nebula, we are confronted with the problem of how a small and very untypical fraction of this matter was segregated from the original mix and then organized into the planetary system during the process of contraction. Two ideas have been proposed as possibilities: (1) Poorly defined "blobs" of planetary material were left behind as the nebula contracted. (2) The planetary material was expelled much later in ringlike form after the sun had reached a size considerably smaller than the orbit of Mercury.

According to the first concept the planets and sun formed simultaneously; the material of the individual planets may have been segregated into their respective orbits and perhaps even into large solid aggregations before the sun reached the status of a self-luminous body. This is important because the presence of a central source of radiant energy could scarcely fail to have had very significant effects. If the idea of simultaneous formation of sun and planets is correct, the rings or blobs left off by the contracting protosun might preserve essentially the composition of the ancestral nebula at successive stages. Some astronomers are inclined to believe that the giant planet Jupiter may, in fact, be a sample of the original material of the solar system. This might also be true of Saturn, which is largely gaseous, and of Uranus and Neptune, which are only slightly more dense than water.

The inner planets—Mercury, Venus, Earth and Mars—are a different matter. They are small, dense, and obviously rich in the heavier elements—the end products of a long series of change and segregation. It is generally agreed that the final material for the inner planets must have resulted from the removal of light material from the original mix. An important consideration in this and any theory is the difference in behavior of dust and gas. Gas would tend either to fall into the sun or to dissipate rather rapidly into space. Dust particles by comparison would be likely to take up various elliptical orbits around the sun. An important factor to be considered at this stage is the impact of the *solar wind,* a general term for the stream of radiation and particulate matter that arises from the sun. The solar wind is very powerful and is able to push gaseous material outward, as shown by the fact that tails of comets always flare away from the sun. At an earlier stage in its development, the sun was probably larger and hotter with correspondingly greater

Figure 7.9   Jupiter, showing banded atmosphere and the large red spot. (Hale Observatories.)

radiation. The probable sweeping away of gaseous material from regions near the sun is of great importance in explaining planet formation. The heavier materials that were left behind in the winnowing process became aggregated into the dense inner planets. How these residual materials were eventually gathered into large individual bodies is not clear. Random motions and countless collisions would probably bring fragments of different sizes together, and the larger ones would constantly grow at the expense of the smaller ones. At first, when these masses were small, and gravitational effects were weak, there may have been other processes capable of holding the pieces together. Probably much of the solidifying process was caused by the freezing of such volatile substances as ammonia, water, and even hydrocarbons. Comets are known to be composed of rocklike pieces held together by frozen constituents of various kinds. The more volatile compounds are released to form a comet's spectacular tail as it approaches, and is heated by, the sun.

Just what other chemical and physical processes may have occurred in the dust cloud or protoplanet that was to become the earth can be known only in a theoretical way. Some investigators have supposed that electrical and magnetic effects were much in evidence and that lightning may have caused fusion of certain constituents and a building up of complex compounds. Geologists believe that actual physical remnants of this stage of planetary evolution may be preserved in the form of small rounded bodies called *chondrules,* which are found in certain meteorites called *chondrites.* These small spheres are usually glassy and have a chemical composition similar to that of earthly rocks such as pyroxene. Chondrules and other finer material including liquid and gaseous components constitute satisfactory raw material for building the earth. An aggregation of such material has been compared to a popcorn ball in allusion to the loosely packed structure that must have existed at this stage. However, as the body that was to become the earth grew in size, the interior would be compacted and eventually highly heated. The emergence of the earth as a single body opens another phase of history that will be discussed in Chapter 8.

Is Earth Unique?

Now that we have summarized what is inferred about the early history of the earth, a brief mention of the topic of the possible prevalence of similar planets is appropriate. If any planets revolve around suns other than ours, they are relatively too small to be visible by earth-based telescopes. So direct observation is of little use in proving or disproving the existence of other planetary systems. But the study of stars that resemble the sun has produced some surprising clues having mainly to do with the matter of rotation.

A basic observation is that the sun spins very slowly at about 2 kilometers per second; a point

Figure 7.10   (a) Internal structure of a gray chondritic meteorite that fell in Indiana. (b) An enlargement of a portion of this meteorite showing the spherical, variously colored chondrules. Specimen is about 10 centimeters in maximum diameter; chondrules are about 0.25 centimeters in diameter. (Smithsonian Institution.)

(a)

(b)

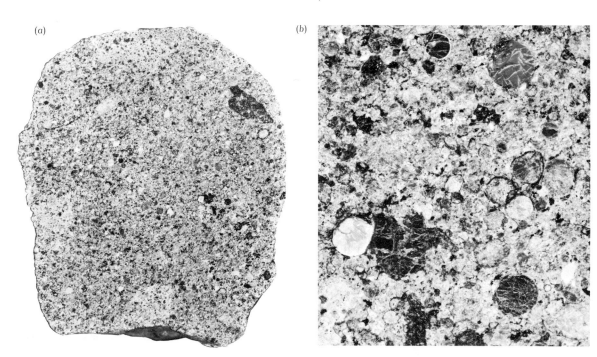

on the equator takes 25 days to complete one revolution. As previously mentioned the angular momentum of the solar system resides chiefly in the planets. Somehow during the course of evolution the sun lost much of the angular momentum that it might be expected to possess, whereas the planets seem to have gained more than might be expected. The many theories that have been proposed to account for this cannot be discussed here. The important fact is that many "slow suns" have been discovered throughout the universe. These slowly spinning stars are also relatively cool like our sun. The argument emerges that any star that resembles our sun in all its basic physical characteristics might also resemble it in possessing a family of planets. The argument is strengthened when a characteristic such as slow rotation can be related directly to the presence of planets. There are billions of slowly spinning stars in our galaxy alone, and some astronomers speculate that millions of these might have planets with intelligent life. It is only fair to state that this opinion is not shared by all scientists. The question "Are we alone?" cannot be answered with assurance from the available scientific evidence.

Earth and Moon— Partners in Space

Among the most spectacular and significant feats of mankind has been the exploration of the moon. The moon's surface has been studied not only by means of earth-based instruments and unmanned space vehicles but also by means of manned landings and actual specimens of moon rock. Much has been written concerning these exploits, and the present short discussion must be confined to those aspects that might be considered as bearing on the history of the earth. In other words, what have been the mutual developmental relations of the moon and earth? That they are at times only about 362,000 kilometers apart and appear to have been even closer to-

gether in the past ensures that their histories must be interwoven in many ways.

Three theories have been proposed as explanations for the moon as a satellite of the earth: (1) the moon formed from the same nebular cloud in near proximity to the earth and at about the same time; (2) the moon originated independently of the earth at some other locality either within, or external to, the solar system and was captured by the earth when the two approached each other; (3) the moon is composed of material that was once part of the earth, material that was forcibly ejected and separated from the earth at a remote point in time. Is the moon a child, a sister, or a visiting distant relative of the earth?

Before the moon landings, scientists were convinced that surface exploration and sampling would reveal evidence to establish one or another of these theories of the origin of the moon. Such hopes appear not to have been realized; evidence is equivocal and proponents of each theory have found new evidence favorable to their respective beliefs.

Certain facts are generally known and accepted. Physical constants for the moon are (earth = 1): mass, 0.0123; surface area, 0.074; surface gravity, 0.1645; volume, 0.02; and density, 0.60. The overall density of 3.36 is nearly the same as the outer part of the earth. If there is an inter-

Figure 7.11   Basaltic moon rock brought back by the Apollo 12 mission. Basalt is a relatively heavy igneous rock that has solidified from liquid lava. It is common on earth as well as on the moon. Specimen is about 5 centimeters long. (NASA.)

Figure 7.12 *Low-level view of the crater Coperni-cus, showing numerous minor craters and the hum-mocky topography created on the lunar surface by successive meteoritic impacts over long time periods. (NASA.)*

nal segregation or stratification similar to that of the earth it is much less evident; the lack of strong electrical or magnetic effects precludes the presence of a molten core or even a large metallic core of any kind. The chemical composition of lunar rocks is similar to, yet different from, the composition of earth rocks.

The age of lunar samples yields significant clues. There are many specimens from 3 to 3.7 billion years old but not many that are older or younger. One important specimen returned by Apollo 15 has an age of 4.15 billion years. Appar-ently, an extensive period of melting about 3.5 billion years ago produced the great lava fields, or *maria* (plural of the Latin *mare*, "sea"). The 4.15-billion-year-old sample escaped this melting but is still not a piece of the original lunar mate-rial. The interval between 3.7 billion years and the formation or solidification of the moon is not

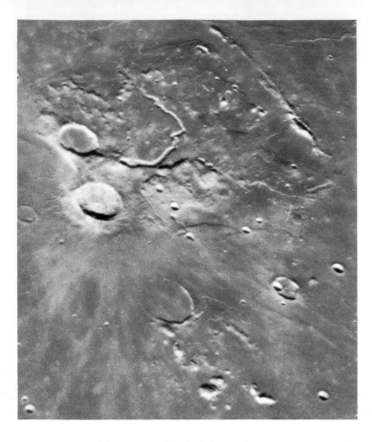

Figure 7.13 Typical lava plain, or mare, on the moon. Note partly inundated prelava crater and protruding mountains in the lower center. Postlava features include craters of various sizes, rills (elongate narrow depressions), and possible faults. (NASA.)

contention that the moon has never been completely molten. The great fields of lava may have been generated near the surface, perhaps by the impact of infalling asteroids or comets.

Apollo 15 photographs reveal a considerable number of relatively new volcanic craters, and there is evidence that water vapor is currently escaping from the interior of the moon. These facts indicate ongoing subsurface activity, but they do not obscure the fact that the major features of the moon have been shaped by external forces. The almost total lack of water or evidences of water in the past assures scientists that the history of the moon must of necessity be greatly different from that of the earth.

The fact that the creation of the large lunar lava fields about 3.7 billion years ago coincides with the age of the oldest rocks on earth hints at some cosmic connection, perhaps one of cataclysmic magnitude. In fact, there may be some truth to the chilling theory that both the moon and the earth were subject to supercolossal meteoritic impacts between 3.5 and 4 billion years ago. The earth at that time may have been emerging from a previous molten state, or it may have been cool while the moon was hot but not molten. In any event the meteoritic collisions could have triggered a partial melting of the moon and indirectly caused the melting of the entire earth. A strange fact is that the large lava fields are all on the side of the moon that faces the earth. A shower of massive impacts could account for the fact that on the earth there is no historical rock record for the period preceding the impact episode, whereas some moon rocks and perhaps even surface features predate it. Considering the fact that there are thousands of heavy asteroids plunging through nearby space and that there probably were many more of these in the past, the idea of a devastating bombardment is entirely plausible.

Scientists who speculated for years about the composition of the moon and eagerly awaited actual specimens of lunar rock brought back by

yet well understood. Very little melting and reconstitution of lunar material seems to have taken place since the event 3.7 billion years ago.

Currently much attention is being given to the thermal history of the moon. To what extent has its history been dominated by internal, heat-driven forces such as those that affect the earth? We are not yet sure whether or not the interior has shells or layers like the earth, but the weakness of electrical and magnetic fields rules out a heavy metallic core. This lends weight to the

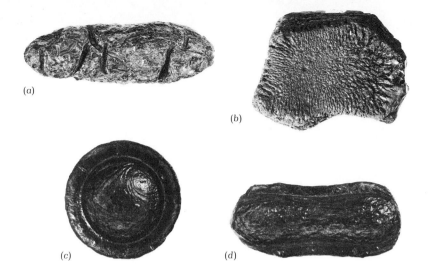

(a)

(b)

(c)

(d)

Figure 7.14 Four tektites showing variations in shape and surface markings; (a) and (b) are from the Philippine Islands, and the other two are from southern Australia. Specimens range from about 2 to 6 centimeters in length. (Smithsonian Institution.)

the Apollo missions were surprised to learn that much moon material may have already arrived on earth by natural means and has been in collections for many years. This material consists of small glassy bodies known as tektites. The impact

Figure 7.15 Iron meteorite from Henbury, Australia. (Center for Meteorite Studies, Arizona State University.)

of large meteorites on the lunar surface supposedly splashed this material into space. It is only fair to state that some investigators do not follow this theory and are inclined to believe that tektites are somehow derived from earthly sources.

Meteorites

Meteorites can be divided into two chief classes: the *stony* and the *metallic*. The stony meteorites are superficially similar in appearance to some igneous and metamorphic rocks and may escape detection due to the resemblance.

The metallic meteorites are composed chiefly of nickel and iron. A number of them have been found to contain diamonds. The physical state of these materials has led investigators to assume that the metallic meteorites crystallized very slowly under intense heat and pressure, probably in the interior of a planet or planets similar in composition to the earth.

Meteorites contain elements, compounds, and minerals identical to those found on earth, including a number of radioactive isotopes from which radiometric ages may be calculated. It seems very significant that *almost all meteorites prove to be*

Figure 7.16 Looking down into Meteor Crater near Winslow, Arizona. The depression is about 1,100 meters across and the deepest part is about 110 meters below the rim. Thousands of fragments of iron meteoritic material have been gathered in the vicinity of this crater. (U.S. Geological Survey.)

about 4.5 billion years old, the same age as the earth. The breaking up of the parent body came later, and may have been due to a collision of the planetary bodies that took place as recently as 100 to 200 million years ago.

The question of whether or not meteorites contain living organisms is much debated. There can be no doubt that the carbonaceous chondrites do contain chemical compounds, including hydrocarbons similar to those found in living things on earth, but these compounds may not occur in organized self-duplicating spores or bacteria, such as some investigators maintain. Final conclusions must await further study. It is certain that no meteorites with sedimentary structures or recognizable fossils have been reported. Even on the earth, the amount of rock that has been modified by sedimentary processes amounts to only a small fraction of 1 percent, and if the earth were to be converted into fragments, we would find very little of the sedimentary material among the debris.

A SUMMARY STATEMENT

Modern views of the universe are distinctly evolutionary. Processes operating to produce the solar system have tended toward progressive segregation and concentration of the heavier elements out of an original "mix" composed chiefly of hydrogen and helium. Of the various ideas proposed to account for the cosmos, the evolutionary, or big-bang, theory is most acceptable. The beginning was the sudden explosion of an extremely dense mass that contained all the matter of the universe. The first galaxies were composed of the outward moving material (chiefly of hydrogen and helium) of this virginal fireball. Self-luminous stars, such as the sun, were formed when matter became sufficiently concentrated to initiate nuclear reactions.

Planet formation probably accompanies the development of a large number of luminous stars similar to the sun. The most acceptable theory of the origin of the solar system is the dust-cloud hypothesis, according to which the heavier constituents of an originally diffuse cloud of matter aggregated into protoplanets circling the central sun. The heavier elements were concentrated as lighter ones were driven away from the sun by radiation pressure.

The moon presents special problems. It is similar to, yet different from, the earth. The weight of present evidence favors a theory that the moon formed about the same time as the earth and from the same cloud of matter, but other theories are still supported. In general, the protomoon received slightly lighter and more siliceous material than the earth.

Meteorites are broadly classified as either stony or metallic and resemble earth and lunar material in composition and age. Tektites are small glassy objects of uncertain origin. They may have been propelled from the surface of the moon by the impact of large meteorites or asteroids; some scientists believe they are of earthly origin.

FOR ADDITIONAL READING

Abell, G., *Exploring the Universe*. New York: Holt, Rinehart & Winston, 1969.

Asimov, Isaac, *The Universe*. New York: Avon, Discus Books, 1966.

Baldwin, R. B., *The Moon—A Fundamental Survey*. New York: McGraw-Hill, 1965.

Bergamini, D., *The Universe*. New York: Time Inc., 1962.

Coleman, James A., *Modern Theories of the Universe*. New York: New American Library, 1963.

Hoyle, Fred, *Astronomy*. Garden City, N.Y.: Doubleday, 1962.

Page, T., and L. W. Page, eds., *Neighbors of the Earth: Planets, Comets, and the Debris of Space*. New York: Macmillan, 1965.

Shapley, H., *Galaxies*. New York: Holt, Rinehart & Winston, 1960.

Toulmin, Stephen, and June Goodfield, *The Fabric of the Heavens*. New York: Harper & Row, 1965.

STRUCTURE AND MECHANICS
OF THE EARTH

The earth has been called "a pebble in the sky," an allusion to the fact that it is essentially a solid, stony body. Another writer has compared the earth to a soft-boiled egg—both have a thin, hard, rather brittle shell or *crust,* a thick envelope of solid material making up the white of the egg and the *mantle* of the earth, and an internal, partly liquid sphere that is the yolk of the egg and the *core* of the earth. The soft-boiled-egg analogy emphasizes the fact that part of the core of the earth is liquid, not solid. The actual structure of the earth as visualized from extensive studies

Earth as seen from the moon. (NASA.)

and theoretical considerations is shown in Figure 8.1. Each of the major solid components—core, mantle, and crust—have peculiar properties that must be considered in connection with the past history and present condition of the earth.

The Core

The core of the earth begins about 2,900 kilometers below the surface and continues for about 3,480 kilometers to the center of the globe. Evidence indicates that there is an outer core and an inner core, the inner portion having a radius of about 1,320 kilometers. Studies of the behavior of earthquake waves have convinced scientists that the outer portion is liquid, whereas the inner zone is solid. Compression waves (those that compress and dilate rock as they travel forward) drop sharply in velocity when they enter the outer core and gain speed again when they penetrate the inner core. Shear waves (those that shake rock sideways as they travel forward) stop altogether at the outer core. Because compression

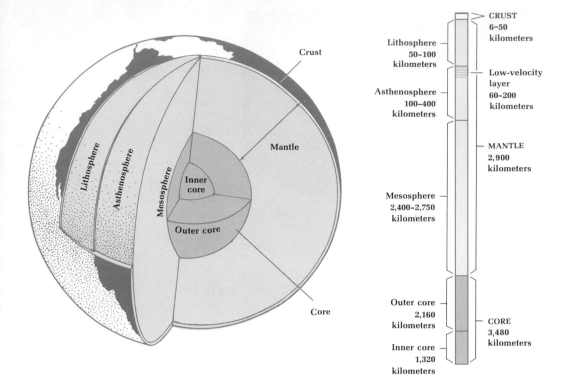

Figure 8.1   The structure of the earth. Bar graph is out of scale for clarity.

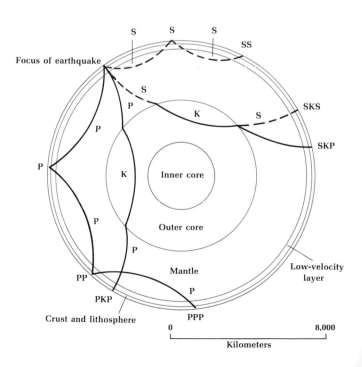

Figure 8.2   A cross section of the earth illustrating the paths of earthquake waves. Compressional waves (P, solid lines) are refracted sharply when they reach the core, as is shown by the PKP trajectory. Shear waves (S, dashed lines), as such, end at the core, but they may be converted into P waves, cross the core as compressional waves (K), and re-emerge in the mantle again as P waves and S waves (thus, SKP and SKS). The waves may also be reflected back into the earth at the surface (trajectories PP, PPP, and SS). (Adapted from The Interior of the Earth, U.S. Geological Survey.)

waves travel through solids and liquids, whereas shear waves travel only through solids, the molten state of the outer core is considered certain.

The core of the earth is estimated to have a specific gravity of about 10.5 (water = 1). This relatively high figure is arrived at in the following manner: The specific gravity of the earth as a whole is 5.5; the crust is relatively light (2.8) and the mantle only a bit heavier (4.5). In order to arrive at the 5.5 figure for the entire earth, scientists must assume that the core has a density of 10 or 11.

The extremely high density of the core indicates several possible compositions. In terms of ordinary matter, the core's density is comparable to that of the heavier metals: lead (11.3), nickel (8.9), gold (19.3), uranium (18.7). For several reasons, geologists theorize that the core is composed of an alloy of iron and nickel, the iron making up about 80 to 85 percent of the total. This theoretical choice has been strongly influenced by what is called the meteorite analogy. Many meteorites are composed of a mixture of iron and nickel that shows evidence of having solidified under high temperature and pressure. Most astronomers consider these metallic meteorites to be debris from the core of a disrupted earthlike planet that once occupied a position between Mars and Jupiter. It should be pointed out that another large class of meteorites, the stony meteorites, has properties similar to those of the mantle of the earth (see Figure 7.10).

The dense, heavy, and partly liquid core within the earth tells us something about the history of this planet. A heavy central core surrounded by concentric shells of successively lighter materials is generally thought to result from gravitational stratification in a liquid sphere. "Liquid" in this connection means molten and hot.

It should be noted that the earth is evidently unique among the inner planets in its internal structure. Measurements made from space vehicles show that Venus, Mars, and the moon have

Figure 8.3   A section of a metallic meteorite found in 1860 near Cleveland, Tennessee. The crystalline structure indicates formation under conditions of high temperature and pressure. (Smithsonian Institution.)

very weak electrical and magnetic fields, an indication that these bodies lack a heavy metallic core. Geologists take this to mean that Venus, Mars, and the moon never passed through a molten stage as did the earth, or at least were never as thoroughly liquefied as the earth. This is an extremely important consideration in connection with the origin of water and the atmospheric gases.

The previous paragraphs emphasize the importance of heat in the formation of the earth. In fact, if we understood the earth's thermal history, the rest would be relatively easy. The earth probably began its existence as a cold body. It is believed that the basic constituents were in the form of dust, small droplets, crystals, or grains that were swept together into a heterogeneous mass and thus became a planet. A number of processes that are capable of generating heat in such an aggregation can be visualized: the impact of infalling meteorites, solar radiation, tidal friction caused by the moon, and heat produced by disseminated radioactive elements.

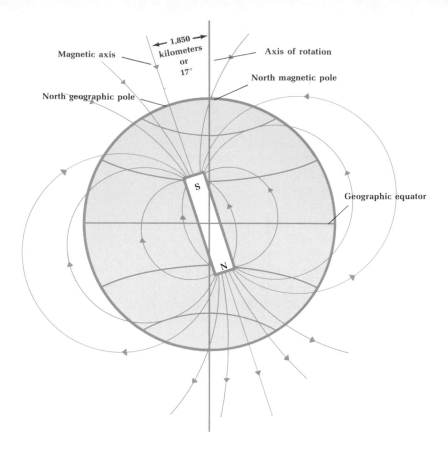

Figure 8.4 *The earth reacts as though it contained a giant magnet, and its magnetic field extends far out into space. Note that the magnetic poles do not coincide with the geographic poles.*

Radioactivity appears to have been the most powerful and consistent energy source. Radioactive elements are known to generate heat as they disintegrate, and radioactive uranium, thorium, and potassium are present in the earth today and were even more abundant in the past.

Under certain assumptions that need not be detailed here, the radioactive elements could have generated sufficient heat for the melting of the earth about 500 million years after its formation. Once the required heat level was reached, melting would have been rapid. Previously disseminated iron possibly descended rapidly to form the core, displacing in the process a corresponding volume of lighter material. Even the surface may have been stirred and engulfed. Once melted the earth remained in a molten state

for another lengthy period, perhaps 500 million years. No matter what the reason, no earth rocks older than 4.1 billion years have been found. During the lengthy period before solid rocks originated, deep-seated convection currents were probably segregating the earth into its major shells.

The immediate effects of the original formation of the core of the earth may never be fully known, but its subsequent influence is better understood. The core is responsible for most of the magnetic and electrical fields that extend throughout the earth and beyond it into space. That the earth behaves like a giant magnet is well known to everyone who has seen a compass or a dip needle. The earth may be an electromagnet rather than a permanent magnet, because the interior is too hot to maintain permanent magnetism. Electrical currents produced by fluid motions within the core or between the core and mantle are sufficient to create the magnetic field. This is the *dynamo theory* of the earth's magnetism.

As we saw in Chapter 6, polarity reversals are a very puzzling feature of the magnetic field. These changes appear to take place at lengthy but irregular intervals; geologists estimate that the last reversal took place about 690,000 years ago. Like other magnetic and electrical phenomena displayed by the earth, polarity reversals must derive chiefly from the core. However, it is thought that they might be triggered by meteoritic impacts or by unusually strong earthquakes.

In the absence of actual observations concerning what happens during a magnetic reversal, there has naturally been a great deal of speculation. It is possible that the strength of the magnetic field gradually declines, reaches zero, and then increases—but with opposite polarity. Regardless of the speed with which they take place, magnetic reversals may have had drastic effects on past life. Scientists know that a weakening of the earth's magnetic field will consequently reduce the strength of the *magnetosphere,* a region from 1,000 to 64,000 kilometers above the earth that traps electrically charged particles approaching the earth from outer space and thus acts as a shield against dangerous particulate radiation. Therefore, the amount of radiation reaching the earth's surface during a polarity reversal would be several times the normal amount and would have a notable impact on living cells. Radiation is considered to be the chief cause of mutations in living things—the greater the dose of radiation the greater the mutation rate. Mutations are mostly harmful; and if too many took place in a short time, the results would be disastrous.

The removal of any significant life group, for example, dinosaurs, gives opportunity for the expansion of others into vacated environments. There is thus a possibility of large-scale turnover. If this line of reasoning is correct, there is a distinct possibility that times of magnetic reversals have been critical ones for living things. The core of the earth, remote as it seems, may thus have had a profound influence on the evolution of surface life. Some authorities believe we are approaching another period of reversal; but because this ominous event is not scheduled until about the year 4000, its prediction has not caused undue alarm.

The Mantle

The massive spherical shell that surrounds the core of the earth is called the mantle. It is about 2,900 kilometers thick and makes up 84 percent of the earth by volume and 67 percent by weight. It is convenient to think of the mantle as the original, or average, material from which the rest of the earth has been derived—the core represents material that has settled downward, whereas the crust, oceans, and atmosphere have risen to the surface. Knowledge derived from the study of earthquake waves indicates that a number of discontinuities within the mantle divide it into perhaps as many as ten subshells.

Except for the very outermost part we have little direct information about the composition of the mantle. Minerals rich in silica, magnesium, and iron are probably dominant, with iron increasing downward and magnesium upward. Fragments believed to have been brought up from the mantle have been found in surface lavas and intrusive bodies. These displaced rock types include very dense and iron-rich materials such as eclogite, peridotite, gabbro, and pyroxene. These are all more dense than the basalt of the oceanic floor but are not greatly different from it in chemical composition.

This constitutes evidence that the outer mantle is also stratified according to density. It is significant that many deep-seated rock types

Figure 8.5   *A diamond in the matrix of basic igneous rock. The diamond and associated material have arisen from a site of formation about twenty-eight kilometers within the earth's crust. (Christensen Diamond Products.)*

have lighter near-surface equivalents with the same chemical composition. Thus magnesium, iron, and silica may combine in various ways to form *enstatite,* with a density of 3.1, *olivine,* with a density of 3.4, or *spinel* with a density of 3.6. Diamond is a well-known high-pressure equivalent of graphite. It is evident that depth of burial can bring about density differences with no change in chemical composition.

Heat, like pressure, appears to increase downward although not at a steady rate. Temperatures of 1000°C have been calculated for the crust–mantle boundary and 5000°C for the mantle–core boundary. The upper mantle is relatively hotter than might be expected in comparison with the lower mantle. A plausible explanation for this is the distribution of radioactive uranium, thorium, and potassium. Ions of these elements are too large to become part of the crystal lattices of the heavier silicate minerals such as olivine and pyroxene. As the deep-seated silicates solidified,

the radioactive elements were left behind to move upward and become concentrated in the last liquid fractions where they finally crystallized with feldspars and micas or remained as films between other mineral grains. In any event, by one means or another, the radioactive elements have been concentrated near the surface of the earth, thereby causing the upper levels to be hotter than they might otherwise be.

The upper mantle and crust are subdivided into a number of shells that are different chiefly because of the reactions of their constituents to heat. Three major zones are recognized: the *mesosphere, asthenosphere,* and *lithosphere.* The mesosphere is a zone of solid, dense material that makes up the great mass of the mantle below about 400 kilometers.

The asthenosphere is a zone of soft, hot, and easily flowing material above the mesosphere and below the lithosphere both of which are hard and rigid. Its thickness ranges from 100 to 400 kilometers. How a hot, plastic, semiliquid layer such as this can exist sandwiched between harder layers so near the surface of the planet is not fully understood. High heat and low pressure do not fully explain the situation, for these factors apparently can melt solid material at even greater depths. The essential agent is probably water, which does much to lower the melting point of silicate minerals such as those in the upper mantle and thus plasticizes rocks made up of them.

The upper region of the asthenosphere is designated by students of earthquake waves as the *low-velocity layer.* Above this is the outermost zone of the solid earth, the lithosphere, with a thickness ranging from 50 to 100 kilometers. At the top of the mantle (base of the crust) is another marked boundary, the *Mohorovičić discontinuity* (Moho for short). The behavior of earthquake waves suggests that rocks below the Moho are more dense, contain less gas and liquid, and are richer in iron than rocks immediately above. The Moho is not a boundary between liquid and solid or even between brittle and plastic rocks.

It probably marks a transition from less dense to more dense material, the change in density being caused primarily by higher pressures without any change in chemical composition of the material. Above the Moho is the crust of the earth, which is from 6 to 50 kilometers thick.

The Crust

We live on the crust of the earth and as a matter of course know more about this part of the earth than we do about the deep interior. The term "crust" originated when it was generally believed that the interior of the earth was entirely molten. Although scientific ideas about the interior have changed, the term "crust" is still useful in designating the outer shell that has different properties from the deeper mantle. The distinction between crust and mantle is not as important as it once seemed, because of the discovery that the upper mantle is attached to, and reacts with, the crust in an intimate way. A more significant discontinuity is that between the upper mantle and the lower mantle, which, as we have mentioned, is called the asthenosphere.

We are well acquainted with the outermost part of the crust from direct observation of surface rocks and from extensive mining and drilling. Analysis of crustal rocks shows that they have an average specific gravity of 2.8. Two distinct but locally intergrading divisions of the crust are recognized. The lowermost, called the *sima* (from *si*lica and *ma*gnesium), is chiefly basaltic and has an average density of 2.9; it occurs in a continuous world-circling layer. The outermost division, called the *sial* (from *si*lica and *al*uminum), has an average density of 2.6. The sial is not a continuous layer but is concentrated chiefly in the continents. It is generally agreed that the sima was derived or segregated from the underlying mantle and that the continental material was in turn produced by the basaltic layer. The whole process is complicated but entirely feasible; there is nothing in the upper layers that could not have

had a source lower down. Basalt is the common denominator of crustal studies, and the composition and manner of formation of many varieties of this material have received a great deal of attention.

The common basalt of the ocean floors, called *tholeiite,* has a low content of alkali metals (sodium and potassium), but basalt-rich volcanic piles, such as the Hawaiian Islands, and continental basalt fields are rich in these elements. This lighter and generally younger basalt appears to have been produced through the melting of tholeiite. Water is intimately associated with near-surface reactions and has an important effect in lowering the temperature at which basalt melts and in transporting volatile materials including the metals of important ores.

Once basaltic rock has reached either the ocean floor or the surface of continents and islands, it will invariably undergo further transformations that increase the silica content, decrease the specific gravity, and lighten the color. On the sea bottom there will be additions of water and deposits of clay derived from weathering of the land; calcareous and siliceous material will be added by lime-secreting and silica-secreting organisms, respectively. A mixture of typical oceanic sediments and basalt flows when melted gives an andesitic volcanic rock, which has soda lime feldspars predominating over alkali feldspars. Further melting and mixing eventually produces light-colored rhyolite, trachyte, and highly siliceous ash and pumice of various kinds. These last-named igneous rocks are rare in ocean basins but common on continents. They may be transformed by metamorphism into gneiss or granite, which along with the parent rock will weather to give ordinary sedimentary sandstone, siltstone, and clay. The topic of how these materials become continents will be treated in the next chapter.

New evidence concerning the structure of the earth has thoroughly revolutionized concepts about this planet. In the past the continents and ocean basins were regarded as stable and permanent, but today the transitory nature of the surface features is emphasized. The recognition that the asthenosphere exists and provides a zone of decoupling between the brittle lithosphere and the interior of the earth is probably the most productive concept to arise in geology in the last century. The fact that the exterior of the earth has a great deal of mobility provides new and revolutionary solutions to many long-standing problems of historical geology.

Continental Drift and Sea-floor Spreading

The idea that the continents may have once been joined together goes back at least to 1620 when Francis Bacon speculated about the meaning of the parallel shores of Africa and South America. The first scientific treatment of the problem came in papers by F. B. Taylor in America (1908) and Alfred Wegener in Germany (1910). Wegener asserted that all land areas were once part of a single supercontinent called *Pangaea*. He supposed this landmass to have split apart and the fragments to have reached their present position by a process called *continental drift*. Another theory posits two original continents, *Laurasia* in the north and *Gondwana* in the south.

Wegener's theory proved to be very productive, for it gave rise to an immediate storm of criticism and was followed by literally thousands of papers for and against the idea. The fact that the continents can indeed be joined together in a reasonably good "fit" was considered to be good geologic evidence for continental drift. The chief argument against drift was more mechanical than geologic—no force or mechanism seemed capable of moving the continents. Because everyone had to combat a mental image of the solid continents plowing along through the solid crust, it is small wonder that the idea was resisted by almost everyone. Some of the arguments for and against continental drift will be

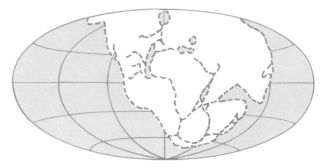

Late Carboniferous

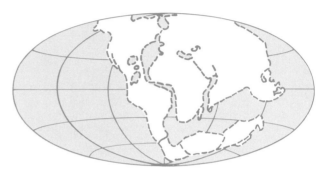

Eocene

Early Pleistocene

Figure 8.6 The concept of the original great landmass, Pangaea, and its break-up into present continents as proposed by Alfred Wegener. (Adapted from W. L. Stokes and S. Judson, Introduction to Geology: Physical and Historical, Prentice-Hall, Inc., 1968.)

discussed at appropriate intervals in succeeding pages.

The long-standing impasse over a sufficient mechanism was broken in the 1960s with information derived not from the lands but from the ocean basins. The basic breakthrough was the charting of a tremendous earth-circling chain, or system, of oceanic "mountains." These are more properly referred to as a "ridge and rift" system because the structure, composition, and arrangement is totally unlike that of continental mountains. The oceanic mountains are composed almost entirely of flows of basalt, are relatively narrow, and have steplike sides. A most puzzling feature is the presence along many stretches of a narrow steep-sided valley or rift along the crest of the ridge. This valley is the site of intense earthquake activity and is bordered by many volcanoes, some of which rise above sea level. Iceland sits on the ridge and is sliced by active faults in line with the trench system.

The placement of the ridge system with relation to the continents is significant. The *mid-Atlantic ridge,* as the name suggests, bisects the Atlantic almost perfectly and is therefore essentially parallel with the coastlines of the Americas and of Europe and Africa. Elsewhere the pattern is not so perfect.

Discovery of the oceanic ridge system was followed immediately by speculation as to its origin. It is clearly a zone of tension. The ocean floor literally splits apart with the formation of the great rifts and fractures. As these form they are "healed" by being filled with basalt rising from within the earth. The obvious mechanical

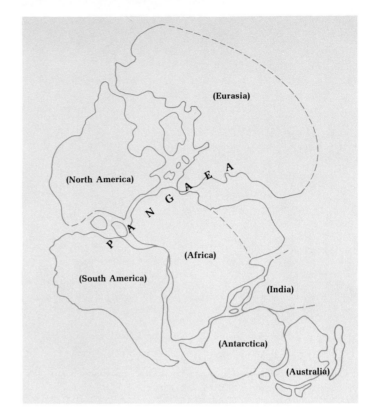

Figure 8.7 *Modern view of Pangaea.*

Figure 8.8 (a) *Pattern of the earth's great folded mountain chains and island arcs. One zone circles the Pacific Ocean; the other traverses southern Asia and Europe. They intersect in the East Indies at a "crossroads" of intensified geologic disturbances. (b) The world-circling mid-ocean ridge system bisects the oceans and has a total length of about 64,000 kilometers. It is the longest single geologic structure on earth.*

(a)

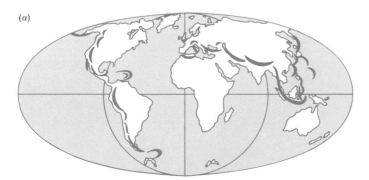

(b)

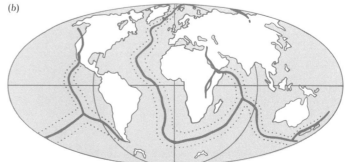

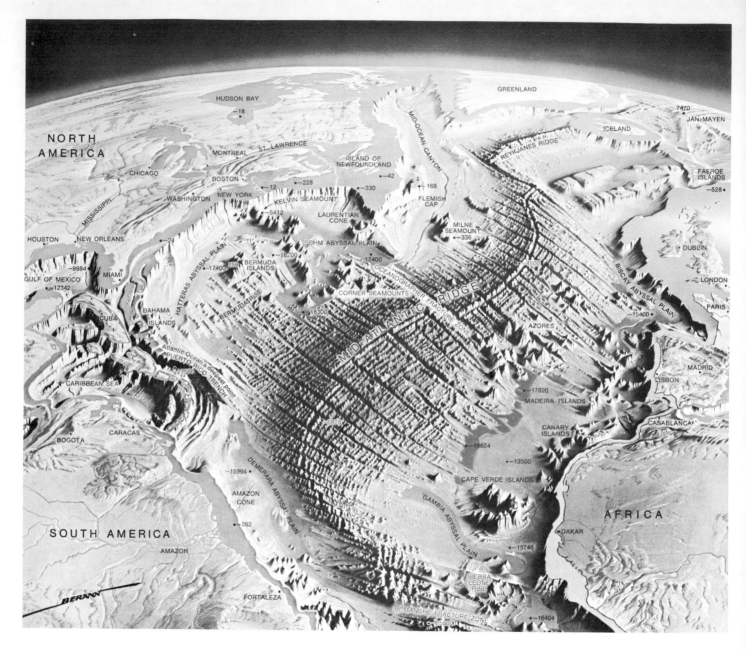

Figure 8.9 A topographical painting by Heinrich Berann of the floor of the Atlantic Ocean, with emphasis on the mid-Atlantic ridge and associated structures. Numbers indicate feet above or below sea level. (Photograph courtesy of Alcoa.)

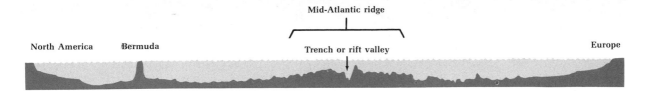

Figure 8.10 Configuration of the Atlantic Basin along a line from Cape Henry to Rio de Oro. Maximum depth of the ocean is about 5,500 meters. The mid-Atlantic ridge and trench are well defined. (Adapted from Bruce C. Heezen et al., The Floors of the Oceans, Geological Society of America, Special Paper 65, 1959.)

necessity of moving previously formed material aside to make room for new material from beneath is basic to the concept of sea-floor spreading. As new ocean bottom is being created, older strips are moved aside and are eventually hundreds or even thousands of kilometers from their place of origin.

Figure 8.11 A simplified diagram illustrating the production and spread of oceanic crust and the origin of belts of material with contrasting magnetic properties. Molten material rises from below into the axis of the ridge where it takes on the magnetic characteristics that prevail when it cools. The solidified basalt is divided at the ridge crest into equal bands that are slowly carried away from the crest in both directions and become the floor of the ocean. A record of variations in magnetic intensity and of the reversals of polarity are preserved on either side of the crest and can be mapped in the form of parallel bands. (Adapted from F. J. Vine, "Sea-floor Spreading—New Evidence," Journal of Geological Education, Vol. 17, 1969.)

In 1963, a spectacular piece of scientific synthesis led to a method for identifying separated bands of basalt so that rates and distances of movement could be calculated on the basis of the magnetic properties of the basalt strips. We discussed this method of correlation in detail in Chapter 6. The important fact here is that newly formed oceanic basalt contains a record of the magnetic field that prevailed at the time of its formation. Because samples of basalt can be dated by radiometric means, geologists can determine just when any magnetized strip was solidified. Maps of great tracts of ocean bottom that show not only the magnetic properties but also the ages of the elongate strips produced by sea-floor spreading have been prepared. Bear in mind that an emerging body of fluid basalt is separated into two bands at the mid-oceanic ridge, so that duplicate records are created. Any map of the patterns on one side of an active ridge will have a mirror image on the other.

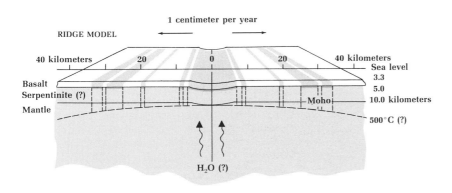

If we know the distance that a strip of basalt has moved since its formation, we can determine how rapid the movement has been. Hundreds of careful measurements in all the major oceans show rates of movement ranging from 1.5 centimeters to 15 centimeters per year; the average rate is perhaps about 2.5 centimeters per year.

Slow as this seems by ordinary standards, it is fast geologically speaking.

The establishment of sea-floor spreading as a fact not only constitutes evidence that continents have moved apart but also shows how this could have happened and how rapidly it was accomplished.

Plate Tectonics

The mere widening of the ocean basins cannot completely explain the present surface geography of the earth. The ocean basins cannot widen indefinitely, and the continents cannot be pushed around in an indiscriminate manner without producing chaotic surface effects. Obviously, crustal material must be destroyed to make room for that produced at the mid-oceanic ridges. The sites of destruction are now known to be the elongate trenches of the deep ocean. Here tongues of crustal material are plunging downward to be remelted and assimilated into the general circulation. This process is called *subduction*. The modern explanation of how crustal material is created, moves, and is destroyed is much wider in scope than the old idea of continental drift. Continental drift presupposes moving continents and stable oceans; the new idea requires dynamic movement of both the lands and the sea bottom. The term *plate*, or *global, tectonics* is used to describe the processes and mechanisms involved.

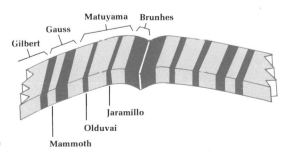

Figure 8.12 (a) A diagram showing polarity reversals over the past 4 million years. The polarity epochs are dominated by either a normal or reversed magnetic field. Within these long intervals, there may be shorter periods during which an opposite field prevailed temporarily. These short intervals are called "events." (b) A diagram indicating how the polarity record is correlated with any sea floor that has been actively spreading during this time. (Adapted from Allan Cox, "Geomagnetic Reversals," Science, Vol. 168, 1969, Fig. 4. Copyright 1969 by the American Association for the Advancement of Science.)

Plate tectonics designates the motions and reactions of the large, relatively thin, shell-like blocks of the earth's crust. These plates are tightly fitted together and cover the entire earth like the cracked shell of a hard-boiled egg. Plates may pull apart, press together, or slide past each other on broad fronts. They can even be created and destroyed in a gradual way. How all these apparently haphazard or opposing reactions are possible has now been explained in a satisfactory but preliminary way.

The theory of how the plates can move was hinted at in our discussion of the asthenosphere and lithosphere. The plates are part of the lithosphere and they are able to slide on the plastic asthenosphere. Six major plates are recognized and named after their dominant geographic features: Eurasian, Pacific, Australian, American, African, and Antarctic. Minor blocks that fit between the angles and curves of the larger blocks have also been outlined and named.

The American plate carries North and South America and the western half of the Atlantic Ocean. The African plate carries the eastern half of the Atlantic Ocean, all of Africa, and part of the Indian Ocean west of the mid-Indian ridge. All of Europe and Asia, except for India and

Figure 8.13 *Basaltic lava flow encrusted with a deposit of manganese. This photograph was taken in the Rift Mountains, about 2,700 meters below the surface of the Indian Ocean. (Official photograph, U.S. Navy.)*

Figure 8.14 *The lithospheric plates, with boundaries between the plates shown by the heavy line.*

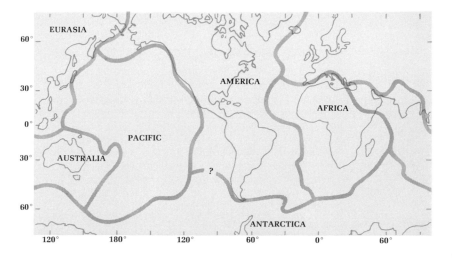

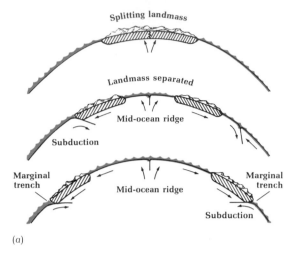

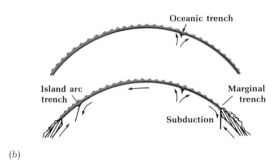

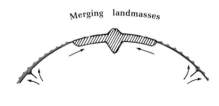

(c)

Figure 8.15 Simple geometric possibilities within the framework of sea-floor spreading, emphasizing the role of the continents. (a) and (b) illustrate the stages and possible configurations resulting from the initiation of rifting and spreading beneath continental and oceanic crust, respectively. (c) illustrates the possibility of two continental blocks coming together over a downcurrent. (Adapted from F. J. Vine, "Sea-floor Spreading—New Evidence," Journal of Geological Education, Vol. 17, 1969.)

Arabia, are carried by the Eurasian block. The Pacific is the only entirely oceanic block and is overridden on the east by the western margin of the United States. Boundaries of the Australian and Antarctic blocks are less well known.

Because the individual plates are strong and rigid, geologic activity within them is relatively minor. Around the borders the situation is quite different. These regions are characterized by intense geologic activity, as evidenced by active folding, faulting, volcanic activity, and earthquakes. This is explained by the pushing, pulling, or sliding motion of the plates. Each of these actions creates large-scale surface manifestations that are geographic as well as geologic in expression. Results of plate movement are shown perfectly well on any globe or world map. Curving mountain chains, island arcs, oceanic deeps and trenches, submerged ridges and scarps, lines of active volcanoes on land and under the ocean— all are results of the interaction of the massive mobile plates that encase the earth.

The destruction of the edges of moving plates is more dramatic than the formative processes in the mid-oceanic ridges. The zones of destruction are generally marked by a downward plunge of one plate and a corresponding overriding action of another. If the collision takes place within an oceanic basin, elongate trenches result (Tonga Trench in the southwestern Pacific). If the action occurs along the offshore edge of a continent, there will be not only curving ocean trenches but also island arcs with active volcanism and strong earthquake zones (Kuril-Japanese ocean trench and island arc system). If the collision of plates is near a relatively abrupt coast, oceanic deposits and even large volcanic piles may be literally scraped off and piled against the edge of the continent (Andes in South America).

Reaction at continental borders differs greatly depending on the relative rigidity and rates of movement of the colliding blocks. If an oceanic plate with a blanket of marine sediments is forced to plunge downward, it must ultimately melt or be metamorphosed. Its heavier constit-

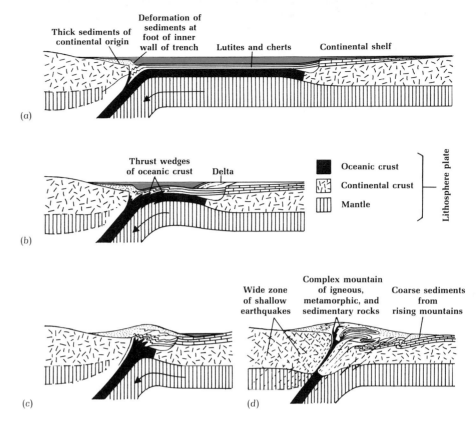

Figure 8.16 New information derived from the study of plate tectonics tells geologists a great deal about the processes connected with mountain building. In this four-part illustration, the complex mountain system of igneous, metamorphic, and sedimentary rocks in the final stage has been created by the collision of two lithospheric plates. (Adapted from John F. Dewey and John M. Bird, "Mountain Belts and the New Global Tectonics," Journal of Geophysical Research, LXXV (1970), 2642.

uents will sink; lighter ones will rise. It is the lighter rock materials plus water vapor and other gases that break forth in the chains of active volcanoes of island arcs or mountains parallel to the nearby trenches. It appears possible that one continental block may strike another in such a way that it is not immediately forced steeply downward into the mantle but instead passes underneath the opposite block so as to raise it into high plateaus and complex elevated ranges. The collision of India and southern Asia seems to illustrate this action. Finally, if two equally rigid and buoyant blocks with fringing deposits of relatively soft sediment approach each other there may be folding of the sediments into spectacular mountains such as the Alps. In the case of the Alps, the African block appears to have forcibly shoved material of the Alps northward an estimated 290 kilometers or more.

Major zones of faulting are created when two plates slide past each other. A famous example is the San Andreas fault in California where the Pacific block is moving northwesterly with respect to the American block. At a few places a ridge system actually impinges on a continent. The East Pacific rise veers gradually eastward from a position in the southern Pacific Ocean and strikes the western United States by way of the Gulf of California. The Great Basin of the west-

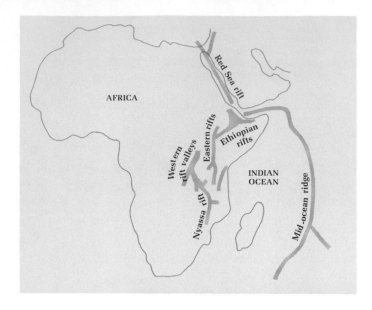

Figure 8.17 *The Indian Ocean rise and the African rift systems. This region is marked by intense volcanic and earthquake activity.*

Figure 8.18 *A radar map of part of the San Francisco peninsula, with an index map identifying important features, including the San Andreas fault zone. Along this zone, the Pacific crustal plate (lower portion of map) is moving actively northward in relation to the American plate. (U.S. Geological Survey.)*

ern United States displays a great many unusual features that have been caused by the fact that the American block has forcibly overridden the rise.

The Indian Ocean rise strikes into the coast of Africa and is responsible for the depressions occupied by the Red Sea and the Gulf of Aden. A northerly trending branch explains the Dead Sea depression. Southward, branches from this system strike into the African continent to create the spectacular rift systems of East Africa. Volcanic and earthquake activity are intensified wherever the rift system slices into a continent or island. Iceland, which sits astride the mid-Atlantic ridge, is perhaps the most geologically active island on earth. It contains many volcanoes and faults. The entire island is built of flows and fragmental materials discharged by volcanoes.

The reality of the movement of large plates of the outer earth cannot be doubted. That the plates float upon, and are buoyed up by, the hot, plastic asthenosphere is obvious, but what causes them to move is not well understood. Any successful theory must take into account the fact that the causative agents are not permanently localized. They appear to spring up, gain mo-

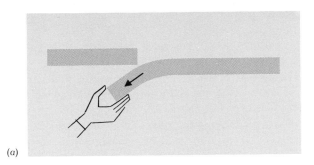

(a)

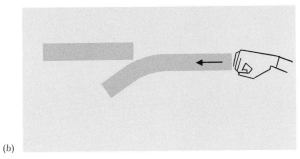

(b)

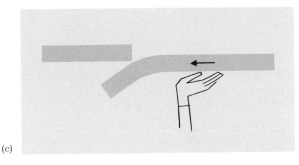

(c)

Figure 8.19 Three possible explanations for the movement of the lithosphere plates. A plate may be (a) pulled by gravity, (b) pushed by rising magma, or (c) carried along by moving currents.

Figure 8.20 Paleomagnetic measurements of rocks in North America and Europe show the paths followed by the magnetic poles of these two continents from Precambrian times to the present. (Adapted from Allan Cox and R. R. Doell, "Review of Paleomagnetism," Geological Society of America Bulletin, LXXI (1960), 758, Fig. 33.)

mentum, and die down over long time periods. The pattern today is different from that of the past. For example, the present-day Pacific plate appears to have been formed by the fusion of at least three ancient ones.

Some scientists believe that the currents are purely convectional features driven by heat arising from within the earth. They occur in a pattern that results in the most effective dissipation of the rising heat. According to this theory, the lithosphere blocks with attached continents ride passively about under the influence of powerful internal forces like close-spaced blocks of ice in the polar sea. Another interpretation is that surface conditions exercise a strong influence. Blocks of material exposed for a long period at the surface or under the ocean might become cooler and denser; the blocks would then slide downward on low gradients and be engulfed by the hotter material beneath. A return flow of material to complete the circulation would be brought about by a lateral displacement in the plastic asthenosphere. In simplest terms the problem is one of deciding whether the lithosphere or the asthenosphere is the active element.

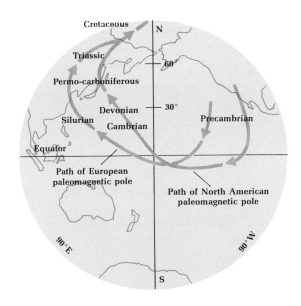

No matter what the final decision may be, it is obvious that heating and cooling must be the ultimate causes.

Polar Wandering

It is a well-known fact that cold climates with large glaciers are associated with polar regions, while warm climates are characteristic of the equatorial zones. Hence it is logical to conclude that a landmass bearing evidence of continental glaciers must have been near a polar position, just as Antarctica now is. But the currently tropical areas of Africa bear evidence of a great ice age, and Antarctica carries a record of a previously warm climate. This contrast between present and past climates must somehow be explained.

Continental drift and plate tectonics give one possible answer. Motions of the individual lithospheric plates may have carried various continental areas into polar regions and then out of them again, thus explaining drastic climatic changes. There are two other possible explanations however: (1) the entire lithosphere of the earth may have slowly shifted, or (2) the earth itself may have gradually shifted its axis of rotation. Both of these actions would have had the same effect as the movement of the individual plates.

If there has been a basic shift of the entire lithosphere and crust independent of the movement of the individual plates, a record of its effects would be preserved in the climatic record of the various regions of the earth. Thus if a succession of landmasses passed into, and out of, polar regions they should acquire and carry with them a record of glaciation appropriate to the times of their location in a polar situation. The successive positions of the pole would then be marked by a wide, potentially continuous path of glacial effects across the earth. This zone of effects is called the polar wandering path, and

the general concept derived from it is known as *polar wandering.*

Unfortunately the term "polar wandering" gives an erroneous impression that we hasten to correct. If we see a line on a piece of paper, it is natural to assume that a pen has moved over the surface. But the same line may have been drawn by moving the paper while the pen point remained stationary. Thus the so-called polar wandering is not an active movement of the poles over the earth; the poles remain steadfastly pointing toward the same regions of space while the continents and seas move through their influence.

As we said, polar wandering could also have been achieved by a rolling motion of the entire planet. (We could tape the paper to the top of a desk and move the desk beneath the pen point.) However, the overall shape of the earth seems to argue against this theory. The equatorial bulge and the flattening of the polar regions appear to be the direct result of rotation and could not easily be altered as the axis of rotation slowly changed position. A slipping of the entire lithosphere upon the plastic zone beneath it seems to be a much more logical explanation for apparent polar wandering.

Paleomagnetic data have also shown that the movement of individual lithospheric plates must also be considered in any interpretation of polar wandering. Thousands of pole positions have been calculated from specimens collected throughout the world. It is known that for any specific time the location of the poles as given by specimens from one continent may be different from the location supplied by specimens from another continent. This is considered to be evidence for large-scale movements of individual continents—the rocks with their contained magnetic orientation have been shifted so that they no longer point where they originally did. Most of the data can be reconciled by shifting the continents backward along certain paths until they occupy the positions they did at the time of magnetization. When this is done, it becomes

Figure 8.21 Eruption of Irazu volcano, Costa Rica. In addition to the fine ash that has covered the slopes of the volcano a great deal of water vapor and other gases are being emitted. Much of the water of the earth is thought to have originated from volcanic eruptions. (U.S. Geological Survey.)

evident that there must have been both slippage of the entire lithosphere and movement of individual plates. Research is continuing on all aspects of the subject.

Oceans and the Atmosphere

Resting upon the surface of the solid earth (the lithosphere) is a discontinuous sheet of water (the hydrosphere) and layers of gas (the atmosphere). The liquid and gaseous outer layers are arranged according to density and have a shell-like structure comparable to that of the solid earth. Water is essentially incompressible and homogeneous, but the atmospheric constituents vary, are compressible, and exhibit distinct spheres or layers. The origin of these outer shells of liquid and gas must be explained in any review of the origin of the earth.

Perhaps the earth should have been called "water," for it appears to be characterized more by its extensive water areas than by dry land. Oceans and seas occupy 71 percent of the surface, a total of 361 million square kilometers. Lakes, streams, glaciers, and snow fields are other manifestations of water that modify the so-called dry land. The swirling patterns of white clouds shown in photographs of the earth taken in outer space are dramatic evidence of water in the atmosphere. The total volume of water in the oceans is estimated at 1.4 billion cubic kilometers, and if the lands were leveled out, water 2.5 kilometers deep would cover the globe.

How can we explain the origin of the oceans, especially when water as such seems to be rare elsewhere in the solar system and universe? Water cannot exist as a liquid except within rather narrow temperature limits. No other visible planet displays extensive sheets of water, although water vapor exists on Mars and Venus and ice derived from water probably constitutes large parts of the major planets. Water cannot exist for long, if at all, on smaller bodies such as the moon or Mercury, because the water molecules gradually escape the pull of gravity and pass into outer space.

How was water produced in large quantities in the first place and how has it persisted for so long? In prescientific times, the traditional view was that the lands and oceans appeared rather suddenly in ready-made working order. Later on, when the concept of a molten earth held sway, scientists visualized a primitive atmosphere containing all the earth's potential water in the form of dense clouds of steam and vapor. When the earth cooled sufficiently, torrential rains fell and formed boiling pools and eventually oceans and seas. This dramatic picture has been greatly modified by recent observations and experiments.

The modern view is that water and atmospheric gases came not from without but from within the earth. As the earth cooled and solidified, many gaseous and volatile constituents were forced upward and escaped to the surface, a process known as *degassing*. Hot springs, geysers, and volcanic emanations indicate that degassing is still going on. Volcanic gases consist mainly of water, carbon dioxide, sulfur, and nitrogen. The water, of course, can be formed within the earth by the combination of hydrogen and oxygen.

This theory for the origin of the oceans and seas goes far in explaining what is known about the earth's present condition. For example, it explains the slow accumulation of carbonate rocks through time. Limestone ($CaCO_3$) and dolomite ($MgCO_3$) have as an essential component the gas carbon dioxide, which is supplied chiefly from the atmosphere. If the total supply of carbon dioxide had been present in the beginning we might expect much limestone and dolomite early in the earth's history and less later on. The fact is that there are few carbonate rocks in the earlier geologic record but the proportion increases later on.

The composition of the earth's atmosphere also helps to substantiate the theory of the origin of the earth's water. In our atmosphere there is relatively more water vapor, carbon dioxide, chlorine, sulfur, and nitrogen than is apparently present elsewhere in other planetary atmospheres. On the other hand, the noble gases *argon*, *krypton*, and *xenon* are rarer on earth than they are on the other planets. This set of circumstances is neatly explained by the theory of planetary degassing. It is probable that as the chemically less reactive noble gases emerged from the interior of the earth they immediately dissipated into space because they could not form compounds that would retain them. The high temperature of the earth itself and the effects of the solar wind (radiation pressure) were sufficient to drive the noble gases into space. By comparison the chemically reactive excess volatiles were able to form heavy compounds and were thus retained on or near the earth's surface. The subsequent release of oxygen and nitrogen to build the present atmosphere is another story which must be put off for later consideration (Chapter 10). At this point suffice it to say that the theory of the origin of the earth's water also explains why the atmosphere has its present composition.

A SUMMARY STATEMENT

On the basis of density, the earth is stratified into three major solid portions—the *core, mantle,* and *crust.* The differentiation into shells is best explained as a consequence of early melting.

The core begins about 2,900 kilometers below the surface and is probably composed chiefly of iron and nickel. The inner core is solid, but the outer core (about 2,160 kilometers thick) is molten. This shell possesses a degree of internal motion which is thought to generate the earth's electrical and magnetic fields.

The mantle, about 2,900 kilometers thick, seems to have been the parent material for the core and for the crust. It is thought to consist of dense minerals rich in silica, magnesium, and iron. Lying in the mantle from 80 to 160 kilometers below the surface is a hot, plastic layer called the *asthenosphere.* This shell, in effect, separates the outer shell (*lithosphere*) from the lower mantle. The upper mantle is attached firmly to the crust but is separated from it by a zone of physical change called the *Mohorovičić discontinuity.*

The lithosphere is divided into six major, close-fitting, mosaiclike plates that more or less float or slide on the asthenosphere. Material for the plates is created along one or more edges where the plates border a world-circling fracture zone. Along this zone basalt rises and solidifies as primary crustal material. This newly created basaltic lava moves away from the fracture zone at a slow rate in both directions and becomes the ocean bottom material. Plates are destroyed when they plunge downward into oceanic trenches or at continental margins. When plates with superincumbent continental material approach or collide, mountain ranges are created. Movement of the plates is the long-sought mechanism for continental drift, but it is still not known whether the lithosphere plates are pulled, pushed, or dragged along their courses.

Another type of movement, known as *polar wandering,* affects the earth. The entire lithosphere, independent of internal plate movements, may slide or migrate so as to bring different areas into the polar regions. Thus it is not the poles but rather the continents that wander; hence polar wandering is more apparent than real.

The watery hydrosphere and multilayered atmosphere are chiefly products of degassing. When entirely molten, the earth was virtually free of both water and atmosphere. Gaseous and liquid components are considered to have emerged gradually from the interior and accumulated under the force of gravity. The whole earth system is still in a state of evolutionary flux and change.

FOR ADDITIONAL READING

Brancazio, Peter J., ed., *The Origin and Evolution of Atmospheres and Oceans.* New York: Wiley, 1964.

Cox, Allan, "Geomagnetic Reversals." *Science,* Vol. 168, 1969, 237–245.

Munyan, Arthur C., *Polar Wandering and Continental Drift.* Society of Economic Paleontologists and Mineralogists, Special Publication No. 10, 1963.

Pinney, Robert A., ed., *The History of the Earth's Crust: A Symposium.* Princeton University Press, 1968.

Takuchi, H., S. Vyeda, and H. Kanamori, *Debate About the Earth.* San Francisco: Freeman, 1967.

THE PRECAMBRIAN

The age of the earth is generally regarded as about 4.5 billion years. Within this immense period of time a number of events stand out as natural mileposts of historic significance: the solidification of the first igneous rocks to form a permanent "crust"; the formation of the first sedimentary rocks, an indication of erosion and deposition by water; the first appearance of life, heralding a fairly stable environment; the formation of recognizable continents containing granite and sedimentary rocks; the "explosive" proliferation of oceanic life at the beginning of the Cam-

Going-to-the-Sun Mountain in Glacier National Park, Montana, is carved from Late Precambrian sedimentary rocks. (Montana Highway Commission.)

brian Period; and, finally, the migration of life from the waters to the land.

One of these events is generally accepted by geologists as being of paramount historical import—the appearance of abundant fossils at the beginning of the Cambrian Period about 600 million years ago. This event in effect initiates the interval that may be studied with the aid of fossils and divides it from the preceding interval, which must be studied virtually without them. The beginning of the fossil record is analogous to the invention of writing in human history.

Fossils provide a means of deciphering and subdividing the rock record just as writing helps immeasurably in clarification of human events.

Subdividing the Precambrian

A number of schemes have been proposed to designate the interval before the appearance of abundant fossils. The most commonly used term is the self-explanatory word *Precambrian*. The Precambrian itself can be subdivided on the basis of the widely used system that separates all past history into five great eras, the last three of which are included in the modern time scale as described in Chapter 6. The five eras in chronological order are the following: *Archaeozoic, Proterozoic, Paleozoic, Mesozoic* and *Cenozoic*. The last three designate progressively advancing stages of life—ancient, middle, and recent. Because Archaeozoic and Proterozoic also designate ancient and primitive life, there is obvious overlap with the term Paleozoic.

In practice, because evidences of life are so few as to be totally useless in making a distinction between the two older eras, Archaeozoic has come to designate the highly metamorphosed and presumably older part of the Precambrian, whereas Proterozoic designates the less altered and younger part. According to these definitions, which really have no time connotations, a rock formation might be Archaeozoic in one place and Proterozoic not far away. Also, a twofold division of the exceedingly long Precambrian is understandably imprecise. There is a rock record of at least 3.6 billion years before the Cambrian began, an interval judged to be six times longer than all succeeding time. To recognize three major divisions within a span only one-sixth as long as another for which only two similar divisions are recognized is obviously less than ideal.

In spite of its obvious deficiencies, many geologists still use the fivefold scheme with the understanding that the Archaeozoic includes any older, highly metamorphosed Precambrian rocks to which they cannot apply ordinary means of study and analysis.

Other schemes are being tested. One of these proposes that the lengthy span of time for which there is little or no fossil record in the rocks be called the *Cryptozoic* ("hidden life"). All later time and rocks would be designated as the *Phanerozoic* ("evident, or visible, life"). These great divisions, being aggregations of eras, would be designated as *eons*. This scheme is useful when attention is focused on the history of life.

Another plan, followed officially by the U.S.

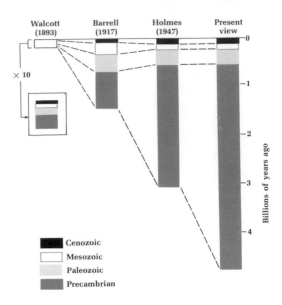

Figure 9.1 *Changing view of the magnitude of Precambrian time in the twentieth century. (Adapted from H. L. James, "Problems of Stratigraphy and Correlation of Precambrian Rocks . . . ,"* American Journal of Science, *258-a (1960), 105.)*

Geological Survey, is to classify Precambrian rocks in terms of relative age as Early Precambrian, Middle Precambrian, and Late Precambrian. This plan is based on the reasonable assumption that the ages of most ancient rocks will eventually be known, after which formal names can be applied. Russian geologists, working along similar lines, use the terms Precambrian I, Precambrian II, Precambrian III, and Precambrian IV.

No matter which method of worldwide correlation is eventually adopted, there must be local divisions with local names. When these are known, they can be integrated later into broader schemes. Fortunately, there exists a physical, objective basis for working out the succession of events and geologic history of each of the major Precambrian areas of the world. The key to local classification is the fact that broad elongate belts thousands of square kilometers in extent have similar ages throughout and are consequently thought to have originated as separate units, or *provinces*. Each province is regarded as having resulted from a sequence of events divisible into three broad phases: (1) the *geosynclinal* or *depositional,* (2) the *orogenic* or *mountain-building,* and (3) the *epeirogenic* or *erosional.* Together these phases constitute a great geologic, or geotectonic, cycle. The sequence is also sometimes referred to as a geosynclinal cycle, even though "geosynclinal" properly applies only to the first stage. The concept of the geosyncline, one of the great unifying ideas of geology, should be briefly discussed.

Figure 9.2 *The evolution of a geosyncline. (a) Initial downwarping near the continental margin creates an elongate depression in which sediments begin to accumulate. A volcanic ridge separates the outer (eugeosynclinal) tract from the inner (miogeosynclinal) belt. (b) Subsidence continues with additional sedimentation and enlargement of the volcanic ridge. (c) Compression of the sediments (caused generally by collision of moving lithosphere plates), intrusion of granite, and uplift create a mountain chain from the former geosyncline. Erosion goes on concurrently with the uplift. See text for more complete discussion.*

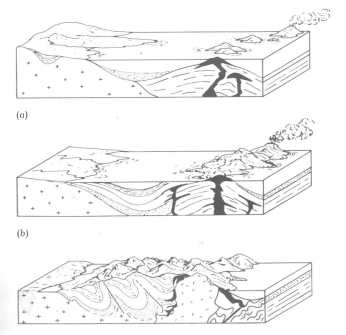

(a)

(b)

(c)

Concept of the Geosyncline

A geosyncline is a great elongate downfold of the earth's crust. Lengths are usually measured in terms of hundreds or thousands of kilometers, widths in terms of scores or hundreds of kilometers. Although downwarping may be initiated on dry land or even in the deep ocean basins, most geosynclines have appeared near the margins of continents. This localization seems to be an expression of the fact that the zone between the oceans and continents is one of intense geologic activity, discontinuity, and weakness.

The downwarping of an elongate tract, no matter where it is located, creates space into which sediments may be carried and deposited. Naturally a geosyncline is filled with whatever materials are available. The nature and intensity of local geologic activity is therefore important in the filling process. A trough lying immediately adjacent to a rapidly rising volcanic range will be filled with coarse, poorly sorted igneous detritus, whereas one next to low-lying sedimentary terrain may receive only fine-grained limy sediments precipitated primarily from ocean water. Geosynclines have been classified in many ways, but no scheme has gained universal acceptance. However, two terms seem valuable in distinguishing geosynclines that lie on the inner (landward) margins of the continents and those that lie on the outer (seaward) margins. The outer type of geosyncline is called a *eugeosyncline;* the inner type is a *miogeosyncline.* Usually these take the form of two parallel troughs. Eugeosynclines are rich in siliceous and volcanic rocks, such as lava, tuff, chert, argillite, and ash. Miogeosynclines are marked by well-washed sandstone, limestone, dolomite, and shale. Volcanic and other igneous manifestations seem to characterize the middle and late phases of geosynclinal history and make major contributions in most regions.

The sedimentary, or depositional, phase of a

geosynclinal cycle continues as long as downsinking is dominant. Sinking may be so slow that deposition is sufficient to keep the basin full of sediment. In fact, one early observation in favor of the geosynclinal theory was that most of the sediments found in the Appalachian Mountains originated in shallow water. How else but by slow downsinking could thousands of meters of sediment accumulate in a shallow-water environment? In other geosynclines, such as the one that preceded the Alps, downsinking was more rapid than deposition and characteristic deepwater formations resulted.

The depositional phase of a geosyncline, on an average, may take several hundred million years. The Appalachian Geosyncline, for example, received sediment for about 300 million years during the Cambrian, Ordovician, Silurian, Devonian, and Mississippian Periods. The deposits in geosynclinal belts range roughly from 9,000 to 30,000 meters. Any deposit that sinks to greater depths seems to be melted and destroyed by the heat of the earth.

Although geosynclines have been recognized as major geologic features of the earth, the mechanics of their formation have only recently been satisfactorily explained. Obviously the most effective way to create downfolds or upfolds in any bendable material is by pressing from one or both sides. Geologists also realized from the

Figure 9.3   The Alps of the Bernese Oberland, Switzerland. The Alpine chain is the uplifted and highly deformed strata deposited in the Tethys Geosyncline. (Swiss National Tourist Office.)

first that in most cases this lateral pressure is directed from the ocean toward the land. The concept of plate tectonics and sea-floor spreading neatly explains this situation.

Continued lateral pressure will eventually cause the contents of a geosyncline to be squeezed and deformed. The relatively weak and unconsolidated contents within the trough are then forced to rise above their former levels. At first there will be only great open folds, but if compression continues these folds will become higher and more closely compressed. Because rock in large masses is inherently weak, the folds will soon begin to topple and slide downward in great faulted masses of amazing complexity. It is thought that mountain ranges such as the Alps represent practically all of the contents of a geosynclinal trough that has been squeezed entirely out of its original form. Not to be ignored is the fact that erosion begins to attack any rising areas and will continue to be effective throughout all the subsequent history of a mountain system. In the more active phases, eroded material usually becomes mixed with, or buried under, the folded and faulted older rocks as they rise to form mountains.

The compressional, or mountain-building, phase of the geosynclinal cycle is not as lengthy as the depositional phase. The rise of mountain ranges, although slow by human standards, is rather rapid geologically speaking. As a matter of fact, mountain building serves as a sort of punctuation mark between the longer, quiet episodes.

With the end of active compression and mountain building, the geosynclinal cycle enters its final stage. This has been called the epeirogenic phase in allusion to the fact that the entire affected area, including the mountain ranges and their borderlands, tends to be uplifted en masse. It is this phase that has brought such systems as the Appalachians, the Alps, and the Rockies above sea level and welded them to the adjacent continental borders. The dominant surface activity of this phase is erosion. Running water and

Figure 9.4 *Air view of rocks and structures in the Precambrian shield near Great Slave Lake, Canada. The great fault separates Precambrian granite on the left from Precambrian sediments on the right. All these rocks were once buried under thousands of meters of material, and the site was probably once a complex mountain range. The body of water is McDonald Lake. (National Air Photo Library, Surveys and Mapping Branch, Department of Energy, Mines, and Resources, Canada.)*

glacial ice gradually reduce mountain ranges to low hills and eventually to level plains. The erosional phase of the cycle is a lengthy and perhaps indeterminate one. To level and remove a full-scale mountain system usually requires two, three, or four geologic periods, in other words, several hundred million years. It may be worth noting that the erosional products from one mountain system will be moved elsewhere to fill another geosynclinal tract and initiate another grand cycle of sedimentation, uplift, and erosion.

The concept of the geosyncline illustrates the importance of the role of the geologic cycle in understanding and classifying Precambrian rocks and events. Briefly, one geologic cycle in its ideal and complete form would embrace the geosynclinal, orogenic, and epeirogenic, phases just de-

Figure 9.5 Exposure of Precambrian gneiss, Brim Creek, British Columbia. Gneiss is a common rock type of the early stages of the earth's history. Although the composition may show a wide range of variation, the contorted, banded appearance is usually displayed. (Canadian Geological Survey.)

scribed. The completion of such a cycle might require from 300 million to 500 million years, which is long enough to constitute a sizable major subdivision of earth history. Thus a rough measure of the past history of any continental area can be known from the number of cycles it displays. For the time being this is indeed the best guide we have for dealing not only with the Precambrian but also with later time and rocks.

A geologic cycle can be identified only by means of extensive field studies, geologic mapping, and age dating. We must recognize that the evidence of the extensive and intensive geologic processes that take place during a geologic cycle cannot be permanently or entirely erased. True, a deeply eroded lowland is all that may remain of a former geosyncline and subsequent massive mountains, but it is still possible to know a lot

about a mountain system from its "stump and roots."

The accumulation of reliable radiometric dates makes it possible to outline the geographic area of each cycle. Remnants of sedimentary rocks, lava flows, and igneous intrusions can be studied and integrated into the comprehensive framework that the concept of a cycle provides. In this process we are aided greatly by what we can learn about modern conditions. We have for examination geosynclines being filled with sediment (Gulf Coast of the United States), mountains in process of formation (the Coast Ranges of North and South America), and older mountains being eroded (the Alps, Rockies, and Appalachians). The geosynclinal concept has proved valuable even though it must now be viewed in a different context than the one in which it was proposed.

Distribution of the Precambrian Rocks

A brief survey of the distribution of Precambrian rocks is essential before we can understand the problems of their origin and evolutionary meaning. We shall avoid for the time being the knotty problem of possible continental drift in Precambrian times and shall describe the great Precambrian areas in terms of their present position on the surface of the earth and their relation to the continental masses of which they are a part. These extensive regions of durable ancient rocks that have been eroded to gently convex shieldlike profiles are called shields, or shield areas. For this discussion, we will ignore the wide bands of younger sediments and mountain ranges and ocean basins that now separate the individual shields. Also, because of lack of evidence, we can say little about any internal movements that may have taken place among the shields during the long interval between the time that they were probably clustered together and the onset of the great breakup about 200 million years ago.

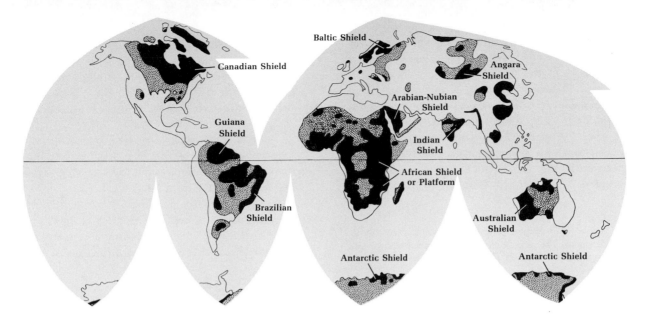

Figure 9.6 *The chief Precambrian shield areas of the world. Black indicates Precambrian rocks at the surface; stipple designates Precambrian rocks under shallow sedimentary cover. (Adapted from W. L. Stokes and S. Judson,* Introduction to Geology: Physical and Historical, *Prentice-Hall, Inc., 1968.)*

North America. In North America, Precambrian rocks are exposed chiefly in the Canadian Shield, an area of 7.2 million square kilometers including part of Greenland. The shield area is not centrally located but is displaced toward the northeast where it borders the Atlantic Ocean. Elsewhere around its edge the Precambrian rocks dip gently downward and disappear under a cover of younger sedimentary rocks. In what is called the covered shield of the north-central United States, Precambrian rocks are occasionally seen in uplifts such as the Black Hills and Ozark Mountains. They are also observed in the cores of the Rocky Mountain ranges or in deep canyons such as the Grand Canyon.

The oldest rocks of the Canadian Shield make up a roughly oval heartland stretching from Greenland to Wyoming, an area that constitutes the Superior-Wyoming Province. Here are found vast masses of granite cutting preexisting lavas and sediments, all mostly well over 2.5 billion years old. Rocks from a small area in southwestern Minnesota have been dated at 3.5 billion years, and an even greater age (3.98 billion years) is indicated for a considerable tract in south-central Greenland. The Greenland rocks are currently the oldest known on earth. Constituent rocks of the Superior-Wyoming Province are rich in silica and low in carbonates; they could have originated from previous sediments. A smaller area with rocks similar in age and composition to the Superior-Wyoming region centers around Great Slave Lake in northern Canada. Although these early islandlike masses have survived from the time of formation, they have certainly been much reduced in size by erosion as they supplied material for sedimentary deposits of surrounding lowlands and seas.

Following the *Kenoran Orogeny,* the term given to the activity that created the Superior-

Wyoming Province, a variety of igneous and sedimentary layered rocks accumulated in near proximity to the original nucleus. These rocks include siliceous volcanic flows, arkoses, and graywackes. There are also a few beds of dolomite. Eventually this sequence was intruded by a second generation of granitic masses with the result that two new provinces, the Churchill to the north and the Central to the south, were created. The Churchill Province makes up most of the exposed Canadian Shield, whereas the Central Province occupies much of the north-central United States and is mostly covered. The great disturbance that created these provinces, called the *Hudsonian Orogeny,* culminated roughly 1.8 billion years ago.

A third great episode of rock formation and mountain building produced a new belt of continental rock that wraps about the eastern margins of North America from Labrador to Mexico. The name applied to this disturbance is the *Grenville Orogeny,* and the belt of new rock is known as the Grenvillian. The foundation of Greenland, too, dates from this episode, and the ages of min-

Figure 9.7 Major geologic provinces of North America. The numbers in parentheses indicate the approximate age of the provinces in billions of years.

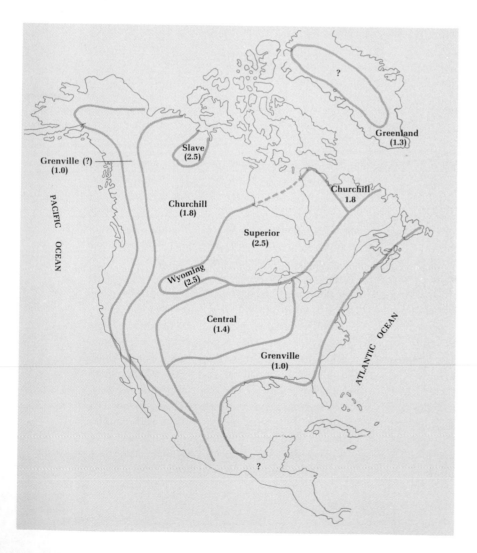

Figure 9.8   Gigantic faces of United States presidents in Mount Rushmore National Park.
The faces are carved from rock of Precambrian age. (U.S. National Park Service.)

erals gathered from this province average about 1 billion years.

Obviously there have been later orogenies and additions to the continent, such as those that produced the Appalachians and the Rocky Mountains. These too have enlarged the continent in the same way that the Precambrian orogenies did before them.

Europe. Most of northern Europe is underlain by Precambrian rocks. They are found in many mountain ranges and make up the surface of the Baltic Shield, which is found chiefly in Sweden and Finland. The Precambrian lies under a thin blanket of sediments east and southeast of the Baltic Shield, an area known as the Russian plat-

form. Precambrian rocks are also found in northern Scotland and neighboring islands and in various areas of the continent. Included are the Alps, the Vosges, Brittany, the Massif Central of France, northwest Spain, Sardinia, and the Carpathians.

The European Precambrian has been intensively studied, and many concepts bearing on the origin of metamorphic and igneous rocks have been developed by European geologists. Four episodes of sedimentation separated by periods of granite formation are recognized in the Baltic Shield, but newly applied radiometric dating methods have cast doubt on the true extent of the two older subdivisions. The most intense and widespread effects occur in the third, or *Karelidic,*

Figure 9.9 *Looking down the inner gorge of the Grand Canyon in Arizona. The dark-colored, unstratified rock in which the river is running is the Vishnu Schist of Precambrian age, about 1.5 billion years old. Note light-colored veins in the walls. The stratified cliff above the Vishnu rests on a great unconformity that is of Cambrian age. (U.S. Bureau of Reclamation.)*

cycle, which culminated about 1.8 billion years ago. The youngest group of Precambrian rocks is little metamorphosed and contains vague fossil impressions. In northern Norway, beds of tillite resting on glaciated rock surfaces occur. Russian geologists, who know these late Proterozoic rocks chiefly through borings, call them the *Rhiphaean* System; in Norway they are known as the *Eocam-*

brian, or *Sparagmitian.* Precambrian rocks yield important mineral products, and in the Baltic Shield these are chiefly iron and copper.

Asia. Precambrian rocks are exposed at the surface or lie under shallow cover over a vast area of north-central Siberia. The central area in which the ancient rocks are actually visible is called the Angara Shield; a much wider area, thinly veneered with younger rocks, is called the Siberian platform. Between the Siberian and Russian platforms lie the Ural Mountains and the West Siberian lowlands, which are sites of important basins and geosynclines of later date.

Precambrian rocks appear in the cores of many of the great ranges of southern Asia. There are also large disconnected patches, not usually referred to as shields, in China, Korea, and Southeast Asia. Precambrian areas in India and the Arabian peninsula are best considered in connection with Africa with which they are geologically related.

South America. About half of South America has Precambrian rocks at, or relatively near, the surface. Three large areas are recognized: the Guiana Shield bordered by the Orinoco River, the Atlantic Ocean, and the Amazon Valley; the Central Brazilian Shield, including most of the southern drainage of the Amazon; and the Coastal Brazilian Shield along the eastward bulge of the continent.

The Precambrian rocks of South America are usually divided into three divisions, Early, Middle, and Late Precambrian. The oldest rocks are gneisses and schists about which little is known. The Middle Precambrian, a time of active deposition in geosynclinal belts, produced great thicknesses of sedimentary and igneous rock and was closed with a period of granite formation and metamorphism. The Late Precambrian consists of less metamorphosed sediments such as quartzite, slate, phyllite, and conglomerate. Important economically are iron-bearing formations in the younger rocks.

Figure 9.10   The Baltic Shield as seen in Finland. The region was heavily glaciated and the Precambrian rocks are covered in many places by lakes, moraines, or forests. (Finnish Tourist Board.)

Figure 9.11   Pão de Açúcar (Sugarloaf Mountain), an imposing monolith eroded from hard Precambrian rock, rises well above the massive formation that juts into the harbor area at Rio de Janeiro, Brazil. (Pan American Airways.)

The South American shields have been mostly emergent since Precambrian time, but the Amazon Valley, which lies between two of them, has been repeatedly submerged by shallow seas since the early Paleozoic.

Africa. Africa has more extensive exposures of Precambrian rocks than any other continent. The continent is in fact a succession of shields and minor resistant blocks that have not yet been sorted out and named in a systematic way. Precambrian rocks make up much of the continent south of the Sahara and also the western bulge that fronts the Atlantic Ocean. The Arabian-Nubian Shield is partly in Arabia and partly in Egypt, having been split by the relatively recent Red Sea rift. Some geologists refer to Africa as a platform and think of the entire continent as essentially Precambrian in nature.

The oldest rocks so far discovered are in Barberton Mountain Land of South Africa. In this area, piles of metamorphosed but still recognizable layered rocks having a thickness of about 19 kilometers and ranging in age from 3 to 3.6 billion years have been found. One date of 4.1 billion years is recorded for a pebble of granite taken from a conglomerate within this section. This specimen, if correctly dated, is the oldest earth rock so far identified and may possibly be a fragment of the original crust.

Australia. Rocks of Precambrian age are found over wide areas of Australia, especially in the western half of the country. These rocks yield rich stores of metallic minerals and are being intensely prospected at the present time. A striking feature is the great volume of sedimentary and volcanic rocks up to 2.3 billion years old that are relatively unmetamorphosed and are comparable in their unaltered appearance with rocks of much younger age elsewhere.

Australian geologists divide the Precambrian into the *Archaean* and *Proterozoic.* The Archaean, as defined by Australian geologists, includes rocks over 2.5 billion years old; it is complex and

has not been subdivided except locally. The Proterozoic has been divided into three systems, from oldest to youngest, the *Nullaginian, Carpentarian,* and *Adelaidean.* These subdivisions include much basaltic lava, quartz-rich sediments, and cherty formations. These rocks occur in scattered patches with thicknesses up to 15,000 meters or more. The Adelaidean System is famous for the occurrence of many well-preserved primitive plant and animal fossils, which are described in the next chapter.

Antarctica. Although most of Antarctica is covered by ice and snow, enough of the bed rock is exposed around the margins and in protruding

Figure 9.12 *Precambrian rocks, dated at more than 2.6 billion years, exposed in a limestone quarry, southwestern Rhodesia. These rocks yielded the fossils shown in Figure 10.10. (NASA, photograph by K. A. Kvenvolden.)*

Figure 9.13 *Ayers Rock, an eroded remnant of Precambrian rock in the central desert region of Australia. Said to be the world's largest monolith, it is 335 meters high and about 9 kilometers in diameter at the base. (Qantas Airways Ltd.)*

Figure 9.14 *Exposure of Precambrian gneiss and schist, Transantarctic Mountains, Antarctica. (Institute of Polar Studies, photograph by John Gunner.)*

mountains to give a general idea of the geology. East Antarctica, the larger rounded part, is almost entirely of Precambrian age. Most of the exposures are dated at between 500 million and 1.5 billion years old; earlier Precambrian time is apparently not represented. There are many intrusive igneous rocks, and the effects of metamorphism and disturbances are much more intense than are shown by rocks of similar age in Australia.

Problem of the Original Crust

The major shield areas show evidence of having developed along similar lines. If rocks older than about 2.5 billion years are present in a particular shield, they are preserved in compact areas usually termed *cratons*. A craton according to this usage is a long-standing stable nucleus within a shield area. Cratons are characterized by great masses of granite surrounding and incorporating highly-metamorphosed volcanic rock, called

greenstone. Greenstones were originally chiefly basaltic or andesitic lava flows. Cratons of the various shield areas are being intensively studied to determine if they might be fragments of the original solid crust of the earth that have remained intact since solidification.

Evidence is not conclusive as to whether the original crust was basaltic or granitic. If the first persistent shell of solid rock formed directly from a well-mixed molten substratum, it would have been heavy, iron-rich, and basaltic. If, however, there was a period of gradual enrichment of lighter siliceous elements in the surface layers before the final solidification, the original crust may have been granitic. We observe that the ocean beds that are forming today are basaltic, but this may be due to the fact that great volumes of light material have already been extracted from it to form the present continents, oceans, and atmosphere. As a result, the subcrustal reservoirs have been depleted and are no longer producing silica-rich crustal material.

Those who have studied the situation in the field in Africa, Australia, and South America are not in agreement as to whether the greenstone belts in general are resting on granites or have been engulfed by them. In other words we cannot yet be sure whether a particular granite is younger or older than the greenstones with which it is in contact.

ORIGIN OF CONTINENTS

There are two separate but related problems concerning the origin of the continents. One, which was briefly discussed in Chapter 8, concerns the origin of the light, silica-rich material that makes up the continental rocks; the other has to do with the concentration of this material in the relatively high-standing landmasses of the present continents. The discovery of areas that contain very old rocks—perhaps even parts of the original crust—tells little about the shape and spacing of the first continents.

This is a difficult topic and there are many unanswered questions. Were there lithospheric plates encasing the earth in the Precambrian similar to those now in existence? If so, what was their configuration? Many geologists are inclined to believe that plates have existed from the time when the crust solidified, but they may not have been as thick, as individually extensive, and as rigid as they are today. As we attempt to reconstruct older and older configurations, the patches of continental material seem to become more numerous but smaller in size. These small segments also appear to have moved about in a more free and lively fashion than do modern continents. Volcanic arcs, subduction zones, oceanic trenches, geosynclinal troughs, and mobile oceanic belts appear to have been in existence and were the source of belts or tracts of continental material as visualized under the general theory of plate tectonics. These smaller subcontinents gradually coalesced to form larger masses. The region extending from Southeast Asia through the East Indies to New Zealand, which is very active geologically, is cited as a modern example of what went on during earlier stages of continent building. In other words, we may be witnessing the creation of a continent in the southwest Pacific.

Regardless of what previous world geography may have been, there is good evidence that the most ancient Precambrian rocks, those over 800 million years old, were once in two clusters, one generally north of the present equator, the other south of it. This arrangement becomes apparent when the present continents are assembled to make the best overall fit according to the continental-drift hypothesis. Figure 9.15 shows the Precambrian areas and continental outlines thus assembled. There are large areas of Precambrian rocks younger than 800 million years old that do not fit entirely within the clusters of older rocks, but their distribution does not alter the evidence that the oldest known rocks were concentrated for a long time in a belt lying almost entirely on one side of the earth. We cannot be sure that this was the original situation; perhaps there was a previous world-circling continental crust that was somehow disrupted and transferred to one hemisphere.

If the observations mentioned above are true, the present distribution of Precambrian rocks or shield areas has been strongly influenced by con-

tinental drift. They appear to have been in close proximity, if not in actual contact, at one time, but now they are widely separated. There may have been a sequence of minor motions between these two arrangements during which separation and collisions may have taken place, but we are not yet able to chart these events. In any event, building of the present continents got under way on a large scale at least 2.6 billion years ago and has continued at an uneven rate ever since.

Figure 9.15 A predrift reconstruction showing the continental blocks (color) isolated in two restricted areas about 1.7 billion years ago. The blocks are transected and surrounded by belts of younger rocks.

Evolutionary Development
of Later Precambrian Rocks

In general, the older a province or region is geologically, the more disturbed and metamorphosed are its component rocks. Many cubic kilometers of ancient rocks that were once sediments have been metamorphosed almost beyond recognition or have even been melted entirely and converted to igneous rock types. Progressively younger rocks naturally retain more of their original structure and composition, and ordinary methods of study can be applied to great areas of younger Precambrian rock. Within the interval from about 2 billion years to 500 million years ago, several significant and recognizable rock types are observed to have widespread distribution and are thought to reflect worldwide environmental processes. These rock types, in approximate order of their appearance are the following: (1) banded iron formations of subaqueous origin and associated calcareous algal growths called *stromatolites,* (2) iron-bearing red beds of continental origin, (3) coarse, unsorted, bouldery deposits of glacial origin called tillites, and (4) mineral salts such as gypsum and anhydrite. The last category is confined mainly to rocks later than Cambrian.

The banded iron formations are of more than academic interest—they constitute the world's greatest commercial source of iron ore. Large mines and attendant industries have grown up in connection with deposits in the United States, Canada, Brazil, Australia, and South Africa. Many deposits with an original, or primary, iron content of about 30 percent have been enriched or concentrated by natural geologic processes to grades of 70 percent or more.

Although opinions differ and research is continuing, it is agreed that the origin of banded iron ore was in bodies of water, possibly freshwater lakes, under certain peculiar processes that are not now operating. Because much of the iron is ferric (the oxidized form, Fe^{+++}) it is evident that there must have been abundant oxygen in many water bodies of Late Precambrian time. A source of dissolved oxygen is indicated by the abundant algal remains that were forming at the same time and often in the same environment. The algal fossils are preserved in layered, hemispherical masses that have been described as cabbagelike, fingerlike, or moundlike. These are not the actual remains of algae but are successive, thin, calcium carbonate platforms laid down by algae that were growing in thin mats upon the platforms. Stromatolites are somewhat analogous to the dead calcareous reefs left by corals. The close association of oxygen-producing organisms and oxygen-consuming sediments seems logical. The important point is that oxygen was at this time abundant enough to infuse many extensive water bodies and influence the type of sediment being deposited. The essential chemical reaction appears to be $4FeO + O_2 \rightarrow 2Fe_2O_3$.

The deposition of banded iron formations gradually declined and came to a close about 1.5 billion years ago. With this decline there came a corresponding increase in land-laid iron-bearing rocks of the type commonly called red beds. None of these are known in formations older than about 1.8 billion years. An early ex-

Figure 9.16   Specimen of Precambrian banded iron ore from the Iron River district, Michigan. (U.S. Geological Survey, photograph by H. L. James.)

Figure 9.17 *Iron mines in the Mesabi Range, northeastern Minnesota. Note the overburden of waste material above the dark-colored iron ore. (M. A. Hanna Company and U.S. Bureau of Mines.)*

ample is the Nankoweap Formation of the Late Precambrian of the Grand Canyon. In red-bed formations, the iron content is relatively low, a few percent at most. However this iron is in the ferric state, thus indicating wide availability of oxygen in the atmosphere. The sequence of events again seems to be logical: after enough oxygen had been produced by photosynthesis to satisfy the chemical requirements within the oceans and lakes it began to diffuse into the atmosphere and influenced surface geologic processes. The most obvious effect was the creation of continental red beds that remain as permanent records of this important change in the atmosphere. A less obvious but no less important action was the creation of the ozone layer, a zone rich in O_3 that encircles the earth about 25 kilometers above the surface. The importance to subsequent life is the ability of ozone to absorb incoming shortwave radiation that is harmful to

all unprotected protoplasm. This will be discussed further in the next chapter.

Near the close of the Precambrian, another event of worldwide significance is recorded. This was an extensive period of glaciation—an ice age—that is indicated in the rock record by the presence of the distinctive, ice-deposited type of material called tillite. Tillites of various types together with other glacial evidence of this age are widely distributed; they have been found and authenticated for every continent except Antarctica. This episode was one of the three great glacial periods known to have affected the earth and it may have been the most intensive and extensive. It is not surprising that we have found no obvious explanation for this ice age—we have

Figure 9.18 *An exposure of some of the earth's oldest red beds. The Hakatai Formation of Late Precambrian age appears in the cliffs immediately above the Colorado River. Light-colored cliffs forming the point at the upper left are composed of the basal Cambrian Tapeats Sandstone. (Courtesy of Donald Baars.)*

Figure 9.19 *Mining gold-bearing quartz veins in the Homestake Mine, Lead, South Dakota. This is the most productive gold mine in the United States. The ore is in highly contorted Precambrian rocks. (U.S. Bureau of Mines.)*

none for the one we have recently passed through. In any event, an ice age is definitely a phenomenon related to the oceans and atmosphere and is not a localized affair. Perhaps changes in atmospheric constituents such as carbon dioxide are involved. From a practical standpoint this ancient ice age provides a rough time marker for correlating between the continents.

The class of sediments that geologists term *evaporites* is almost totally absent in the Precambrian. This includes gypsum ($CaSO_4 \cdot 2H_2O$), anhydrite ($CaSO_4$), halite ($NaCl$), and salts of potassium and magnesium. Several explanations for this absence are possible. Gypsum and anhydrite need oxygen to form and would not be expected in the early oxygen-poor environment. The more soluble minerals such as ordinary salt may have been formed but they have not survived to the present. They may have been dissolved or literally squeezed out of the rock formations in which they were deposited. Such materials are very plastic and easily displaced by earth movements. In any event, no significant evaporite formations are known in the Precambrian. That they do begin to appear in the Cambrian signifies some important change in the chemistry of the oceans.

A SUMMARY STATEMENT

The lengthy period that elapsed between the formation of the earth and the appearance of abundant fossils about 600 million years ago is known as the *Precambrian*. It may also be referred to as the *Cryptozoic* ("hidden life"), in contrast to subsequent time, which is called the *Phanerozoic* ("visible life").

Fossils are of little use in subdividing the 3.6 billion years of the Precambrian interval. Currently the best means of classification is based on geologic cycles. A typical cycle involves the following: (1) deposition of great volumes of underwater sediment in downsinking belts called geosynclines; (2) compression of sediments with resulting folding and emergence; and (3) erosion of the emergent rocks and stabilization of the belt as part of a continent. Details vary, and the type of sediment depends on local sources. The magnitude of intrusive and extrusive igneous contributions ranges from small to very great. Affected areas can be large mountain systems such as the Rockies or Appalachians. Duration of a geosynclinal cycle ranges between 300 million to 500 million years.

All the continents have a Precambrian *shield* area made up of granitic intrusions and downfolded masses of altered sediments that are considered to be the remnants of geosynclinal fillings. There is good evidence that the shield areas were in close proximity during their early history, as would be expected according to the concept of continental drift.

A typical shield may show three to six geological cycles. In North America, for example, the oldest portion has a central position extending from Laborador to Wyoming. Following outward are four successively younger belts. Important Precambrian areas are the following: the Canadian Shield in North America, the Baltic Shield in Europe, the Angara Shield in Asia, the Australian Shield, the Indian Shield, and the Guiana-Brazil Shield of South America. Africa is dominantly Precambrian and has a complex history.

Most shields are rich in ores of gold, silver, nickel, copper, uranium, lead, and zinc. Iron ore is especially characteristic of the Late Precambrian. Toward the closing stages of the interval, red, iron-bearing sediments appear, and widespread evidence of glaciation characterizes the latest Precambrian of all the large landmasses.

FOR ADDITIONAL READING

Clark, T. H., and C. W. Stern, *The Geologic Evolution of North America.* New York: Ronald Press, 1969.

Cloud, Preston, ed., *Adventures in Earth History.* San Francisco: Freeman, 1970.

Engle, A. E. J., "Geological Evolution of North America." *Science,* Vol. 140, 1963, 143–152.

Kay, Marshall, *North American Geosynclines.* Geological Society of America, Memoir 48, 1951.

Rankama, Kalervo, ed., *The Geologic Systems: The Precambrian.* New York: Interscience Publishers. Series includes: Vol. 1, Denmark, Norway, Sweden, and Finland (1963); Vol. 2, Spitsbergen and Bjoroya, British Isles, Greenland, and Canada (1965); Vol. 3, India, Ceylon, Seychelles Archipelago, Madagascar, and the Congo, Rwanda and Burundi (1967); Vol. 4, Southeastern United States, Southcentral United States, Northwestern United States, Mexico (1970).

THE BEGINNING OF LIFE ON EARTH—
EARLIEST FOSSIL EVIDENCE

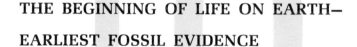

The problem of when and how life appeared on earth is currently receiving a great deal of attention. As might be expected, the problem is complex, and so scientists in many different fields have become involved. Astronomers, geologists, chemists, biologists, and others have all made significant contributions. Clues are being sought at tremendous expense by means of current space explorations, and at the same time earth-based astronomers continue to delve into matter and energy relations in near and remote regions

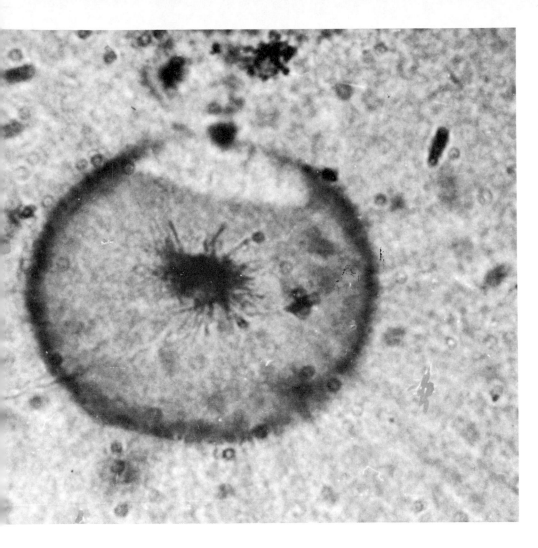

of the universe. The techniques of chemistry and biology have combined in a number of new fields, such as biochemistry and molecular biology, that focus on the composition and activities of living things. Similarly, there are expanding fields of study under the headings of biophysics and even bioengineering. With the aid of new instruments, including electron microscopes capable of enlargements up to 100,000 diameters, it is now possible to view small subcellular entities as an aid in understanding the basic structure and function of living matter. However, because of obvious limitations we must confine the present discussion of the appearance of life to those aspects which most concern geology.

What Is Life?

Life must be defined. On first thought it would appear easy to distinguish the living from the nonliving. We say that some organized aggregations of matter are alive because they possess certain characteristics, among which are the following: (1) *internal chemical activity* by which nutrients are used for growth, repair, and generation of energy; (2) the ability to produce new individuals like themselves, in other words, the power of *reproduction;* (3) the capacity to *respond to outside stimuli;* and (4) spontaneous movement of some sort (*locomotion*). One or the other of these characteristics may be found in certain nonliving entities, but the concurrence of all four generally indicates a living organism. But these criteria are not easily applied in all cases; many entities, usually very small, are found at the borderline of life and nonlife, and it is among these beings that we naturally look for clues to how life began.

Is There an Ultimate Origin?

Scientists are not likely to find out exactly when and where life first appeared on earth. No paleontologist seriously expects to discover fossils of the earth's first living inhabitants; they were surely much too small and fragile to leave recognizable remains. Aside from this, there is the strong probability that the origin of life may not have taken place on earth; perhaps life began in interplanetary or interstellar space before the formation of the earth or on remote earthlike planets at many times and places.

The idea that life is not confined to this earth is widely accepted by scientists. Based on our present knowledge, it is easier to believe that life is universal than to believe that it is not. If we accept this thinking, it is evidently best to refer to the *appearance* of life on earth, rather than to the *origin* of life on earth.

Recent studies of interstellar matter have provided data that must be considered in connection with the origin and prevalence of life in the universe. Not only simple chemical elements but also simple compounds such as the hydroxyl radical (OH), ammonia vapor (NH_3), water vapor

(H_2O), carbon monoxide (CO), hydrogen cyanide (HCN), and formaldehyde (CH_2O) have been discovered. The compounds are relatively abundant and occur in cloudlike concentrations comparable in mass to the sun. Needless to say the presence in space of abundant molecules that are associated with life suggests that the elemental beginnings of living things might be going on. In this connection, it is thought that the planet Jupiter may be characterized by conditions similar to those prevailing on earth during its earlier stages. Jupiter may in fact be an immense natural laboratory where the synthesis of molecules basic or essential to life might currently be taking place.

If life is widespread, even universal, a natural question is how it came to be thus distributed. Either it arose independently wherever it is found or it has spread from one or a few places of origin throughout the universe. The crucial question here is whether or not living material can travel across interplanetary or interstellar space. We shall omit from consideration the real or imagined feats of modern space science and of science fiction and question whether very simple organisms or organic molecules could survive a haphazard and unprotected journey through the perils of space. Astronomers and biologists feel that certain primitive life forms or the key molecules necessary to initiate life may have arrived in meteorites. As was mentioned in an earlier chapter, certain types of meteorites, known as carbonaceous chondrites, contain a variety of carbon-rich compounds. Some investigators claim that these compounds are aggregated into specific shapes like simple organisms. A chondritic meteorite that fell in southern Australia in September 1969 contains amino acids in considerable amount and variety. This suggests but does not prove the existence of extraterrestrial life.

The theory that organic molecules or living things could be widespread in space received a setback when no traces of such material were found on the moon. However, conditions in the

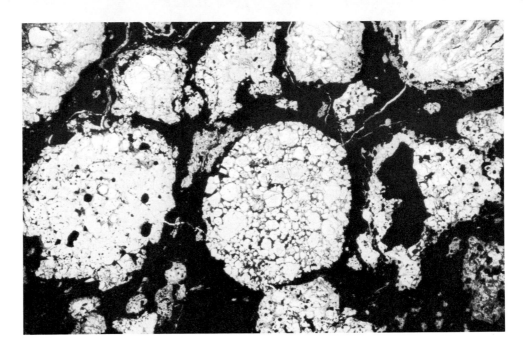

Figure 10.1　Enlarged view of a carbonaceous chondrite, a type of meteorite rich in asphaltlike material. The rounded bodies, known as chondrules, are in the carbon-rich matrix and measure about 1 millimeter in diameter. (Smithsonian Institution.)

distant past could have been more favorable for "seeding" the moon and earth than at present. The concept of panspermia, that life is present throughout the universe, certainly cannot be discounted as groundless.

Prerequisites of Life

We usually refer to earth life as "life as we know it," thus implying that there may be other forms we do not know about. Earth life is essentially based on water and the chemical reactions of carbon within the temperature range of from 0° to 100°C. In spite of much imaginative science fiction and theoretical chemistry, there seems to be very little possibility for any other workable life systems. This probability is so strong that the search for life beyond the earth is not really for life as such but for earthlike planets where life as we know it might exist. Just how earthlike a planet need be to support life is not positively known, but there are certain standards or limits that are understandably restrictive.

If a planet is to be able to support life, it must measure up favorably on three basic criteria: *star*, or *sun*, *size*, *planet size*, and *temperature*. Its sun must be warm enough and must have a sufficiently long life-span. Stars ranging in mass from about one-half to one and one-half that of the sun would fulfill these requirements. The energy of such stars is maintained by hydrogen conversion, and their expectable term of existence is from 5 to 30 billion years. If our evaluation of the history of the earth is correct, a period of at least 3 billion years is required for life to emerge and evolve to the level of man. Stars only slightly larger than the sun have life-spans that are too short to admit of evolution to this stage. For example, a star having twice the mass of the sun would have an effective life-span of only about 2 billion years.

The criterion of planet size is important be-

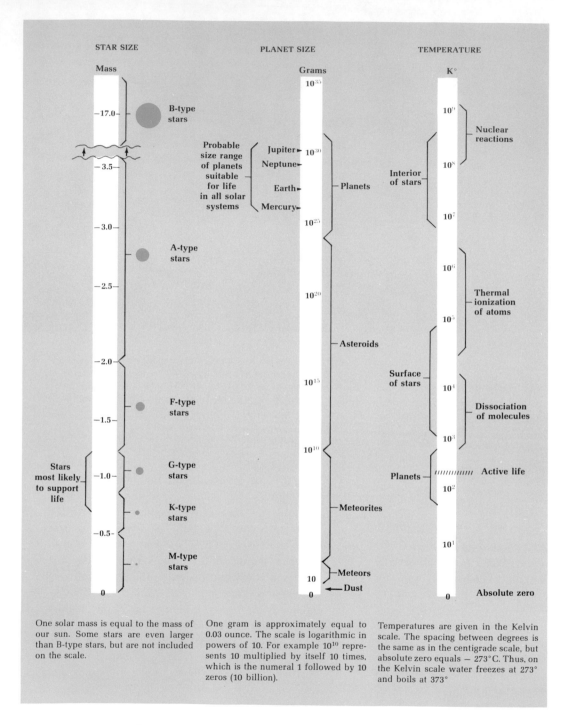

STAR SIZE

Mass

—17.0— B-type stars

—3.5—

—3.0— A-type stars

—2.5—

—2.0—

—1.5— F-type stars

Stars most likely to support life —1.0— G-type stars

—0.5— K-type stars

0 M-type stars

PLANET SIZE

Grams

10^{35}

Probable size range of planets suitable for life in all solar systems

Jupiter ▶ 10^{30}
Neptune ▶
Earth ▶
Mercury ▶ Planets

10^{25}

10^{20} Asteroids

10^{15}

10^{10} Meteorites

10 Meteors
0 ◀ Dust

TEMPERATURE

K°

10^{9} Nuclear reactions

10^{8}

Interior of stars 10^{7}

10^{6} Thermal ionization of atoms

10^{5}

Surface of stars 10^{4} Dissociation of molecules

10^{3}

Planets Active life

10^{2}

10^{1}

0 Absolute zero

One solar mass is equal to the mass of our sun. Some stars are even larger than B-type stars, but are not included on the scale.

One gram is approximately equal to 0.03 ounce. The scale is logarithmic in powers of 10. For example 10^{10} represents 10 multiplied by itself 10 times, which is the numeral 1 followed by 10 zeros (10 billion).

Temperatures are given in the Kelvin scale. The spacing between degrees is the same as in the centigrade scale, but absolute zero equals — 273°C. Thus, on the Kelvin scale water freezes at 273° and boils at 373°

Figure 10.2   Three requirements for life. (Adapted from Walter Sullivan, We Are Not Alone, rev. ed., McGraw-Hill Book Company, 1966.)

cause it affects the gravity of the planet. A habitable planet that would be suitable for life as we know it should be large enough, and thus have sufficient gravity, to retain an oxygen atmosphere and small enough to allow hydrogen to escape. On the basis of this, we are probably safe in saying that Jupiter is too large and Mercury is too small. The fate of the water molecule is critical. The moon is a dramatic example of an essentially waterless body, Mars has very little water, and Mercury probably none. Of course all earthly organisms are adapted to the earth's gravity, and we do not really know how much (or how little) gravity a watery, carbon-rich organism could withstand. In any event, existence would not be possible for anything light enough to fly into space or so heavy that it would be immobilized or crushed by its own weight.

The range of temperature found in the universe is a subject about which a great deal is known. At one extreme is the absolute zero of space, a total absence of heat. On the other hand, the interior temperature of stars can reach 1,000,000,000°C. As was mentioned earlier, life can exist only within a very narrow temperature range. On a cosmic scale this zone is rare and would occupy a section only 0.1 centimeter long on a thermometer that was about 100 kilometers long. But the search for habitable planets is surely not hopeless. As anyone knows, a person can be comfortable either close to a small fire or farther away from a big one. In terms of life, this means we might find a habitable planet close to a warm star or farther away from a hot one.

ESSENTIAL CHEMICAL ELEMENTS

All the chemical elements essential to life must have been present on earth before life could begin. Regardless of any intangible constituents, every organism seems to require at least twenty chemical elements and perhaps traces of several others. Hydrogen, carbon, nitrogen, oxygen, phosphorous, and sulfur make up at least 95 percent of all protoplasm; other less prominent constituents are potassium, sodium, magnesium, calcium, and chlorine. Most organisms also require traces of iron, copper, zinc, manganese, molybdenum, boron, fluorine, silicon, and iodine. A certain amount of substitution among the trace elements appears possible, but, in general, any deficiency of those mentioned is serious.

The mere presence of the essential elements is not enough to ensure the well-being of life—all those named are present on the moon but not in right proportions. On earth those elements that constitute water and the atmosphere are notably abundant and are particularly important because they help to circulate and distribute the other elements. Without the assistance of currents of wind and water there is little possibility that the essential elements would be available to life in all its locations. The effects of the wind help to distribute particles of matter to all parts of the earth. Every part of the ocean is stirred by currents that bring oxygen and other dissolved or suspended material to marine organisms. Running surface water gathers particles from diverse regions and builds up the fertile soils of deltas and floodplains. Glacial ice also picks up and distributes rock material and is partly responsible for the fertility of such regions as the central United States.

Even the interior of the earth is in a state of flux and movement. Igneous activities such as the deep-seated melting of rock in one place and its emergence as lava through volcanoes are incidental manifestations of inward unrest. All the active geologic processes of the earth that are currently in operation have apparently been going on for several billion years and must be counted of great significance in the maintenance of life.

WATER

The elements essential to life are universally present in seawater. Water makes up 80 to 90 percent of all living matter and is the most important component of protoplasm. Because water is the universal solvent and circulates freely over the entire globe, it is able to take all elements into suspension or solution and carry them from areas in which they are abundant to regions where they are absent or rare.

Water and its dissolved constituents must have been present not only when life appeared but also throughout all subsequent periods of time. There is good evidence that seawater accumulated rather rapidly during the early stages of the earth's history and has remained almost constant in quantity for the past 3.5 billion years. Have the oceans always been salty? Although the old view that the original ocean was fresh and that it gradually became salty over the ages has given way to the current belief that the ocean has always been as salty as it is now, there is evidence of very significant fluctuations and variations in both the composition and the physical properties of seawater. At certain favorable periods in the past, great amounts of salt and other constituents have been moved from the ocean into sedimentary beds on land. Extensive temperature changes have also been brought about in the oceans during past ice ages.

Figure 10.3 Eruption of the volcano Kilauea, Hawaii, 1924. The great clouds emitted from the volcano consist largely of steam and water vapor. (U.S. Geological Survey.)

Until recently the continents and ocean basins were regarded as permanent unchanging features of surface topography. Under this arrangement, many geologic processes were thought to have only one-way effects, the most important of which was the permanent loss from the land to the deep oceans of vast volumes of sediment including elements such as phosphorus and potassium, elements essential to land life. Today, the theory of sea-floor spreading and plate tectonics has led to a totally different concept. The essential feature of plate tectonics is that all ocean bottoms are moving in such a way that they impinge against the continents. Thus, oceanic oozes and deep-water deposits are scraped off the ocean bottom and welded to the land. In this way, thousands of cubic kilometers of sediments, particularly calcareous and siliceous oozes and volcanic products, have been added or returned to the continental margins.

In this process of continental and oceanic collision, a great deal of material is also forced downward, melts, and is converted into igneous material. When this molten material reaches the surface, it is piled up in volcanic mountains or lava flows. In due time it will be eroded and will become sediment and soil on which land life can subsist.

GEOLOGIC CYCLES— CARBON AND HYDROCARBONS

Because many essential substances are in short supply, they must be used over and over again. They are released on the death of one generation for use in a succeeding one. Thus numerous cycles operate in nature whereby the same elements are used and reused by living things and not permanently locked up in the rocks or indestructible compounds. Some of the most important of these cycles involve carbon and hydrocarbons.

Carbon is available from many sources and becomes an essential part not only of living things but also of the common minerals that constitute limestone, chalk, and dolomite. The carbon cycle is complex and important. Weathering, animal respiration, and volcanic action prevent carbon from being locked up in compounds that are unavailable to living things. Although carbon makes up only 0.09 percent of the crust of the earth, it constitutes about 18 percent of the total mass of plant and animal tissues. The known compounds of carbon exceed those of all other elements combined and are the basis of the science of organic chemistry. The compounds of carbon and hydrogen—the *hydrocarbons*—are important almost everywhere. Coal, oil, and combustible natural gas, the important natural fuels, are classed as hydrocarbons.

Hydrocarbons are abundant in the atmospheres of Jupiter, Saturn, and Uranus. At the low temperatures prevailing on these outer planets, such compounds as ethane, ethylene, and acetylene are in liquid form. Although the atmosphere of the earth now contains only small traces of the hydrocarbons, they may have been major constituents in earlier stages. Hydrocarbons are easily destroyed by oxygen, which converts methane to carbon dioxide.

The Problem of Spontaneous Generation

The point of the foregoing discussion is that conditions favorable to life probably came into existence at a remote period in the past. This raises a most important question: Will life appear spontaneously if all known favorable factors— essential chemical elements, water, energy sources, tolerable pressure and temperature —prevail in a planetary environment? The important word in this question is "spontaneously," and the subject of spontaneous generation is a key concept that must be treated at some length.

The subject of spontaneous generation has had a curious history. The ancients believed that living organisms could arise spontaneously and fully formed from other organisms of entirely

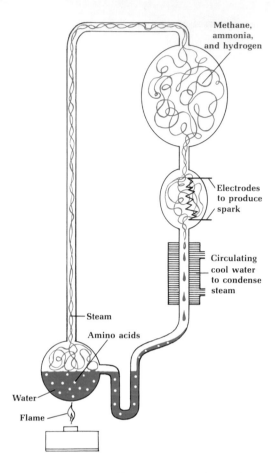

Methane, ammonia, and hydrogen

Electrodes to produce spark

Circulating cool water to condense steam

Steam

Amino acids

Water

Flame

Figure 10.4   S. L. Miller's apparatus for producing amino acids under primitive earth conditions. (Redrawn from Miller.)

different nature or even from nonliving matter. They noticed that tadpoles became frogs and that caterpillars turned into butterflies, and so they felt that it was possible for worms to arise from horsehairs, geese from barnacles, and sparrows from falling leaves. The fact that maggots appear in decaying matter and that mice emerge from old rags seemed proof enough that even dead material has power to bring forth life. The best minds of the Middle Ages fully supported these ideas.

Louis Pasteur demonstrated that spontaneous generation of this type is impossible. Pasteur's ingenious experiments and clear-cut results are still hailed as triumphs of the scientific method over mysticism and superstition. His conclusion that no type of spontaneous generation is possible has proved to be shortsighted, premature, and somewhat of an embarrassment to succeeding scientists. If we discount the theory of spontaneous generation entirely, then the only alternative is supernatural creation. There are no other possibilities. Science therefore finds itself reexamining the possibility of spontaneous generation.

PRELIFE CHEMICAL EVOLUTION

It is impossible to duplicate the past exactly, but scientists have been able to devise experiments that match some of the conditions that once existed. Thus, we may attempt to determine experimentally just how far the synthesis of carbon-rich molecules could have proceeded under prelife conditions. We know that cells consist of complex molecules that differ from nonliving material only in degree of complexity and not in the kind of material contained. On the basis of present evidence, we know that living material apparently can originate only from previous living material, but there is a possibility that it might rarely arise by some other method. Science is not seeking the spontaneous generation of fully formed organisms such as man but only of the simplest of living entities. Evolution appears fully capable of all further developments leading to complex organisms.

Some relatively simple but very significant experiments have produced surprising results. In 1953, S. L. Miller, then a graduate student at the University of Chicago, devised an apparatus for circulating steam through a mixture of ammonia, methane, and hydrogen (Figure 10.4). This mixture was then subjected to a high-energy electrical discharge. He discovered that a number of amino acids, including glycine and adenine, could be synthesized by this means. Adenine, a basic unit of the genetic material DNA and RNA, is very resistant to destruction once it is formed. It has also been created by bombarding a mixture of gases with high energy radiation similar to that produced by natural radioactive material. Complex molecules have been formed in hot-water solutions of simple basic materials, and synthesis of significant compounds has also been achieved by ultrahigh sound waves, by heating and drying, and by ultraviolet radiation.

The essential point in a geologic context is that these experiments start with materials and conditions that were prevalent on the primitive earth. Natural lightning supplies electrical discharges, thermal springs provide steam and water at all temperatures, strong radiation arises from natural minerals and arrives in the form of cosmic rays from space. Ultraviolet and other forms of nonparticulate radiation pour in from the sun and other extraterrestrial sources, shock waves

and pressure phenomena are provided by faulting and by meteoritic impacts.

If we assume that moderately complex compounds are present along with energy sources sufficiently strong to synthesize them into still more complex forms, will life emerge spontaneously? Other factors, favorable and unfavorable, need to be considered before an answer can be given. One of these is the nature of the atmosphere. We have seen that the early atmosphere, the one that prevailed when life supposedly appeared, was relatively poor in oxygen. In the absence of this chemically powerful element, a great variety of carbon compounds could have formed and remained in existence without being

Figure 10.5   Hot spring, Yellowstone National Park, Wyoming. The first appearance of life on earth may have been in a localized pool such as this where the right combination of chemical elements and energy sources happened to coincide. (Union Pacific Railroad.)

destroyed by oxidation. In a strong reducing (oxygen-poor) environment, highly organized amino acids, sugars, carbohydrates, and other carbon compounds could have accumulated in water solutions and would have constituted what is popularly called the primeval "soup." Under varied situations that might easily be imagined, portions of this solution might be isolated and subjected to temperature changes, evaporation, and exposure to radiation. Segregated portions might have dried to the consistency of slime or ooze, a condition that is thought by some to be even more favorable for the generation of life than dilute solutions.

The world into which life emerged was different from the present earth in yet another way—it was free from predators. There were no organisms of any sort relentlessly seeking food wherever it could be found. Any energy-rich organic material that appeared might remain in existence until it was broken up by natural means and would thus, for some period of time at least, be available for chemical reactions. When radically new forms of life enter previously unoccupied environments, they are of course temporarily free from intense competition.

That the original environment of life was different from the one now prevailing provides an answer to the anti-evolutionists' contention that if the spontaneous appearance of life were possible once, it should be going on today. Life, it seems, has a way of burning its bridges behind it. Evolution is a one-way process and never repeats itself.

Undoubtedly the greatest obstacle to understanding or accepting a spontaneous synthesis of living material is the element of chance. Most people are unwilling to accept the origin of life as an accident, for this would mean that their own existence as human beings might also be accidental. Such thinking is obviously emotional and nonscientific. However, for reasons that are entirely different many scientists also reject the idea of spontaneous biogenesis. They point out that the statistical improbability of all the essential elements coming into proximity and uniting to form a living thing are so small that the event would never take place in the time and space available.

The problem of synthesizing one simple protein of about 300 amino acids has been cited. It is assumed that the protein must be synthesized by a gene with at least 1,000 nucleotides (the basic chemical units of which genes are built) in its chain, the nucleotides being the four basic ones—adenine, thymine, guanine, and cytosine. A chain of 1,000 nucleotides made of the four basic units might exist in any of $4^{1,000}$ ways, but only one will form the protein being sought. The chances that the correct sequence would be achieved by simple random combination is said to be so small that it would not occur during billions of years on billions of planets, each covered by a blanket of concentrated watery solution of the necessary amino acids.

While this thinking carries a ring of mathematical finality, it too is erroneous. We should be reminded that the process need not be accomplished in one step starting with simple elements or even with simple compounds. It may start with very large molecules suitable for a role in a larger entity. Just as automobiles are made by assembling units already made up of many parts, so might the origin of the first living things have been on a piecemeal basis. Some biologists assert that there is good reason to suppose that the smallest duplicating entity need be only a small strand of RNA.

Also, it is dangerous to say that an unlikely or highly improbable event can never take place. According to one view, any possible event, no matter how improbable, must happen if enough time is allowed. Furthermore, only *one* successful synthesis is needed; it matters not how many unsuccessful tries are made. "Chance" is a word of many meanings as anyone can learn by consulting a dictionary. Chance may be interpreted as opportunity, probability (hence statistical certainty), possibility, fate, luck, or accident. To say a thing is determined by chance is not the same as saying it is accidental, lucky, or unplanned. To assert that there is no chance in the origin of the human body and personality is to show a profound ignorance of genetics and the reproductive process.

Experimental work aimed at duplicating what took place at the beginning of life has yielded encouraging results. The general public is intensely interested in what is going on, and any steps leading to "life in the test tube" receive wide publicity. Some scientists are convinced that life has already been artificially synthesized. Many, if not most, biologists believe it can and will be. The artificial synthesis of a simple living entity would be a triumph for science, but this event would be far different from the creation of a fully operating higher animal such as man. Until laboratory synthesis is achieved, we must accept the appearance of life as a rare event of the greatest significance that "happened" in remote times under conditions not yet fully understood.

ORGANIZING THE CELL

Once a self-duplicating molecule was in existence the entire subsequent history of the earth was profoundly affected. The stores of energy-rich chemicals that had accumulated in the primeval soup could be utilized in an orderly meaningful way. Loose amino acids that were attracted to the first reproducible molecule were built into an exact replica of it. The replica split off, and each half attracted new components and began a potentially endless lineage of descendants. The process has been likened to a fire or an explosion because it could spread rapidly until all the material on which it fed was consumed. An ultimate shortage of raw material was inevitable, and this led to the first competition among living organisms. Those molecules that were able to capture more of the available food supply were more likely to survive and reproduce.

At this stage, we may suppose that *enzymes* began to be incorporated into the operating molecules. Enzymes are types of proteins that speed up chemical reactions or enable them to proceed at lower temperatures than they otherwise could. Hundreds of different types of enzymes have been synthesized and used by living things in the process of evolution. The benefits are obvious— from the beginning those organisms that were able to secure beneficial enzymes in one way or another were favored over their less successful contemporaries. The most successful primitive forms became more than simple strands of DNA-like material.

With the incorporation of more and more enzymes, living things took on a cellular structure, and living material came to be enclosed within a protective cell wall. It is beyond the scope of this book to discuss in detail the various components of cells, and so we shall be content to say merely that a cell is the smallest structural aggregate of living matter capable of functioning as an independent unit. There are two basic types of cells: *procaryotic* and *eucaryotic*. Procaryotic cells have a very simple structure with no nucleus and are found only in bacteria and blue-green algae. They are generally characteristic of unicellular organisms, although chains and clumps of cells are possible. Eucaryotic cells are found in all animals and plants and are much more complex. They have a nucleus and are composed of several elements that perform various functions.

Advances from simple to complex cell structure apparently took place during prefossil time, and so we obviously have no evidence of this development. Biologists feel that some of the components of eucaryotic cells may represent once free-living entities that entered into the structure of a cell for their own and the cell's benefit. Thus, chloroplasts, one of the components of eucaryotic plant cells, benefit their host by stimulating the production of energy-rich compounds by means of photosynthesis and receive protection and other benefits in return.

Some scientists consider the advance from procaryotic to eucaryotic cells to be the greatest evolutionary advance of all time. The time-worn phrase "amoeba to man" is an obviously misleading description of evolution. The amoeba is tremendously complex and can exist only because of a great deal of prior evolution.

The struggle for survival in the final analysis embraces all means and methods to obtain energy or to escape being converted into it. If our reconstruction of the earliest living things is correct, we must conclude that they led a passive existence, floating aimlessly in the seas among a variety of carbon-rich but lifeless molecules. By means of purely random movements, the living organisms occasionally contacted the nonliving molecules. The two entities became attached by chemical bonds and there was a consequent in-

Figure 10.6   Enlarged model of a cell showing the complex structures where life's basic functions originate. (Photo by Ezra Stoller, the Upjohn Co.)

crease in size and duplication of structure. This purely passive method of keeping alive was adequate for a time and continued as long as there were sufficient stores of suitable carbon compounds in the environment. Only when the original energy sources were depleted or when local living space became crowded did competition begin. At this stage Darwinian evolution took over.

If it were not for the process of photosynthesis, all life might have disappeared from the earth at an early stage of evolution. Photosynthesis was the only method of obtaining life-supporting energy that was capable of operating on a sustained basis because it utilized the one inexhaustible energy source—solar radiation.

In green (chlorophyll-bearing) plants, photosynthesis takes place when sunlight is utilized to convert carbon dioxide and water into energy-rich carbon compounds, with the release of oxygen as a by-product. The simplified reaction in chemical terms is as follows:

$$6CO_2 + 12H_2O \xrightarrow[\text{radiation}]{\text{solar}} C_6H_{12}O_6 + 6O_2 + 6H_2O$$

When properly combined, as in the respiration, digestion, and assimilation of animals, the carbon compounds and the oxygen yield the energy that supports animal metabolism. Organisms that were most effective in combining oxygen and food were favored in the struggle to survive and were able to displace their less adaptable associates.

Although photosynthesis became the basic means of obtaining energy for practically all plant and animal species, it is not universally used by all organisms. A large number of simple organisms live in the complete absence of light and degradable organic material. These so-called "chemical eaters" obtain energy from common minerals such as gypsum or hematite and appear to be dead-ended insofar as further evolution is concerned.

Although the basic raw materials of photosynthesis are widely distributed, other elements

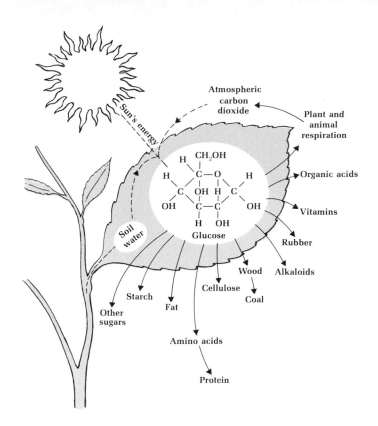

Figure 10.7 *The green plant uses the energy of the sun to produce many valuable products. (Adapted from A. Galston,* The Life of the Green Plant, *2nd ed., Prentice-Hall, Inc., 1964.)*

that are needed in small quantities are not necessarily available in all regions. Only if these *trace elements* are spread about by geologic processes can photosynthesis operate with maximum efficiency. The unavailability of raw materials for photosynthesis limited the spread not only of plant life but also of animal life that used the plants as a source of food. Animals cannot spread beyond their plant resources. We may state a general principle: *No life form can spread unless it is preceded by a suitable food (energy) source.* Present-day life is not as restricted by this rule as was primitive life; basic food supplies suitable to almost any type of organism are now available

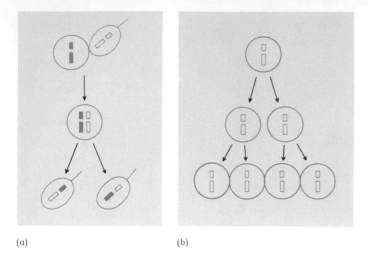

(a)   (b)

Figure 10.8   Sexual reproduction (a) compared with asexual reproduction (b). Individuals produced by the union of male and female display characteristics from both parents, and the combinations in a large population provide new variations in great numbers. Asexual reproduction provides no such variety; all the offspring resemble the parent from which they were derived. The small bars in this illustration represent the chromosomes of the reproductive cells.

almost everywhere. Man can exist off the land almost anywhere because of the contributions of previous inhabitants.

Other Advances of Early Life

The emphasis of modern biology has shifted to the molecular levels of organization away from a former concentration on complete organisms. What has been learned serves to emphasize the fundamental importance of various subcellular entities. The new knowledge has naturally raised questions as to how these smaller components may have arisen and evolved. Answers to the question of origin cannot be obtained by direct observation and must depend on theoretical considerations and study of presently existing organisms. In any event it is obvious that most of the structures and functions below the level of the cell must have been attained before multicellular organisms appeared and hence before abundant fossils could exist.

Some of the advances of early life have been mentioned—the appearance of replicating molecules with a genetic code, the acquisition of enzymes, organization of living material into cells and the development of energy-gathering photosynthetic systems. Several other advances, associated with the higher, more successful forms of life, are very important and should be mentioned at this point.

REPRODUCTION

Reproduction is a basic function of living things. Duplication of an organism is accomplished either sexually or asexually. Sexual reproduction involves some degree of union of cells and exchange of parental genetic material; asexual reproduction involves simple cell division with no union or exchange of genetic material.

The importance of sexual reproduction arises from distinct advantages that it gives to any species. It provides a spectrum of variability that helps an organism to adapt to changing conditions. An asexually produced organism receives essentially the same inheritance as its predecessor in a chainlike line. Beneficial mutations may occur, but they are not shared with others of the same species. A sexually propagated organism also inherits its parent's characteristics, but in the case of these organisms any new and beneficial mutation is shared immediately with others in the same group and may soon spread throughout the entire population. The sexual process provides relatively more varied raw material on which natural selection may operate—in a word, it speeds up the evolutionary process.

Primitive beginnings of sexual reproduction with limited exchange of genetic material are evident among the lowest procaryotic algal and bacterial organisms. Sexual reproduction apparently had its beginning early in the history of life and must be reckoned as one of the most significant advances of all time.

DEATH AS A BENEFICIAL DEVELOPMENT

Asexually produced unicellular organisms are, in a sense, immortal. They may die by accident or starvation, but as long as conditions are favorable the individual subdivides and loses identity but the constituent living material does not die. In sexually produced organisms, all body cells perish at death, and only a minute quantity of genetic material can be said to live on with the next generation.

What is the advantage of death? Consider the alternative—eternal existence. If any species were immune from natural deterioration and death, it would fill the world in a relatively short time. All living space would be occupied. Food would be exhausted and essential elements would be tied up permanently in living bodies. Starvation, in the broad sense, is the price to be paid for eternal life in a world where space and resources are limited. The important fact is that without death there could be no further progress. Whatever level an organism reached would be fixed forever. As it is, death removes the less perfect and makes way for something better. Implications are that death became an important factor with the development of sexual reproduction rather early in the history of the world.

Beginning of the Fossil Record

It may be helpful to recall that fossils are defined as the remains or evidences of an ancient organism preserved in materials of the earth's crust. This definition includes more than objects with definite form and shape. Disaggregated particles of carbonaceous material as well as chemical residues of materials known to have been synthesized by living things are genuine evidences of life and may be considered in the broad sense as fossil material. Petroleum is a mixture of organic liquid hydrocarbons and thus indicates the prior existence of life. Systematic search has been made in sediments for organic molecules such as phytane and pristane, which are related to chlorophyll. Starting with relatively young oil-rich rocks in which recognizable fossils are abundant, the examination has proceeded backward to sedimentary rocks about 3.6 billion years old. So far, rocks of all ages have yielded molecular fossils that suggest the presence of photosynthetic organisms at a very early date. At no place in a long series of analyses has a point been reached where the *residues* or products of living things are replaced by the more simple *predecessors* of life.

OLDEST FOSSIL ORGANISMS

Well-preserved remains of minute organisms have been found in sedimentary rocks billions of years old. Such preservation is possible mainly because of the physical properties of the mineral called *chert*, which in pure form is silicon dioxide (SiO_2). This material precipitates rapidly as nodules or thin beds in seawater, hardens quickly, and becomes a very durable, noncompressible rock; therefore, this mineral is ideal for the preservation of simple, semirigid organisms such as bacteria and fungi. When geologists suspect that a sample of chert contains microscopic remains, they cut it into thin sections and study it under high magnification.

AFRICAN DISCOVERIES

The oldest known, structurally preserved objects that have been interpreted as organisms are from carbonaceous chert of the Onverwacht Series of the Barberton-Badplace region of the eastern Transvaal, South Africa. An igneous intrusion that cuts the chert beds is dated at 3.2 to 3.3 billion years, a minimum date for the age of the remains. There are algalike spheroids, filamentous structures, and carbonaceous fragments. That these are true fossils is indicated by their close association with chemical compounds not

(a)

(b)

known to originate independently of living things.

In the same sequence of rocks but several hundred kilometers distant are found the earliest known *stromatolites*. As mentioned in Chapter 9, these are distinctly layered accumulations of calcium carbonate having a rounded, cabbagelike or branching fingerlike form. These are not actual

Figure 10.9 Oldest suspected fossil organisms. (a) A thin section of an algalike form. (b) A similar body freed from the matrix by etching in acid. These are from the 3.2-billion-year-old Onverwacht Series of South Africa. Scale is given in microns (1 micron = 0.001 millimeter). (Courtesy of Albert E. J. Engle.)

Figure 10.10 (a) Oldest known stromatolites, from the Bulawayan Group of southwestern Rhodesia. Age is more than 2.6 billion years—probably between 2.8 and 3.1 billion. Portion shown is about 9 centimeters long. (b) Stromatolites from the Precambrian rocks of Glacier National Park, Flathead County, Montana. The photo shows the weathered upper surface of a number of colonies. (a, NASA, photograph by K. A. Kvenvolden; b, U.S. Geological Survey.)

organisms, but rather the deposits laid down by algae. That they are organic is shown by their resemblance to deposits created by living algae, by the discovery of scattered algal bodies within them, and by the absence of any other known mode of formation. Stromatolites are found in the Early Precambrian but do not become common until the Middle Precambrian. By Late Precambrian they are abundant enough to serve as specific guide fossils for rocks in which they occur.

Another significant assemblage of primitive organisms has been discovered in the Fig Tree Series, which is about 1,000 meters above the Onverwacht and in the same region. The Fig Tree fossils include rodlike, spheroidal, and threadlike forms of unknown relationships.

THE GUNFLINT FLORA

A much more extensive assemblage, the first Precambrian microflora to be described, was reported from the Gunflint Iron Formation of Ontario, Canada, in 1954. The age as determined by radiometric means is about 2 billion years. The flora is diverse and includes threadlike, rodlike, and spheroidal species identical in form to present day bacteria and blue-green algae. There are also abundant star-shaped and umbrella-shaped species of unknown affinities. Eight new genera with a large number of species have been named from the Gunflint Formation. All the forms are regarded as procaryotic (no organized nucleus), and none seem to be animals. Extensive chemical analysis of the containing rocks has produced unmistakable hydrocarbons of organic origin as well as finely disseminated carbonaceous material.

OTHER PRECAMBRIAN FOSSILS

The oldest eucaryotic organisms (with organized nuclei) so far discovered are from the Beck Springs Dolomite of eastern California. The estimated age is 1.3 billion years. Another important

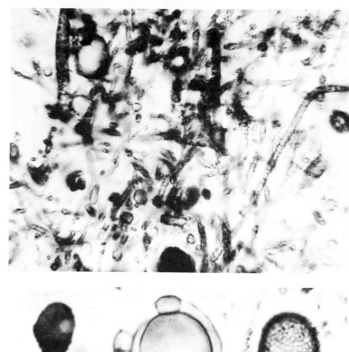

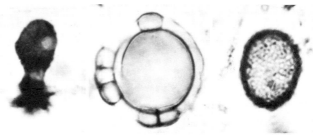

Figure 10.11 *Primitive algae and fungi from the Gunflint Chert near Schreiber, Ontario, Canada. All are highly magnified. The age is between 1.7 and 2.1 billion years. (Courtesy of Elso S. Barghoorn.)*

and especially well preserved assemblage of plants, found in the Late Precambrian Bitter Springs Formation of central Australia, has been determined to be approximately 1 billion years old. The fossils are preserved as organic residues in laminated black chert. Blue-green algae are abundant, and colonial bacteria, funguslike filamentous organisms, spheroidal green algae, and other cellular forms are also present. Fossil evidence indicates that the first eucaryotic organisms developed from certain procaryotic cells about 1.5 billion years ago.

The Bitter Springs cherts have yielded thirty

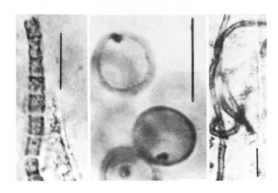

new species of plant microfossils. Over half of these are blue-green algae, some of which show no significant differences from living forms of the same families. No animal life has been recognized.

Life Creates the Second Atmosphere

A short digression is necessary at this point to comment on the interaction between early life and its surroundings, an interaction that appears to have resulted in significant changes in the total environment. It may be recalled that the atmosphere in which life first appeared was probably rich in methane, ammonia, hydrogen, and water, whereas the present atmosphere has great amounts of nitrogen and oxygen. What changes could have given rise to the highly oxygenated atmosphere essential to present-day organisms? Two chief processes are known that could produce free oxygen: (1) disassociation of water vapor and carbon dioxide by solar ultraviolet radiation and (2) photosynthesis. The first pro-

Figure 10.12   Microscopic plant fossils from the Late Precambrian Bitter Springs Formation, central Australia, approximately 1 billion years old. The fossils are preserved in chert, which can be cut into thin sections to reveal specimens such as these. Both spheroidal and filamentous forms are represented. The spheroidal forms (center) show a nucleus and indicate that the development of eucaryotic cells had taken place by this time. In each illustration the scale represents 10 microns. (Courtesy of Elso S. Barghoorn and J. W. Schopf.)

Figure 10.13   Some ways in which the atmosphere influences living things.

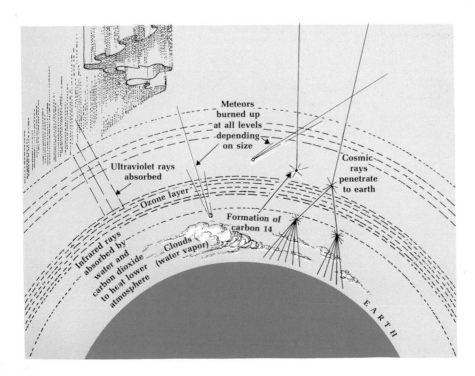

cess, properly called *photodissociation,* is not considered adequate to provide oxygen in the amounts necessary to create the observed effects. Photosynthesis, on the other hand, seems to be entirely capable of bringing about substantial changes in the atmosphere and in other earth materials with which it can react.

Release of oxygen by photosynthesis seems to have commenced over 3 billion years ago, and in a 2 billion year interval the process completely altered the chemistry of the earth. Oxygen reacts readily with other elements and compounds. The first oxygen probably reacted with iron dissolved in seawater and produced iron oxide. The Late Precambrian was the greatest known period of extensive iron ore production throughout the earth. Oxygen also reacts with ammonia to form nitrogen and water and it will make carbon dioxide out of methane. By such reactions the old atmosphere was changed, and surface minerals were reduced to their oxides. After these events, free molecular oxygen began to accumulate and became available for animal respiration. It has been suggested that animal life began in local areas of the ocean where algae were abundant and later spread almost everywhere when free oxygen was also widely available. This may be the reason why animal life diversified with apparent suddenness about 600 million years ago.

A second important product of excess oxygen was the ozone layer, which forms a shell around the earth about 25 kilometers above sea level. As we said earlier, ozone absorbs most of the sun's ultraviolet rays and prevents them from reaching the earth's surface. These rays are deadly to living cells—without the ozone layer, life could exist only under water, under rocks, or within thick shells. It is reasonable to suppose that life was capable of moving into shallow water and ultimately onto land when the ozone layer became increasingly effective as a shield. This may be the explanation as to why land plants and animals both came on land together about 350 million years ago.

ADVANCED FOSSILS

The earliest evidence of metazoan (many-celled) animals consists of trails and burrows. Important as these fossils are they are so imperfect that we are forced to refer to them only as traces of worms or wormlike organisms. Zoologists recognize as many as eleven phyla of living wormlike animals, but these are mostly soft-bodied and at best leave only elongate, winding trails and burrows. In spite of this poor record, much can be inferred regarding the earliest of animal fossils. Worms possess a number of very significant adaptations; they have varied means of locomotion, a head end containing a brain and rudimentary sense organs, bilateral symmetry, and a means of taking food into the body where it is digested and from which it is excreted. These are all rather advanced characteristics not possessed by the majority of living things. In explaining the evidence of highly evolved characteristics at an early geologic date we assume a long period of evolution in which these characteristics were produced and selected.

The most varied and distinctive trails and burrows so far discovered come from the Precambrian of Australia. These consist of sinuous trails, meandering, closely spaced "grazing" patterns, evenly spaced grooves or indentations, and rows of ovoid structures interpreted as fecal material. All seem to have been created by elongate, wormlike animals, some of which may belong to annelids, ancestral mollusks, or coelenterates. Specimens from the Torrowangee Group of New South Wales are so consistently well defined that seven species have been described. The rocks are thought to be Late Precambrian in age, within the range of 1 to 1.5 billion years old.

Much more satisfactory than trails and burrows are fossils showing complete shapes. A considerable number of these are now known from Precambrian rocks at widely scattered spots on the globe. The most important locality thus far

Figure 10.14   Unidentified fossil organisms from probable Late Precambrian rocks, south-eastern Newfoundland. Impressions only are preserved. Both the branching and circular impressions are tentatively regarded as coelenterates, distantly related to modern corals and jellyfish. (Courtesy of S. B. Misra.)

reported is the Ediacara Hills in southern Australia. Here the fossils occur as impressions in sandstone. Many hundreds of specimens have now been collected and have been classified into the following basic forms: rounded impressions with radiating grooves resembling jellyfish; stalk-like fronds, with grooved branches, similar to living sea pens; elongate, segmented, wormlike impressions, with a horseshoe-shaped head and about forty segments, that closely resemble annelids; flattened, nearly round, wormlike impressions; oval impressions, with T-shaped grooves, that seem to be unrelated to any living organisms; and curious circular impressions, with bent arms that radiate outward, also unlike any known living thing. Among these forms there are unquestionably representatives of two living phyla, the coelenterates and the annelids; arthropods are probably present also. A great variety of trails and burrows and possible plant remains are also found in the Ediacara region.

Another locality with fossils similar to the Ediacara fauna is in southern Newfoundland. Casts and molds of a variety of organisms are described in terms of outline as spindle-shaped, leaf-shaped, round to lobate, and dendritelike. Some of these are definitely coelenterates, but most cannot yet be assigned to specific zoological groups. The age of this locality is in doubt, but the preponderance of evidence indicates that it is Late Precambrian.

With the discovery of the abundant Australian and Newfoundland fossils, interest has been re-

Figure 10.15 Biological evolution was preceded by a long and complex chemical evolution that led to the formation of molecules of amino acids, sugars, and nucleic acids. These somehow acquired the organization and behavior of living systems, first as cells that derived their energy from the breakdown of chemical substances and later as cells that could use photosynthesis to harness the energy of the sun. The emergence of eucaryotic cells signaled the beginning of a period of rapid organic evolution. (After Barghoorn, 1971.)

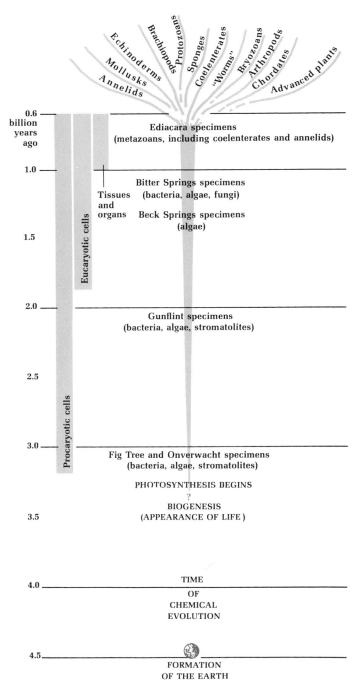

ABUNDANT WORLDWIDE FAUNAS

Echinoderms
Brachiopods
Protozoans
Sponges
Coelenterates
"Worms"
Bryozoans
Arthropods
Chordates
Advanced plants
Mollusks
Annelids

0.6 billion years ago

Ediacara specimens
(metazoans, including coelenterates and annelids)

1.0

Tissues and organs

Bitter Springs specimens
(bacteria, algae, fungi)

Beck Springs specimens
(algae)

Eucaryotic cells

1.5

2.0

Gunflint specimens
(bacteria, algae, stromatolites)

2.5

Procaryotic cells

3.0

Fig Tree and Onverwacht specimens
(bacteria, algae, stromatolites)

PHOTOSYNTHESIS BEGINS
?
BIOGENESIS
(APPEARANCE OF LIFE)

3.5

4.0

TIME
OF
CHEMICAL
EVOLUTION

4.5

FORMATION
OF THE EARTH

vived in previous less spectacular and hence un-appreciated finds. Similar fossils had been found in England and South Africa, but these were so few in number and so poorly preserved that scientists were hesitant to place much weight on them.

Although doubt has been expressed about the correct age assignments of the early metazoan fossils just described, it is generally agreed that they are Late Precambrian. Certainly they are distinctly different from the trilobite-dominated faunas of the Cambrian. Evidence indicates that diversification into modern phyla was well under way, and further discoveries can be expected. Of significance is the fact that most, if not all, Precambrian animals so far discovered are soft-bodied and shell-less, a fact that may account for the general scarcity of remains. The abrupt changes that are recorded at the beginning of the Cambrian will be discussed in the next chapter.

A SUMMARY STATEMENT

That life may exist on planets other than earth is admitted as probable by many scientists. A few meteorites have been found to contain carbonaceous residues, amino acids, and even minute objects resembling simple organisms. This and other astronomical evidence indicates that space contains at least some of the basic essentials for life.

Advanced forms of life such as we have on earth might develop on any planet of proper size if the essential requirements of living things were present. The prevailing temperature range would have to be suitable; water would have to be present; and conditions would have to remain stable long enough for evolution to produce advanced life forms. These conditions might well exist on planets attending the plentiful stars that are similar to our sun.

If all prerequisites are present, there appears to be no barrier to the spontaneous appearance of simple beginnings of life. Whether or not life on earth was spontaneously generated cannot be proved, but, in one way or another, life appeared and has left evidence of its existence in sedimentary rocks more than 3 billion years old.

In chronological order, currently known evidences of Precambrian life are the following: (1) formless organic residues, stromatolites, and structurally recognizable algalike spheroids and filamentous structures from the Onverwacht Series, Barberton-Badplace region of the eastern Transvaal, South Africa, dated at at least 3.2 billion years; (2) spheroidal and rodlike bodies resembling bacteria from chert of the Fig Tree Series, also in the Barberton region, dated at about 3.2 billion years; (3) an abundant microflora (at least eight genera comprising twelve species) of bacteria, blue-green algae, and organisms of unknown affinities, together with hemispherical stromatolites almost surely of organic origin in the Gunflint Iron Formation of western Ontario, all of which are from 1.6 to 2 billion years old; (4) a flora of diverse types, including the first evidence of nucleated cells, from the Beck Springs Dolomite of eastern California, estimated at about 1.3 billion years old; (5) an abundant, well-preserved flora including blue-green algae, colonial bacteria, probable fungi, filamentous organisms, and spheroidal green algae from the Bitter Springs Formation, central Australia, dated at about 1 billion years; (6) trace fossils, trails, grooves, and symmetrical indentations probably made mostly by elongate, wormlike animals, from New South Wales, Australia, estimated to be 1 billion years old; (7) the so-called Ediacara fauna, consisting of distinct impressions of a variety of animals, some confidently classified as coelenterates (jellyfish and sea pens) and annelid worms, together with animals of unknown affinities, found best-preserved in southern Australia, but also in Newfoundland, England, and South Africa and dated at about 700 million years.

FOR ADDITIONAL READING

Barghoorn, Elso S., "The Oldest Fossils." *Scientific American,* Vol. 224, No. 5 (May 1971), 31–42.

Cloud, Preston, ed., *Adventures in Earth History.* San Francisco: Freeman, 1970.

Margulis, Lynn, *Origin of Eukaryotic Cells: Evidence and Research Implications for a Theory of the Origin and Evolution of Microbial, Plant, and Animal Cells in the Precambrian Earth.* New Haven: Yale University Press, 1970.

Oparin, A. I., *The Origin of Life,* 3rd ed., trans. Ann Synge. London: Oliver and Boyd, 1957.

Rutten, M. G., *Geologic Aspects of the Origin of Life on Earth.* New York: American Elsevier, 1963.

Schopf, J. William, "Microflora of the Bitter Springs Formation, Late Precambrian, Central Australia." *Journal of Paleontology,* Vol. 42, May, 1968.

11

THE EARLY PALEOZOIC PERIODS

The appearance of abundant fossils at the beginning of the Cambrian Period introduces added complexity and much detail to the story of the past. Because of this, we must condense our discussion of the Cambrian and succeeding fossil-rich periods and generalize a great deal in order to keep within reasonable space and time limits. Although there is a strong tradition and many good reasons for devoting a full chapter to each of the standard periods, this method is time-consuming, tends to be repetitious, and may obscure the more important underlying themes of

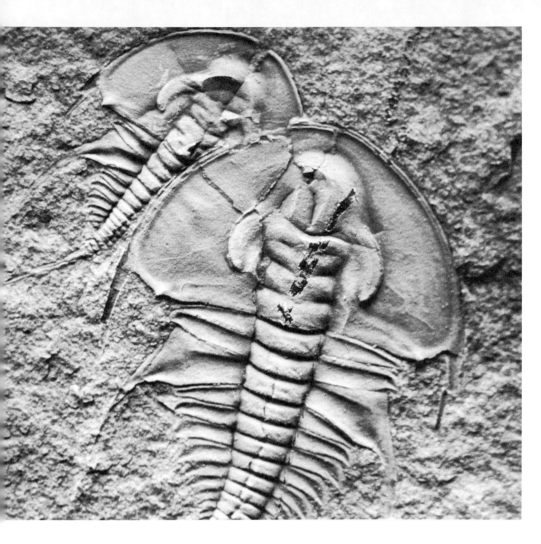

the subject. In line with this reasoning, this and the following two chapters discuss a combination of successive periods rather than a single one at a time.

This chapter treats the combined Cambrian, Ordovician, and Silurian Periods, or the early Paleozoic Era, with a total length of 190 million years. The next chapter covers the late Paleozoic Era (Devonian, Mississippian, Pennsylvanian, and Permian Periods), with a combined duration of 180 million years. The Mesozoic Era (Triassic, Jurassic, and Cretaceous Periods), which lasted about 162 million years, constitutes another convenient unit and will be discussed later in Chapter 13.

Major Patterns of Fossil-bearing Periods

It is obvious that these three- or four-period intervals are of approximately equal duration; a closer look shows that each interval encompasses a roughly similar sequence of events. At the be-

Figure 11.1 Angular unconformity between Precambrian and Cambrian rocks, northeastern British Columbia. (Geological Survey of Canada.)

ginning of a typical major cycle, the oceans are found transgressing inward over the continental margin and interior lowlands. Mountain building and igneous activity are at a low ebb. Climates are mild, and fine-grained marine sediments dominate the sedimentary record.

After they have reached a position of maximum flooding, the shallow seas begin to retreat as crustal disturbances elevate the land. Climates diversify, and glaciers and deserts become common. Continental types of sediment are produced in abundance, and marine deposits are correspondingly restricted. Thus, one cycle ends and the stage is set for the next to commence.

Like most cycles that historians recognize, these geologic repetitions are neither perfect nor universal. They are best thought of as generalized patterns for the earth as a whole. Some geologists refer to the times of greatest emergence as *land periods* and to the times of submergence as *sea periods*. We may consider a combined land-sea interval as a great cycle because it tends to repeat a certain pattern of events.

European and Russian geologists designate the events of the early Paleozoic as *Caledonian*, those of the late Paleozoic as *Hercynian*, those of the Mesozoic as *Kimmerian* and those of the Cenozoic as *Alpine*. These names emphasize chiefly physical events such as mountain building rather than sedimentary rocks or fossil forms. Best current estimates give the Cambrian a duration of 100 million years, the Ordovician 70 million years, and the Silurian 20 million years. The great disparity in length of these periods is another reason for not treating each in a separate chapter.

Beginning of the Cambrian

The early stages of continental growth have been described in Chapter 9. The ancient continental cores or nuclei that formed during Precambrian time had many important influences upon subsequent history. These preexisting land areas furnished the erosion products of which later formations are composed and also determined the positions of basins, geosynclines, and mountain ranges that were yet to form. Because each of the major continents had one or more nuclei, there is a certain basis of uniformity with which to begin the story of later time.

During the latest stages of the Precambrian, the lands appear to have been relatively high and locally mountainous, and the oceans were confined well within their basins. The many clear indications of extensive glaciation near the beginning of Cambrian time that are found in Greenland, Australia, South Africa, India, China, Canada, and the United States have already been mentioned. There appears to be a logical connection between high continents, cold glacial climates, deep contracted oceans, and scarcity of fossils. The underlying cause of all these conditions may have been a period of widespread adjustment in the earth's crust during Late Precambrian time. When this disturbance had subsided, the Cambrian began; the continents were

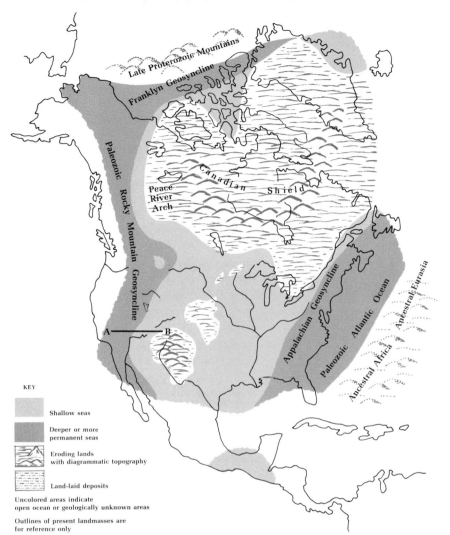

Figure 11.2 Reconstructed paleo-geography of the Cambrian Period.

KEY

Shallow seas

Deeper or more permanent seas

Eroding lands with diagrammatic topography

Land-laid deposits

Uncolored areas indicate open ocean or geologically unknown areas

Outlines of present landmasses are for reference only

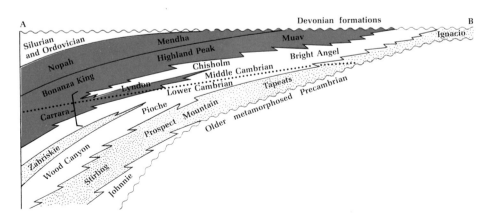

Figure 11.3 Cross section showing Cambrian formations from southeastern California to southwestern Colorado. Length of section is about 960 kilometers, and position is shown by line A-B in Figure 11.2. The section illustrates the encroachment of Cambrian seas inward across the continent. The sandy formations (Stirling, Prospect Mountain, Tapeats, and Ignacio) represent near-shore and beach deposits that started to form much earlier to the west and were completed later to the east. The dotted line, which is the boundary between Lower and Middle Cambrian deposits, shows that while beach material was being deposited to the east, deep-water limestone was forming to the west (Nopah, Mendha, etc.).

lowered by erosion, glaciers melted, the oceans warmed and flooded inland, and marine life flourished and spread widely. The slow rise of sea level may have stemmed from a number of causes, such as the release of water from glaciers, the dumping of sediment from the land, or changes in the shape of the ocean floor or even of the entire globe.

Whatever the explanation may be, the early Paleozoic was a time of extensive flooding of land areas, especially in the Northern Hemisphere. Beginning as shallow embayments, the

Figure 11.4   Evidence of the Caledonian mountain-building disturbance, Yorkshire County, England. Horizontal Carboniferous limestone overlies steeply dipping and eroded Silurian beds. (Crown copyright, Geological Survey photograph. Reproduced by permission of the Controller of Her Britannic Majesty's Stationery Office.)

seas crept inward and at times covered more than half the present land areas.

Even though the statements just made may be generally valid, they do not help in defining the precise time at which the Cambrian commenced or in drawing a boundary in the rocks that represents this time. It should be understood that the problem of placing a mappable boundary between the latest Precambrian and earliest Cambrian exists only where uninterrupted deposition was going on from one period into the other. There are many places where this is the case and no worldwide or even continentwide criteria have been discovered. It is generally agreed that the Cambrian began with the appearance of certain primitive arthropods, the trilobites, but the first trilobites are different in different areas. In the absence of diagnostic guide fossils, it is customary to draw the base of the Cambrian at some local and arbitrary plane, such as an unconformity or a conglomerate bed. Geologists will probably have to be satisfied for the time being with an approximate position for this important datum.

Paleogeographic Reconstructions

THE CALEDONIAN REVOLUTION

The science of geology had its inception in northwestern Europe, and one phase of the formative period was the designation of the type sections of the Cambrian, Ordovician, and Silurian Systems in Great Britain. Further investigation has shown that the Cambrian as originally described at the type locality must now be regarded as Ordovician. However, by means of studies of many contiguous areas in Wales, England, Scotland, Norway, Denmark, and Sweden, a good representation of the early Paleozoic systems has been pieced together. The countries just named were part of the subsiding *Baltic,* or *Caledonian, Geosyncline,* which trended northeastward along the northwest border of Europe at this time. At

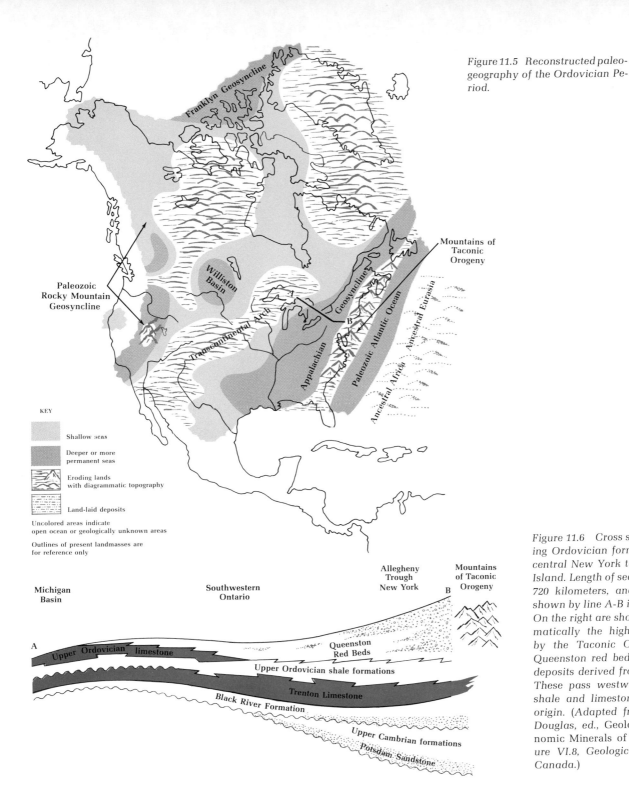

Figure 11.5 Reconstructed paleogeography of the Ordovician Period.

KEY

Shallow seas

Deeper or more permanent seas

Eroding lands with diagrammatic topography

Land-laid deposits

Uncolored areas indicate open ocean or geologically unknown areas

Outlines of present landmasses are for reference only

Franklyn Geosyncline

Williston Basin

Paleozoic Rocky Mountain Geosyncline

Transcontinental Arch

Appalachian Geosyncline

Mountains of Taconic Orogeny

Paleozoic Atlantic Ocean

Ancestral Africa

Ancestral Eurasia

Michigan Basin

Southwestern Ontario

Allegheny Trough New York

Mountains of Taconic Orogeny

A

Upper Ordovician limestone

Queenston Red Beds

Upper Ordovician shale formations

Trenton Limestone

Black River Formation

Upper Cambrian formations

Potsdam Sandstone

B

Figure 11.6 Cross section showing Ordovician formations from central New York to Manitoulin Island. Length of section is about 720 kilometers, and position is shown by line A-B in Figure 11.5. On the right are shown diagrammatically the highlands raised by the Taconic Orogeny. The Queenston red beds are deltaic deposits derived from the uplift. These pass westward into fine shale and limestone of marine origin. (Adapted from R. J. W. Douglas, ed., Geology and Economic Minerals of Canada, Figure VI.8, Geological Survey of Canada.)

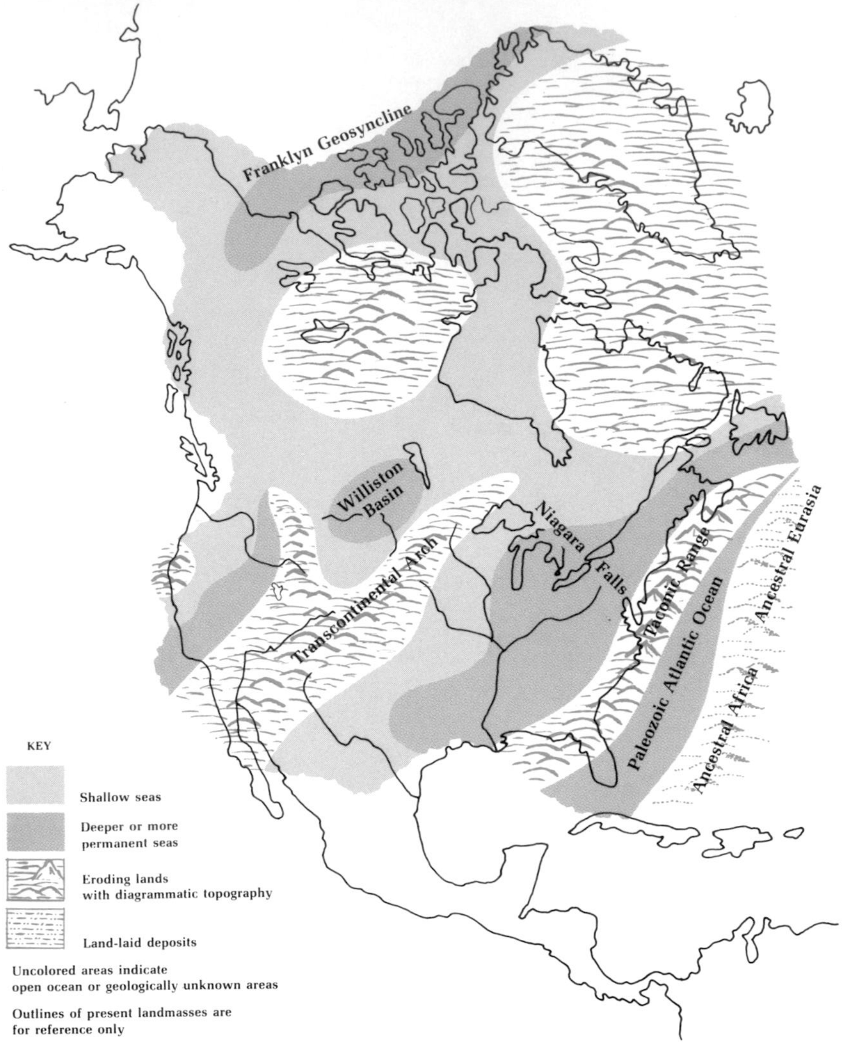

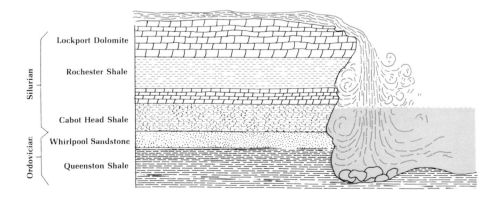

Figure 11.7 Reconstructed paleo-geography of the Silurian Period.

Map labels:

Franklyn Geosyncline

Williston Basin

Transcontinental Arch

Niagara Falls

Taconic Range

Paleozoic Atlantic Ocean

Ancestral Africa

Ancestral Eurasia

KEY

Shallow seas

Deeper or more permanent seas

Eroding lands with diagrammatic topography

Land-laid deposits

Uncolored areas indicate open ocean or geologically unknown areas

Outlines of present landmasses are for reference only

Figure 11.8 stratigraphic column:

Silurian
- Lockport Dolomite
- Rochester Shale
- Cabot Head Shale

Ordovician
- Whirlpool Sandstone
- Queenston Shale

Figure 11.8 Classical exposure of Late Ordovician and Silurian rocks at Niagara Falls. See Figure 11.7 for geographic relations of this section.

least 12,000 meters of early Paleozoic rocks accumulated in the Welsh segment of this trough.

The type, or standard, Cambrian System in the Caledonian belt has fossil trilobites and a variety of rock types appropriate to the varied environments of a subsiding trough. The thinner marginal deposits of Denmark and Sweden are still practically horizontal and unmetamorphosed and contrast strongly with equivalent geosynclinal, well-metamorphosed beds of Norway.

As the Caledonian Geosyncline continued to subside, Ordovician and Silurian deposits followed those of Cambrian age. Sediments of these two later periods are well represented in Wales, the type area, and elsewhere along the trend of the geosynclinal trough. The Ordovician is characterized by black shale facies with graptolites; shell-bearing facies with abundant corals, brachiopods, and trilobites; facies of mixed sandstone and conglomerate with few fossils; and, finally, thick volcanic facies of ash, tuff, and lava. The graptolites have proved to be the best guide fossils for subdividing both the Ordovician and Silurian into zones. Also worthy of note are the Late Silurian (*Ludlovian*) beds of northwestern Europe, which yield the best-known remains of early primitive vertebrates and eurypterids.

The Silurian came to an end in northwest Europe with an important and widespread mountain-building episode, the *Caledonian Revolution*. This was very intense in Norway, where rocks of Precambrian, Cambrian, Ordovician, and Silurian age were metamorphosed almost beyond recognition. Only the occasional discovery of vague fossils enabled geologists to interpret the history correctly. Granitic bodies were intruded, and the area was cut by great thrust faults that further complicated the picture.

Farther south, in Great Britain, the effects were less intense, but faulting and folding give clear indication of mountain building in Wales and Scotland. During all these disturbances there is evidence of an extensive landmass off the present northwest coast of Europe. This has been called the *Old Red Sandstone continent,* or *Atlantis.* Current theories of continental drift indicate that this landmass was in reality North America and Greenland. The sediments of the unstable seaway between the continental blocks were compressed to form the Caledonian Ranges. The Caledonian Revolution formed more or less rigid areas that have never been invaded by the seas since Silurian time. Similar intense effects are noted in the Silurian of Greenland.

THE TETHYS SEAWAY TAKES FORM

The long, relatively narrow tract of the earth's crust that includes the area of the present Mediterranean Sea and the great ranges of the Alpine-Himalayan chain is the world's most extensive geosynclinal belt. Geologists call this ancient tract the *Tethys,* or *Mesogean, Seaway.* In its western reaches this seaway lay between the African and Baltic shields; to the east it was open south of the Asian block until the close approach of India in the Late Cretaceous. Tethys was thus in a position of instability and unrest, and with the passage of time the belt subsided and accumulated a record of all the geologic systems. During the early Paleozoic it was not as active as the Caledonian trough, but deposits of the Cambrian, Ordovician, and Silurian are known across its entire length from Spain to Southeast Asia.

Rocks of the early Paleozoic are widespread in China and contain abundant fossils. The oldest known fossils of Japan and of Indonesia are of Silurian age. There is little or no evidence of sedimentary deposits along the western Pacific or East Indies before this time.

EURASIA NORTH OF TETHYS

The vast plains of northern Russia and most of Siberia are underlain by varied formations of Cambrian, Ordovician, and Silurian age. Generally these rocks are deeply buried, but they appear at the edges of the Baltic and Angara shield areas, in the Ural Mountains, and in the ranges of central Asia. There are many similarities with rocks of the same age in North America, and the fossils prove occasional shallow-water connections across and between the northern continents.

The Cambrian of the northwestern Russian platform is described as soft shale or clay, indicating that it has not been disturbed or deeply buried since deposition. A broad seaway is thought to have separated the Baltic and Angara shield areas, but Cambrian sediments are not particularly thick or extensive in the Ural Mountains. In the interior of Asia the Cambrian contains salt and gypsum beds, which are among the earliest deposits of this type known.

The Ordovician System is generally thicker and is at least as extensive as the Cambrian. Ordovician rocks are found throughout the length of the Urals, indicating geosynclinal subsidences in this area. There were very strong mountain-building movements in the Ordovician along much of the western and southern borders of the Angara Shield. These produced a variety of eruptive rocks and large granitic intrusions. A comparison with similar events in eastern North America is suggested. These disturbances began in the Ordovician and continued into the Devonian and are considered as part of the Caledonian orogenic cycle by most European geologists.

The Silurian is generally less extensive than the Ordovician, but it reaches a thickness of 3,600 meters in the Ural Mountains and in Turkestan. Silurian volcanic rocks are frequently found, but signs of mountain building are less evident than in the Ordovician.

NORTH AMERICA

North America has the most complete and extensive record of the Cambrian, Ordovician, and Silurian Periods of any continent. Seas of these periods spread far inland and at times covered practically all of the Canadian Shield. In addition there were geosynclinal belts that almost encircled the continent. The abundant fossils that are preserved in the troughs and shelf areas give a remarkably complete picture of the life of the early Paleozoic.

Commencing with the Cambrian, the North American continent may be thought of as being divided by a sort of backbone into two great provinces—one to the southeast, the other the northwest. The backbone, or *transcontinental arch* as it is properly called, extends from the Mojave Desert area of California to Lake Superior and into Labrador. This feature is really a part of the ancient Precambrian nucleus of the continent that happened to have a northeast trend.

Across the summit of the arch the geologic systems are thin and discontinuous; they thicken away from it in both directions. In general, the fossils of the western formations are related to the Pacific Ocean, those on the east to the Atlantic or Gulf of Mexico. This natural subdivision of the continent serves as a convenient basis for the following discussion of geologic history.

Eastern North America. That portion of North America lying east and southeast of the transcontinental arch displays the most extensive outcrops of early Paleozoic rocks to be found on earth. The interior regions, chiefly the drainage area of the Mississippi River, are essentially a continuation of the Canadian Shield with a thin cover of Paleozoic sediments. A number of broad structural domes and basins surrounded by wide bands of Paleozoic sediments characterize this region. The major uplifts (with the age of the oldest rocks exposed in their respective central areas) are the Ozark Mountains, Precambrian; the Cincinnati Arch, Ordovician; the Wisconsin Dome, Precambrian and Cambrian. The major basins (with ages of youngest rocks in their central regions) are the Illinois Basin, Pennsylvanian; the Michigan Basin, Pennsylvanian; and the Allegheny Basin, Permian. These domes and

Figure 11.9 *Flat-lying Cambrian sandstone, Pictured Rocks, Michigan. (U.S. National Park Service.)*

basins began to form primarily in Ordovician time and have been accentuated by later forces; not however by intensive mountain building.

Bordering the interior plains are mountain systems born of Paleozoic geosynclines. Along the eastern margin from Newfoundland to Georgia is the *Appalachian geosynclinal belt.* The *Ouachita Mountain system,* now largely buried, arose from a geosyncline that lay along the southern margins of the continental nucleus. These troughs received sediments from the continental interior and occasionally from nearby lands beyond present continental borders.

In the Cambrian there was little sediment from sources outside the continent itself, and the geosynclines were strongly onesided. Volcanic rocks are very rare in the Cambrian, but by the Middle Ordovician volcanic action was contributing much material to the nearby troughs. Volcanic rocks mixed with coarse erosion products are especially characteristic of the outer, eugeo-synclinal belt in northern New England and the Maritime Provinces of Canada. These rocks reach 4,500 meters in New Hampshire and 6,000 meters in eastern Newfoundland.

By contrast, the inner, miogeosynclinal belts and shelf areas were receiving ordinary shallow-water deposits. The Lower Cambrian is chiefly sandstone swept from the continental interior. Later in the Cambrian and continuing into the Ordovician came the deposition of great volumes of dolomite and limestone. These calcareous rocks reach a thickness of 3,000 meters in Alabama and about 2,000 meters in Oklahoma. The immense number of fossil shells and other organic debris that characterize these rocks attest to the importance of organisms in creating the blanketlike limy beds that lie one upon another in the continental interior and miogeosynclines. Incidentally, it is supposed that the material that is now dolomite was initially limestone—the chemical substitution of calcium for magnesium took place immediately after deposition and before the material had entirely solidified.

The early Paleozoic carbonate sheets effectively sealed off much of the Precambrian crystalline outcrops so that coarse sandy sediments were not available to erosion until laid bare in subsequent mountain building of late Paleozoic time.

The outer or seaward zones of the bordering geosynclines beyond the limy deposits received variable proportions of volcanic material and fine-grained siliceous shale. During the Cambrian there was surprisingly little volcanic activity in North America, and the outer belt was inhabited by a variety of silt-loving trilobites. During the Ordovician, increasing amounts of dark-colored shale with fossil graptolites were laid down. In recognition of the almost worldwide association of dark shale and graptolites in the Ordovician and Silurian Periods, the term *graptolite facies* has come into general use. The term contrasts with *shelly facies,* which designates the contemporaneous shell-bearing limy rocks.

It has been suggested that the fine-grained sediments making up the outer shale belt represent material blown seaward from the continental interior. The lack of coherent soil and vegetation at this time would certainly have allowed large-scale erosion and transport of fine particles by the wind.

As we have already mentioned, the Atlantic border began to show deep-seated unrest in Middle Ordovician time. Earth movement and volcanic activity mounted in intensity and culminated near the close of the period in the *Taconic Disturbance,* so named from intense effects in the Taconic Range of Vermont and New York. The offshore landmasses at this time shed coarse-grained sediment westward and brought the great lime-depositing interval to a close. Sandy sediments spread across Pennsylvania and New York as lands were elevated to the east; this aggregation of coarse deposits is known as the *Queenston Delta.*

The highlands created by the Taconic Disturbance continued to supply coarse sediments throughout most of the Silurian, but by the end of the period the seas had cleared again and limestone was deposited. Not to be overlooked are the extensive salt-bearing beds of the eastern interior, which also accumulated near the close of the Silurian. Through a peculiar combination of landlocked basins and restricted seaways a vast, almost "Dead Sea" type of environment came into being in western New York, Pennsylvania, Ohio, and Michigan. The greatest salt thickness (about 500 meters) is in the center of the Michigan Basin. Salt from this formation is extensively mined at many places. The most famous exposure of Silurian rocks is at Niagara Falls; the rocks here are older than the salt-bearing series.

Although the Taconic Disturbance of North America was formerly considered to have been separate and distinct from the Caledonian Disturbance in Europe, it is becoming apparent that these early Paleozoic episodes may not be as clear-cut and distinct as once supposed. Radiometric dating and refinements of the time scale have tended to fill the gaps to show a more or less continuous series of disturbances, active now at one place and now at another. Because these effects center around the north Atlantic in western Eurasia, Greenland, and eastern North America, the general term Caledonian Disturbance may well be applied to all of them, with local names for the separate pulses or phases.

Western North America. During the early Paleozoic the Pacific Ocean washed the margins of the transcontinental arch and the Canadian Shield. There was no Rocky Mountain chain but its geologic forerunner, the Rocky Mountain Geosyncline, was a dominant feature. This great curving trough begins in southern California, extends into Nevada, Utah, and Idaho, and then back toward the Pacific through Montana, Alberta, and Alaska. By contrast with eastern North America there is practically no evidence of an offshore western borderland analogous to Appalachia in early Paleozoic time. Essentially all of the earlier deposits of the trough came from seawater or from the wearing down of low interior lands. The hypothetical borderland, Cascadia, evidently is not needed and for the early Paleozoic at least the Rocky Mountain Geosyncline was essentially one-sided. Another significant difference is the scarcity of volcanic products in the western trough. Here the sedimentary history of the early Paleozoic was unbroken by strong mountain building or volcanic effects.

The Cambrian began with the encroachment of shallow seas at the northwest and southwest parts of the geosyncline. The initial deposits were chiefly clean quartz sands washed from the continental interior. With the passage of time and the steady transgressions of the seas the sediments became finer and more calcareous. The amount of limestone and dolomite deposited in the inner, miogeosynclinal belts is even greater than that deposited in the eastern United States. The combined Cambrian-Ordovician carbonate section reaches 4,500 meters in western Utah. Much of the spectacular mountain scenery of the Alberta Rockies is carved from Cambrian strata. Late in the Cambrian, the seas had flooded inward until the transcontinental arch was reduced to a number of islands.

During the Ordovician the Rocky Mountain trough shows a clear division into a miogeosynclinal carbonate belt and a eugeosynclinal siliceous belt. During this period the east and west margins of the continent reached a high degree of symmetry and similarity. In both east and west the inner limestone facies passes into the outer graptolite facies rather abruptly, but the structural complications are greater in New England than in Nevada. In general, throughout the western miogeosyncline the early Ordovician is limy and the later part is dolomitic. Between them is a widespread sandstone or quartzite phase similar to the St. Peter Sandstone of the midcontinent. Thin deposits record the eastward spread of shallow seas onto and occasionally across the transcontinental arch. The Williston Basin in North Dakota and south-central Canada began to form in the Ordovician and received sediments of all subsequent Paleozoic periods.

The Silurian is a short period (20 million years) and its deposits are correspondingly thin. A single sheet of dolomite in the Rocky Mountain Geosyncline gives a relatively incomplete record of the time. That the seas did occasionally spread farther inland is known from small isolated patches in Wyoming and Colorado, but it is certain that no very thick or extensive deposits were laid down.

Early Paleozoic rocks are widespread in northern Canada, including the Arctic Islands. Thin deposits on the mainland thicken northward into the Franklyn Geosyncline, where as much as 5,500 meters of early Paleozoic marine beds accumulated. Sediments are varied, but, as elsewhere in North America, carbonates are dominant.

THE SOUTHERN HEMISPHERE

Compared with North America and Eurasia, the southern continents have a very incomplete and scattered sedimentary record of the early Paleozoic periods. This deficiency is probably due to the relatively emergent conditions of the southern landmasses, which favored erosion rather than deposition.

The most complete representation of the Cambrian, Ordovician, and Silurian Systems in the Southern Hemisphere is in Australia. In the eastern part of the continent there was steady deposition in the northerly trending *Tasmanian,* or *Tasman, Geosyncline.* The types of fossils found in these beds indicate that migration of organisms to and from distant lands was possible. In the Early Cambrian, for example, a peculiar and diagnostic trilobite, *Redlichia,* is found in Australia, eastern Asia, and India. Cephalopods of the Australian Ordovician are much like those of western North America, and the graptolites of the Ordovician and Silurian are almost identical to those from Wales and the United States. The migration routes allowing these similarities are obscure.

Early Paleozoic strata are rare in that part of Africa lying south of the equator. Ordovician fossils are known from the extreme southern region, but there are no fossils of Cambrian or Silurian age. Most of Africa was probably well above sea level in the early Paleozoic, just as it now is. That part of Africa north of the equator has thick accumulations of Cambrian, Ordovician, and Silurian rocks laid down in shallow embayments of the Tethys Seaway; these are naturally closely related to contemporaneous rocks of southern Europe.

Although the early Paleozoic rock record of South America is much better than that of Africa, it is still greatly inferior to that of North America. Fossiliferous Cambrian deposits are very rare, but deposits in Argentina have yielded Middle Cambrian trilobites almost identical with those of the Baltic region. By contrast, the Ordovician is fairly well represented along the entire Andean belt, in places by deposits of great thickness. Graptolite-bearing shales are common and many of the forms are related to those of eastern North America and the Baltic region of Europe. A short period of glaciation is recorded during the period in western Argentina.

Early Silurian seas spread across wide areas of South America, mainly east of the Andean belt in the Amazon trough and across northern Argentina. In the Middle Silurian the seas retreated to the Andean belt; no late Silurian deposits are known. A glacial deposit of Silurian age is said to be present in northern Argentina and Bolivia.

Antarctica has yielded fossils of Early, Middle, and Late Cambrian age, including well-preserved trilobites, brachiopods, and echinoderms related to Australian and Asiatic forms. No Ordovician or Silurian rocks or fossils have yet been identified.

Life of the Early Paleozoic

Layered calcareous structures secreted by algae are among the oldest known fossils. Bacteria over 3 billion years old have also left unmistakable remains in siliceous rocks and iron ores. Considering this very early origin of plant life and its rather simple requirements, it is strange that the transition from seawater to land took so long. Plants had evidently drifted and diversified in the watery environments for long periods before becoming adjusted to land life. The early seaweeds were soft and nonresistant and for the most part quickly disintegrated to shapeless fragments or carbonaceous residues. Although the Cambrian Period has been called the *Age of Seaweed,* the implications of the term are not supported by strong fossil evidence.

A variety of fossil spores have been reported from Precambrian and Cambrian sediments, especially in Russia; these seem to pertain to water-living plants that have left no other significant remains. Ordovician plant fossils appear to be entirely algal, and it is not until the Silurian that other types appear. An aggregation of plant fossils from Lower Silurian rocks of Maine contains several types that are preserved in an erect position and are probably very primitive land plants. They lack true vascular (conductive) tissue but have tiny spines along the stems. Spores that may have been produced by this early land flora have recently been found in Lower Silurian rocks in New York State.

It has been suggested that plants could not thrive in the open atmosphere until the middle of the Paleozoic because of the strength and intensity of the ultraviolet rays of the sun, which at present are screened out by the ozone layer of the atmosphere. The ozone layer in turn had to be built up from excess oxygen liberated by plants in the photosynthetic process. Plants lived

Figure 11.10 *Indeterminate seaweed, probably a green algae, from the Wheeler Shale, Middle Cambrian, Utah. (Courtesy of Richard A. Robison.)*

Figure 11.11 *Trails of a large invertebrate, thought to be an annelid worm, on ripple-marked sandstone of Cambrian age. (Smithsonian Institution.)*

partly shielded by shallow water until the protective ozone layer permitted both plants and animals to emerge on land.

INVERTEBRATES

The Appearance of Abundant Marine Fossils. With the beginning of the Cambrian Period, fossils began to be preserved in great numbers. Although we do not understand why fossils of marine invertebrates suddenly became plentiful, we are inclined to suppose that it was at least partly because of the increasing production of shells and skeletons by animals that had formerly been without them. The scattered remains of both plants and animals that have been found in Precambrian rocks are mainly of soft-bodied organisms, such as jellyfish, that could not leave abundant fossils.

The Cambrian is characterized by a marked increase in the number of animals that use calcium to build their shells—the *mollusks* and *echinoderms*, which use calcium carbonate almost exclusively; and the *brachiopods* and *arthropods*, which secrete mixtures of calcareous and phosphatic material. The increased use of calcium by shell-building animals seems to coincide with the formation of thick beds of lime-

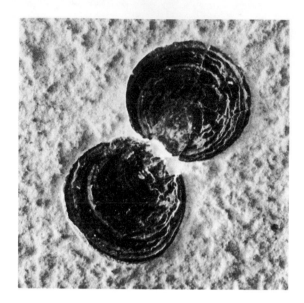

Figure 11.12 Acrothele, a typical inarticulate brachiopod, from the Middle Cambrian. The small, two-piece shell is preserved as it lay open on the Cambrian seabed.

suggestion has been made that shells evolved as a protective measure against harmful radiation rather than against living predators. The reasoning is that the ultraviolet radiation of the sun would have been much stronger without the ozone layer, which, as we have mentioned, may not have formed until an excess of oxygen had been built up by plants.

The secretion of shells and the building of skeletons required animals to expend additional energy and raised other serious problems. Every animal with a shell, for example, must, at least in theory, decide whether a light but mobile shell is better than a heavy but immovable one. A shell may provide protection, but it requires energy to construct and move; animals without shells save energy but are vulnerable to their enemies.

stone and dolomite. The prevalence of limy formations and calcareous shells during the Cambrian and Ordovician may be related to a general warming of the ocean, for calcite is deposited and secreted more easily in warm water than in cold water.

Shells are chiefly protective and skeletons are mainly for support. Skeletons are, therefore, chiefly internal structures, whereas shells are external. In some animals the same structure may serve both as a shell and a skeleton. We assume that some shells evolved in response to an animal's need for protection against neighboring predators. The appearance of skeletons indicates that many animals that originally were rather small had grown larger and heavier. Small animals can be supported by water and are not harmed by gravity, but large animals must have strong supporting structures or they will be crushed by their own weight. The interesting

THE CAMBRIAN—
AGE OF TRILOBITES

The trilobites were crawling or swimming arthropods with light, jointed skeletons. The student should read more about them in Chapter 20 at this point. Their remains clearly show that they must have been mainly bottom feeders capable of actively searching out favorable food and living places. The brachiopods by contrast are immovable and have a two-piece shell that closes to protect the animal inside. The earlier brachiopods resembled two shallow saucers fitted face to face and held together by a number of small muscles. They were oval, round, or tongue-shaped, and their shells were mostly phosphatic. They are called inarticulates because their shells lacked definite hinge structures. More advanced brachiopods developed hinge structures and are called articulates because their shells can open and close like a lady's compact. All brachiopods, living or extinct, are fixed permanently in one spot by a sort of root or fleshy extension of the body. The trilobites could go wherever food was most abundant, but the brachiopods had to depend on the movement of water to bring food within reach of the feeble inflowing currents that

Figure 11.13 *A well-preserved Middle Cambrian trilobite (Alokistocare harrisi) from the House Range, western Utah. Specimen is about 2 centimeters long. (Courtesy of Richard A. Robison.)*

were generated by certain coiled structures within their shells. Trilobites squandered some energy dragging their shells about; brachiopods conserved energy by remaining in one spot. But because both animals survived and prospered, each mode of life must have had its advantages.

A puzzling group of Cambrian organisms, the *Archaeocyatha* ("ancient cups"), were the earliest known reef-forming animals. They left conical calcareous fossils up to 10 centimeters long. Although they resemble both sponges and corals, they are sufficiently different to merit being included in a phylum by themselves. Archaeocyathids had very wide distribution and were especially abundant in Australia and Siberia. Their fossils gave the first clues to the presence of Cambrian rocks in Antarctica. The group is confined almost entirely to the Early Cambrian. So far as is known, this is the first phylum to become extinct.

Trilobites make up about 60 percent of all Cambrian fossils; brachiopods constitute 10 to 20 percent; and the remainder includes archaeocyathids, protozoans, sponges, worms, gastropods, echinoderms such as cystoids, cephalopods, and

Figure 11.14 *The blind trilobite* Peronopsis *represents an important but short-lived group. The agnostid trilobites, of which this is an example, are characterized by a peculiar dumbbell appearance with the head and tail of similar size and outline. The lack of eyes and the shovel-like shape may have adapted these creatures for burrowing in the muddy sea bottoms. The agnostids lived through the Cambrian Period but died out in the Ordovician. Among the many species are a number of excellent guide fossils including some that seem to have reached all the major continents. Length of specimens is about 1 centimeter.*

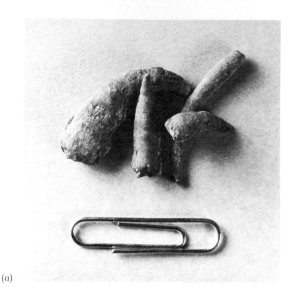

(a)

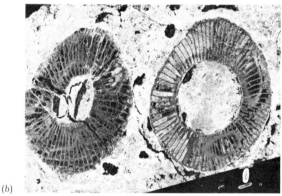

(b)

Figure 11.15 (a) Archaeocyathids from Cambrian rocks of Nevada. The exact zoological classification of archaeocyathids is unknown; they show affinities both with corals and with sponges. (b) Cross-sectional views of two specimens in Lower Cambrian rocks of the Yukon Territory. Cross sections are about 1 centimeter across. (a, Smithsonian Institution; b, courtesy of V. J. Okulitch.)

Figure 11.16 A primitive echinoderm (Gogia granulosa) distantly related to present-day crinoids. From the Middle Cambrian, near Calls Fort, northern Utah. Specimen is about 7 centimeters long. (Courtesy of James Sprinkle.)

arthropods other than trilobites. The trilobites and brachiopods so completely dominate the Cambrian fossil record that all other forms are interesting chiefly for the information they yield about the primitive beginnings of their respective lines. Among the groups not yet positively identified in Cambrian rocks are *bone-bearing animals, pelecypods, bryozoans, true corals, starfish,* and *sea urchins.*

In view of their abundance, it is not surprising that the trilobites are excellent guide fossils in identifying and tracing Cambrian rocks. They have left casts and impressions, unaltered chitinous remains, petrified replicas, and permineralized replacements in shale, sandstone, and limestone on all the continents. Although their

Figure 11.17 *Typical Ordovician brachiopods.* (a) Platystrophia *and* (b) Dinorthis. *The largest specimen measures about 3 centimeters across.*

skeletons could easily fall apart, the separate pieces, especially the heads and tails, are varied and relatively easy to identify. The experts can use these fossil fragments to subdivide Cambrian rocks wherever they are found. In North America, for example, the Cambrian has been divided into twenty-six zones, all but four based on specific trilobites. Locally, and for shorter periods of time, certain trilobites are valuable guides in later periods but are most useful in identifying subdivisions of the Cambrian.

We assume on the basis of available evidence that trilobites inhabited shallow shelves and could migrate or spread along offshore areas until stopped by deep water or by land barriers. Specific trilobites became adapted to different environmental conditions—some to sandy and some to limy bottoms, as shown by the consistent occurrence of certain types in sandstone or limestone. During the Paleozoic a number of migration routes or open seaways seem to have been available to the trilobites. Many trilobite faunas of the Early Cambrian were widespread, or cosmopolitan, indicating that they migrated freely.

But the Middle Cambrian faunas are rather restricted. Middle Cambrian faunas of New England and western Europe are similar, whereas Appalachian faunas of this period resemble collections found elsewhere in North America, Argentina, and Russia.

Brachiopods are less important than the trilobites as Cambrian guide fossils, chiefly because individual species are more difficult to identify.

ORDOVICIAN FAUNAS—
ALL MAJOR PHYLA IN EXISTENCE

During the Ordovician Period conditions continued to be favorable for marine invertebrates. New groups joined those that had appeared in the Cambrian, and by the close of the Ordovician all major animal phyla capable of leaving fossils were in existence. The shallow seas were almost literally filled with a great variety of invertebrate forms adapted to varied environments and food sources. Calcite continued to be the chief construction material of shells and skeletons, and calcareous shells were locally so abundant that entire formations are composed of them. The appearance of abundant oil and gas resources in Ordovician rocks may indicate increasing organic productivity of the ocean waters during this period.

Trilobites reached the height of their development during the Ordovician and assumed a great variety of shapes and sizes. Ordovician trilobites tended to be either smooth and rounded or bristled with nodes and spines. A number of forms probably took up a free-floating existence at this time. The brachiopods also were numerous and varied. Ordovician brachiopods were mostly articulates, less than 3 centimeters across, and were generally of calcareous composition. The majority lacked spines and ornaments other than simple ribs.

The *bryozoans* appeared early in the Ordovician and increased tremendously during that period. They are extremely small animals that always grow in composite masses or colonies composed of multitudes of single individuals. They construct colonies of calcite or other material in the form of twigs, branches, crusts, mounds, or networks. The individual animals are microscopic, which explains why biologists first thought they were plants and called them bryozoa, or "moss animals." The bryozoans were the first group to exploit thoroughly the possibility of community existence. Existing specimens suggest that their food has always been small particles strained from surrounding water.

The *graptolites* are another group of colonial animals that became common in the Ordovician. Some types were fixed like small shrubs to the sea bottom, and others floated freely in the upper levels or were attached to seaweeds. They are

Figure 11.18 Ordovician bryozoans. The large specimen at top is Constellaria, so named for the small star-shaped elevations that cover the surface. The small, twiglike specimens are Eridotrypa; the small pores are the living chambers of extremely small, individual animals. The Constellaria specimen measures about 6 centimeters across.

Figure 11.19 Ordovician cup, or horn, corals. The illustration shows several individuals of the common genus Lambeophyllum. The fossils represent only the stony framework in which the soft, polyplike body of the animal was fixed. The opening of the top specimen measures about 2 centimeters across.

extinct, and we know very little about their relationships to other animals. Their skeletons were composed of chitinous material and were light enough to float but too thin to afford much protection. Graptolites existed in many forms and went through distinctive evolutionary stages during the Ordovician and Silurian.

The *cephalopods* (described in greater detail in Chapter 20) also flourished during the Ordovician. Their variously shaped, chambered shells were buoyant enough to permit the animals to move about rapidly, and their keen senses probably made them the most advanced of all marine invertebrates. Varieties with straight shells up to 4 meters long were probably the largest animals of the period.

Another group, the *crinoids*, also began to leave an abundant fossil record during the Ordovician. Crinoids are echinoderms with plantlike stems and roots and a flowerlike crown or head. The stem enabled the animals to keep their food-collecting devices well above the ocean bottom. Complete crinoid specimens are rare; the heads, or calyxes are composed of many small plates that usually fall apart. The most common remains are the round, flat, disclike segments that make up the stem.

Figure 11.20 *Ostracods from the Swan Peak Formation of Ordovician age, southern Idaho. Notice the impressions of a trilobite tail and brachiopod shell.*

Figure 11.21 *Specimens of the Ordovician graptolite* Climacograptus *on a piece of shale from the Vinini Formation, central Nevada.*

Figure 11.22 A well-preserved crinoid (Eucalypto-crinites crassus) of Silurian age. (American Museum of Natural History.)

During the Ordovician, *corals* became increasingly common, both colonial as well as solitary types. The corals capture food from the surrounding water with the aid of stinging cells and threadlike structures they shoot at prey.

THE SILURIAN—
HEYDAY OF THE BRACHIOPODS

The Silurian was a relatively short period and its life represents an orderly outgrowth from the Ordovician. Shallow seas still spread widely over the continental areas and provided the chief en-vironments for life. Apparently few animals besides the graptolites were adapted to existence in the open oceans, and nonmarine or even brackish water invertebrates are not positively known.

The chief invertebrates of the Silurian were brachiopods, trilobites, and graptolites. The mollusks including cephalopods, pelecypods, and gastropods were increasing as were the corals, bryozoans, and crinoids. Toward the period's end *eurypterids* became common.

More families of brachiopods have been identified in the Silurian than in any other period. There was an increase of larger forms with complex internal structures and roughly five-sided outlines (*Pentamerus*). *Spiriferoid* forms with wing-shaped shells appeared suddenly and launched a long and successful career. Trilobites were still abundant but had definitely passed their zenith and few new genera were produced. Spiny and smooth forms were characteristic.

The graptolites of the Silurian were superficially more simple than their Ordovician predecessors. Most are classed as monograptids with living chambers (*theca*) ranged along one side of a single threadlike support. From twenty to twenty-two worldwide zones based on graptolites have been recognized in the Silurian. Corals proliferated and began to build extensive reefs; those of the Late Silurian of the east-central United States are outstanding. Also contributing to the Silurian reef structures were great numbers of the lime-secreting crinoids and bryozoans. A peculiar coral with a chainlike cross-section (*Halysites*) is very characteristic of Silurian rocks.

The mobile and predaceous cephalopods increased in importance; the *nautiloids*, with simple, unfolded internal partitions and a variety of shell shapes, represented this group. Judged by their obviously fierce appearance and inferred predaceous habits, the rulers of Silurian seas were the eurypterids, or "water scorpions." They were the largest animals of the time, some reaching a length of 2 meters. Another arthropod group, the ostracods, with small bean-shaped, bivalved shells, was also increasing.

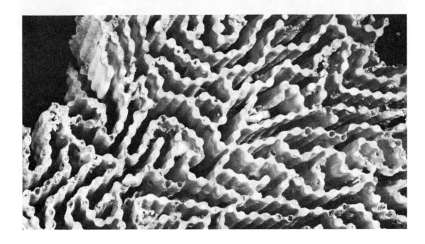

Figure 11.23 Halysites, *the chain coral. This specimen was etched from limestone of Middle Silurian age from western Utah.*

Figure 11.24 *A well-preserved eurypterid (Eurypterus lacustris) from Silurian rocks of western New York. Length is about 20 centimeters. (Buffalo Museum of Science.)*

Although it is customary to consider vertebrates as more highly organized and specialized than invertebrates, it is impossible to make a fully accurate classification of all animals on the basis of whether or not they possess a backbone. Instead of the term "vertebrate," it is more precise to use the term "chordate," which designates not only the vertebrates but also their relatives that are not strictly invertebrates. Many living chordates do not have a bone in their bodies and are of such insubstantial construction that they probably cannot leave fossil remains. Undoubtedly, a variety of such forms lived and died in the past and left few traces or none at all; some of them may have given rise to today's vertebrates.

Three groups are of interest in this regard: the primitive echinoderms, the graptolites, and the conodonts. There is much good evidence that echinoderms (starfish, sea urchins, and so on) are related to the chordates. Certain primitive forms such as the *carpoids* (Figure 11.25) are considered to show this connection. They are, in other words, close to the dividing line between the echinoderms and chordates. We have already described the graptolites as invertebrates in the preceding section, but it is only fair to state that many investigators believe them to be primitive

If either graptolites, or conodonts are chordates, we can say that this phylum was present in the Cambrian. Ignoring these problematical forms, the geologic history of vertebrates begins with the appearance of bone. This hard, durable, easily fossilized substance assumes many forms and is ideal for the construction of internal skeletons.

The Earliest Vertebrates. The earliest known fossils attributed to vertebrates are small jaws with sharp pointed teeth discovered in rocks of Early Ordovician age in Missouri. Aside from this find, the first really abundant remains occur in the Middle Ordovician Harding Sandstone of Colorado. This formation contains many fragments of bone and enamel, but no complete specimens have yet come to light. The bone fragments occur intermingled with shells of invertebrates and with the tiny conodont teeth, which

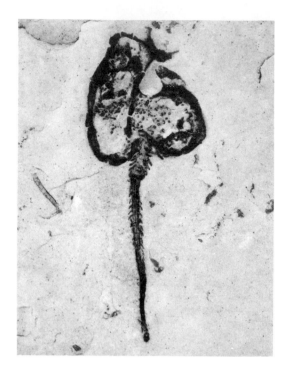

Figure 11.25 *A primitive type of echinoderm known as a carpoid. Not yet named, this tadpole-shaped animal has many characteristics similar to those of lower chordates, but the resemblances may be of no genetic significance. This specimen, from the Middle Cambrian of northern Utah, is about 2 centimeters long. (Courtesy of Richard A. Robison.)*

chordates, members of the subphylum *Hemichordata* to be exact. The conodonts are among the most puzzling of all fossils, for they are represented only by curious, very small toothlike objects. The body shape is known to be bilaterally symmetrical from rather vague outlines discovered in rocks of Pennsylvanian age. Conodonts range from the Middle Cambrian to the Triassic.

Figure 11.26 *One of the earliest known conodont specimens (Proconodontus) from the Late Cambrian of Utah. The specimen is about 0.2 centimeter long. (Courtesy of James F. Miller.)*

do not appear to belong to the same animals as the bones. It is possible that these early fishlike vertebrates may have lived in fresh water and that their remains later were washed into the sea and mixed with ordinary marine forms.

So far, Late Ordovician rocks have yielded very little information about vertebrates, but impressive evidences of the group are found in Silurian rocks of northwestern Europe, where a number of strange fishlike forms are represented. Here, for the first time, the real nature of the earliest vertebrates is revealed. They have elongate, bilateral, fishlike proportions and are mostly not over a few centimeters long. Some, known as *ostracoderms,* have the heads and foreparts encased in bone; others are covered with queer scale patterns, and a few were apparently devoid of any protective covering except skin. Ordovician and Silurian fish seem to have been mainly mud-grubbers; most lacked teeth and jaws.

However obscure their beginnings may have been, these early fish possessed the potential for becoming masters of the sea. By slow degrees and the evolution of many improvements, they achieved a dominant position in the ocean, which has never been successfully contested by the invertebrates. Jaws appear to have developed from gill supports that lay behind and near the

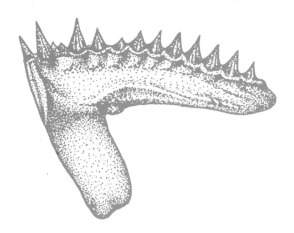

Figure 11.27 A possible early chordate. Specimens of this toothed, jawlike object are found in Lower Ordovician rocks of Missouri. Exact classification remains in doubt.

mouth, while teeth may have been derived from modified toothlike structures that are similar to the tiny pointed scales that make the skin of modern sharks feel so rough. Marvelously well preserved fossils of the brain cavity and nerve canals reveal that the nervous system and sense organs of fish were already well developed by

Figure 11.28 A complete specimen of Rhyncholepis, a primitive fish, from Silurian rocks near Oslo, Norway. The length is about 12 centimeters. (Courtesy of A. Heintz.)

Silurian time, and there can be little doubt that certain other bodily systems were also relatively efficient and highly organized. These characteristics gave the early fish great advantages: they could seek and capture large and active prey, including animals with thick skins or heavy shells, such as the abundant and ever-present trilobites and the floating graptolites. They could also move rapidly to escape their arthropod enemies, the eurypterids. In spite of the poor fossil record, it is evident that vertebrates made many of their most important advances in early Paleozoic time.

Evolutionary Status of Early Paleozoic Life

THE AGE OF MARINE INVERTEBRATES

The early Paleozoic was predominantly an age of sea life. Disregarding the soft-bodied forms, about which we know relatively little, it is evident that all major groups of invertebrates were established during the Cambrian and Ordovician Periods. No new phyla and very few classes of invertebrate animals have appeared since that time. Although we have ample proof that the shallow *epicontinental seas* (those lying on the continents) were teeming with invertebrate life, we have not learned a great deal about the deep ocean and the dry land. Deposits laid down in deep water have not been positively identified, but the graptolites may indicate something about the type of life that floated in the oceans. Early Paleozoic continental sediments are relatively scarce, but all evidence indicates that the lands were barren of life with the exception of certain very primitive plants and perhaps a few arthropods, such as scorpions.

Compared to their descendants and living relatives, the inhabitants of the shallow early Paleozoic seas were relatively primitive and inefficient. They had not yet developed the most effective ways to capture, use, and protect their energy reserves, although the fundamental resources of sunlight and aquatic plants were obviously abundant. In the first place, fixed forms predominated. There were a number of floaters but only a handful of animals that could move about efficiently under their own power. This distribution indicates that most groups found their food by sifting or straining either mud or seawater. Mud-grubbers include the trilobites and most worms; among the food-sifters, or filter feeders, are the brachiopods, bryozoans, crinoids, sponges, and corals. All these animals are at the mercy of ocean waves and currents; they may be buried by sediment, overturned by storms, killed by temperature changes, and starved by stagnant conditions. Their most favorable habitats are warm, shallow seas that are not agitated by violent waves or strong currents. The early Paleozoic was obviously a very favorable time for this type of existence.

The graptolites were passive floaters. They found in the upper levels of the ocean a wide-open environment and for a while were the dominant form in these realms. But they too were at the mercy of the waves and were occasionally destroyed en masse, as their closely packed remains indicate. They appear to have been wholly defenseless and gradually declined as fish and other predators proliferated.

The swimmers, crawlers, and burrowers, with relatively efficient methods of locomotion, are represented in the early Paleozoic by the trilobites and a few mollusks. These animals were able to escape unfavorable natural conditions and could also pursue and capture suitable prey. Even so, by comparison with their descendants and living relatives, they were slow, clumsy, and inefficient.

The early Paleozoic seas also bore a high proportion of shelled or armored forms. The possession of a shell has obvious advantages, but it is not entirely an unmixed blessing. Few heavily shelled animals have remained dominant forms of life, for the shell as a defensive mechanism may actually inhibit or retard its possessor. As we shall see in studying subsequent intervals, the forms with heavy and cumbersome shells tend to be less successful in the struggle for survival than lighter or shell-less forms.

The organs and methods of locomotion appear to have been poorly developed in early Paleozoic animals. Although it is impossible to judge the speed and agility with which these extinct creatures may have moved, it is certain that few of them possessed well-developed locomotive organs. Legs and fins are rather scarce, and those animals that did possess them were mostly weighed down by armor. The cephalopods evidently developed a jet-propulsion type of locomotion in the Ordovician, but it is doubtful if it was as efficient as it now is among such shell-less animals as the squids.

Another clue to the rather passive life of the early Paleozoic is the rarity of teeth or even biting mouth parts. There are conodont teeth but these were probably not used in biting or chewing. The only really effective biting organs seem to be those of the eurypterids.

The degree to which reefs were in evidence during the early Paleozoic is significant. We have already suggested the ways in which marine organisms are dependent on, and at the same time are at the mercy of, moving water. The great bulk of marine life must live within the upper hundred meters or so of water, where light penetrates and food is provided by plants. This same zone, however, is subject to the most violent wave action and is constantly being agitated by the action of currents. Few individual food-sifting animals can maintain their stability in this zone, except in quiet interior seas where wave action is at a minimum. The outer fringes of the continents and the islands of the open ocean provide the best environment for any organism able to build structures that can survive the pounding surf. There is a real advantage to the individual in being able to cooperate in building solid and continuous structures that not only withstand the waves but may even control them.

We do not mean to say that there were no animals capable of building reefs or community structures during the early Paleozoic. The first form to begin to exploit the possibility were the *archaeocyathids,* a rare group of organisms that were extinct by Middle Cambrian time. Its so-called "reefs" were so small and incoherent, however, that they scarcely deserve the name. We find the first reefs of any real consequence in the Silurian. Some of these structures were several hundred kilometers long and were solidly made of corals and other marine organisms. But these reefs were built up in quiet interior seas and were, therefore, quite different from the kind we encounter today in open oceans.

The typical marine invertebrate of the early Paleozoic led a sedentary existence, was largely without sense organs or means of offensive or defensive action, possessed a shell immovably fixed in place, fed itself by sifting mud or seawater, and was largely at the mercy of waves and currents. Here and there were creatures with other, more advanced tendencies, the forerunners of animals whose story we shall tell in succeeding chapters.

Economic Products of the Early Paleozoic

The study and evaluation of economic mineral deposits constitutes an important subdivision of geology. For present purposes we must confine our attention to the historical aspects of the subject—that is, to the times and manner in which mineral deposits were formed and their relation to other geologic events. Disregarding building materials and ordinary stone products, which are available almost everywhere, it is convenient to classify mineral products into the following groups: (1) those originating by or through igneous activity, (2) those originating by sedimentary

Figure 11.29 A slate quarry in the Pen Argyl Formation of Ordovician age, Northampton County, Pennsylvania. (U.S. Geological Survey.)

processes and not affected by organic influences, and (3) those originating through the influence of organic materials or organisms.

The early Paleozoic is not particularly rich in metallic minerals because igneous action (which is usually connected with their deposition) was not strong. A few important sedimentary deposits with metallic minerals should be noted. Economic deposits of sedimentary iron occur in Ordovician and Silurian rocks in northern Norway, Great Britain, Germany, northeast Siberia, Spain, and the eastern United States. The Clinton Formation, which is the chief iron producer in the Appalachian region, is a sedimentary deposit that extends for hundreds of kilometers from New York State to Alabama. It is possible that certain bacteria were precipitating iron-rich deposits at this time.

Commercial deposits of copper are found in Late Cambrian and Ordovician sediments of Siberia, and there is phosphate-bearing ore in the Cambrian of central Asia.

Among the deposits associated with intrusive igneous rocks are platinum and gold in the Ural Range; copper, nickel, titanium, and chromium in areas of Caledonian disturbances in northern Norway; gold in New South Wales and Tasmania; and copper and gold in eastern North America. Important commercial deposits of salt are found in the Silurian of western New York and the Great Lakes region and in the Cambrian of central Asia.

A great deal of oil and gas is derived from Ordovician rocks in North America. As a matter of fact, in the United States the Ordovician ranks third among the systems in known oil and gas yield and reserves. The largest Ordovician fields are in the mid-continent region. An extensive deposit of oil shale is found in the Cambrian of Esthonia. Plants had not yet become abundant enough for the formation of significant coal beds.

A SUMMARY STATEMENT

The Paleozoic Era began with the appearance of abundant fossils and a slow submergence of the continental margins. During the Cambrian, Ordovician, and Silurian Periods, the major geosynclines were occupied almost continually by shallow seas and the shield areas occasionally were flooded. The Cambrian was generally quiet, with volcanism and mountain building at a low ebb. Sediments from this period are chiefly fine grained, and there are thick accumulations of limestone, dolomite, and shale.

During the Ordovician, the eastern margin of North America was affected by the Taconic mountain-building disturbance. Other areas bordering the North Atlantic were in a state of geologic unrest during the Ordovician and Silurian. Major effects during the Silurian were in northwest Europe, where the previously downfolded Baltic Geosyncline was compressed and intruded by granites (Caledonian Revolution). The Southern Hemisphere and Pacific borderlands were geologically quiet.

The early Paleozoic was an age of marine invertebrates. All the major phyla were in existence and the seas were teeming with life. The vertebrates appear as fossils in the Ordovician, and fish increased in the seas during the Silurian. The first land vegetation appeared in the Silurian, but there were no forests of inland vegetation sufficient to leave coal beds.

FOR ADDITIONAL READING

Augusta, J., and Z. Burian, *Prehistoric Animals*, trans. G. Horn. London: Spring Books, 1956.

Barnett, Lincoln, *The World We Live In*. New York: Time Inc., 1955.

Berry, W. B. N., and A. J. Boucot, *Correlation of the North American Silurian Rocks*. Geological Society of America, Special Paper 102, 1970.

Clark, Thomas H., and Colin W. Stearn, *The Geologic Evolution of North America*, 2nd ed. New York: Ronald Press, 1968.

Dott, R. H., and R. L. Batten, *Evolution of the Earth*. New York: McGraw-Hill, 1971.

Dunbar, Carl O., and Karl M. Wagge, *Historical Geology*, 3rd ed. New York: Wiley, 1969.

Fenton, C. L., and M. A. Fenton, *The Fossil Book*. Garden City, N.Y.: Doubleday, 1958.

Holland, C. H., ed., *Cambrian of the New World*. New York: Wiley, 1971.

Kay, Marshall, *North American Geosynclines*. Geological Society of America, Memoir 48, 1951.

———, and Edwin H. Colbert, *Stratigraphy and Life History*. New York: Wiley, 1964.

King, Phillip B., *The Evolution of North America*. Princeton, N.J.: Princeton University Press, 1959.

Kummel, Bernhard, *History of the Earth*, 2nd ed. San Francisco: Freeman, 1970.

McAlester, A. Lee, *The History of Life*. Englewood Cliffs, N.J.: Prentice-Hall, 1968.

Raup, David M., and Steven M. Stanley, *Principles of Paleontology*. San Francisco: Freeman, 1971.

12 THE LATE PALEOZOIC PERIODS

The late Paleozoic is divided into four periods with names and estimated durations as follows: Devonian, 60 million years; Mississippian, 25 million years; Pennsylvanian, 40 million years; and Permian, 55 million years. During this 180-million-year interval there appear to have been important movements of the continental masses but they did not break apart entirely as they did later in the Mesozoic Era. The most intense effects were in and among the northern landmasses, especially those bordering what is now the North

Atlantic and Arctic oceans. In addition to the movements and reactions between the individual blocks or plates, there were extensive floodings of shallow seas upon each of them. During the Late Devonian and Early Mississippian there was a notable preponderance of limy sediments. In the Late Mississippian and the Pennsylvanian, many former seaways became swamps in which coal beds accumulated. The Permian is characterized by an abundance of red, land-laid sediment and deposits of salt and gypsum.

Before discussing some of the important advances made by life forms during the late Paleozoic, we shall take a brief look at some key geographical developments.

Paleogeographic Reconstructions

The exact geography of the continental plates and intervening seaways and geosynclinal belts during the late Paleozoic is far from fully understood. Although it is fairly certain that all southern continents were closely associated in one coherent mass, the northern continents appear to have been somewhat separated from one another and from the southern aggregation. It seems certain that the Pacific was much wider than at present, whereas the Atlantic, if it even existed, was narrow and discontinuous. An ancestral form of the Arctic Basin may have been in existence, but this too is uncertain. As is the case with many other matters concerning the history of the earth, future evidence must be awaited.

As an explanation for many relevant facts, it is assumed as a working hypothesis that Gondwana and Laurasia came into forcible contact in the Late Mississippian and the Pennsylvanian Periods. This brought about intense foldings that created the Ouachita and Wichita ranges in Oklahoma, Arkansas, and Texas, the Appalachian Range of the southeastern United States, and the Hercynian and Varascian ranges (the so-called Paleozoic Alps) of Europe, Asia, and northern Africa. Disturbances of this interval have long been referred to as the Hercynian Revolution in Europe and the Appalachian Revolution in North America.

It appears likely that the Siberian and Russian shield areas approached each other and collided

in the Permian or Early Triassic. This created the Ural Mountains and other folded structures that have been eroded and buried under the Siberian Lowland. The trend of mountains formed in the Pennsylvanian and Permian conforms with what would be expected from massive movements of crustal plates described above. Details must be worked out by careful mapping and study of mutual time and space relations.

TETHYS

As it had during the early Paleozoic, the Tethys Seaway continued to act as a partial barrier between lands of the Southern and Northern hemispheres. What is now central and southern Europe, the Mediterranean Sea, and Africa north of the Sahara was occupied by a succession of shifting seas, coal swamps, lakes, and sandy lowlands, which received a great variety of sediments. This belt occupied an unstable position between the rigid Baltic Shield to the north and the African platform to the south, and was in a state of constant unrest. By contrast, most of

Figure 12.1 The arrangement of the continental landmasses at the end of the Permian. (Adapted from Robert S. Dietz and John C. Holden, "Reconstruction of Pangaea," Journal of Geophysical Research, Vol. 75, 1970, p. 4943.)

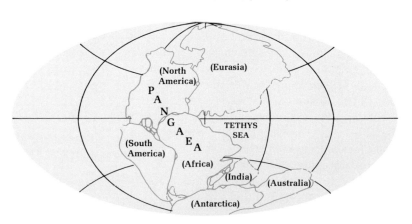

central and southern Africa remained stable and above water.

Farther east, where the Himalayan and associated mountain ranges now rise, the Tethyan belt lay along the margins of the continent. India had not yet arrived by drift from the south. Tethys presented a barrier to land plants and animals and could be crossed only occasionally during the late Paleozoic. At the same time, it was frequently possible for marine organisms to migrate along its length from one extremity of the great Eurasian continent to the other. Correlating by fossils along the Tethys Seaway is, therefore, fairly easy.

We should take particular notice of a number of offshoots of the Tethys Seaway. One of these occupied the site of the present Ural Mountains and extended well into the Arctic region. This seaway had already received deposits during every early Paleozoic period, and with the passage of time representative types of Mississippian, Pennsylvanian, and Permian rocks accumulated to give the Urals a fairly complete Paleozoic record. This area is in fact the world's standard of reference for the Permian System.

NORTHWESTERN EUROPE

Early in the Devonian Period, land-laid sediments began to fill the depressions between and around the Caledonian folded mountains that had formed in the Late Silurian. These land-laid sediments are preserved in several areas of Great Britain, where they are known as the Old Red Sandstone. This formation consists of sandstone with occasional shale and conglomerate beds. Although the total area now covered by these beds is small, they are of historical interest because their study and classification has had a great influence on geologic thought. Scattered throughout the Old Red Series are beds containing fish that were at home in brackish or fresh water, eurypterids, and fragments of land plants. These fossils were among the first to become well known to the general public, and they helped direct attention to the fact that the rocks have a definite story to tell about past conditions. From the fossils and from the evidence of wind action, torrential floods, and recurrent droughts in the area, we assume that the rocks accumulated in upland lakes and shallow arms of the ocean where climatic extremes were common.

South and east of the land-laid Old Red deposits in southern England and the northeastern parts of continental Europe are limestones of marine origin. An excellent display of fossiliferous Devonian rocks that has become a world standard is found in Belgium and Germany.

The Carboniferous (combined Mississippian and Pennsylvanian) of northwestern Europe is of great economic importance because of the limestone and coal deposits it contains. Deposits of the Carboniferous are mostly in disconnected basins between uplifts created by the Hercynian Revolution.

During the Permian, western Europe was dominated by shifting seas in which a variety of evaporites (salt, potash, gypsum) were deposited. Unfossiliferous sandstones are also common. The whole sequence has much in common with contemporaneous deposits in the United States.

THE HERCYNIAN OROGENY
IN THE EASTERN HEMISPHERE

During the Late Pennsylvanian and Early Permian a great system of mountains known as the *Hercynian chain* was formed in Europe and Asia. It seems significant that the dominant trend of this system was east-west, the same direction taken by nearly contemporaneous structures in North America. The region affected included southern Wales, northern and central France, southern Germany, Bohemia, and parts of Russia and central Asia. Volcanic outpourings and granitic intrusions accompanied the folding, and many important ore deposits were formed. This ancient mountain chain was subsequently deeply eroded; portions sank and were flooded by Tethys and still later became involved in the

Figure 12.2 Devonian limestone exposed in a quarry in the valley of the Orneau River, southern Belgium.

the continent. Lava and volcanic products were emitted on a large scale, and glaciers descending from the uplifted ranges dumped their characteristic deposits into adjacent seas. Because the Devonian of southern Africa is more intensely folded than overlying beds, we assume that the late Paleozoic orogenies also may have affected this area.

The Urals were compressed and uplifted during the closing stages of the Paleozoic, and the site thereafter was covered only by thin beds mainly of continental origin. At frequent intervals, seas spread northeastward across central Asia through what is now China.

EASTERN NORTH AMERICA—
THE APPALACHIAN REVOLUTION

It is known that the Paleozoic geologic histories of eastern North America and western Europe are similar. The concept of continental drift goes far in explaining this. For decades American geologists have found it necessary to recognize a hypothetical landmass, which they called *Appalachia,* off the eastern margin of the Atlantic coast. At the same time European geologists found evidence for a landmass (Old Red Sandstone continent) off the northwest coast of Europe, as was noted in Chapter 11. Because there is evidence that Europe and North America may have been much closer together, these hypothetical landmasses are no longer needed to explain geologic history. It is possible, however, that a considerable strip of land between the continents may have been destroyed during or after the separation. In any event the Hercynian Revolution of Europe and the Appalachian Revolution of America must be regarded as closely related phases of the same great disturbance.

Whatever landmass lay off the eastern coast of North America was apparently alternately uplifted and eroded and gave rise to a number of large river systems that carried sediments westward to the geosyncline. The thickest Devonian section (4,000 meters) is exposed in the

great Alpine and Himalayan chains. An important expression of Hercynian Orogeny in China, called the *Tungwu Revolution,* culminated in Early Permian time. The trend of this chain was northeasterly, and the foldings were accompanied by lava flows. The last Paleozoic disturbance in China was the *Suwanian,* near the close of the Permian.

One of the most violent disturbances to affect Australia began during the Mississippian and continued intermittently until the Permian. Areas affected were chiefly in the southeastern part of

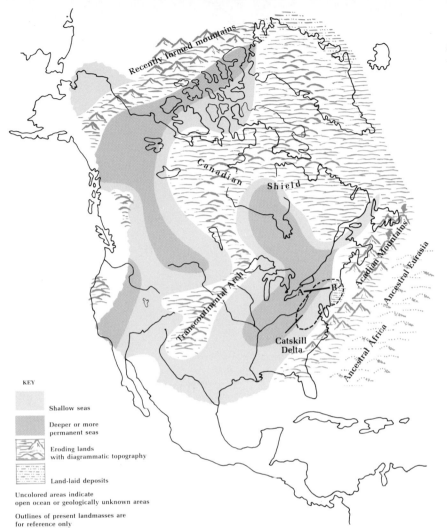

Figure 12.3 Reconstructed paleogeography of the Devonian Period.

KEY

Shallow seas

Deeper or more permanent seas

Eroding lands with diagrammatic topography

Land-laid deposits

Uncolored areas indicate open ocean or geologically unknown areas

Outlines of present landmasses are for reference only

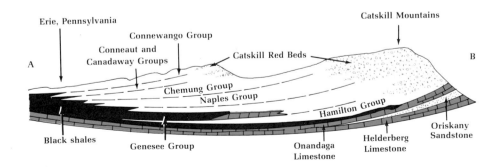

Figure 12.4 Cross section of Devonian sediments from western Pennsylvania to the Catskill Mountains. Length of section is about 400 kilometers, and position is shown by line A-B in Figure 12.3. This section illustrates the internal structure of the great Catskill Deltas, especially the gradation from coarse red beds to the east to fine black shales to the west. (Adapted from Chadwick and Kay, 16th International Geological Congress Guidebook 9a, 1933.)

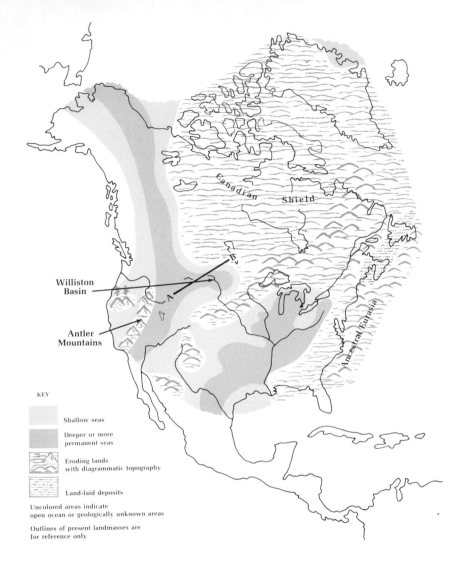

Figure 12.5 Reconstructed paleogeography of the Mississippian Period.

Williston Basin

Antler Mountains

Canadian Shield

Ancestral Eurasia

KEY

Shallow seas

Deeper or more permanent seas

Eroding lands with diagrammatic topography

Land-laid deposits

Uncolored areas indicate open ocean or geologically unknown areas

Outlines of present landmasses are for reference only

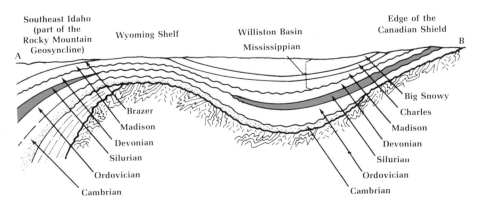

Figure 12.6 Restored cross section from southeastern Idaho to southern Manitoba at the close of the Mississippian Period. Length of section is about 1,300 kilometers, and position is shown by line A-B in Figure 12.5. Most of the formations of the Williston Basin are either limestone or dolomite. The irregular wavy contact lines signify erosion surfaces or interruptions in deposition during which seas were absent from the area. This type of section is typical of interior basins.

Catskill Mountains and adjacent territory. Here are the deeply eroded remnants of a large delta, with coarse sediments to the east that grade westward into finer marine beds. Elsewhere in the geosyncline and across most of North America, open seas prevailed during most of the Devonian, and limestone and fine-grained shale beds were deposited.

The close of the Devonian was marked by a localized disturbance, the *Acadian Orogeny,* which affected territory in, and adjacent to, New England and eastern Canada (Acadia). Intense folding and metamorphism of older rocks, the extrusion of lava, and the intrusion of granite accompanied this activity. In general, its effects were stronger in the northern areas and died out gradually southward. The Acadian Disturbance affected much of the same territory that was involved in the Taconic Disturbance, which had occurred earlier.

Mississippian time in the Appalachian Geosyncline and over most of North America was relatively quiet; but a great deal of coarse debris, including a large deltaic accumulation, now known as the *Pocono Formation,* was brought in by rivers from the east. Extensive shoals and wide mud flats lay over the area affected by earlier disturbances, which would henceforth remain dry land. Most of the rest of North America was covered by wide shallow seas in which limestone was the dominant sediment. The period gets its name from the general area near the junction of the Mississippi and Missouri rivers, where the formations are for the most part very limy, with many caverns and other evidences of solution.

WESTERN NORTH AMERICA

Geologists generally believe that the Appalachian trough turns rather abruptly to the northwest somewhere under the edges of the Gulf of Mexico. The succession of rocks found in the Ouachita-Wichita uplifts in the Texas-Oklahoma region are similar to formations in the Appalachians. In the Texas-Oklahoma region we find evidences of intense orogeny in the Late Mississippian and again at several times during the Pennsylvanian. Another great mountain chain,

Figure 12.7 Large quarry in Mississippian Salem Limestone near Oolitic, Lawrence County, Indiana. (Indiana Geological Survey.)

known as the *Ancestral Rockies,* was formed in the Utah-Colorado-New Mexico area. The ancestral Rockies trend in the same general direction as the folds in Texas, Arkansas, and Oklahoma, and were formed during the Middle Pennsylvanian. The large number of east-west folds that came into existence between the middle of the Mississippian and the Late Pennsylvanian are broadly considered by most geologists to be an expression of the intense Appalachian Revolution.

The western and southwestern part of the United States was occupied by a succession of shifting seaways and uplifts during the late Paleozoic. Devonian and Mississippian downsinking and deposition followed the northerly trend of the great Rocky Mountain Geosyncline, and there were substantial interior seas in the Williston Basin in the heart of the continent. A major disturbance, called the *Antler Orogeny,* affected the western border of North America, particularly Nevada and Idaho, with pulses of mountain building in the Late Devonian and the Mississippian.

A major change in the pattern of geologic development commenced in the Late Mississippian

Figure 12.8   Monument Valley, Utah. The imposing monoliths are erosional remnants of Permian De Chelly Sandstone resting on red siltstone, also of Permian age. On top of some of the monuments are small remnants of Triassic formations. (Utah Travel Council.)

Figure 12.9   El Capitan and Guadalupe Peak, Culberson County, Texas. The cliffs are eroded in fossil reef material of Middle Permian age. (U.S. Geological Survey.)

and the Pennsylvanian. Prior to this time, the Appalachian and Rocky Mountain geosynclines had been receiving most of the sediments. In the later Paleozoic periods a number of major basins came into being in the southwestern part of the continent, and thick Pennsylvanian and Permian deposits were laid down in Texas, Oklahoma, Arkansas, Kansas, Colorado, New Mexico, Utah, and Nevada. The most complete and continuous accumulation of Pennsylvanian-Permian rocks is in Utah, where one formation, the *Oquirrh,* is over 9,000 meters thick.

Recent exploration of the Arctic regions has shown an extensive rock record of late Paleozoic time. The Devonian reaches great thickness and is chiefly dolomite and shale. The Mississippian,

Pennsylvanian, and Permian have less volume. Evaporite (salt and gypsum) deposits occur in the Pennsylvanian and Permian, but coal is not known.

The various orogenic movements that affected the borderlands of the Canadian and Baltic shields during the late Paleozoic supplied much coarse sediment. Prior to this period, limestone, dolomite, and shale predominated; afterward, the sediments contained a higher proportion of sandstone and conglomerate, both in the marine and in the continental sections. It was also a time of

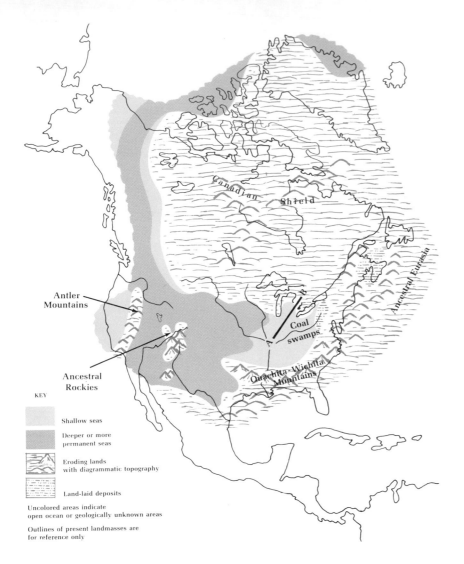

Figure 12.10 Reconstructed paleogeography of the Pennsylvanian Period.

Antler Mountains

Ancestral Rockies

Coal swamps

Ouachita-Wichita Mountains

Canadian Shield

Ancestral Eurasia

KEY

Shallow seas

Deeper or more permanent seas

Eroding lands with diagrammatic topography

Land-laid deposits

Uncolored areas indicate open ocean or geologically unknown areas

Outlines of present landmasses are for reference only

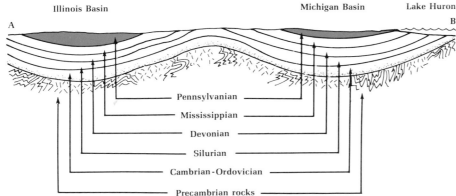

Illinois Basin

Michigan Basin

Lake Huron

A

B

Pennsylvanian

Mississippian

Devonian

Silurian

Cambrian-Ordovician

Precambrian rocks

Figure 12.11 Restored cross section across the Illinois and Michigan basins. Length of section is about 880 kilometers, and position is shown by line A-B in Figure 12.10. These two basins sank gradually during the Paleozoic and received sediments of each of the pre-Pennsylvanian periods. The Pennsylvanian has been covered by little or no sediment since its deposition and the numerous coal beds it contains lie near the present surface.

formation of gypsum, salt, and red beds. The Appalachian Geosyncline and its western extensions probably ceased to exist as open seaways during the Pennsylvanian, for the areas of Permian deposition follow a distinctly different pattern.

THE SOUTHERN HEMISPHERE

The geologic record of each of the southern continents is remarkably similar, a fact that argues strongly that they were once united in one great landmass. Compared with the same interval in the Northern Hemisphere, fossils are relatively scarce, but, on the other hand, a variety of physical events and processes are clearly recorded that have no counterparts in northern lands.

At the extreme southern tip of Africa a section of marine Devonian rocks about 2,000 meters thick indicates a temporary subsidence. About half of South America was covered by Devonian seas, including the present Andes chain and the Amazon Basin. Some of the fossils are related to North America, some to other southern continents. Devonian rocks with primitive plants and brachiopods have been found in Antarctica, indicating shallow, warm seas at that time. Devonian rocks are also widespread in Australia; they are some 6,000 meters thick in the Tasmanian Geosyncline.

ICE AGE
IN THE SOUTHERN HEMISPHERE

Some of the most interesting deposits of the late Paleozoic are associated with glacial action. Glacial sediments are very characteristic, including, as they do, the poorly stratified and heterogeneous material called *tillite,* which represents material picked up and deposited by ice. Among the constituents of tillite are smooth and flattened pebbles that have been pushed across bedrock while held fast in the ice. And where the ice has ground forward, the bedrock below is also polished, scratched, and abraded. Signs such as

these prove that a great ice age gripped the Southern Hemisphere during the late Paleozoic. This ice age is difficult to explain because the evidence of ice action is centered in what are now tropical and semitropical regions. Many geologists find it easier to believe that the continents bearing this evidence have shifted to their present position from a former location near the South Pole, rather than that thick ice caps could have formed and spread in tropical or semitropical lands.

Tillites of Middle Devonian age, dated by associated fossils, occur in western Argentina; but the glaciers there were probably rather local. Evidence of more extensive ice sheets occurs in the Pennsylvanian of much of southern South America. In Australia, where marine beds with fossils are found interbedded with glacial deposits, it is possible to distinguish Permian and possibly Pennsylvanian glaciations. The *Dwyka Tillite,* which covers many thousands of square kilometers in South Africa, is regarded by geologists as a product of a large continental glacier and has been dated as Late Pennsylvanian. An extensive tillite that resembles the Dwyka in almost every way has recently been discovered in Antarctica. Although geologists have not evaluated all the evidence, it appears that continental glaciation in the Southern Hemisphere reached a maximum in the Late Pennsylvanian and Early Permian with minor effects both before and after this time.

The trend of scratches and grooves in the bedrock indicates the direction in which the glaciers moved. In South Africa and South America the glaciers moved principally from east to west— some away from the Equator. In central Africa and Madagascar other deposits show that the ice moved northward, well within the tropics. Most surprising has been the discovery of great beds of glacial debris in northern India, where the direction of movement was northward. Evidences of glaciation are widespread also in Australia and Tasmania, where the ice moved from south to north.

Figure 12.12 Reconstructed paleogeography of the Permian Period.

Red sandstone

Canadian
Shield

Appalachian Mountains

Ancestral Eurasia

Ancestral Rockies

A
B

KEY

Shallow seas

Deeper or more
permanent seas

Eroding lands
with diagrammatic topography

Land-laid deposits

Uncolored areas indicate
open ocean or geologically unknown areas

Outlines of present landmasses are
for reference only

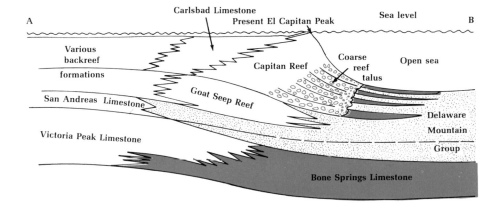

Carlsbad Limestone

Present El Capitan Peak

Sea level

A   B

Various
backreef

formations

Capitan Reef

Coarse
reef
talus

Open sea

San Andreas Limestone

Goat Seep Reef

Victoria Peak Limestone

Delaware

Mountain

Group

Bone Springs Limestone

Figure 12.13 Restored cross section through the Permian rocks of the Guadalupe Mountains of west Texas. Length of section is about 64 kilometers, and position is shown by line A-B in Figure 12.12. The section portrays the growth of the great reef complex forward and upward with the relative rise of sea level. (Adapted from P. B. King, U.S. Geol. Survey Prof. Paper 215, 1948, and N. D. Newell et al., The Permian Reef Complex of the Guadalupe Mountains Region, Texas and New Mexico, W. H. Freeman and Company, copyright 1953.)

Figure 12.14 *Aerial view of the Cape Peninsula, Union of South Africa. Stratified rocks of early Paleozoic age make up Table Mountain, which rises above Cape Town. (South African Information Service.)*

Figure 12.15 *Arrangement of the continents around the South Pole during the late Paleozoic. The white areas indicate the location of present-day glacial deposits, and arrows show the movement of ice. (Adapted from Warren Hamilton and David Krinsley, "Upper Paleozoic Glacial Deposits of South Africa and Southern Australia," Bulletin of the Geological Society of America, Vol. 78, June 1967.)*

THE BEGINNING
OF THE KAROO SYSTEM

The *Karoo*, a geographical part of the Cape Province of South Africa, gives its name to a sequence of rocks that contributes much to an understanding of the geologic history of the Southern Hemisphere and of the land life of the time. The Karoo System reaches a thickness of 6,000 meters and was deposited almost entirely under continental conditions. Although the Karoo deposits are now preserved only in scattered patches, they were probably once united in one huge downsinking basin. Deposition started early in the Pennsylvanian and continued without interruption through the Permian and Triassic into the Jurassic. At the base is the Dwyka Tillite, the previously mentioned deposit of glacial origin. This reaches a thickness of 700 meters and is composed of a heterogeneous mass of rock fragments resting, in places, upon scratched and polished surfaces across which the glacial ice has slowly traveled.

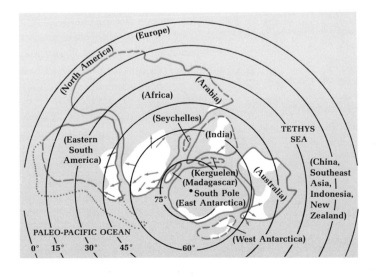

Figure 12.16   Glacial striations of Dwyka age on the Otavi Sandstone about 21 kilometers west of Ohopoho, Kaskoveld, southwest Africa. (Geological Society of South Africa, photograph by E. S. W. Simpson.)

Above the Dwyka lies the *Ecca Series* which contains the best coal of Africa and a variety of freshwater fossils. Among these remains are the first fossil leaves of the *Glossopteris flora*, which became widespread on all southern continents (see discussion later in this chapter). Following the Ecca is the much thicker *Beaufort Series*, famous for the abundance and variety of its vertebrate remains. There are a few fish and amphibians, but mostly reptiles, both herbivorous and carnivorous. Scores of species have been described, and the series has been split into a number of zones, each characterized by its particular life forms. These fossils give the most complete and comprehensive picture of the early evolution of the reptiles. The Permian-Triassic boundary lies within this series. The Mesozoic portion of the series will be described in the next chapter.

Life of the Late Paleozoic

PLANTS

The plant kingdom advanced and diversified tremendously during the late Paleozoic. The most important developments took place on the continents, where the spread of vegetation was amazingly rapid, considering how barren the long preceding periods had been. The development of *vascular tissue* (tissue that conducts sap and nutrients) and of cellulose as a building material made possible the growth of large upright trees capable of withstanding high winds and the pull of gravity. Wood in the broad sense became the chief plant tissue in response to the need for a strong, flexible, yet porous conducting medium to lift the leaves toward the sunlight and at the same time provide an effective pathway for water and nutrients from the roots.

During the Paleozoic, plants with seeds began to replace those with spores, and there was a growing independence from water as the transporting medium for spores. Wind pollination became much more effective as plants pushed their limbs upward into the air. Seeds, which are essentially an additional food supply that gives the plant embryo a greater chance to survive, appeared in the Mississippian Period. During the Pennsylvanian and Permian, certain plants acquired the ability to survive in semiarid or cold regions. Although fossils from upland areas are scarce, there are a few, such as from the *Hermit Shale* (Permian) in the Grand Canyon, that show that desert-type vegetation was in existence during the late Paleozoic.

Devonian Floras. Our knowledge of Early Devonian land plants is rather imperfect, although remains have been discovered in Europe, North America, China, and Australia. These early plants have been known collectively as the *psilophytes flora,* from a common genus, *Psilophyton.* It is interesting that two simple plants that appear to be descendants of this primitive land flora still exist in the tropics. In general, these earliest known land plants lacked roots and leaves and bore spores at the ends of simple branching stems.

More advanced types of vegetation appeared in the Middle Devonian. The best-known plants of this age are from a silicified peat bog found in the Old Red Sandstone at Rhynie, Scotland. Here, well-preserved representatives of three genera of vascular plants occur. Two of these ancient plants, *Rhynia* and *Hornea,* had no

Figure 12.17 Model of the Middle Devonian plant Rhynia. The simple, leafless stems are about 7 centimeters high. (Field Museum of Natural History, Chicago.)

leaves and roots and were only a few centimeters high. They had spores suitable for dispersal on land and a central strand of conducting tissue. The third genus, *Asteroxylon,* was somewhat larger than the other two and had scalelike leaves.

There is a distinct break between the flora of Middle and Late Devonian time, marked by the decline of the psilophytes flora and the rise and broad dissemination of a new group called the

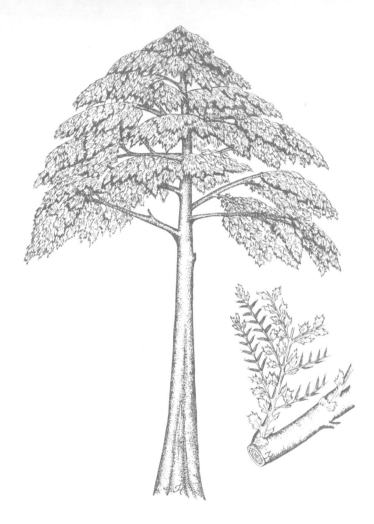

Figure 12.18 An artist's conception of the appearance of Archaeopteris, *dominant genus of Late Devonian floras. (Drawing by Naoma E. Hebbert.)*

perse their spores or seeds more efficiently. They were mainly lowland types and grew along shores and valley bottoms. The fossil record they left behind is fairly complete and, in addition, their modern descendants tell us a good deal about their structure and habits. There were scouring rushes, or horsetails (today's *Equisetum* is a descendant), seed ferns (*Archaeopteris, Protopteridium*), and lycopods (spore-bearing plants) such as *Protolepidodendron* and *Archaeosigillaria,* which were destined to give rise to immense forests in the following period. The most primitive gymnosperm, *Callixylon,* is found in Late Devonian rocks.

These Late Devonian plants produced the earliest known fossil forests. Near Gilboa, New York, numerous stumps up to slightly over 1 meter in diameter have been uncovered. In one quarry at least eighteen stumps were found in a space of about 5 square meters, all in an upright position and with their roots in shale beds and their trunks surrounded by sandstone. Geologists estimate that these trees were 9 to 12 meters high.

By the close of the Devonian, plants had acquired the adaptations necessary to spread over much of the face of the earth. During the Mississippian Period plants evolved steadily, but conditions governing growth and fossilization were somewhat unfavorable, although true ferns, club mosses, and horsetails were rather common. The first known seed plants, the pteridosperms, are found in rocks of earliest Mississippian age.

The Carboniferous Coal-forming Swamps. The Carboniferous Period was ideal for swamp vegetation and became the great period of coal formation. The fossil record is exceptionally complete because mining operations have yielded a wealth of specimens. Among the most important contributors to the coal beds were the so-called *scale trees, Lepidodendron* and *Sigillaria,* both lycopods. *Lepidodendron* had a slender trunk and a crown of forking branches that soared skyward to heights of over 30 meters. The leaves were lance-shaped and were arranged in

Archaeopteris flora. Fossil remains of *Archaeopteris* have been found in such widely scattered areas as Russia, Ireland, Ellesmere Island to the north of Canada, and Australia. It had large fernlike leaves with clusters of spore-bearing organs; some regard it as a possible link between the ferns and seed-bearing plants. As a whole, the dominant Late Devonian plants were larger than their Middle Devonian relatives and had branching root systems, stronger stems, and could dis-

spirals around the branches. When the leaves were shed, they left a characteristic pattern of leaf scars on the branches, which gives the tree its popular name. Only about 10 percent of the trunk of *Lepidodendron* was actual woody tissue; it was therefore not as sturdy as most living trees. In *Sigillaria* the leaf scars are in vertical rows.

Also inhabiting the coal forest were many jointed plants called *Sphenopsida.* The leaves of these plants radiated in whorls around the joints, as in the modern horsetail, *Equisetum.* Largest of the jointed plants was *Calamites,* which reached a height of over 30 meters and a diameter up to 1 meter. These trees had a hollow or pithy central cavity that often filled with sediment and preserved a cast of the inside of the tree.

In addition to the important coal-forming plants just described, there were minor types that were destined to give rise to important vegetative groups in later periods. True ferns made up much of the undergrowth on the forest floor. The gymnosperms (wind-pollinated, nonflowering plants), which were to dominate the Mesozoic, were represented by at least five orders: the *Pteridospermae, Cordaitales,* and *Bennettitales,* which are extinct; the *Ginkgoales,* represented by one living species; and the *Coniferales,* which make up a key segment of present-day vegetation.

Figure 12.19   Pennsylvanian plant fossils. (a) An impression of the trunk of Lepidodendron rimosum, *a scale tree, showing the characteristic diamond-shaped leaf scars. (b) Various types of foliage preserved in nodules found at Mazon Creek, Illinois. Three of the specimens represent halves of nodules that can be fitted together. Remains of insects and small vertebrates are occasionally found in nodules from this locality. (Field Museum of Natural History, Chicago.)*

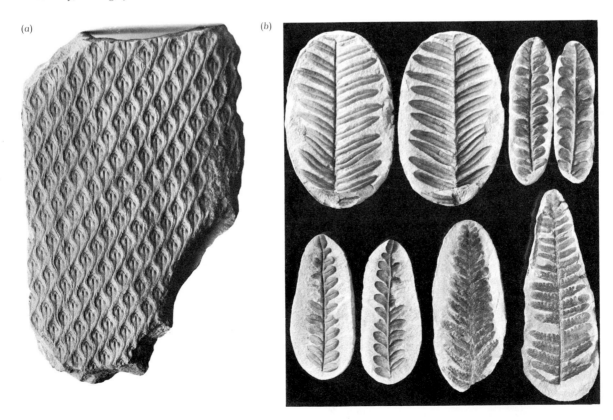

(a)

(b)

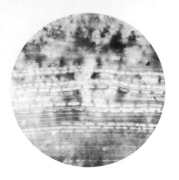

Figure 12.20 *Highly magnified, thin-sections of coal showing remnants of cell structure in compressed and altered vegetative material. (U.S. Bureau of Mines.)*

Coal as Fossil Vegetation. Coal is the most important by-product of past life known to man. It has accumulated in prodigious quantity at many times and places when the right biological and physical factors have happened to coincide. All evidence indicates that coal consists of altered plant remains. Every stage in the process of alteration has been thoroughly studied. Commencing with peat, which is easily studied in the process of formation, the successive ranks of coal are *lignite, sub-bituminous, bituminous,* and

Table 12.1 *Constituents of Cyclothems*

Top
10. Shale with ironstone concretions
9. Marine limestone
8. Black shale with black limestone concretions or layers
7. Impure, lenticular, marine limestone
6. Shale
5. Coal
4. Underclay
3. "Freshwater" limestone
2. Sandy shale
1. Sandstone, locally unconformable on underlying beds
Base

anthracite. The rank depends on the conditions to which the plant remains were subjected after they were buried—the greater the pressure and heat, the higher the rank of the coal. Higher-ranking coals are denser, contain less moisture and volatile gases, and have a higher heat value than lower-ranking coals. Obviously, more than a heavy growth of vegetation is required for coal formation. The plant debris must be buried, compressed, and protected from erosion and from intensive metamorphism.

The complete story of coal formation involves not only a study of the coal forests themselves but also a consideration of the types of sediment in which the coal is buried and an understanding of the processes acting on the plant material after it has been buried.

Coal and Cyclothems. Even if all the biological, geographic, and climatic factors were favorable, coal still could not form unless plant debris was submerged and buried by sediments. More than a mere downsinking of the vegetated land is involved. Simple uninterrupted subsidence would permit only one bed of coal to be buried, and the area thereafter would become permanently covered by the sea or by deep-water sediments. The importance of the geologic factor becomes apparent when we realize that many coal beds lie one above another in the same area. It is not uncommon to find five or ten superimposed beds, and in some places as many as fifty beds can be counted, one above another.

A regular sequence of repeating or cyclic sediments of which coal is a normal member is called a *cyclothem.* Its constituent units are shown in Table 12.1.

Cyclothems, with their individual coal beds, are repeated one above the other many times in the coal-forming areas. Regular repetitions of rock types without coal are not technically called cyclothems. They may form in shallow water or arid basins and are termed *rhythmites.* These too are common in the Pennsylvanian Period and must be related in some way to cyclothems. A number of possible mechanisms have been proposed to account for the alternate flooding and emergence that produced the cyclothems. All theories agree that there must be a prolonged regional downsinking of coastal areas. One school of thought maintains that the downsinking is not steady, but alternately slows down and speeds up in giant pulsations. During the pauses, coal swamps flourish; during more rapid downward movements, the sea spreads inland, burying and compressing the vegetation of the former swamps. An opposing school of thought attributes the cyclic effects to the rise and fall of sea level superimposed on the sinking lands. During emergent, or low-water, periods forests flourish; during flooding they are covered with water and sediment. Supporters of this idea point out that the period of coal formation in the Northern Hemisphere was the time of glaciation in the Southern Hemisphere and that alternate periods of freezing (withdrawal of water from the sea) and melting (restoration of water to the sea) could have caused changes in sea level. The whole problem is in an interesting state of speculation, and no one theory is accepted by all students of coal geology. What we do know is that a peculiar combination of apparently unrelated factors came into existence and operated over a period of many millions of years to produce many raw materials and a large part of the coal resources on which we depend for a great deal of our energy.

Here let us turn for a moment to the geologic and geographic environments in which the coal-forming vegetation grew. Wide expanses of level ground at or near sea level are necessary before continuous swamps can form. Elevated or hilly terrain may support ample vegetation, but it obviously cannot become a site of coal deposition.

During the Pennsylvanian Period, large areas of Europe and North America were ideally suited for the spread of swamp vegetation. The Appalachian Geosyncline had been filled to overflowing with a variety of sediment, and there were few patches of high ground in the interior of the

Figure 12.21 *Slab of Permian siltstone with fossil leaves of* Glossopteris browniana, *Transantarctic Mountains, Antarctica. (Institute of Polar Studies, photograph by David H. Elliot.)*

continent. An unbroken expanse of lowland extended from New York to Texas. In Europe and Asia the coal-forming areas were not so extensive, and many were situated in interior basins not connected with the sea; nevertheless, the delicate balance that existed between sedimentation and earth movements permitted coal to form. Climates of the coal-forming areas appear to have been moist and warm but not always tropical. A great deal of coal formed under temperate conditions and some even in close proximity to glaciers.

The Permian Glossopteris *Flora.* An important and much-studied aggregation of fossil plants appeared in Gondwana during the Permian. It is dominated by the genus *Glossopteris* from which it takes its name—the *Glossopteris flora.* The leaves have a distinctive tongue-shaped outline and have numerous fine, branching veinlets. *Glossopteris* is found in the Permian of India, Australia, South Africa, South America, Madagascar, the Falkland Islands, and Antarctica. Although there are numerous fossil leaves and the plant is known to have borne simple seeds it is not known whether it was treelike or not. About forty different species have been described, and from the association with glacial deposits it is thought that *Glossopteris* may have been adapted to cool or temperate conditions.

Glossopteris is especially important because it provides evidence for continental drift. At a very early date biologists realized that this plant could not have spread to its present fossil localities across thousands of kilometers of open water. This argument still holds true and has been greatly strengthened by evidence from other sources.

Figure 12.22   A spectacular slab of fossil crinoids from Mississippian rocks near Legrande, Iowa. Preservation of entire individuals such as these is very rare. (Iowa State Department of History and Archives.)

The invertebrates of late Paleozoic time show few spectacular advances and reveal less of evolutionary significance than either the plants or vertebrates of the same period. In general, it was a time of intense competition in the seas, with a consequent gradual elimination of less effective types and a flowering of more progressive ones.

Aquatic animals had already achieved most of their basic adaptations before the late Paleozoic and had settled down, as it were, to strong competition involving mainly minor changes. As a rule, there was a greater number of relatively mobile creatures in the ocean than there had been in earlier periods. Legs, swimming organs, and fins were more in evidence, and more animals were able to crawl or burrow. Even sluggishly moving animals such as starfish, echinoids, and sea cucumbers gradually replaced the fixed members of their phylum typified by the cystoids and blastoids. Crawling mollusks were on the increase, and the cephalopods already had developed their coiled, many-chambered, buoyant shells to high levels of mobility.

At the same time, there is greater evidence of both defensive and offensive mechanisms and structures. Brachiopods were abundant, with a preponderance of thick-shelled and highly spiny types. As if to balance the scales, many species of sharks with heavy, flat teeth, well adapted for shell crushing, cruised the seas. Slow-moving animals with weak skeletons, such as trilobites, inarticulate brachiopods, and graptolites, were either extinct or in decline as the more mobile or better-protected forms flourished.

Food-gathering and food-protecting techniques improved. Passive feeders, which depended on the vagaries of passing currents for their sustenance, were gradually replaced by animals able to move about or at least able to gather food by self-generated currents. Virtually all brachiopods of the late Paleozoic possessed more powerful food-gathering mechanisms than their ancestors. Particularly efficient in this respect were the stalked crinoids, whose food-gathering arms branched and subdivided to comb and filter wide areas of water. As the crinoids flourished, their relatives, such as cystoids and blastoids, with less efficient food-collecting systems, declined and disappeared.

Finally, we should mention the evolutionary advances made by the reef-building and colonial animals. These creatures were able to secrete interlocking structures of great size, in which both plants and animals cooperated. The Permian reefs of Texas are examples. Here, by presenting a sufficiently strong front to the waves, organisms were in a position to find food and living space that might otherwise go unclaimed.

Although late Paleozoic marine species were more advanced than their predecessors, they still had not achieved maximum efficiency or security, as we shall see later.

The late Paleozoic saw the rise and fall of many important groups of invertebrate animals. Important debuts included the *fusulinids* in the Mississippian, cephalopods with moderately crinkled septa in the Early Mississippian, cephalopods with highly crinkled septa in the Middle Permian, land gastropods in the Pennsylvanian, spiny brachiopods in the Late Devonian, and insects in the Middle Devonian.

The corals, bryozoans, brachiopods, gastropods, and pelecypods managed to maintain a relatively stable level of existence and were common and abundant during the late Paleozoic. These groups contribute the vast bulk of the marine fossils, and many thousands of species of each have been preserved. Within each group there were significant trends, marked by the extinction and replacement of individual families. Among the important coral genera were *Halysites* (Late Ordovician to Early Devonian), *Favosites* (Early Ordovician to Permian), *Chaetetes* (Ordovician to Jurassic), and *Syringopora* (Silurian to Pennsylvanian). An important group of late Paleozoic bryozoans were the lacy

Figure 12.23 *Blastoid calyxes from Mississippian rocks of the midwestern United States. The large specimen and the five smaller ones at right belong to the genus Pentremites; the two middle specimens are Globoblastus. The large specimen is about 2 centimeters long.*

Figure 12.24 *Fenestellid bryozoan, Upper Mississippian rocks, central Utah. Complete specimen (a) is an entire colony and illustrates why this is known as the "lace fossil." Enlarged view (b) shows the individual living chambers. Bryozoans are intensively and exclusively colonial animals. Specimen measures about 9 centimeters across.*

(a)

(b)

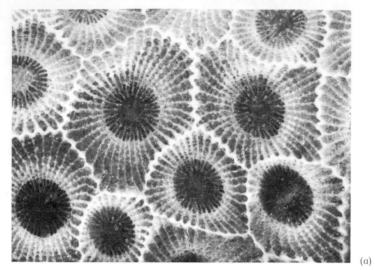

(a)

(b)

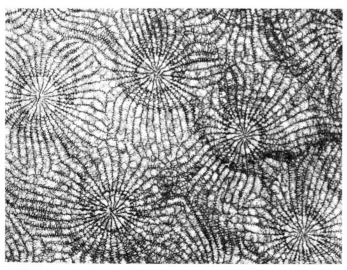

(c)

forms, or *fenestellids* (Silurian to Permian), which left delicate fossils representing colonies of many individual animals. Also prevalent was the curious screw-shaped bryozoan *Archimedes* (Mississippian to Permian). The most distinctive brachiopods during the late Paleozoic were *spirifers* (Silurian to Jurassic), characterized by pointed, wide, and winged shells, and the very spiny *productids* (Devonian to Permian), some of which reached 30 centimeters in diameter. Pelecypods were represented by a variety of forms but were still subordinate to brachiopods in late Paleozoic seas. Gastropods of many types abounded locally but made no special advances, except for the evolution of certain air-breathing species.

The eurypterids (Ordovician to Permian), which have been aptly described as "sea scorpions," were the largest known arthropods of the late Paleozoic. Their jointed external skeletons ranged in size from a few centimeters to over 2 meters. They reached the apex of their development in the Silurian and Devonian and gradually declined thereafter. We may assume that they competed with the early fish, especially during the Ordovician and Silurian Periods.

Although the fossil record is incomplete, we can trace the broad outlines of the evolution of land-living arthropods during the late Paleozoic. The first known insects are from Middle Devonian Rhynie Chert, which yielded the well-preserved plants mentioned earlier. Over four hundred forms of insects are known from Lower and Middle Pennsylvanian rocks. Many were large by present-day standards, and the record is held by a dragonfly with a 71-centimeter wingspread found in Belgium. Spiders, cockroaches, centipedes, and scorpions appear to have been com-

Figure 12.25 *Cut and polished specimens of three Devonian colonial corals.* (a) Hexagonaria, (b) Prismatophyllum, *and* (c) Billingsastrea. (c, *courtesy of W. A. Oliver, Jr.*)

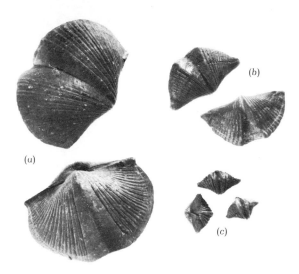

Figure 12.26 Spiriferoid brachiopods of the late Paleozoic periods. (a) Paraspirifer, from the Middle Devonian, (b) Mucrospirifer, common in the Appalachian region, and (c) Punctospirifer kentuckiensis, of Pennsylvanian age, common in the mid-continent.

Figure 12.27 Spine-bearing brachiopods (productids) from Late Permian rocks of the Khisor Range, West Pakistan. Specimens have been etched from the enclosing limestone with their fragile spines almost intact. These samples are about 1 centimeter across. (Courtesy of Richard E. Grant, Smithsonian Institution.)

mon. The attainment of flight evidently occurred during the Mississippian Period. Of interest is the recent discovery of an insect wing in Permian rocks of Antarctica.

Guide Fossils. Among the great variety of invertebrates inhabiting the late Paleozoic seas, it is not surprising to find a number of groups useful as guide fossils. In the Devonian the species that have furnished the more useful, widely distributed, and distinctive forms are the trilobites, corals, brachiopods, conodonts, and ammonoids. Brachiopods were extremely varied, and some genera spread from continent to continent. The same is true of Mississippian, Pennsylvanian, and Permian forms.

Beginning in the Late Devonian, the coiled cephalopods with complex internal structures began to be abundant and widespread. About ten worldwide zones are recognized in the Devonian, seven in the Mississippian, six in the Pennsylvanian, and five in the Permian.

Excellent guide fossils are furnished by the toothlike conodonts. These reach a high point of diversity and usefulness in the Late Devonian, in which twenty-nine zones and subzones are recognized. With minor exceptions these zones

(a)

(b)

Figure 12.28   Conodonts of Pennsylvanian age. (a) Lonchodina, (b) Hindeodella. Specimens are about 0.16 centimeter long. (Courtesy of Dwayne D. Stone.)

marine Pennsylvanian and Permian rocks are the protozoans called *fusulinids*. Their small, spindle-shaped skeletons resemble grains of wheat, oats, rice, or rye. They have left their skeletons abundantly in limestone, sandstone, and shale, indicating a cosmopolitan type of existence. Although simple in outward appearance, their internal structure is very complex, consisting of numerous coils separated by curving partitions into a great

(a)

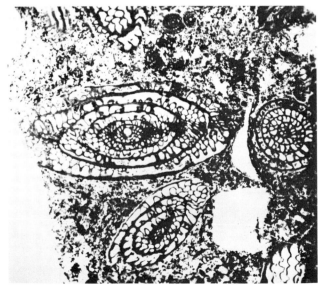

(b)

occur in proper order in Late Devonian rocks of all major continents. Because the Late Devonian is estimated to have had a duration of about 20 million years it is possible to divide the rocks into zones of about 750,000 years, which is extremely precise in geologic work. Conodonts are also useful in all other portions of the late Paleozoic but in a less spectacular way.

In North America the best guide fossils for

Figure 12.29   Fusulinids. (a) A collection of wheat-shaped specimens of Triticites. (b) A highly magnified, thin section showing the complex internal structure. In the center is a specimen cut along the long axis; to the right of it is a specimen cut at right angles to the long axis to show the spiral coiling and chambers.

many chambers. The internal details are characteristic for the different species, and because the fusulinids evolved rapidly, we can distinguish the fossils of one interval or formation from those of succeeding or preceding ones.

Fusulinids appeared during the Mississippian as very small, flat, coil-shelled creatures and increased in size and complexity during the Pennsylvanian and Permian. The last representatives were as much as 7 centimeters long and 1 centimeter thick. Because fusulinids are generally quite small and tend to occur packed together in dense masses, they are frequently brought up in cores from wells and are very useful in correlating certain oil-bearing formations.

Figure 12.30 An ostracoderm (Cephalaspis lyelli) from the Old Red Sandstone of Scotland. The crescent-shaped head shield is characteristic. Length about 17 centimeters. (By permission of the Trustees of the British Museum, Natural History.)

VERTEBRATES

The Devonian Period is known as the *Age of Fishes*, because, for the first time, the fossil record reveals the existence of numerous and varied fish forms that represent a distinct evolutionary advance over the contemporary Paleozoic invertebrates. In fact, so many different kinds of fish appeared during the Devonian that paleontologists have had a difficult time classifying them, and we can mention only a few here.

As marine life diversified, competition increased. The less fit were eliminated; the more advanced survived and improved. Among those eliminated were the jawless ostracoderms, a few of which had lingered on from the Silurian. These creatures were the first large group of vertebrates to become extinct. Forms that enjoyed a temporary success and then disappeared before the close of the period were the *antiarchs* (spiny sharks) and the *arthrodires* (joint-necked fish). The antiarchs were sharklike in shape, with several pairs of spiny fins and various types of scales; the arthrodires had jointed necks that permitted the skull to be moved up and down. One of the arthrodires, *Dinichthys,* from the Late Devonian rocks of Ohio, was 9 meters long, probably the largest animal of the time.

Figure 12.31 Fossil fish from the Old Red Sandstone. Geologists feel that specimens containing packed remains such as these suggest wholesale death in pools that dried up during dry periods. (Crown copyright, Geological Survey photograph. Reproduced by permission of the Controller of Her Britannic Majesty's Stationery Office.)

Success of the Fish. During the Late Devonian two classes of fish began to establish their superiority: the *bony fish* (*Osteichthyes*) and the *cartilage fish* (*Chondrichthyes*). The bony fish, representing the most numerous, varied, and successful of the aquatic vertebrates, include the

Figure 12.32 Well-preserved fossil of the Late Devonian crossopterygian fish Eusthenopteron foordii. This fish is near the line leading to the first land-living amphibians. (Courtesy of Erik Jarvik.)

Figure 12.33 Internal skeleton of the forelimb or pectoral fin of Eusthenopteron foordii. The basic elements of the walking limb of land vertebrates are plainly evident. (Courtesy of Erik Jarvik.)

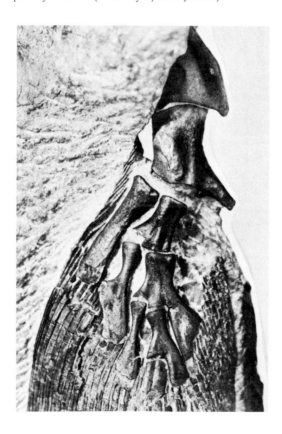

vast majority of living fish and many extinct forms. They are adapted for life in both salt and fresh water and have lived in practically all water environments on earth. Their story is mainly a chapter in later geologic time.

Included in the Osteichthyes class is a less numerous and rather unimpressive group, the *nostril-bearing fish (Choanichthyes),* which can take in air through their nostrils as well as through their mouths. This group includes the Devonian ancestors of the modern *lungfish,* which are all characterized by large, flattened leaflike teeth and the ability to survive dry periods by burrowing in the moist beds of streams or lakes. In addition to the lungfish, the Choanichthyes include the *crossopterygians,* or *lobe-finned* fish, which are in the direct line of evolution from fish to land-living vertebrates and are now represented by the solitary "living fossil," *Latimeria.* In these fish the fin is a solid, muscular structure with a central axis of bones—a structure that proved crucial in the drive to mount the land.

Out of the Water. Late in the Devonian Period certain lobe-finned fishes succeeded in leaving the water and establishing themselves on land, a step made possible by a great many structural and functional modifications. The lobe-fin, with its axis of internal bones, had to be converted into a walking limb, the lung was adapted to breathe air just as it now does in the lungfish, and the circulatory and excretory systems were modified along much the same lines as in present-day tadpoles. These processes are only partly indicated by fossil evidence, but the change in skeletal framework is clearly shown in some exceptionally well-preserved remains. The earliest known amphibians, called *ichthyostegides,* of which the genus *Ichthyostega* is a typical example, are found in Late Devonian rocks of Greenland. *Ichthyostega* was about 60 centimeters long and possessed a strange mixture of newly acquired characteristics and older traits inherited from its fish ancestors. Its tail was long and still retained a fringelike fin; its body was sprawling, and the creature must have presented an ungainly picture moving over the land. The animal's legs and feet were too weak to support its body for any length of time, and it may have returned frequently to the water. Its fishlike skull can be compared bone for bone with the skulls of its fish relatives that never left the water. It possessed many pointed conical teeth with a peculiar structure that occurs also in its crossopterygian ancestors. This tooth structure, known as *labyrinthodont,* is characterized by deep infolding of the enamel. All in all, the fossil record of the skeletal changes that accompanied the important transi-

tion from water to land is quite complete. Taken in connection with what we can learn from still living forms such as salamanders, lungfish, and *Latimeria,* we can reconstruct the process with considerable accuracy.

We cannot, however, be sure just why the crossopterygians invaded the land. Certainly competition in the water had become very severe, and the land offered food and protection to any animals that could live out of water. It seems logical that they made their first approach to land across the moist and sandy beaches where food, cast up by the tides, was generally available. Here too the opportunity or necessity of digging in the moist sand favored the development of stout limbs with toes. The enticement of live food in the forests in the form of various arthropods eventually led the first pioneers to abandon the shores and to forsake the water completely.

Success of the Amphibians. During the Mississippian and Pennsylvanian Periods, *amphibians* became the dominant land animals. The Mississippian, with its extensive shallow seas, was not so favorable to their expansion as was the suc-

Figure 12.34 Drawing and reconstruction of the earliest known amphibian (Ichthyostega). *The specimen, discovered in the uppermost Upper Devonian of East Greenland, is about 1 meter long.* (Courtesy of Erik Jarvik.)

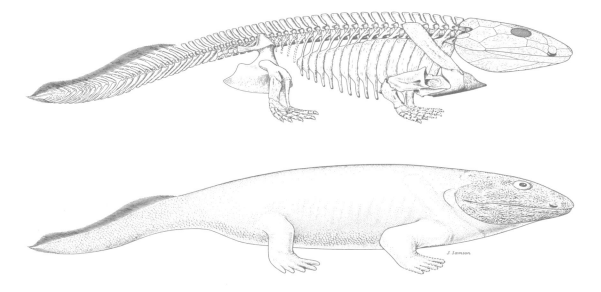

Figure 12.35   Skeleton of Trematops, a sprawling, flat-headed amphibian from the Early Permian of Texas. (Field Museum of Natural History, Chicago.)

Figure 12.36   Restoration of Paracyclotosaurus, a 3-meter-long amphibian from the Permian of New South Wales, Australia. (By permission of the Trustees of the British Museum, Natural History.)

ceeding Pennsylvanian, with its great swamps, mild climates, and luxuriant forests of coal-forming plants. Such surroundings were obviously ideally suited to the amphibians' way of life, and most of our knowledge about Pennsylvanian vertebrates comes from specimens found

in coal-bearing rocks. From these remains we can reconstruct many types of amphibians—some with lizardlike shapes, some resembling snakes, and some much larger, with contours similar to those of crocodiles and salamanders. We find not only the full-grown animals, but also many remains of immature or tadpole stages. Associated with the bones of the amphibians are the fossilized remains of the creatures they preyed on: spiders, centipedes, scorpions, and a variety of winged insects. In the world of the swamp forests, the amphibians ruled supreme, preying on any of the lower forms of life with which they came in contact.

Challenges of Life on Land. The gradual drying and cooling that characterized the Permian Period caused a corresponding decline in the number and variety of amphibians. Those that survived were confined mainly to watercourses in dry regions, and we find their remains chiefly in red-bed types of sediments. These later amphibians were adapted to a large extent for life out of water, but they were awkward and inefficient by comparison with later land animals. As we look at their skeletons, we are impressed by the clumsy sprawling legs, heavy tails, and immense flattened heads. Their skeletons tell us that one of the amphibians' chief foes was the force of gravity. Merely to raise their bodies off the ground must have required intense physical effort, and really rapid locomotion was out of the question.

Many adaptive changes in the structure of the backbone strengthened it for its new role of supporting the weight of the amphibian's body. The changes involved mainly the development of interlocking devices and processes for muscle attachment. These variations in the structure of the backbone serve as a basis for classifying the amphibians. The ancestral amphibians had rather deep, fishlike heads, about as wide as they were high. With the passage of time the skull gradually flattened until the entire head was many times wider than it was thick. This curious

Figure 12.37   *Reconstructed scene depicting the ancestral reptile Petrolacosaurus. Specimen is from the rocks of Late Pennsylvanian age near Garnett, Kansas. Fossils of the other plants and animals depicted in the illustration were found in the same deposit. (From F. E. Peabody, courtesy of the University of Kansas.)*

adaptation eventually enabled the animal to open its mouth by raising its upper jaws and skull while its lower jaw lay flat on the surface on which the animal rested. The usual method of chewing by lowering and raising the lower jaw is obviously inefficient if an animal must raise its entire body in order to permit the lower jaw to operate. Because virtually all groups of late amphibians developed broad, flat skulls, we suppose that the same principle of conservation of energy was acting on them all.

The term *stegocephalian* (roof-headed) is applied to the larger flattened amphibians of the Pennsylvanian, Permian, and Triassic Periods. Although this type of amphibian lingered on into the Triassic, it was already declining at the close

of the Paleozoic. Frogs and toads were products of a later time, and investigators have found no trace of their precise ancestors among Paleozoic fossils.

The Coming of the Reptiles. All evidence indicates that reptiles developed from amphibians some time during the Mississippian or Pennsylvanian. A few isolated bones that some geologists assign to reptiles have been found in Mississippian deposits, but the first complete and authentic reptile skeletons have been taken from rocks of Pennsylvanian age. As you might expect, the distinction between the earliest reptiles and their amphibian ancestors is not very clear, and the fine points of classification must be left to special-

Figure 12.38 *The oldest known vertebrate egg. Although crushed and distorted, this object has been identified as an egg by the structure of the shell material. It was found in Early Permian red beds of Texas. (Courtesy of A. S. Romer, Museum of Comparative Zoology at Harvard College.)*

Figure 12.39 *Reconstruction of the reptile Lycaenops from the Permian beds of South Africa. The skeleton, although mammal-like, exhibits many reptilian characteristics. Notice the presence of two bones in the lower jaw, the small brain case, the absence of external ears, the ribs on the neck vertebrae, the small scapula and pelvis, the absence of a kneecap, and the long, heavy tail. (Reproduced with permission from Edwin H. Colbert, Evolution of Vertebrates. Copyright 1955 by John Wiley & Sons, Inc.)*

ists. The earliest known reptile is *Romeriscus* from early Pennsylvanian rocks near Port Hood, Nova Scotia. The remains are small and fragmentary, but the identification appears to be correct.

Another important link in the history of the reptiles was discovered in Late Pennsylvanian rocks near Garnett, Kansas. Here, the complete, well-preserved, 61-centimeter skeleton of a small, lizardlike reptile was found in natural association with a large number of other plant and animal fossils. This animal, called *Petrolacosaurus*, had a delicate, well-constructed skeleton, sharp teeth, and the general appearance of an alert, swift-moving animal. In comparison with contemporary amphibians, *Petrolacosaurus* shows clearly the improved adaptations to land life that the reptiles had by this time achieved.

Not to be overlooked is the one unique and basic attainment that separates reptiles from amphibians—the ability to produce an egg that can be laid and hatched out of water. Just how did the reptiles develop this capability? On this point the fossil record is silent, but it is significant that a fossil egg, the oldest known, has been discovered in Permian rocks of Texas. The ability to produce an egg complete with food, water, and oxygen supply encased in a protective but porous shell entails many basic adjustments that involve intricate chemical and mechanical processes.

Long before the Permian Period ended, the amphibians had sharply declined and reptiles were spreading over the continental areas. Re-

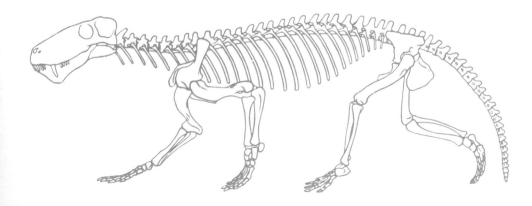

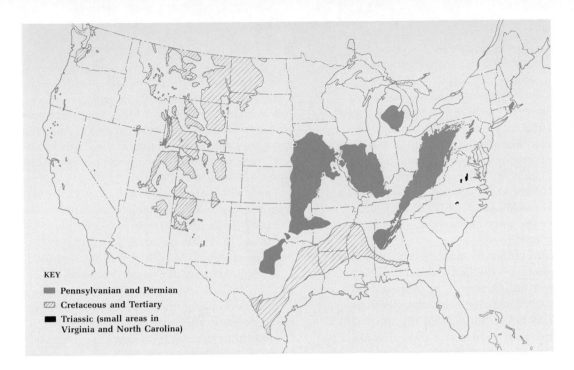

KEY

▰ Pennsylvanian and Permian
▨ Cretaceous and Tertiary
▪ Triassic (small areas in
  Virginia and North Carolina)

Figure 12.40   Coal-bearing regions of the contiguous forty-eight states. Geologic age of the deposits is shown by the legend. Alaska also has large reserves; Hawaii has none. (U.S. Bureau of Mines.)

mains of late Paleozoic vertebrates are scattered over all the great continents, but the most complete and continuous record is found in the Karoo Basin in South Africa. The most characteristic animals of the Karoo belong to the *mammal-like reptiles (Therapsida)*. This group, which includes animals of various sizes, developed adaptations similar to later mammals. Some had well-differentiated teeth, the limbs were pulled in, not sprawling, there were fewer ribs, the tail was smaller, and the whole skeleton was light and well constructed. These were all reptiles, however; the transition to mammals took place during the Triassic, and it was many millions of years before the reptiles yielded their supremacy to the mammals.

In general, the best records of Mississippian and Pennsylvanian vertebrates are found in the United States, England, and Europe, and the best records of Permian land life are found in South Africa, China, and Russia.

Economic Products of the Late Paleozoic

Mineral deposits of all types were formed during the late Paleozoic periods. Especially significant are the vast stores of fossil fuels: coal, oil, and gas. Coal of Mississippian age is mined in Russia, and Pennsylvanian coal supports heavy industry in the eastern United States, Great Britain, and western Europe. Permian beds are the chief source of coal in India, Australia, South Africa, and China, and are also quite important in the Soviet Union.

Oil and gas are plentiful in late Paleozoic rocks, especially in North America. The Devonian is a major source of oil in Canada, and important Mississippian, Pennsylvanian, and Permian pools are found in interior parts of the United States. Important oil and gas pools recently found in late Paleozoic rocks beneath the North Sea will be of great importance to western Europe, particularly Great Britain. It has been estimated that reserves of Paleozoic oil account for about 10 percent of the world's total, which indicates just how extensive the later Mesozoic and Cenozoic contributions have been.

The extensive mountain-building activity and granitic intrusions of the late Paleozoic, which we usually call the Hercynian or Appalachian orogenies, brought many important metalliferous deposits into existence throughout the affected areas. Tin, lead, zinc, copper, and silver deposits were formed in western Europe, and the iron, copper, chromium, nickel, and chromite ores of the Urals date from this period. Large deposits of sedimentary copper occur in lower Carbon-iferous rocks of south-central USSR. Gold, lead, silver, and other metals were deposited in central Asia, along with tin and tungsten ores in the Malaya-Burma region and tin, zinc, antimony, and mercury deposits in China. Important ores of various metals were deposited also in Australia and New Zealand. By contrast, there are apparently very few metal deposits of this age in the Western Hemisphere.

The late Paleozoic was a time of extensive salt deposition, especially in the northern continents. Ordinary table salt, *halite,* is found in the Pennsylvanian of Colorado and Utah, and in the Permian of Texas, New Mexico, and Kansas. It is accompanied by the more valuable *potash salts* in the western localities. Salts of various kinds are also mined from Permian rocks in England, Germany, and Russia.

Important reserves of phosphate were laid down in Late Permian (*Phosphoria Formation*) rocks in Idaho and adjacent states and form the current foundation for a flourishing mineral fertilizer industry.

A SUMMARY STATEMENT

Sediments of late Paleozoic age occur on all the continents, including Greenland and Antarctica. The northern continents were flooded by extensive shallow seas in the Devonian and Mississippian Periods. In the Pennsylvanian and Permian there is a higher proportion of land-laid sediments, including extensive coal beds. The record of marine sedimentation for the Southern Hemisphere is fragmentary except in Australia. Signs of glacial action in the Pennsylvanian and Permian have been found in Africa, Antarctica, Australia, South America, and India. This feature, taken in connection with similarities in fossil forms and igneous activity, is considered by many geologists to indicate that all southern continents were then joined in one super landmass called Gondwana.

Marine life diversified and expanded. Brachiopods, especially spirifers and productids, were numerous. Echinoderms such as blastoids and crinoids reached their zenith. Fusulinids were prolific in the Pennsylvanian and Permian. Extensive reefs with varied inhabitants were widespread. Sweeping exterminations affected many families at the close of the era.

The late Paleozoic was a time of evolution and expansion of land life. Forests appeared in the Devonian and became dominant features of the landscape in succeeding periods. Amphibians appeared in the Devonian and were common in the coal swamps. Reptiles appeared in the Late Mississippian or Early Pennsylvanian and had become adapted to life in the continental interiors by the end of the Permian.

FOR ADDITIONAL READING

Newell, Norman D., *et al., The Permian Reef Complex of the Guadalupe Mountains Region, Texas and New Mexico.* San Francisco: Freeman, 1953.

Watson, D. M. S., *Paleontology and Modern Biology.* New Haven: Yale University Press, 1951.

Additional references for the material in this chapter can be found at the end of Chapter 11.

THE MESOZOIC ERA

13

The Mesozoic Era, or "time of middle life," lasted an estimated 162 million years and is divided into three periods as follows: Triassic, 25 million years; Jurassic, 65 million years; and Cretaceous, 72 million years.

During this lengthy interval there were drastic alterations in the distribution of continents and ocean basins through sea-floor spreading and continental drift. Meanwhile, ordinary geologic processes of erosion and sedimentation continued on the continental blocks as they drifted

about, and plants and animals evolved and migrated as various opportunities or necessities arose.

Paleogeographic Reconstructions

During the Permian Period, just before the beginning of the Mesozoic, all the continents were apparently joined together in a single north-south arrangement stretching almost from pole to pole.

Northeastern North America lay against Europe, the bulge of Africa fitted against the eastern and southeastern United States and northeastern Brazil, the bulge of South America was against the west coast of Africa, Antarctica made contact with the southern capes of South America and Africa, and Australia joined Antarctica farther east. India fitted between East Africa and Australia. A large triangular embayment of the Pacific Ocean lay between southern Asia and northeastern Africa. This imposing geographical feature was the Mesozoic expression of the Tethys Seaway.

By the end of the Triassic Period, about 25 million years later, North and South America had separated by several hundred kilometers, and a rift of similar width had appeared between Africa and Antarctica on one hand and India and Antarctica on the other. Australia remained attached to Antarctica but India was fully detached from its previous niche.

During the Jurassic, there was even more displacement between North and South America, and the North Atlantic began to assume significant magnitude as Europe was detached from North America. The splitting-off of North America had reached northward to about the vicinity of southern Greenland by the close of the Jurassic Period. South America and Africa had just begun to separate. India was fully adrift and headed for southern Asia leaving behind it the ever-widening ancestral Indian Ocean. Antarctica, with Australia still attached, was swinging away to the southeast. Tethys almost completely separated the northern and southern landmasses from each other.

Seventy million years later, by the end of the Cretaceous Period, the arrangement of continents and oceans had changed considerably. The South Atlantic had reached a width of about 2,800 kilometers, and Europe and North America were no longer in contact except at the extreme north, where Greenland still linked them together. Antarctica had left South Africa, thereby creating a seaway several hundred kilometers wide, and a rift was developing between Australia and Antarctica. India, moving northward, was narrowing the gap with Asia. Tethys had narrowed because of the northward movement of Africa, and the deeper parts of the geosyncline were where the Alpine-Himalayan chain was due to appear.

By the close of the Mesozoic there were great bodies of water—the Atlantic, the Arctic, and the Indian oceans—that had not been in existence when the era commenced. Primary evidence for these water bodies is the presence of marine formations of Mesozoic age around their margins. It is interesting to note that all the oceans were formed at the expense of the Pacific Basin. There was no great increase or decrease in the total volume of water or the area of land; changes

Figure 13.1 *The positions of the continental landmasses at the end of the Triassic. (Adapted from Robert S. Dietz and John C. Holden, "Reconstruction of Pangaea,"* Journal of Geophysical Research, *Vol. 75, 1970, p. 4946.)*

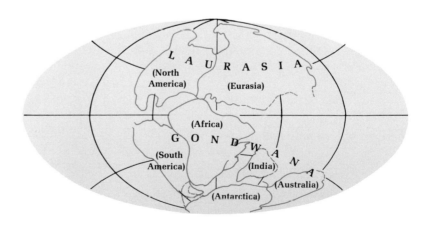

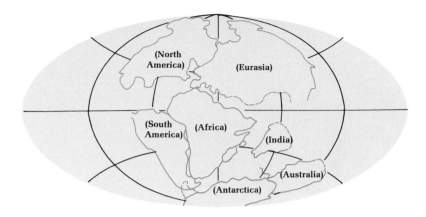

Figure 13.2 *The continental drift dispersion of the continents as of the Late Jurassic. (Adapted from Robert S. Dietz and John C. Holden, "Reconstruction of Pangaea,"* Journal of Geophysical Research, *Vol. 75, 1970, p. 4947.)*

resulted from the major rearrangement of surface features of the earth. The effects of sea-floor spreading and continental drift during the Mesozoic were much greater than any that had occurred before that time, and these movements set the pattern that has continued into the present.

Climatic conditions, while diverse, appear to have been less extreme than during the late Paleozoic. The great glaciers of the previous ice age had melted away and the high mountain ranges of Europe and North America were being lowered by erosion, with consequent amelioration of climate in the Northern Hemisphere. The South Pole apparently lay in the south-central Pacific, whereas the North Pole was alongside what is now Japan in northeastern Asia. Perhaps the location of the poles over or near open water inhibited the accumulation of glacial ice, so that the entire span of the Mesozoic was relatively more mild and equable than succeeding or preceding intervals.

TETHYS—THE GREAT LIMY SEAWAY

The greatest geosynclinal belt on earth, known as Tethys, lies south of the Baltic and Angara shields and north of the African platform and Indian block, as we indicated in Chapter 11. The history of its earlier depositional phase has been outlined in previous chapters; its dramatic elevation into great mountain ranges is yet to come. During the Mesozoic, Tethys was still downsinking and was receiving great thicknesses of sediment. Historically, the Triassic, Jurassic, and Cretaceous Periods were first named and studied intensively in or near the Tethyan belt, and many important concepts were originated there.

By comparison with rocks of similar age elsewhere, the sediments of the Tethys Seaway are exceptionally rich in the carbonate rocks (limestone and dolomite). Deposition of limy sediment was favored by the warmth of the Tethys Seaway, which lay almost parallel to, and not far from, the equatorial regions. Constant supplies of seawater came from the open oceans, particularly to the east. Lime-secreting organisms, both plants and animals, flourished and left their fossil remains and precipitation products in abundance. Great reefs were common; the thick limestone and dolomite masses of Triassic age in the

Figure 13.3 *The Rock of Gibraltar is composed of a mass of Lower Jurassic limestone uplifted during Tertiary mountain-building disturbances. It is honeycombed with ancient caves and artificial excavations. In one of the caves the first known skull of Neanderthal man was found. (British Information Service.)*

South Tyrol Dolomites are chiefly coral and algae. The Jurassic likewise has reefs of corals and sponges. The Cretaceous, as shown by its name (signifying chalk), was a time of tremendous limy accumulations in both Europe and Asia.

In its western part the Tethys Seaway was bounded on both sides by well-defined continental areas. Other sites of Mesozoic marine deposits are mainly marginal to continental masses, bordered by shield areas on only one side. The deeper portions of the Tethys trough lay across southern Europe, but there were occasional shallow embayments to the north across central Europe and Russia, and to the south in North Africa.

The Tethys Seaway appears to have continued across the Himalayan area, southeastward across southeastern Asia and into the East Indies. It was unconfined on the south until the approach and subsequent collision of the Indian plate in the early Cenozoic. Scattered exposures of Mesozoic rocks are found from the eastern part of the Celebes and on through Ceram, Tanimbar, and Timor. The geosyncline may have extended as far as western Australia, where it has been called the *Westralian Geosyncline.* The same great zone is thought to be represented in the Western Hemisphere by the Caribbean region between the North American and South American plates.

Triassic, Jurassic, and Cretaceous deposits of the Tethys Seaway are now uplifted in many great mountain ranges, including the Alps, Pyrenees, Apennines, Atlas, Carpathians, Caucasus, Pamirs, and Himalayas. It is in these regions that we can read the record of the previous history of Tethys.

EURASIA NORTH OF TETHYS

Much of Europe north of the Alps has Mesozoic rocks at or near the surface. The highly populated regions of France, England, and Germany are underlain by these rocks. The Triassic, with the type section in Germany, consists of red beds and limestone; the Jurassic is mixed and has nonmarine and marine facies. The Cretaceous makes up the famous White Cliffs of Dover and many other calcareous outcrops. Among the mineral resources are gypsum and natural gas in the Triassic, iron ores in the Jurassic, and gas in the Cretaceous.

Over most of the area between the Ural Mountains and the east coast of Asia there is no lengthy interruption between the Triassic and Permian rocks. There was, however, a lapse in sedimentation between the Early and Middle Triassic. After this time most of northern Asia was free of marine invasions and received instead a great variety of land-laid sediments. Southward toward the Tethys belt the amount of marine sediment, chiefly limestone, increases until it makes up whole ranges in the Crimea and Caucasus.

Notable in the Early Triassic of central Siberia are great fields of basaltic lava and intrusions of the dark igneous rock *dolerite*. Extensive marine beds of the same age occur in China and eastern

Figure 13.4   Cliffs of chalk of Cretaceous age on the coast of Selwick's Bay, Yorkshire, England. Glacial deposits overlie the chalk beds. (Crown copyright, Geological Survey photograph. Reproduced by permission of the Controller of Her Majesty's Stationery Office.)

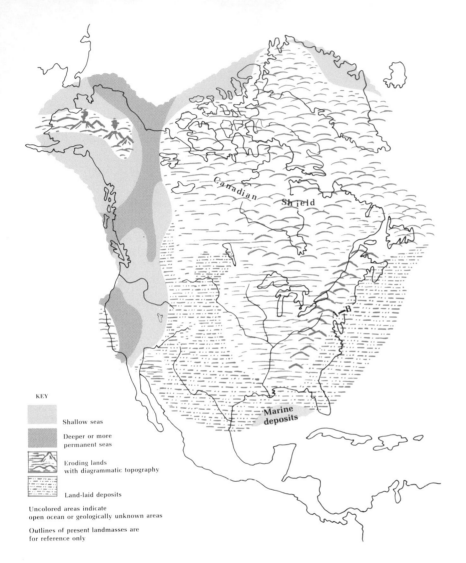

Figure 13.5 Reconstructed paleogeography of the Triassic Period.

KEY

Shallow seas

Deeper or more permanent seas

Eroding lands with diagrammatic topography

Land-laid deposits

Uncolored areas indicate open ocean or geologically unknown areas

Outlines of present landmasses are for reference only

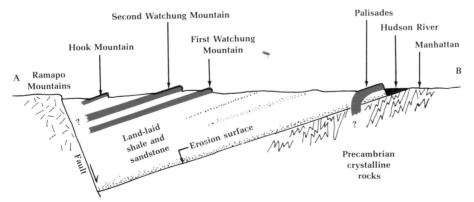

Figure 13.6 Generalized geologic section from Manhattan to the Appalachian Plateau to show the downfaulted Triassic rocks of the New Jersey Lowland. Length of section is about 40 kilometers, and position is shown by line A-B in Figure 13.5. Igneous rocks darkened; the Palisades are carved from a diabase dike (intrusive), the Watchung Mountains are flows of basaltic lava. (Adapted from Christopher J. Schuberth, Geology of New York City and Environs. © 1968 by Christopher J. Schuberth. Reproduced by permission of Doubleday & Company, Inc.)

Siberia. The Middle Triassic is absent and the Upper Triassic is poorly represented. The Jurassic was a time of interior lakes and alluvial plains when many important coal beds were deposited in the USSR and China. Locally, Cretaceous rocks contain coal, bauxite (aluminum-rich rock), and iron ores. In Mongolia the continental beds yield well-preserved dinosaur bones and eggs.

Midway through the Cretaceous, the seas advanced inland far beyond any previous Mesozoic invasion. Much of Russia, western Siberia, and central Asia was flooded.

Figure 13.7   The Palisades of the Hudson River near Alpine, N.J. The vertical cliff is columnar diabase of Triassic age. (U.S. Geological Survey.)

EASTERN NORTH AMERICA

The Triassic and Jurassic are poorly represented in the eastern half of North America. The only significant deposits are nonmarine red beds of Late Triassic age that occur in a dozen narrow downfaulted basins from Nova Scotia to South Carolina. These red formations, known collectively as the *Newark Group,* consist of shale, sandstone, and conglomerate, with great sheets of dark lava and intrusions of dolerite. The famous Palisades of the Hudson are representative of this type of formation. Freshwater fish and dinosaur tracks are common fossils. These rocks are fairly well known because they crop out near centers of population.

No Jurassic is known to appear on the surface east of the Mississippi River, but thick formations, some with oil pools, have been found by drilling in states marginal to the Gulf of Mexico.

The Cretaceous is well represented as a marginal belt along the Atlantic and Gulf coasts; this belt is under the Atlantic from Long Island northward, but it is over 900 kilometers wide in the Mississippi Valley. The Lower Cretaceous appears at the surface only in a narrow belt from northern New Jersey to Maryland; elsewhere it is covered by the Upper Cretaceous. The Cretaceous is mixed marine and nonmarine, and the marine facies die out landward in such a way as to prove that there was an Atlantic Ocean at this time and that the east coast of North

America had essentially its present form when the Cretaceous formations were laid down.

The Appalachian Mountains, the Canadian Shield, and other central areas, were undergoing erosion throughout the Mesozoic. At the beginning of the Triassic, the Appalachians must have been an imposing, youthful-to-mature range; by the Cretaceous nothing remained but the planed-off, deeply eroded stumps. Most of the material removed in the Triassic and Jurassic was carried beyond the present shores of the Atlantic and Gulf coasts and may in fact make up the deeply buried foundations of the continental shelves. That part eroded in the Cretaceous is still near at hand in visible or near-surface formations.

WESTERN AND NORTHERN NORTH AMERICA

A number of shallow marine invasions washed over the western portions of North America during the Mesozoic. Conditions were such that the marine formations alternated with land-laid deposits of the interior, and thus a record of both land and sea life is preserved. During the Mesozoic a strip of land up to 650 kilometers wide was

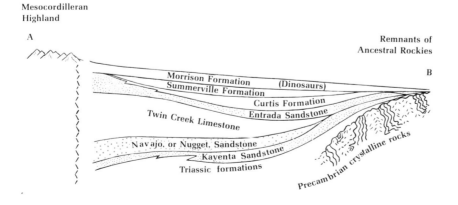

Figure 13.8 Reconstructed paleogeography of the Jurassic
Period.

Canadian Shield

Mesocordilleran
Highland

Remnants
of Ancestral
Rockies

Appalachian Mountains

Modern Atlantic Ocean (?)

KEY

Shallow seas

Deeper or more
permanent seas

Eroding lands
with diagrammatic topography

Land-laid deposits

Uncolored areas indicate
open ocean or geologically unknown areas

Outlines of present landmasses are
for reference only

Mesocordilleran
Highland

A

Remnants of
Ancestral Rockies

B

Morrison Formation   (Dinosaurs)
Summerville Formation
Curtis Formation
Twin Creek Limestone   Entrada Sandstone
Navajo, or Nugget, Sandstone
Kayenta Sandstone
Triassic formations

Precambrian crystalline rocks

Figure 13.9 Restored cross section from central Utah to west-
central Colorado at the close of
the Jurassic Period. Length of
section is about 240 kilometers,
and position is shown by line
A-B in Figure 13.8.

formed by sedimentation, by igneous action, and by the accumulation of material scraped from the ocean bottom. This strip was welded to the western margin of the preexisting continental mass as a permanent addition.

In Early Triassic time, the seas spread inward from the Pacific to central Utah and Wyoming, while broad low floodplains extended well beyond into the central part of the continent. This was the last time marine waters crossed the western United States from the Pacific, because a narrow uplift, the *Mesocordilleran Highland,* arose from the former seabed in the Middle Triassic to block out the western seas.

For a while the protected interior area behind the highland was the site of accumulation of desert sands on a large scale, but eventually, with continued downsinking, the seas entered a pass to the north in Canada or Alaska and again flowed into the west-central parts of the continent. During the Jurassic Period this interior seaway was occupied by at least two successive shallow embayments up to 1,300 kilometers wide. These reached as far south as southern Utah and east to the Black Hills. They failed, however, to connect with other embayments spreading inward from the ancestral Gulf of Mexico and from California. One of the last events of the Jurassic

Figure 13.10 *Erosion forms carved from continental formations of Late Jurassic age near Lake Powell, south-central Utah. (U.S. Bureau of Reclamation.)*

Figure 13.11 Cretaceous formations near Grand Junction, Colorado. The light-colored slopes are cut on the Mancos Shale, a relatively soft marine formation. The upper vertical cliffs make up the coal-bearing Mesa Verde Group. Rocks of similar composition and appearance are found in the western interior of North America from the Yukon to New Mexico. (U.S. Geological Survey.)

was the accumulation of the *Morrison Formation*, which covers nearly 2 million square kilometers in the Rocky Mountains and western plains. The Morrison is composed of material worn from the Mesocordilleran Highland and spread eastward by sluggish, meandering rivers. Buried in the Morrison are skeletons of gigantic dinosaurs and other Late Jurassic life forms. This formation also supplies most of the uranium ores of the United States.

With continual downsinking during the Cretaceous, shallow seas spread inward on all sides in the greatest flood of the Mesozoic. Although the Mesocordilleran Highland was not submerged, it was surrounded like an island, as seas from the north connected with seas from the south across the central United States and Canada. In this great Cretaceous seaway many hundreds of cubic kilometers of silty mud accu-

mulated; this material now forms extensive soft formations from Minnesota to central Utah. On the shallow shelving deltas and floodplains along the western borders of this sea, successive swamps with abundant vegetation gave rise to valuable coal beds.

The Cretaceous came to a close as the central sea was crowded out by its own deposits. Swamps and lakes marked the final phases, and the stage was set for the great Rocky Mountain Revolution, which affected much of the Pacific borderlands.

Northern Canada has a Mesozoic history much like that of the western United States. The Triassic exists in minor amounts with nonmarine red beds and marine shale. The Jurassic has beds of both marine and nonmarine origin, with little or no limestone. The Cretaceous is also well represented by varied sediments. Notable is the presence of basalt flows ranging up to the Cretaceous in age.

Greenland has richly fossiliferous Triassic rocks, a thick and varied Jurassic section, and Cretaceous beds showing alternate retreats and transgressions of the sea. A very large basalt field of Late Cretaceous or Early Tertiary age is found in east-central Greenland.

SOUTH AMERICA AND ANTARCTICA

South America has marine and nonmarine deposits of each of the Mesozoic periods, but the record is not as extensive as that of North America. The Middle Triassic is well represented in Argentina, and there are Late Triassic marine beds in the Andes and nonmarine deposits in eastern Brazil. Land plants and vertebrates closely comparable to those of South Africa are found in the nonmarine formations of Argentina.

During the Early Jurassic great floods of basaltic lava covering nearly 1 million square kilometers, with an average thickness of 600 meters, were extruded in the Paraná Basin of Brazil. Similar flows are found in northeastern Brazil and below the surface in Argentina. Intrusion of extensive diabase sills and dikes accompanied the surface extrusions of basalt. The Jurassic is represented by marine beds in the mountain chains of western South America, with fossils that correlate with the Jurassic of other continents. Only a few small outcrops of nonmarine rocks are known along the east coast.

Cretaceous rocks are much more extensive. All parts of the system are represented by marine deposits, and deposits of some intervals appear in great thickness in the Andes. The eastern coast and lower Amazon Valley was invaded by shallow seas on several occasions, and there were extensive but thin accumulations of continental rock in interior basins. Only in the Cretaceous do marine rocks become abundant along the eastern margin of South America.

Antarctica consists of an ancient shield area called East Antarctica and a western folded-mountain area called West Antarctica, which is in many ways a continuation of the Andes Range. Recent explorations have greatly increased our knowledge of this remote region and have revealed a fairly complete representation of the Mesozoic subdivisions. The Triassic is represented by continental rocks containing coal and a very significant assemblage of fossil verte-brates, including amphibians and large reptiles. Among the reptiles are forms found in southern Africa, India, and China. The Early Jurassic was a time of great intrusive igneous action with massive diabase sheets covering thousands of square kilometers. Basaltic flows also occur, and in places these rest on sedimentary rocks containing Early Jurassic plant fossils. The Cretaceous is known chiefly on the Antarctic Peninsula, which extends toward South America. Thick marine beds with fossil mollusks are found in this region.

THE PACIFIC OCEAN

Although geologists are confident that there was a Pacific Basin during the Paleozoic and early Mesozoic, the actual known rock record goes only into the Late Jurassic. No fossils older than this have been found in spite of much dredging, boring, and surface collecting throughout the Pacific. The best current explanation for this strange circumstance is the hypothesis of sea-floor spreading, according to which all older rocks have been swept into the continents or oceanic trenches and destroyed.

It can scarcely be coincidence that the borderlands of the Pacific became geologically very active about halfway through the Cretaceous. Practically all the ranges that face the Pacific and the island arcs that rise along its margins date from this time. Cretaceous rocks chiefly of igneous derivation abound in Formosa, the Japanese islands, and Korea. Farther south, deposits of the same age occur in the East Indies and New Zealand. The great granitic masses of the Andes and Rockies are chiefly of Cretaceous age.

Not to be overlooked are the evidences of Cretaceous deposits on some of the 10,000 or so submerged mid-Pacific *guyots,* or *seamounts.* These are thought to be mainly decapitated volcanoes. Cretaceous fossils have been dredged from the tops of these mountains at depths of up to 1,800 meters. It must be inferred that the fossil-bearing guyots were at or near the surface of the oceans during the Cretaceous Period. Incidentally, if enough of them had risen above the water, they could have served as "stepping stones" or routes of migration for plants and animals into, or even across, the Pacific Basin.

Taking into account the facts just mentioned and the extensive flooding that took place on all continents during the Cretaceous, it seems that geologic activity in and around the Pacific reached a high point at this time. As one geologist has put it, "a commotion in the ocean" of some importance is indicated.

Southwestern Pacific. Australia was an emergent land during the Triassic, Jurassic, and earliest Cretaceous, but it did receive and retain much

Figure 13.12 Artist Chesley Bonestell's conception of the topography of the Pacific Ocean bottom. If the water should be removed, the flat-topped guyots, or seamounts, would be revealed. The steep front and fringing deposits of a coral reef are in the foreground. (Courtesy of E. L. Hamilton.)

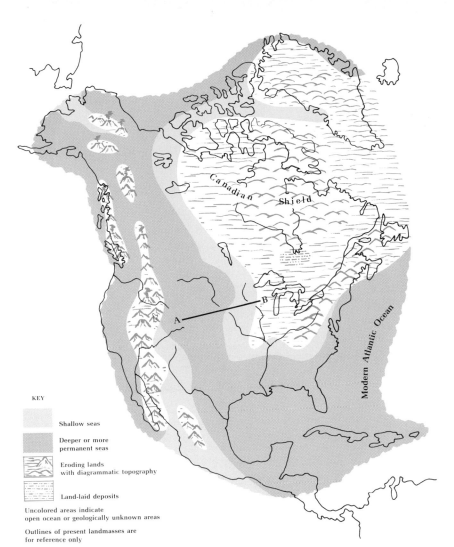

Figure 13.13 Reconstructed paleogeography of the Cretaceous Period.

KEY

Shallow seas

Deeper or more permanent seas

Eroding lands with diagrammatic topography

Land-laid deposits

Uncolored areas indicate open ocean or geologically unknown areas

Outlines of present landmasses are for reference only

Canadian Shield

Modern Atlantic Ocean

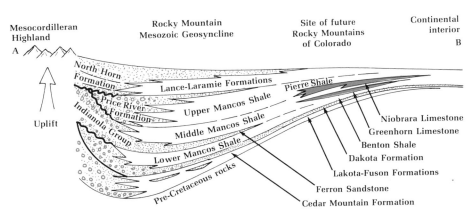

Mesocordilleran Highland

A

Uplift

Rocky Mountain Mesozoic Geosyncline

Site of future Rocky Mountains of Colorado

Continental interior

B

North Horn Formation

Price River Formation

Indianola Group

Lance-Laramie Formations

Upper Mancos Shale

Middle Mancos Shale

Lower Mancos Shale

Pre-Cretaceous rocks

Pierre Shale

Niobrara Limestone

Greenhorn Limestone

Benton Shale

Dakota Formation

Lakota-Fuson Formations

Ferron Sandstone

Cedar Mountain Formation

Figure 13.14 Restored cross section of Cretaceous formations from central Utah to Iowa. Length of section is about 1,450 kilometers, and position is shown by line A-B in Figure 13.13. Uplifts in the area of the Great Basin supplied thousands of cubic kilometers of sediment to the interior of the continent during the Cretaceous. The formations are coarse and conglomerate near the source and become increasingly finer eastward. Minor limestone beds (Niobrara and Greenhorn) are found in the plains region.

Figure 13.15 Upper part of the Karoo System as seen in the Barklay East District, Cape Province, Africa. The gorge is cut in red beds in which Late Triassic vertebrates, including the earliest dinosaurs are found; the light colored ledges are the Cave Sandstone; the uppermost, dark-colored ledges are lava beds of the Drakensberg group which caps and completes the Karoo System. (Geological Society of South Africa, photograph by A. O. Fuller.)

continental sediment. The Triassic yields fossils of freshwater animals and land plants. The Jurassic was a time of extensive swamps and bodies of fresh water, one of which, Lake Wallon, covered at least 700,000 square kilometers and received deposits over 1,500 meters thick. Freshwater organisms and land plants are present, together with dinosaurs. Seas invaded Australia about midway through the Early Cretaceous and converted the continent into two large islands.

New Zealand has a record of Triassic and Jurassic deposition, and a mountain-building episode there in Early Cretaceous time gave rise to

intrusions of granite and later to coarse marginal deposits that interfinger outward with marine beds.

AFRICA AND THE COMPLETION OF THE KAROO SYSTEM

Africa, especially that portion south of the Sahara, was remarkably stable during the Mesozoic. The northern part bordering on the Mediterranean Sea was a part of Tethys and received thick and varied sediments of all the Mesozoic periods. During the Cretaceous a shallow arm of the sea spread across the bulge of Africa and connected with the Atlantic. On the east coast, scattered Jurassic marine deposits are found, and Cretaceous sediments extend many kilometers inland.

Of unusual interest are the continental deposits of the Karoo Basin of southern Africa, whose earlier history has been summarized in the previous chapter. The Karoo System began to accumulate in the Pennsylvanian Period and continued into the Early Jurassic without notable interruptions. The transition from Permian to Triassic takes place in the Beaufort Series and may be located approximately by the appearance of certain types of reptiles. The Triassic part of the Karoo is made up of sandstone, shale, and red beds with abundant fossil bones and footprints. Included are remains of the earliest dinosaur and the earliest known mammal. Fish, crustaceans, and plants also occur.

The closing stages of Karoo history were marked by strong volcanic action that produced the so-called *Drakensberg volcanics*. Both lava (basalt) and intrusive sheets (dolerite) were produced on a large scale. Much of the volcanic rock has been removed by erosion, but remnants totaling 26,000 square kilometers are still preserved in Lesotho (formerly Basutoland). Although fossils are understandably rare, it is con-

cluded that the volcanic activity lasted into Jurassic time.

Like other continents, Africa received Cretaceous deposits about its margins and one shallow seaway even advanced across the westward bulge to make an island of the western Sahara.

Two very widespread, somewhat overlapping geologic disturbances shook the earth during the Mesozoic. These differ in the regions affected, in the types of structures produced, and especially in the varieties of igneous products.

The first disturbance is characterized by the production of dark-colored basic igneous rocks on a large scale. These rocks appeared on the surface as flows of basalt and in the subsurface as dikes and sills of dolerite. The most extensive activity was in the Southern Hemisphere; flows spread one upon another in South America (Brazil and Argentina), South Africa, and Ant-

Figure 13.16   *Thick sheets of igneous rock alternate with sediments to form this imposing escarpment of the Drakensberg Mountains in Natal, South Africa. The age of the sediments and igneous rocks is thought to be Early Jurassic or Late Triassic. (South African Information Service.)*

Figure 13.17 *The glaciated crest of the central Sierra Nevada, California. The granite making up the range was intruded as a number of large batholiths over a long period of time in the Late Jurassic and Early Cretaceous. (Courtesy of Mary Hill.)*

arctica. Simultaneously, intrusions of thick dikes and sills of dolerite occurred beneath the surface. In places, the intrusions seem almost to have lifted and engulfed the overlying sediments. Geologists estimate that at least 300,000 cubic kilometers of igneous rock were produced in South Africa alone. Flows of Triassic lava cover nearly 2 million square kilometers of southern Brazil and constitute the largest known exposed field of volcanic rocks in the world. Erosion of these rocks has produced some striking scenery in parts of South Africa and South America.

Igneous activity continued on a moderated scale during the Jurassic, and in India the volcanic outpourings appear to have been postponed until near the end of the Cretaceous. Over 500,000 square kilometers of the Indian peninsula are covered by flows of basalt about 3,200 meters thick. Known as the *Deccan traps*, these flows

are strikingly similar to others in Abyssinia. Parts of the Northern Hemisphere are now known to have been affected. Dolerite intrusions and basalt flows are found in the Triassic basins of eastern North America. The Palisades of the Hudson (Figure 13.7) are eroded from a thick intrusion in the midst of red Triassic shale. Basalt flows and basic intrusions are known to occur in the Arctic islands of Canada. These may be of Cretaceous age. A large basalt field is also found in Greenland. Last to be noted are the basalt flows and intrusions of central Siberia that cover 1,500,000 square kilometers.

It is tempting to speculate that the widespread appearance of basic igneous material in shield areas or among flat-lying sedimentary rocks not in connection with mountain building represents changes in the crust or upper mantle of the earth and is somehow associated with continental drift, which seems to have occurred at the same time as the igneous action. The great outpourings probably accompanied the lateral movements of the plates upon which they are found. The wide occurrence of red sediments in the Triassic might be due also to the availability of iron, which was supplied on a large scale by the breakdown of the widespread, iron-rich igneous rocks.

The second notable Mesozoic disturbance affected chiefly the margins of the Pacific Basin. It produced some of the greatest mountain ranges of the earth, with attendant intrusion of light-colored granitic rock and extrusion of corresponding types of lava, chiefly *andesite*. Because this disturbance produced the Rocky Mountains, it is sometimes called the *Rocky Mountain,* or *Laramide, Revolution.* Like the basaltic episode, it was spread over a considerable period commencing late in the Cretaceous and dying down in the Paleocene. The Rocky Mountain Revolution provides a rough division between the Mesozoic and the Tertiary.

Life of the Mesozoic

PLANTS

The Mesozoic was a time of transition and change in the plant kingdom. Later Cretaceous vegetation bears little resemblance to that of the Triassic, truly revolutionary changes having occurred during the interval between these periods. By comparison, the changes between Late Cretaceous and present-day vegetation are relatively minor. Emphasizing the distinct nature of the early Mesozoic floras is the fact that plants such as *seed ferns, lycopods,* and *horsetails*—all common in Carboniferous coal forests—are represented in the Mesozoic only by small, insignificant species.

Ferns prospered during the Mesozoic, and their delicate remains are particularly common in continental rocks of Triassic and Jurassic age. Insofar as our understanding of past climates is concerned, it seems significant that types of fern that lived during the Mesozoic in temperate latitudes are now restricted mainly to tropical countries. As a rule, the Mesozoic ferns are more closely related to species now living than they are to Paleozoic types.

Another characteristic group of Mesozoic plants are the *cycads*. The term "cycad" includes not only true cycads, or *Cycadales,* but also certain cycadlike plants known as the *Cycadeoidales,* or *Bennettitales,* which are extinct and are thus referred to as the "fossil cycads." The *Cycadeoidales* have a fossil record extending into the Early Permian. They were gymnosperms that bore true flowers with spore-bearing stamens and a seed-bearing female organ quite similar to flowers of more recent plants. The true cycads, which are now represented by nine genera confined chiefly to tropical regions, appeared in the Pennsylvanian but did not become common until the Triassic. They were so prevalent during the Jurassic that this period sometimes is called the *Age of Cycads.* The group at that time was very cosmopolitan, ranging across all the great continents. They have reproductive parts similar to

Figure 13.18 *A portion of a cycad trunk* (Cycadeoidea marylandica), *about 33 centimeters high, found in Early Cretaceous sediments of Anne Arundel County, Maryland. (Smithsonian Institution.)*

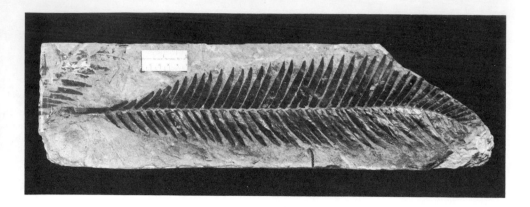

Figure 13.19 *A well-preserved frond of a cycad tree from the Mesozoic rocks of Great Britain. Length about 46 centimeters. (By permission of the Trustees of the British Museum, Natural History.)*

living conifers; there were small, medium, and large species. The typical cycad fossil is a silicified, keg-shaped trunk with a pattern of diamond-shaped indentations marking the position of the leaf bases. The foliage, which is seldom found attached to a trunk, is typically palmlike. Another gymnosperm group, which is now represented by one species, *Ginkgo biloba* (the Maidenhair tree), was prominent during the Mesozoic. Its remains have been found in Mesozoic rocks of North America, Europe, Central America, Malaya, South Africa, and Australia. *Ginkgo* may be the oldest living genus of seed plant, but it is not known to exist in a wild state today.

The early Mesozoic marked a high point in the history of cone-bearing trees. Dominating the landscape were the descendants of primitive conifers, such as *Walchia* and *Lebachia,* which were minor inhabitants of the Carboniferous forests. The cool and dry climate of the Permian and Triassic favored the spread of conifers throughout the world. Especially prominent during the early Mesozoic were members of the family *Araucariaceae,* which now survive only in the Southern Hemisphere. Petrified trunks of araucarians account for most of the fossil remains in the famous Petrified Forest in Arizona.

Here lie trunks as much as 1.5 meters in diameter and over 30 meters long. The first true pines appeared in the Late Jurassic and spread widely during the Cretaceous, as their fossilized remains —cones, needles, and wood—indicate. The *sequoias* appeared during the Jurassic and had become very common by the Cretaceous.

A distinct and rather sudden change in the vegetation of the earth took place during the middle Cretaceous. Earlier, during the Triassic and Jurassic, the most abundant plants had been ferns, various types of cone-bearing plants, and the cycads and their relatives. After the middle Cretaceous, the chief plants were members of the great group known as the *flowering plants,* or *angiosperms.* The angiosperms include plants that are pollinated through floral structures and bear seeds enclosed in an ovary or pod. Angiosperms are divided into two groups: the *dicotyledons,* which have two seed leaves and net-veined leaves; and the *monocotyledons,* with one seed leaf and parallel-veined leaves. Dicotyledons include such diverse plants as oaks, maples, buttercups, cacti, sagebrush, peas, and violets. Typical monocotyledons are grasses, palms, lilies, irises, and orchids. Estimates indicate that there are about 175,000 species of living flowering plants; at least 30,000 fossil species have been found. They have flourished in all climates and include trees, shrubs, and herbs.

The origin of the angiosperms is an unsolved

problem. They apparently appeared first in a restricted area and later spread, becoming almost worldwide by the close of the Cretaceous Period. A peculiar palmlike plant, *San miguelia,* found in Upper Triassic rocks of southwestern Colorado may be the oldest known angiosperm. Leaves of the magnolia, sassafras, fig, and willow are common in Upper Cretaceous rocks. Forests of angiosperms contributed to the formation of Cretaceous coal, and pollen grains of this group are useful in understanding climatic conditions and also in correlating plant groups.

Angiosperms are important not only because they have become the dominant form of plant life on earth, but also because of the influence they exert on animal life. With the flowering plants came a variety of grains, nuts, and fruits, which furnish a food supply that ensures the survival of the plant embryo. Many of these food supplies, such as the fruits of tropical plants, are transitory and must be harvested by animals as they ripen. Others, which are produced by plants growing in temperate and cold climates, are adapted to endure seasons of cold and drought and therefore can be eaten by animals during the winter season. The relatively concentrated nature of the food supply and the small size of most seeds, nuts, and fruits makes them ideal fare for small animals and, unfortunately, for insects as well. It is unlikely that the large Mesozoic reptiles ate food of this sort. They fed on large quantities of coarse herbage and on succulent water vegetation of various types. The mammals and birds, on the other hand, could and did subsist very nicely on the highly concentrated products of flowering plants (or on the insects that fed on these plants). Although the point is perhaps impossible to prove, there must have been a distinct connection between the rise of the flowering plants and the subsequent increase and expansion of the mammals and birds. The relationships between insects and flowering plants will be discussed in the section on invertebrates.

No less important than the origin and spread of the angiosperms on land was the expansion

Figure 13.20   Broken sections of fossil logs in the Chinle Formation of Late Triassic age in Petrified Forest National Park, Arizona. (U.S. National Park Service.)

Figure 13.21   Fossil angiosperm leaves from Late Cretaceous rocks, Clark County, Idaho. These are typical of the plants that became common during the Cretaceous. (Smithsonian Institution.)

of lower forms of plant life in marine and fresh water. The small aquatic algae known as diatoms are not positively known before the Jurassic, but increased in importance and are now a major food source for animal life in the sea. Diatoms secrete skeletons of silica that accumulate to form oceanic ooze or the sediment *diatomite*.

The chalk beds so characteristic of the Jurassic, Cretaceous, and younger ages are composed of extremely small fossils called *coccoliths*, which are generally thought to be secreted by very simple floating algae.

NONMARINE INVERTEBRATES

Invertebrates living on dry land and in fresh water were fairly well established by the Mesozoic. Streams and lakes were well stocked with clams, snails, arthropods such as *ostracods* and *branchiopods*, and small, inconspicuous sponges. All these fed chiefly on water plants with which their remains are usually associated. Air-breathing snails are often found.

The history of insects is particularly important, especially as it relates to the flowering plants. The plants furnish food for the insects in the form of nectar and pollen, in return for which insects serve the vital function of fertilization by transferring pollen from flower to flower. The ...nt of this interdependence is indicated by the fact that over 60 percent of all flowering plants are insect pollinated, and 20 percent of the insects, at least in some stages of their development, depend on flowers for food. This association undoubtedly evolved gradually over the ages, which suggests that many of the compli-

Figure 13.22 *Two types of the chalk-forming fossils known as coccoliths from the Taylor Marl of Cretaceous age, Texas. Magnified about 15,000×. (Courtesy of S. Gartner and W. W. Hay, University of Illinois.)*

cated behavior patterns of insects and the myriad forms of flowering plants are the outcome of a long process of adaptation. The insects chiefly involved in this insect-plant dependency are bees, butterflies, beetles, wasps, and flies.

The food the insects take from the plants is a small price for the plants to pay, because insect-pollinated plants are not obligated to produce as much pollen as plants that are pollinated by other means. Corn, a wind-pollinated plant, may produce 50 million pollen grains per plant, while certain insect-pollinated orchids produce only eight grains.

The ascendancy of insects and flowering plants during the Mesozoic was greatly aided by the relationship that developed between them. The benefits that accrued to birds, mammals, and ultimately to man are incidental to the process. Most of our fruits, vegetables, ornamental plants, and industrial plants are insect pollinated.

The insects, besides being man's chief competitors for food and living space on land, are important in other ways. Although we cannot trace by fossil remains the intricate stages of their adaptations, we should examine some of their more significant achievements. They have developed a number of social orders that, although composed of individuals, function as a unit. This type of social arrangement combines the benefits of individual action and of group solidarity. Nowhere in nature has this ideal been approached so closely as in the case of the bees and ants. Ants not only harvest the natural vegetation, but some species also plant and cultivate special crops. The earliest known ant is a well-preserved individual in amber from Upper Cretaceous rocks of New Jersey. The adaptations of the bees to plants and to one another are becoming well known. Apparently, their small size is the only thing that has prevented insects from becoming the dominant creatures of the earth. Their evolution has been rapid, but once a group reaches a certain level, it tends to stagnate. Many fossil insects found preserved in amber 50 million years

Figure 13.23 The famous White Cliffs of Dover, southern coast of England. The white sediment making up this exposure is predominantly the remains of coccoliths and other calcareous organisms. (British Information Office.)

Figure 13.24 The earliest known fossil ant, discovered in Late Cretaceous deposits of New Jersey. The specimen is preserved in amber. Besides being the first known ant and the first clear indication of the social level of insect organization, it is also an almost perfect link between wasps and ants. Note stinger at end of abdomen. (Courtesy of Frank M. Carpenter.)

old are practically indistinguishable from living forms. They apparently completed all essential phases of their evolution at least that long ago.

MARINE INVERTEBRATES

As a group, Triassic marine invertebrates show the effects of the late Paleozoic exterminations. Apparently the Early Triassic seas continued to be unusually cool or otherwise inhospitable to such animals as corals, sponges, bryozoans, and protozoans. The scarcity of carbonate rock in Early Triassic formations except within the Tethys Seaway is notable. Corals seem to have escaped extinction by a very narrow margin, for few solitary corals and no coral reefs have been found in the Early Triassic. Other groups that were relegated permanently to minor positions were crinoids and brachiopods, which are generally rare in Mesozoic rocks. Two groups that eventually recovered a great deal of the ground they had lost were the protozoans and bryozoans, but the species that became common during the late Mesozoic and Cenozoic were quite different from their Paleozoic predecessors. Protozoans and bryozoans are rare in the Triassic, increase during the Jurassic, and become quite prolific during the course of the Cretaceous.

The mollusks seem to have weathered the critical late Paleozoic disturbance without great losses and came to be the most important shelled invertebrates of the Mesozoic seas. You will recall that this group includes the coiled, single-chambered gastropods, the bivalved (two-shelled) pelecypods, typified by the clams and oysters, and the coiled many-chambered cephalopods. As a general rule, the mollusks gradually grew more varied and numerous during the Jurassic and Cretaceous. Freshwater clams and gastropods were plentiful, and air-breathing snails were locally abundant. By the Cretaceous, we find entire reeflike structures composed of oysters and oysterlike shells. Of special importance are the large conical or twisted shells of a group of pelecypods known as *rudistids,* some

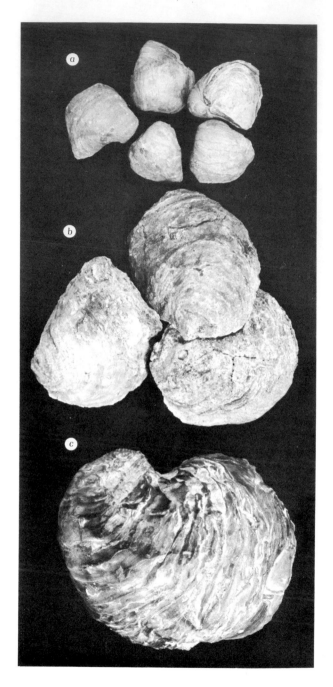

Figure 13.25 Typical Cretaceous pelecypods. (a) Gryphaea, (b) Ostrea (oyster), and (c) Exogyra. *In these forms the shells are of irregular shape and one valve is larger than the other. The Exogyra specimen measures about 10 centimeters across.*

of which reach a length of over 1 meter and a diameter of more than 0.5 meter. Fossil rudistids are distributed in Cretaceous rocks in a world-circling belt along Tethys and in warmer regions of the Western Hemisphere.

The most spectacular, varied, and successful marine invertebrates of the Mesozoic were the cephalopods. Because they were able to swim and crawl and because the shells of some species could float after the animals had died, their remains have been scattered and preserved in many localities. Cephalopod shells make striking fossils, and they are usually beautifully preserved. The coiled varieties range from a few centimeters across to more than 1 meter in diameter. Also included in the cephalopod class are the *belemnites*, squidlike forms whose internal skeletons have the form of a solid, stony, cylindrical object shaped like a cigar. Fossilized belemnite skeletons are particularly abundant, resist erosion, and have been known for centuries as "thunderbolts."

The ammonites (cephalopods whose shells have complex suture patterns) are unequaled as guide fossils for marine Triassic, Jurassic, and Cretaceous rocks. Individual species ranged widely, lived short lives, were independent of ocean-bottom conditions, and, of course, are usually well preserved. Ammonites successfully weathered the critical Permian-Triassic transition, and descendants of Paleozoic forms blossomed in Mesozoic seas over the entire world. Although the group suffered another serious setback at the close of the Triassic and again at the close of the Jurassic, the surviving members recovered rapidly and each time succeeded in repopulating the seas. Only at the close of the Cretaceous did the race suffer final extinction.

The extent to which ammonites have been useful in worldwide correlation is indicated by their selection as guides to many zones and stages. European authorities recognize twenty-nine zones in the Triassic, based mainly on collections from the Alps. A recent summary of the North American Triassic formation shows thirty-five ammonite guide species. Because geologists estimate that the Triassic spanned approximately 25 million years, we can theoretically correlate segments of time less than 1 million years long over the entire earth if necessary ammonite fossils are found.

The Jurassic ammonites are even more diagnostic and have been studied more thoroughly than those of the Triassic. You will recall that the striking shells in the Jurassic rocks of England

Figure 13.26 Some typical Mesozoic cephalopods. Late Cretaceous: (a) Plancenticeras, (b) Baculites, and (c) Scaphites. Early Cretaceous: (d) Cheloniceras and (e) Dufrenoya. Jurassic: (f) Pachyteuthis and (g) Cardioceras. Triassic: (h) Meekoceras and (i) Columbites. The Placenticeras specimen measures about 8 centimeters across.

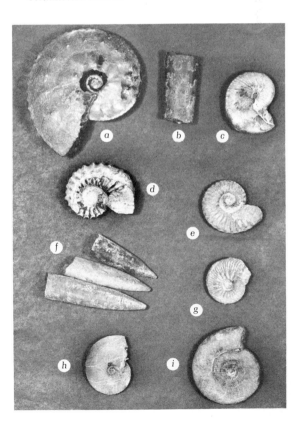

Figure 13.27 A lobsterlike crustacean (Eryon) from Upper Jurassic limestone in the Solenhofen quarries, Germany. Specimen is about 15 centimeters long. (Field Museum of Natural History, Chicago.)

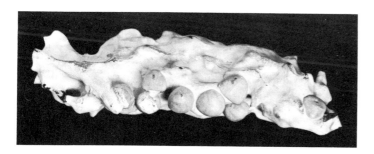

Figure 13.28 A group of echinoids (sea urchins) partly embedded in a mass of chert that evidently solidified about them rather suddenly on the sea bottom. Upper Cretaceous of England. (By permission of the Trustees of the British Museum, Natural History.)

strongly influenced the early development of the science of stratigraphy and correlation. On the basis of their numerous and excellent collections of ammonites, British geologists recognize from forty to fifty zones. The North American Jurassic is relatively incomplete, with only about ten zones having been established so far in the United States and twenty-four in Mexico. Under favorable conditions, then, intervals of somewhat more than 1 million years may be correlated by Jurassic ammonites.

The Cretaceous rocks of Europe cannot be as easily zoned by ammonites as the Jurassic and Triassic. Other types of invertebrates, such as echinoids and brachiopods, are used in conjunction with the ammonites. Perhaps twenty to thirty ammonite zones can be recognized in Europe. Twenty-nine fossil zones have been tentatively recognized in the Late Cretaceous of the United States, of which twenty-six are based on ammonites. The relative length of recognizable intervals again approaches 1 million years. By utilizing other fossils in combination with ammonites, zones spanning about 350,000 years have been identified in Montana.

During the Mesozoic the arthropods evolved and spread in the oceans, as well as on land and in the air. The living space vacated by the trilobites was taken over at least in part by a variety of other crustaceans. The group that includes the familiar shrimps, crabs, crayfish, and lobsters appeared in the Triassic. Over 8,000 living species of this group have been identified. The first true crabs especially adapted for life along the beaches and shallow offshore areas appeared during the Jurassic, along with the earliest barnacles.

Mesozoic echinoderms were represented by a variety of forms, including starfish, sea urchins, crinoids, and sea cucumbers. It is significant that the majority of these creatures were capable of some degree of locomotion.

Protozoans were at a very low ebb in the Triassic but increased in the Jurassic and reached spectacular numbers in the Cretaceous. A recent

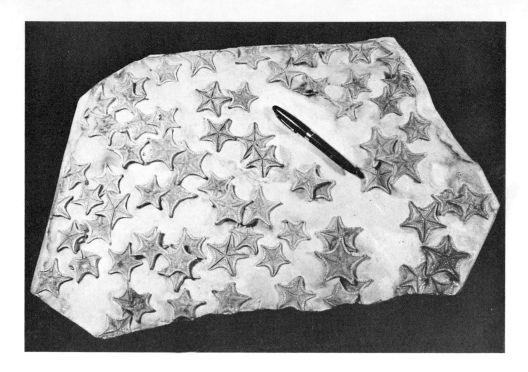

Figure 13.29 Numerous specimens of the starfish Austinaster *from the Cretaceous Austin Formation of Austin, Texas. (Texas Memorial Museum.)*

survey shows nine families in the Triassic, twenty-four in the Jurassic, and thirty-six in the Cretaceous. An important development of the Cretaceous was the appearance and spread of floating forms such as *Globigerina*, a lime-secreting form living in open oceans.

EVOLUTIONARY ADVANCES
AMONG MARINE INVERTEBRATES

The foregoing brief summary indicates that the marine invertebrates of the Mesozoic show many superior adaptations not generally possessed by their late Paleozoic predecessors. The inhabitants of the Mesozoic seas were on the whole more mobile, more intelligent, and more efficient at gathering a greater variety of food. The success of pelecypods such as the oysters and the corresponding decline of the brachiopods can probably be attributed to differences in their food-gathering mechanisms. A brachiopod generates only feeble currents about the margin of its shell, but the pelecypod has a much more powerful

mechanism for drawing water across its gills, where the food is extracted. Side by side, pelecypods prosper at the expense of brachiopods.

Animals that could move about flourished, whereas stationary forms declined. This trend is well illustrated by members of the echinoderm phylum. Immovable forms such as blastoids were extinct by the Permian, and the stalked crinoids had almost disappeared by the end of the Triassic. The crinoids that prospered during the Mesozoic and later were mainly floating forms. Meanwhile, starfish, sea cucumbers, and sea urchins, which move about over the ocean floor, were steadily increasing.

The Mesozoic invertebrates developed many other improvements in food-gathering techniques. Many groups were equipped with teeth and claws, obviously for opening or crushing shells and cutting and tearing flesh. Even such simple animals as the starfish and sea urchin have sharp, powerful teeth. The muscular arms and claws of lobsters and crabs contrast with the weak appendages of the trilobites and king crabs. Many gastropods acquired a rasplike structure studded with minute teeth (*radula*), and the cephalopods, in addition to the radula, are equipped with a birdlike beak with which to tear their victims apart. We do not mean to imply that all these structures were strictly of Mesozoic origin; many were present in the late Paleozoic, but in relatively fewer creatures. Competition for food had clearly favored those animals that could capture and devour large, fleshy prey.

Although the matter of nervous reactions and intelligence is difficult to assess, we should mention that the groups with highly developed brains and sense organs achieved relatively greater success than others not so endowed. The cephalopods, keen-eyed and with quick reactions, multiplied considerably; and even forms such as the starfish, which have rudimentary eyes in the ends of their arms, and *Pecten*, a pelecypod with eyes set in the tissue at the edge of its shell, became common.

The Mesozoic was a time of abundant land life. It is popularly known as the *Age of Reptiles* because of the dominant position this group achieved on land, in the air, and in the seas. Although mammals and birds appeared during the Mesozoic, they were not particularly abundant. Fish continued to evolve and became better adapted to their special ways of life, but the amphibians declined to a level of comparative insignificance and have never recovered.

Fish and Amphibians. Fossil fish are common in many Mesozoic formations of both freshwater and marine origin. The most important advances were made by the bony fish, especially the *actinopterygians (ray-finned fish)*. During the early stages of the Mesozoic most fish belonged to the *Chondrostei,* a group characterized by heavy diamond-shaped scales, asymmetrical tails, somewhat scaly fins, and considerable cartilage in their skeletons. Toward the middle of the era these fish were gradually replaced by the *Holostei,* with lighter scales, more symmetrical

Figure 13.30   A thick-scaled fish (Lepidotus) from the Early Jurassic of Germany. The scales are relatively thick, bony, and inflexible. Specimen is about 70 centimeters long. (By permission of the Trustees of the British Museum, Natural History.)

Figure 13.31 *Graveyard of Buettneria, a giant amphibian, found in Triassic beds near Santa Fe, New Mexico. The massed remains of many specimens probably record the drying up of a pond or lake. The skulls are about 60 centimeters long. (Smithsonian Institution.)*

tails, fins supported by flexible rays, and relatively more bone in their skeletons. This group was in turn displaced during the Cretaceous by the *Teleostei*, flexible-scaled fish with completely bony skeletons and powerful, well-formed fins and tails. This is the group that dominates present-day seas and rivers. The sharks were poorly represented during the early Mesozoic but gradually expanded during the Jurassic and Cretaceous to regain the ground they had lost during the critical late Paleozoic period. The crossopterygians and lungfish held a subordinate position in out-of-the-way surroundings, just as they now do. The Mesozoic was a period of decline for the amphibians. All the large, flatheaded stegocephalians became extinct early in the Triassic, and the group thereafter is represented by the familiar toads, frogs, salamanders, and the legless, wormlike apoda. The first true frogs are found in Triassic rocks of Madagascar, but ancestral forms from which they could have evolved have been discovered in the preceding period. The first known salamanders come from the Late Jurassic Morrison Formation.

Reptiles. Although the amphibians as a group never achieved complete sway over the land, their descendants, the reptiles, became rulers of the earth. About a dozen orders of reptiles were present during the Late Triassic and Early Jurassic, but only four orders exist today. Their basic adaptations permitted them to expand into hitherto unoccupied territory and even to reenter the ocean and compete with the fish.

The few paragraphs that we can devote to the story of the Mesozoic reptiles obviously cannot

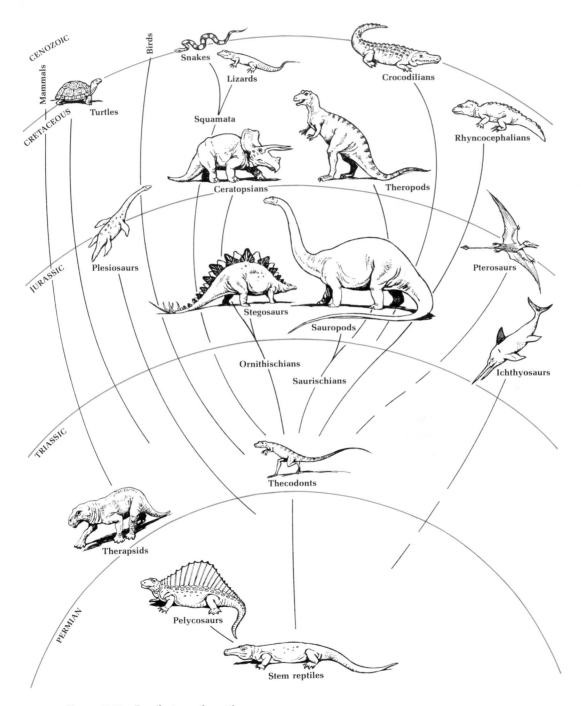

Figure 13.32   Family tree of reptiles.

do justice to the dramatic spread and remarkable achievements of the group. It may help our perspective to remember that the reptiles were dominant for 125 million years, compared with the approximately 65 million years of the Cenozoic, or *Age of Mammals,* and the few thousand years of man's reign.

The Triassic Period was marked by the decline and disappearance of certain of the older orders of reptiles and by the appearance of new and vigorous stocks. Among the Permian holdovers that disappeared in the Triassic were the *protosaurs* and certain mammal-like reptiles, the *therapsids.* The exact nature of these animals need not concern us here, but the newly evolved groups merit more detailed attention. The first turtles appeared during the Triassic and quickly achieved their typical form. The *plesiosaurs* and *ichthyosaurs,* adapted to life in the ocean, also appeared at this time, along with the first crocodiles. Most notable from the viewpoint of evolutionary progress were the *thecodonts,* lightly built, relatively small reptiles adapted to land life. As a group, they are confined to the Triassic, but they gave rise to a remarkable array of descendants that dominated the Jurassic and Cretaceous Periods. Crocodiles, dinosaurs, flying reptiles (*pterosaurs,* or *pterodactyls*), and birds all sprang from thecodont ancestors. One short-lived line of thecodonts, the *phytosaurs,* resembled the crocodile in size, shape, and mode of life, so nearly in fact that they are called the "ecological ancestors" of the crocodiles, meaning that they occupied a similar place in nature but are distantly related.

Dinosaurs. The term "dinosaur" is popularly applied to members of two orders of reptiles, both of which achieved gigantic size and occupied similar environments. The group first appeared during the Middle Triassic as certain slender two-legged forms belonging to the order *Saurischia.* A typical genus is *Coelophysis,* a carnivore found in Late Triassic rocks of New Mexico and adapted for life on dry uplands.

Figure 13.33 *Restoration of the Triassic mammal-like reptile Lystrosaurus. Remains of this animal have been found in China, South Africa and, most recently, in Antarctica. Its presence in these widely separated localities is taken as convincing evidence of the breakup of Gondwana in post-Triassic time. (Reprinted from Edwin H. Colbert,* The Age of Reptiles, *by permission of W. W. Norton & Company, Inc., and George Weidenfeld & Nicolson, Ltd. Copyright 1965 by Edwin H. Colbert.)*

Coelophysis was the forebear of a host of saurischian dinosaurs that adapted to many different terrestrial environments. Some increased in size and seem to have reverted to the four-footed stance, eventually giving rise to the gigantic, long-tailed, long-necked *sauropods.* The sauropods were most common in the Jurassic but existed in restricted localities until the very end of the Cretaceous. Among the well-known sauropods are the ancestral form *Plateosaurus,* from the Late Triassic of Germany; *Diplodocus,* a slender, whip-tailed type from the Jurassic Morrison Formation in the western United States; *Brontosaurus,* a heavier contemporary of *Diplodocus;* and *Brachiosaurus,* the long-necked but short-tailed form, perhaps the greatest land animal of all time.

Those saurischians that retained their two-legged pose and carnivorous habits continued to live side by side with the sauropods and perhaps eventually began to prey on them. This group includes the well-known *Allosaurus,* from the Morrison Formation; *Tyrannosaurus rex,* the

Figure 13.34 Diplodocus, *a giant sauropod dinosaur from the Morrison Formation.* (Smithsonian Institution.)

greatest land carnivore of all time and perhaps the best known of all the dinosaurs; and *Gorgosaurus,* which lived shortly after *Tyrannosaurus* on the Late Cretaceous plains of Canada.

The earliest known member of the second major group of dinosaurs, the *Ornithischia,* thus far discovered is from Late Triassic rocks of the Cape Province, South Africa. This animal, *Lycorhinus,* had a skull about 10 centimeters long and a strange mixture of tooth types perhaps indicating a varied diet. The descendants of the Triassic ornithischians became numerous and varied during the Jurassic and Cretaceous. Here again there are two-legged types, but few members of the group lost the use of their forelimbs to the extent that the saurischian bipeds did. All

later ornithischians were herbivores, and the group adapted to environments ranging from arid deserts to marshes and lagoons. The suborder *Ornithopoda* (bird foot) included many large, water-loving plant-eaters with duck-billed faces. That they existed on tough woody vegetation is indicated by the numerous closely packed teeth that were set in their strong jaws. The duck-billed dinosaurs were most plentiful during the Cretaceous. The suborder *Stegosauria* included heavy four-legged herbivores armed with various types of spikes, plates, and bony protuberances; they were exclusively of Jurassic age. The suborder *Ankylosauria* were heavily armored types known only from the Cretaceous. A fourth order, *Ceratopsia,* included large four-legged species

Figure 13.35 Skeleton of the Jurassic carnivorous dinosaur Allosaurus from the Morrison Formation. Length of skeleton is about 6 meters. (Courtesy of James Madsen, Jr.)

Figure 13.36 Duck bills and neck frills typify Cretaceous dinosaurs. (a) Lambeosaurus, a duck-billed specimen, has a peculiar skull structure that may correlate with underwater feeding habits. (b) Triceratops, the giant horned dinosaur, is a Late Cretaceous ceratopsian from North America. (c) Protoceratops is an Early Cretaceous ceratopsian. This skeleton is from the Gobi Desert. (a and c, Field Museum of Natural History, Chicago; b, Smithsonian Institution.)

(b)

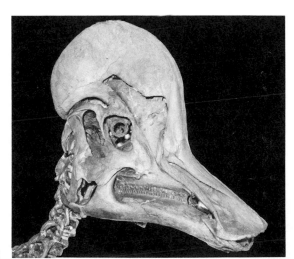

(a)

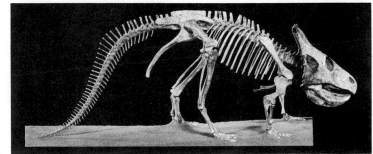

(c)

Figure 13.37 *Dinosaur bones embedded in rock at the Cleveland-Lloyd Quarry in the Morrison Formation, central Utah.*

that had tremendous skulls and a sort of collar or frill that extended backward across the neck and shoulders. Most of the later ceratopsians had one or more forward-pointing horns. This group is known only from the Cretaceous. Geologists believe that it originated in Asia and reached North America late in the Cretaceous.

A list of the inhabitants of any of the larger continents during the Late Triassic, Jurassic, or Cretaceous invariably includes a great number of dinosaurs. These animals spread widely, and even the sauropods reached Australia, a fact suggesting that landmass was connected with the

other continents as late as the Cretaceous. Generally, the various types of dinosaurs occur together in fossil "boneyards," for the carnivorous types appear to have preyed on, and lived among, the herbivores. The most widely distributed forms are the sauropods, but we must remember that their huge bones fossilized readily and when exposed can scarcely escape being discovered. The dinosaurs disappeared at the close of the Cretaceous and as a group are therefore strictly confined to the Mesozoic Era.

Sea Monsters. During the Mesozoic a number of reptilian groups adapted to life in the seas. There were fishlike forms, appropriately called *ichthyosaurs;* paddle-swimmers, mostly with short tails and long necks, the *plesiosaurs;* gigantic long-tailed marine lizards, the *mosasaurs;* and large sea turtles. Of these, only the sea turtles survive; the rest became extinct at or before the close of the Cretaceous, 65 million years ago.

The ichthyosaurs were the first large reptiles to invade the seas, and they became the most fishlike in their adaptations. The earliest members of the group thus far identified are of Triassic age; they became very common during the Jurassic and Cretaceous. They had fishlike limbs, a forked tail, and unusually large eyes, indicating that they may have lived in the dimly lighted depths of the ocean. Some were toothless, whereas others had jaws lined with many sharp teeth; stomach contents, which have been found petrified inside the body of some specimens, indicate that the animals lived on a varied diet, chiefly fish and cephalopods. The largest speci-

Figure 13.38 *An ichthyosaur* (Stenopterygius quadricissus) *with the outline of the body preserved. The specimen is from Lower Jurassic deposits near Holzmaden, Germany, and is nearly 3 meters long. (Field Museum of Natural History, Chicago.)*

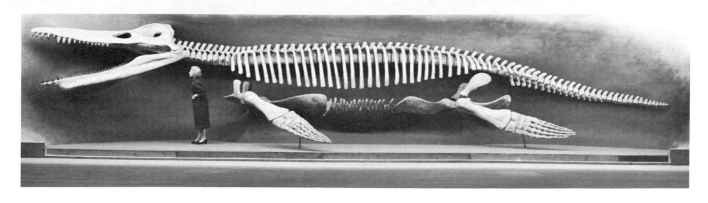

Figure 13.39 *Gigantic 13-meter-long reconstructed skeleton of* Kronosaurus, *a short-necked plesiosaur, from the Cretaceous of Australia. (Harvard University News Service.)*

mens reached a maximum length of a little over 4 meters. All evidence indicates that the ichthyosaurs hatched their young internally and brought them forth into the sea fully formed.

The plesiosaurs appeared in the Triassic and had vanished by the close of the Cretaceous. There were two chief lines of evolution. One, typified by *Alasmosaurus*, had long necks and small heads; the other, including the remarkable *Kronosaurus*, had short necks and large heads. The plesiosaurs had wide oarlike flippers, or paddles, and evidently preyed on fish. Because they could not move as quickly as the contemporaneous ichthyosaurs, they probably depended heavily on the flexibility of their long necks as an aid in capturing prey. The number of vertebrae in the long-necked group increased steadily during their evolutionary history. The last form had seventy-six neck vertebrae, which added up to over half the animal's total length.

The 7-meter-long mosasaurs appeared in the Late Cretaceous but had vanished by the close of the period. With their long, flat tails and sinuous bodies, they were undoubtedly excellent swimmers, and their numerous conical teeth were well suited for capturing large, active prey.

Archelon, the giant sea turtle found in the Cretaceous chalk beds of Kansas, had a flattened body and was over 3 meters long from its nose to the tip of its tail. The bones of the shell were very small, and the covering leatherlike. The estimated weight is 3 metric tons.

Crocodiles, Snakes and Lizards. The earliest known crocodile, *Proterochampsa*, was discovered in Triassic rocks of Argentina in 1958. Crocodiles have succeeded in a modest way through the Mesozoic periods and into the present time.

The first lizards made their debut during the Jurassic but have left few remains. We have already mentioned the large marine forms. The

Figure 13.40 *An artist's conception of* Macroplata, *a plesiosaur about 6 meters long, from the Cretaceous of England. (By permission of the Trustees of the British Museum, Natural History.)*

Figure 13.41 *Fossil skeleton of the oldest known aerial vertebrate* (Icarosaurus siefkeri). *There were no wings; the flight membrane was stretched between extensions of the ribs. The specimen was found in Triassic rocks of New Jersey. (American Museum of Natural History.)*

snakes, which are the last major group of reptiles to appear, are first found in Cretaceous rocks. They are obviously highly modified lizards, with which they are grouped into a single order, *Squamata*. Lizards and snakes have thrived both in wet and in dry areas.

Flying Reptiles. For many millions of years in the late Paleozoic, the only animals capable of flight were the insects. This aerial monopoly was broken in the Triassic when vertebrates took to the air. A gliding reptile from the Newark Group of New Jersey is the oldest known aerial vertebrate. It has been named *Icarosaurus siefkeri* and is a highly modified lizard. In the succeeding Jurassic, the feathered birds and leathery-winged pterosaurs became airborne and were highly successful. The pterosaurs appear in the fossil record slightly before the birds and may have outnumbered them in the Jurassic and Early Cretaceous. The remains of both pterosaurs and birds are rare because both had light and delicate skeletons and were apt to live and die in forests under conditions unfavorable for fossilization.

The pterosaurs ranged from sparrow size to giants with a wingspread of over 8 meters. In general, Jurassic forms were smaller than their Cretaceous descendants and had many small teeth, long tails, and relatively heavier skeletons. *Rhamphorhynchus*, represented by well-preserved skeletons from the Solenhofen quarries in Bavaria, is typical of the group. Later species, such as the well-known Cretaceous form *Pteranodon*, were generally larger and toothless and had short tails and thin, delicate bones. The remains of pterosaurs were preserved only under very special conditions, usually in extremely fine-

Figure 13.42 *Skeleton of* Pterodactylus elegans *preserved in fine-grained Jurassic limestone from the Solenhofen quarries. The specimen is not much larger than a sparrow. (American Museum of Natural History.)*

Figure 13.43   Pterosaur tracks from the Morrison Formation of Jurassic age, Apache County, Arizona. The four-toed print of the hind foot lies ahead of the deep depression made by the "knuckle" of the front foot.

Figure 13.44   Pteranodon, a giant flying reptile. Fossils have been found in Late Cretaceous rocks of the western plains of the United States. (By permission of the Trustees of the British Museum, Natural History.)

grained sediments that accumulated in quiet, shallow water. Paleontologists generally agree that the pterosaurs were gliders, for their remains show no signs of attachments for powerful wing muscles. The discovery of well-preserved footprints with impressions of both the hind feet and front feet has led some investigators to speculate that pterosaurs could walk when necessary and could take to the air after gathering speed on the ground. The wing membrane was stretched between an elongated "little finger" and the sides of the body, and it is evident that damage to such a structure would be difficult to repair and would perhaps permanently disable the animal.

Birds. Birds appear in the fossil record during the Late Jurassic. Two well-preserved skeletons and parts of two others, all with imprints of feathers, have been found in the Solenhofen quarries. These remains are of two genera, *Archaeopteryx* and *Archaeornis*, and belong to animals about the size of a crow. In anatomy, *Archaeopteryx* stands about midway between thecodont reptiles and later birds. These earliest birds had entered many new environments, but their fossils are nevertheless rare. *Hesperornis*, a bird from

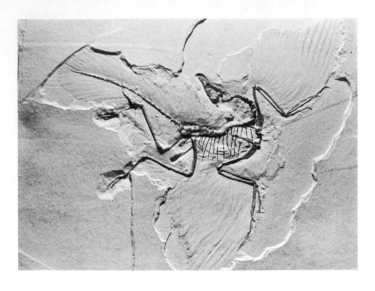

Figure 13.45 *Cast of the original Archaeopteryx skeleton from Jurassic rocks of Bavaria, Germany. (Courtesy of Nelson B. Wadsworth.)*

the Cretaceous chalk beds of Kansas, was about 2 meters long and had a good set of teeth but no trace of wings. It was probably a powerful diver and swimmer.

A comparison of their wings plainly reveals why birds succeeded and pterodactyls failed. The pterodactyl's wing, as we have already mentioned, was formed of membrane, was difficult to repair, and was adapted chiefly for gliding; but a bird's wing permits true flight and is composed of many individual feathers that can be renewed and replaced as needed to keep the wing in a state of constant efficiency.

Mesozoic Mammals. The fossils of Mesozoic mammals are of greatest evolutionary importance even though they are insignificant in quantity when compared with the remains of contemporary reptiles. Bones of early mammals are extremely small and generally scattered. At a few places, however, they are abundant, and thousands of individual bones and teeth have been taken from concentrated pockets and fissures in South Wales. This locality is of very Late Triassic age and is the most prolific and significant source of early mammals so far discovered.

The most important vertebrate found in the South Wales deposits is *Morganucodon,* currently judged to be the earliest mammal. It is less than 10 centimeters long and has many primitive characteristics that suggest relationship with the modern monotremes (platypus, and others). Species of *Morganucodon* have also been discovered in such widely scattered areas as China and South Africa. Contemporaries of *Morganucodon* had also achieved many mammalian characteristics. Of these so-called mammal-like reptiles the *tritylodonts* are best known. They have been

Figure 13.46 *Restoration of Archaeopteryx constructed at the British Museum of Natural History. (By permission of the Trustees of the British Museum, Natural History.)*

found in many localities including a site with many complete skeletons near Kayenta, Arizona.

Fossils of Jurassic mammals have been found both in Europe and in North America, but again remains are relatively rare and consist only of partial skulls, jaws, and teeth. All are rodentlike, about the size of mice and rats or smaller. Several distinct orders are represented. The teeth are variously modified and differentiated into incisors, canines, and molars; the animals seem to have subsisted on a high-energy diet of insects, seeds, and fruit. The typical Jurassic mammal inhabited the undergrowth or the branches of trees and was ever on the alert for the large predatory reptiles with which it could not hope to compete in open combat.

Mammals continued to be scarce during the Cretaceous, and some of the Jurassic forms even disappeared. The best-known late Mesozoic remains are from the Upper Cretaceous of Mongolia and the western United States. Early in the period true *marsupials* (pouched mammals) appeared. *Placentals* (advanced mammals) came in the Late Cretaceous. All in all, the mammals were quite rare during Mesozoic time, but they had already developed the potential to replace the reptiles when the time arrived.

Close of the Mesozoic

No single dramatic event marks the end of the Mesozoic Era. Although a number of gradual changes occurred near the Mesozic–Cenozoic time boundary, geologists have found no evidence of a rapid, simultaneous, world wide physical change. In general, the oceans were withdrawing during this interval, converting former sea bottoms into swamps, floodplains, and tidal flats. At places the withdrawal was accompanied or caused by mountain building.

In the biological sense there were a number of startling changes near the end of the Mesozoic Era. Almost all fossiliferous areas indicate that large and important groups were being extermi-

Figure 13.47 *Jaw and teeth of* Triconodon, *a primitive mammal from the Late Jurassic of England. The specimen is about 2 centimeters long. (Crown copyright, Geological Survey photograph. Reproduced by permission of the Controller of Her Britannic Majesty's Stationery Office)*

Figure 13.48 *An outcrop of a part of the famous Mother Lode near Bear Valley, California. Weathering of a system of veins has released the placer gold for which the western slopes of the Sierra Nevada are famous. The Mother Lode veins were formed during the Cretaceous at a late stage in the emplacement of the granitic batholiths. (U.S. Bureau of Mines.)*

nated both on the land and in the oceans. Dinosaurs and pterodactyls vanished from the lands and in the ocean the last plesiosaurs, mosasaurs, and ichthyosaurs disappeared. Of even greater significance to paleontologists was the extinction of the ammonite cephalopods, the belemnites, the large rudistid pelecypods, and certain ancient lineages of oysterlike pelecypods. Although all these groups did not disappear simultaneously everywhere, their extinction does provide a guide to the end of the Mesozoic Era. In practice, the line between the Cretaceous and the Tertiary is placed above the highest, or last, occurrence of dinosaurs, ammonites, or other key forms.

Economic Products of the Mesozoic

Mineral resources of Mesozoic age are critically important to our modern industrialized civilization. On an average, Mesozoic rocks are richer in oil than either Paleozoic or Cenozoic rocks. According to a recent survey, almost 20 percent of all the world's oil fields tap Mesozoic rocks, and these rocks contain 52.7 percent of the world's known reserves. The prolific Middle East fields draw oil chiefly from Jurassic and Cretaceous rocks. Other areas of Mesozoic production are the Rocky Mountains, the upper Gulf Coast, western Venezuela, and Argentina. Recently, large oil and gas accumulations have been discovered in underwater Cretaceous formations in the Bass Strait between Australia and Tasmania. Large reserves have also been discovered in Triassic rocks of northern Alaska and under the North Sea.

Coal is found in rocks of each Mesozoic period; Triassic coal occurs in the eastern United States, South Africa, and China; important Jurassic coal deposits are found in China and Siberia; and Cretaceous coal is so abundant in the United States and Canada that the system ranks second only to the Pennsylvanian in amount of total

Figure 13.49 The "Big Hole" of the Premier Diamond Mine, South Africa. The hole has been left after removal of the diamond-bearing rock. The diamond-rich rock occurs in the form of vertical cylindrical pipes that are thought to have been exit ways for deep-seated volcanic action during the Cretaceous Period. (U.S. Bureau of Mines.)

reserves. Cretaceous coal is also important economically in northeast Asia.

Metallic mineral products were formed in many places in connection with Mesozoic igneous activity and sedimentation. Copper occurs in scattered deposits in Triassic sandstone and shale in Germany, Russia, and the United States. Jurassic rocks contain important reserves of sedimentary iron ore in England and Alsace-Lorraine. Cretaceous sedimentary iron ore is mined in Siberia. Both Triassic and Jurassic rocks in the western United States contain important reserves of uranium. The famous Mother Lode, source of much of California gold, was formed in the Late Jurassic or the Cretaceous. The salt plugs, or domes, in the Gulf Coast area, which yield sulfur in addition to salt, arose from Jurassic beds thousands of meters beneath the surface. Salt is also found in Lower Cretaceous formations of Brazil and Gabon. The diamond-bearing pipes of South Africa are probably of Cretaceous age and those of Siberia are thought to belong to the Triassic and Jurassic. The diamonds they contain, however, may have been formed at an earlier time and, of course, many stones were freed from their original matrix and redeposited in rocks that were formed much later.

A SUMMARY STATEMENT

The Mesozoic Era dawned with the lands relatively high and with few seas in the interiors of the continents. The Triassic Period was geologically rather quiet, and large volumes of red continental sediments were laid down on the land areas. The succeeding Jurassic Period was without extensive mountain building, the oceans spread more widely over the continents, and a variety of marine and nonmarine sediments accumulated. The Triassic and Jurassic were marked by widespread intrusion and extrusion of basic igneous rocks, especially in the Southern Hemisphere. Proponents of continental drift maintain that the breakup of Gondwana was accomplished chiefly in the mid-Mesozoic.

During the Cretaceous, interior seas spread across the continents in one of the greatest floods of all times. Commencing in middle Cretaceous time, the margins of the Pacific became geologically active; igneous masses were intruded and older geosynclinal deposits were folded and uplifted. The era closed with the continents being uplifted and eroded.

Mesozoic life was in transition. Plants became better adjusted to dry-land environments and to seasonal changes. Pines, ferns, and cycads were dominant during most of the era. During the middle Cretaceous the flowering plants appeared in considerable numbers. Marine organisms slowly recovered from the critical late Paleozoic. Pelecypods, gastropods, and coiled cephalopods multiplied, and toward the end of the era, floating organisms of various kinds increased. On land, the reptiles, represented by dinosaurs, were the dominant vertebrates. Mammals appeared in the Triassic and birds in the Jurassic, but neither group succeeded in becoming dominant during the Mesozoic.

FOR ADDITIONAL READING

Arkell, W. J., *Jurassic Geology of the World.* New York: Hafner, 1956.
Augusta, J., and Z. Burian, *Prehistoric Reptiles and Birds,* trans. M. Schierl. London: Hamlyn, 1961.
————, *Prehistoric Sea Monsters,* trans. M. Schierl. London: Hamlyn, 1964.
Colbert, E. H., *The Age of Reptiles,* New York: Norton, 1965.

Additional references for material in this chapter can be found at the end of Chapter 11.

THE TERTIARY PERIOD

This chapter deals with the *Tertiary Period,* one of the two periods of the Cenozoic Era. Rocks and fossils of this period are abundantly preserved, and they indicate that the physical conditions under which they originated were much like those of the present. Most of the major geographic features of the continents and many topographic details were produced by Tertiary events. Most present-day life forms have their obvious ancestors among the fossils that are found in profusion in rocks of Tertiary age. The geologic time scale in Chapter 6 shows the sub-

Peaks of the Grand Teton Range, Wyoming. The central higher portions are ancient Precambrian rocks. The range was uplifted during the Rocky Mountain Revolution early in the Tertiary Period. (Courtesy of Hal Rumel.)

divisions of the Tertiary, their duration in years, and their relation to the *Quaternary,* the period in which we now live (see Figure 6.2).

North American and British geologists favor the terms Quaternary and Tertiary as Cenozoic subdivisions. Other European geologists are inclined to prefer *Neogene* and *Paleogene.* The important mountain-building activity that created the Alps culminated in the middle Tertiary, and this event naturally serves as a guidepost in Old World classifications. The term *Nummulitic Period* is also used in Europe for the combined *Paleocene, Eocene,* and *Oligocene Epochs* in recognition that *nummulites* (large, disc-shaped fossil protozoans) are common in this interval. We cannot predict which of the proposed terms will be adopted and found best suited for worldwide use.

The Tertiary began with the continents well above water. Of the many Cretaceous seaways, only Tethys and its embayments remained. Even this feature was largely obliterated during the middle Tertiary, Alpine disturbances that elevated the great east–west mountain chains of

Figure 14.1 Craters on the island of Java, Indonesia. Volcanic activity and crustal unrest have characterized the southwestern Pacific for many geologic periods. (Indonesian Information Office.)

Europe and Asia. During the middle and late Tertiary the other continents were practically free of interior seas. The borders of the Pacific continued to be geologically active, and the East and West Indies took on their modern appearancc.

Although the Tertiary was a time of extensive migrations of plants and animals and there was great diversification into orders and lesser categories, no large or radically new groups of plants or animals made their appearance or were exterminated during the period.

Paleogeographic Reconstructions

At the beginning of the Tertiary Period, each of the continents had been almost completely isolated and were separate entities. However, there were a few exceptions to this generality: northern Europe and Greenland were still connected and Australia was still attached to Antarctica. Subsequently, during the 65 million years of the Tertiary the various landmasses drifted into their present configuration. The Atlantic Ocean widened and Greenland split away from Europe. The westward push of the Americas completed the somewhat violent creation of the Rockies-Andes chain, which, as an incidental effect, brought

Figure 14.2 The continental drift dispersion of continents as of the beginning of the Tertiary. (Adapted from Robert S. Dietz and John C. Holden, "Reconstruction of Pangaea," Journal of Geophysical Research, Vol. 75, 1970, p. 4949.)

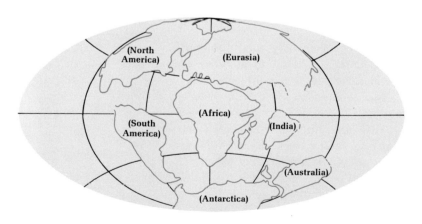

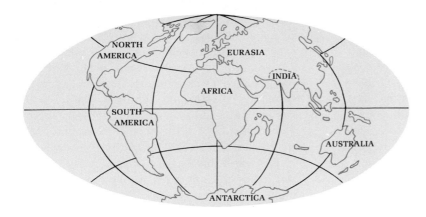

Figure 14.3 The position of continents today. (Adapted from Robert S. Dietz and John C. Holden, "Reconstruction of Pangaea," Journal of Geophysical Research, Vol. 75, 1970, p. 4951.)

about the rejoining of the North American and South American continents by means of the Isthmus of Panama. This made possible the migration of land life between the continents.

India completed its spectacular northward journey and collided with Asia during the middle Tertiary. This together with a general northward movement and counterclockwise rotation of Africa against Europe created the Alpine-Himalayan chain. Finally, Australia parted company with Antarctica and drifted rapidly northward pressing strongly into the developing East Indies.

These motions of the great landmasses also served to outline the present oceans. Insofar as the geologic record is concerned, the destruction of Tethys was a major event. A general vertical uplift of all the continents also took place. This may or may not be a consequence of their drifting but is a basic fact to be taken into account in explaining the diversity of climate, patterns of wind and water currents, and the onset of the Pleistocene Ice Age.

TETHYS—FROM SUBSIDING TROUGH TO MOUNTAIN RANGES

Although the wide and lengthy Tethys Seaway was still in existence at the beginning of the Tertiary, the downsinking and sedimentation phases of its history were drawing to a close. Mountain range after mountain range began to appear in Europe and Asia as compressive and volcanic processes affected the area.

The closing of the seaway and the severe compression of its contained sediments can be explained by the northward movement of the African block toward Europe and by the collision of India with southern Asia.

The closing of the European segment of Tethys evidently resulted from a scissorlike motion that began at the west end and progressed eastward. Thus the Pyrenees had emerged at the close of the Eocene, as had the Atlas Range of northwest Africa. The most important foldings of the European Alps and the Carpathians took place in the course of the Oligocene. In Asia the Himalayan area began to be affected early in the Tertiary, with the most intense phase coming in the middle Miocene. In the Middle East, mountain building commenced in the Cretaceous and continued into the Pliocene, with diminishing marine deposition and corresponding nonmarine accumulations. The entire Alpine Revolution, as it is generally called, seems to have culminated in the Miocene and the total effect was to create the highest mountain ranges from the area of deepest sediments. A number of residual seas in the late Miocene and Pliocene encircled the previously formed mountains and received extensive deposits of sediment from the newly formed ranges. These deposits later became involved in secondary movements and foldings.

Rising ranges cut off large areas from the sea, and these became lagoons with brackish and freshwater deposits. As the Himalayan Range rose, it shed debris southward into a great subsiding trough that was developing upon the newly arrived Indian block. This deposit, the *Siwalik Series,* is 4,500 to 6,000 meters thick and yields a key record of life and environments from late Miocene to early Pleistocene time.

Much of the actual folding and mountain building of the Tethyan belt took place at or below sea level. Later effects served to elevate the mountains as well as the surrounding areas on a broad (*epirogenic*) scale, so that the seas were permanently expelled and broad tracts bordering the ranges became dry land. This effect came chiefly in the Pliocene and early Pleistocene. The Tibetan Plateau, highest in the world, originated as the area was lifted by the underthrust of the Indian block. The Mediterranean Sea is not technically regarded as a remnant of Tethys although it lies along the same trend and covers some of the same territory. Tethys was destroyed by mountain building and the Mediterranean was opened by later movements related to the formation of the Atlantic Ocean.

EURASIA NORTH OF TETHYS

Most of Europe between the Tethys Seaway and the Baltic Shield was occupied by a series of shallow Tertiary seas or lagoons. There was a discontinuous barrier between the northern lagoons and Tethys in the form of the eroded remnants of the Hercynian ranges. The famous succession of deposits in the Paris Basin, where Lyell conceived the subdivisions of the Tertiary, represents deposition in shallow embayments from the Atlantic and contrasts sharply with contemporaneous deposits in Tethys. Southeastern Britain, France, northern Germany, and the Netherlands have extensive Tertiary deposits.

Eastward, the Tertiary seas were less extensive. They did not reach Moscow, but covered much of southern Russia. For a short period in the Oligocene a narrow seaway lying east of the Urals extended from the Arctic Ocean to Tethys. This was the last flooding of Eurasia north of Tethys. Most of Siberia and China was free of marine transgressions and received instead a variety of continental rocks, including coarse conglomerates. Many large basins came into existence and trapped sediments from adjacent ranges. In these basins are well-preserved fossils of contemporary mammals and plant life. The largest land mammal known, *Indricotherium,* a rhinoceros, lived during the Oligocene in central Asia.

In extreme northeastern Siberia the Tertiary is again represented by marine rocks, and there are lava flows and granitic intrusions appropriate to the circum-Pacific zone. Deposits of valuable metals and petroleum are found in Tertiary rocks of northeastern Asia.

EASTERN NORTH AMERICA

A relatively narrow belt of Tertiary sediments fringes the Atlantic and Gulf coasts of North America. The visible part of the belt begins at Cape Cod, widens southward to include all of Florida, and is up to 320 kilometers wide in the lower Mississippi Valley. The interior regions, including the stumps of the Appalachian Range, supplied most of the sediments. Because the coastal plain has been little deformed and because its sediments can be related directly to

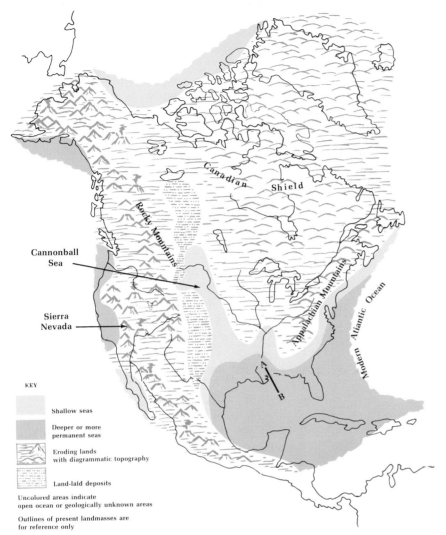

Figure 14.4 Reconstructed paleogeography of the early Tertiary Period.

KEY

Shallow seas

Deeper or more permanent seas

Eroding lands with diagrammatic topography

Land-laid deposits

Uncolored areas indicate open ocean or geologically unknown areas

Outlines of present landmasses are for reference only

Cannonball Sea

Sierra Nevada

Canadian Shield

Rocky Mountains

Appalachian Mountains

Modern Atlantic Ocean

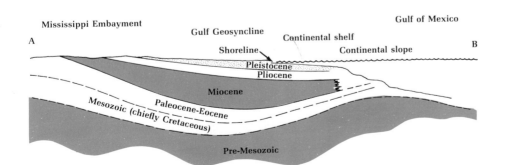

Mississippi Embayment

Gulf of Mexico

Gulf Geosyncline

Continental shelf

Shoreline

Continental slope

A

B

Pleistocene

Pliocene

Miocene

Paleocene-Eocene

Mesozoic (chiefly Cretaceous)

Pre-Mesozoic

Figure 14.5 Cross section from Natchez, Mississippi, into the Gulf of Mexico. Length of section is about 560 kilometers, and position is shown by line A-B in Figure 14.4. This marginal geosynclinal trough has received about 15,000 meters of sediment from the interior of the continent, beginning in the Jurassic and continuing to the present. The bottom of the trough is now considerably lower than the bottom of the Gulf of Mexico. (Adapted from P. B. King, The Evolution of North America, Princeton, N.J.: Princeton University Press, 1959. Reprinted by permission.)

their source areas, we know that the Tertiary was relatively quiet and uneventful insofar as the eastern part of North America is concerned.

As would be expected, the Tertiary rocks thin toward the land and contain more and more marine fossils as they are traced seaward. Coarse sands and silts characterize the northern outcrops, but limy rocks reflecting warmer seas increase southward. Florida has many limy formations with masses of shells and even coral reefs. The oldest formation seen at the surface in Florida is of Eocene age, and the whole state is essentially a product of shallow Tertiary seas.

Not to be overlooked are the Bahama Banks, which resemble Florida but contain even more calcareous material. A well drilled on Andros Island directly east of the Florida Keys passed

Figure 14.6 Columbia River Basalt, Franklin and Whitman counties, Washington. Successive flows are exposed as cliffs in the canyon walls. (U.S. Geological Survey.)

through a continuous sequence of nearly 4,500 meters of Tertiary and Cretaceous calcareous sediments, which suggests that the area has been subsiding for a long time and that deposition of calcareous sediment kept pace with the down-sinking.

WESTERN NORTH AMERICA

The Tertiary brought extensive changes along the Pacific Coast of North America, and it may be said that the Coast Ranges were essentially a product of this interval. At the beginning of the Tertiary, the Pacific extended inward over the site of the present Cascade Range and to the foothills of the Sierra Nevada. Great outpourings of lava, especially in the Miocene, spread widely in Washington, Idaho, and Oregon. The Columbia River basalts cover at least 500,000 square kilometers and are up to 3,000 meters thick. Twenty-three separate flows, one above another, have been counted, and a total of about 150,000 cubic kilometers of lava is estimated to have been produced in the Columbia Plateau.

In California, a succession of temporary embayments accompanied the uneasy shifting of the land. The Sierra Nevada was eroded nearly to sea level by the Miocene, but was then rejuvenated and supplied additional sediments westward to the nearby basins. With the passage of time, by the addition of sediment and by general uplift, a fringing strip of land including the Coast Ranges became a part of the continent. The Cascade Range was uplifted in the late Pliocene, about 4 million years ago. The famous volcanic cones capping the Cascades came later during the Pleistocene. All these features are compatible with the westward thrust of North America against and over the floor of the Pacific.

The Rocky Mountains continued to grow during the early Tertiary. The Great Basin, with its peculiar north-south trending ranges, began to form in the Oligocene, as great numbers of parallel faults cut the crust. The area between the Sierra Nevada and the Wasatch Range subsided unevenly and created basins for the lakes that appeared in the more humid phases of the Pleistocene Ice Age.

Tertiary deposits are widespread in Alaska but rare in northern Canada and the Arctic islands. In Alaska, coal beds and fossil plants are common, but no Tertiary vertebrates have yet been found, hinting that climates were not as favorable as they were farther south.

Continental Sediments. The most complete sequence of Tertiary land deposits known is found in the western United States. Here conditions were especially favorable for the production and preservation of large amounts of sediment. Deposits of the Paleocene Epoch accumulated mainly in the interior basins lying between the ranges created by the Laramide Orogeny. The Paleocene is well represented by formations in the *Big Horn Basin* of Wyoming and the *San Juan Basin* of New Mexico. There are also many thousands of square kilometers of Paleocene rocks in the plains section of Montana and the Dakotas. The rocks are chiefly sandstone and shale, with immense reserves of low-grade coal. The fossil vegetation and animal life indicate that a generally humid, semitropical climate prevailed over the western interior of the United States at this time. Apparently the whole region was several thousand meters lower than it is at present. Paleocene sediments that now lie at relatively high elevations in the Rockies accumulated originally only slightly above sea level.

The Eocene is represented by formations in at least a dozen separate Rocky Mountain basins, but there are practically no sediments of this age in the plains area. The sediments that filled the individual basins were obviously derived from the adjacent uplifted ranges and show considerable variety. Among the most common rock types are thick conglomerates and coarse sandstones laid down by powerful streams. An unusual feature of the Eocene was a system of large interior lakes that formed in the adjoining parts of Colorado, Utah, and Wyoming. This water

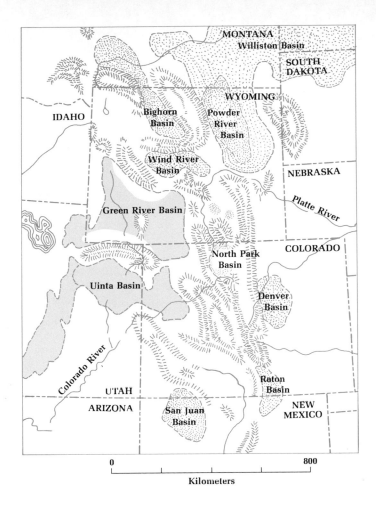

Figure 14.7 *Basins with Tertiary sediments within and near the Rocky Mountains. From these basins come most of the fossil remains that are the basis for reconstructing the early Tertiary life of North America. Areas occupied by lakes for significant periods of time are indicated by light blue. (Adapted from P. B. King, The Evolution of North America, Princeton, N.J.: Princeton University Press, 1959. Reprinted by permission.)*

body, called the *Green River Lake System,* covered over 100,000 square kilometers. Its waters teemed with plant and animal life whose remains have become incorporated in the fine-grained bottom sediments. These organic-rich deposits constitute the world's greatest single oil reserve. If all the contained oil could be recovered, it

Figure 14.8   Eroded cliffs of the Green River Formation of Eocene age, near Rifle, Moffat County, Colorado. Rich beds of oil shale, which yield up to 80 gallons of oil per ton of rock, make up part of the higher cliffs. (Colorado Department of Public Relations.)

Figure 14.9   Reconstructed stages in the development of the Medicine Bow Mountains, Wyoming. Drawn by S. R. Knight. (Wyoming Geological Association.)

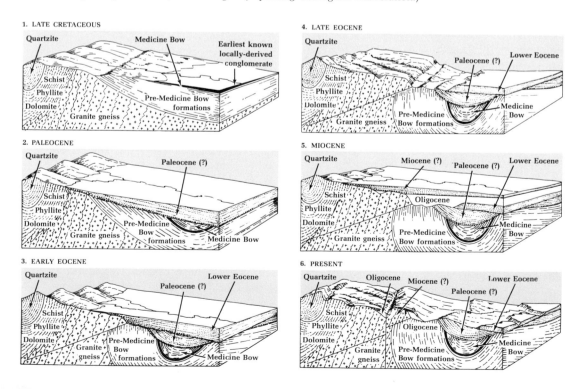

Figure 14.10 *Badlands National Monument, South Dakota. The barren, rugged outcrops are eroded in river-laid deposits of Oligocene age. Fossil mammals are abundant in this formation. (U.S. National Park Service.)*

would exceed all other known reserves in the world. The delicate fossils of insects, leaves, and fish found in the Green River Formation are world famous.

During the period of lake formation, the western United States remained near sea level and a subtropical or humid climate prevailed. Erosion and sedimentation were the dominant geologic processes, and the mountain ranges were worn down and buried in their own debris. The many excellent fossils of primitive mammals and other organisms that occur in the Eocene of the Rocky Mountains provide us with a near-perfect picture of life of the time.

With the coming of the Oligocene Epoch, the geologic quiet was broken by volcanic outbursts that blanketed large areas of the American West with a thick covering of ashes and dust. Explosive volcanoes erupted in many places and were particularly active in the Yellowstone Park area of Wyoming and the San Juan Mountains of

Colorado. Although the volcanic deposits were originally hundreds of meters thick, they have since been eroded away to the point where the topography that existed before the eruptions now lies exposed. There is relatively little Oligocene sediment preserved within the Rocky Mountains proper. The most complete record of this epoch is in the foothills and plains areas of Colorado, Wyoming, Nebraska, and South Dakota. Here we find the famous White River beds, from which come the most perfect and complete fossil mammals to be found anywhere in the world. The fine preservation of these remains is attributable to the frequent overflowing of sediment-laden rivers in the area, which buried not only scattered bones but also the complete skeletons of animals drowned on the floodplains. From the White River badlands have come remains of at least 150 species of fossil animals, including types ancestral to forms still alive and others that are completely extinct. The mammalian life of the west-

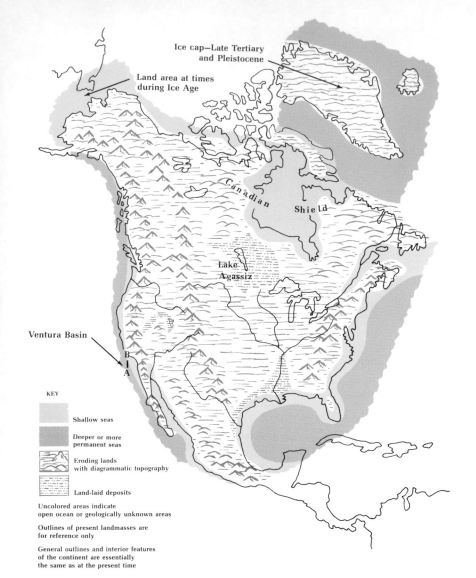

Ice cap—Late Tertiary
and Pleistocene

Land area at times
during Ice Age

Canadian

Shield

Lake
Agassiz

Ventura Basin

B
A

KEY

Shallow seas

Deeper or more
permanent seas

Eroding lands
with diagrammatic topography

Land-laid deposits

Uncolored areas indicate
open ocean or geologically unknown areas

Outlines of present landmasses are
for reference only

General outlines and interior features
of the continent are essentially
the same as at the present time

Figure 14.11 *Reconstructed paleogeography of the late Tertiary
Period.*

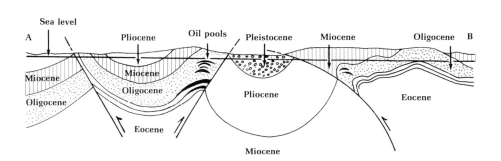

Figure 14.12 *Cross section of
the Ventura Basin, California.
Length of section is about 40 kilometers, and position is shown
by line A-B in Figure 14.11. This
very deep basin has over 13,000
meters of Tertiary sediments in
addition to those from the
Cretaceous. The complex structures are mainly the result of
deformation within the Pleistocene. The Ventura Basin has
produced over 500 million bar-
rels of oil from Tertiary rocks.
(Modified from Reed and Hol-
lister, "Structural Evolution of
Southern California," Bulletin of
The American Association of
Petroleum Geologists, Vol. 20,
1936. Published with permission.)*

ern plains of North America at this time must have resembled that seen today only in central Africa. The Oligocene climate was evidently somewhat cooler and drier than the preceding Eocene.

The succeeding Miocene Epoch witnessed a general uplift over a large part of North America. The entire Rocky Mountain area was raised bodily over 1,000 meters, and the climate became drier and cooler. Miocene deposits are spread out mainly over thousands of square kilometers between the Rocky Mountains and the Mississippi River. Here we find ancient stream channels, lake beds, floodplains, and soils composed of sediments derived from the rejuvenated mountain ranges to the west. Mammal remains are abundant, but the warmth-loving semitropical forms are no longer in evidence. The animal life suggests that extensive grassy plains were spreading over the area, replacing forested sections.

Pliocene deposits are found over much of the same area as are those of the Miocene. The chief source of sediment was still the Rocky Mountain ranges, and the place of deposition was the western plains. Animal and plant remains and the types of sediment deposited indicate that climates were cooler and drier than they were during the Miocene. Mammals adapted to moist, warm climates were no longer represented. Some had migrated elsewhere and some had become extinct. The stage was set for the great Pleistocene Ice Age, which we shall consider at some length in the following chapter.

SOUTH AMERICA AND ANTARCTICA

A fairly complete and varied rock record of the Tertiary Period is found in South America. All epochs are represented and some portions are extremely fossiliferous. Volcanic rocks and intrusions are abundantly represented, especially in the southern part of the Andes; in fact, much of the southern tip of the continent is a product of Tertiary activity.

Continental deposits are much more extensive

Figure 14.13 Andes Mountains as viewed from the east in eastern Bolivia. Mount Mururata on the skyline is 5,800 meters high; it is about 4,000 meters from the bottom of the canyon to the summit. (Courtesy of Eugene Callaghan.)

in South America than are marine deposits. Most of the land-laid sediments came directly from the Andes, which were already present in the Late Cretaceous and continued to rise throughout the Tertiary. The mountains shed large volumes of sediment eastward onto the margins of the shield areas and across the Argentine plains. In places there are abundant deposits of fossil mammals not unlike those of the plains of North America. The belt of sedimentation along the eastern border of the Andes widened and deepened with time and in the Pliocene extended from Venezuela to Patagonia, with extensions down the Amazon Valley.

Marine deposits were most extensive in the Miocene when an area near the mouth of the Amazon and another in central Argentina were

Figure 14.14 *Lake Kivu in the Congo Basin, which is filled mainly with Tertiary sediments. Thick vegetation and deep soil cover hinder geologic explorations in regions such as this. (UNESCO, photograph by Basal Zarov.)*

That portion of Africa bordering the Mediterranean Sea belongs to the Tethys geosynclinal belt. Its general history is not greatly different from that of southern Europe, which has already been discussed. All epochs of the Tertiary are represented by marine deposits, those of the earlier Tertiary being marked by abundant nummulites. Egypt and Libya were inundated by successive shallow seas and were little affected by the Alpine disturbances.

The marine Tertiary deposits of the remainder of Africa occur chiefly as a marginal band of variable width. An embayment several hundred kilometers long entered Nigeria in the early Tertiary, but elsewhere the seas did not penetrate the continent, which was generally well above sea level.

A number of extensive and important interior basins received land-laid sediment during various intervals of the Tertiary. The Congo Basin and its great river system were created in the late Pliocene and have extensive deposits of this age. The enormous Chad Basin in and adjacent to the southern Sahara was occupied by a succession of fluctuating lakes and received varied late Tertiary and Quaternary deposits, including desert sands and volcanic dust. The fossil record proves many humid intervals with abundant animal life.

Much of Africa south of the Congo, especially the Kalahari Desert, is covered by the unusual Kalahari beds, which are mainly loose or partly consolidated red sands. The age is uncertain, and the sands may have been deposited, blown about, and redeposited a number of times. Stone artifacts are frequently found in the shifting sands.

Most important of the continental formations of Africa are those found in and near the rift-valley systems. The rift valleys are elongate, steep-sided depressions, lowered by parallel faults and bordered by large volcanoes. The massive fracturing of the African platform was a by-product of its northward movement and its collision with

inundated. Recent explorations of the sea floor off the Argentine coast have revealed thick sedimentary deposits, apparently swept out to sea from the mainland in Tertiary time.

West Antarctica has a geologic history much like that of southern South America, especially that of the Andean chain. Volcanic eruptions broke forth intermittently, and there was intensive faulting and folding. Deposits of Tertiary age are chiefly found on the Antarctic Peninsula, which points fingerlike toward South America. Fossils are mainly those of plants and marine mollusks. Mild climates are indicated by the vegetation and by the fossil mollusks and brachiopods. No land vertebrates have yet been found. There is evidence of close connections with South America early in the Tertiary, but similarities diminish with time.

Eurasia. The Red Sea and Gulf of Aden were opened to seawater during the middle Tertiary. Within the continent large lakes, such as Lake Albert, Lake Nyasa, and Lake Tanganyika, occupy rifted depressions, and even larger water bodies are recorded for the past. In the Miocene, Pliocene, and Pleistocene, a remarkable variety of fossils were entombed, including the best-known specimens of early human and prehuman creatures. The famous *Olduvai Gorge,* which is discussed in Chapter 16, is near the rift valleys.

AUSTRALIA, THE EAST INDIES, AND NEW ZEALAND

The Tertiary geologic history of the numerous islands making up Indonesia is very complex; it contrasts with that of Australia, and both are different from nearby New Zealand. The three are included together merely for convenience.

Australia has marine, nonmarine, and volcanic rocks of Tertiary age. Outcrops are concentrated chiefly in the southeast quarter of the continent and cover relatively little area. The marine beds are at the margins of the continent and are chiefly of Oligocene and Miocene age. Many scattered lake and river deposits are known; some of these contain important coal beds and many species of fossil plants. The record of vertebrate life is very scanty and casts little light on the early history of the peculiar marsupial fauna.

The East Indies as they now exist are essentially the result of Tertiary geologic events; about three-fourths of the surface of the islands consists of sediments and volcanic deposits of Cenozoic age. The sedimentary rocks are mostly marine, and limy deposits rich in corals, algae, and foraminifera are common. On a few of the larger islands, continental beds with coal are found. There are tremendous volcanic accumulations consisting of flows, tuffs, and breccias. Some of these solidified below sea level.

The history of the region is one of unstable, rapidly changing landscapes. Local oblong basins that filled rapidly and were then uplifted without intensive deformation seem to have been typical. In terms of continental drift much of the unrest came from the northward push of Australia against the East Indies. The Tertiary deposits are very thick; measurements of from 5,000 to 15,000 meters are common.

The most abundant fossils are mollusks, corals, and foraminifera, many of the latter being nummulites in the broad sense and correlating with distant European formations. Vertebrate fossils are known from scattered localities; *Pithecanthropus,* the Java ape-man, is from the Pleistocene of central Java.

Tertiary rocks make up about half the surface of New Zealand. All parts of the period are present and volcanic rocks are abundant. Intense mountain building occurred in the Miocene and Pliocene, and the New Zealand Alps date from this time. Many similarities with the Tethyan belt can be noted. Some geologists regard New Zealand as a youthful continental nucleus.

THE WEST INDIES

The Caribbean Sea, with adjacent lands and islands, presents many geologic problems. The known rock record begins in the Mesozoic, and the vast bulk of exposed material is of Cretaceous or Tertiary age. Volcanic products are common and limestone is the prevalent sediment. There has been continual geologic unrest, and most of the structures are very complex. Some geologists believe the Caribbean area is a much shattered fragment of crustal material caught between the North American and South American blocks. The material seems to have been left behind as the two adjacent continents moved westward. Although the Isthmus of Panama is clearly a part of the great "backbone" of North and South America, it is of fairly recent origin and lacks older configurations.

As would be expected, the curving West Indian island arcs have much in common with the Pacific borderlands, specifically as to the age and type of rocks involved. Another feature of

Figure 14.15 *Grassland. This remnant of undisturbed prairie in Pottawatomie County, Kansas, illustrates the type of vegetation that has dominated the western plains through millions of years of the late Cenozoic Era. (U.S. National Park Service.)*

the area is the close relationships of the Cretaceous and Tertiary fossil forms with those of the Tethys Seaway of the Old World. Obviously many warm-water organisms, including large foraminifera, echinoids, corals, and mollusks, were able to travel across the Atlantic, which may have been narrower at this time. West Indies land life shows little similarity to that of the Old World, however.

Life of the Tertiary

In a general but unmistakable way the life of the Tertiary was modified in response to new and rigorous conditions. As shown by the preceding review of the physical environments, the conti-

nents became relatively high and rugged while the interior and marginal seas withdrew.

Extremes of heat and cold and great variations in precipitation have characterized world climates during the last 60 million years. It seems safe to say that there has been greater diversity in living conditions during this interval than during any other equivalent span of the earth's history. The rigorous surroundings of the Tertiary Period presented many challenging opportunities to living things. There were many harsh environments, but most of these were successfully colonized by life forms. Although the great continental glaciers and the rainless deserts are almost devoid of life, the extent to which animals and plants have invaded other unfavorable habitats, such as high mountains, caves, the depths of the ocean, and the ice-free polar regions, is truly remarkable. In general, the story of Tertiary life is one of response to climatic extremes and topographic diversity, with a premium on adaptations that could overcome cold and seasonal changes.

All major groups of plants are represented by fossils in Tertiary rocks. Even soft and unsubstantial forms, such as mosses, fungi, molds, and bacteria, have left actual remains or indirect evidence of their existence at this time. The descendants of earlier important groups, such as club mosses, horsetails, ferns, ginkgoes, and cycads, also continued to leave a few scattered fossils. Cone-bearing plants, including sequoia, pine, juniper, cypress, fir, and cedar, have left more numerous remains.

The angiosperms, or flowering plants, dominate the plant world and are preserved in great profusion. This large group had become well established in the Cretaceous, and all modern families appear to have evolved by Miocene time. Remains include trunks, branches, leaves, flowers, and pollen grains. Where conditions were favorable, Tertiary plants accumulated to form coal, lignite, and peat.

Grass is the most important angiosperm of the Tertiary. The first fossil grass seeds appear late in the Cretaceous; the family now includes about 500 genera embracing 5,000 species, and there are scarcely any areas where grass will not grow. Grass is generally an outstanding feature of most landscapes and is important geologically because it inhibits erosion and indirectly retards the weathering of rocks. In temperate climates, grass forms extensive turf, meadow, or prairie land. As long as grasslands remain unbroken, they are practically immune to erosion. When the grass is removed, erosion may take over. Grass retards the escape of moisture from the ground, favors the accumulation of humus, and permits soil formation to proceed to great depths. In order to

Figure 14.16   Fossil ginkgo leaf on a slab of Miocene shale from southwestern Montana compared with the leaves from a living tree. As far as can be determined, there are no ginkgo trees growing in a wild state at the present time, but fossil remains are common in Mesozoic and Cenozoic rocks.

appreciate the influence this one family of plants has exerted over vast areas of the earth, we need merely observe that sterile badlands may exist side by side with lush grasslands.

Among the important items of food furnished by the grass family are rice, wheat, barley, oats, rye, corn, and millet. Other important members of the group include bamboo, useful as a food and as a building material; sugar cane, a giant grass that supplies half the world's sugar; and sorghum, which provides grain, forage, straw, and molasses. Although all these plants have been modified by man, their ancestors provided food for plains-living mammals throughout the Tertiary epochs. The grazing habits of many of our large mammals evolved in response to the availability of grass. The grains sustain many rodents, birds, and insects. It is difficult to appre-

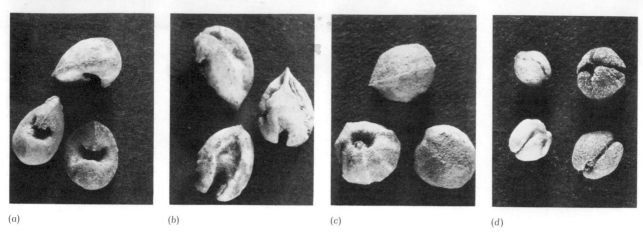

Figure 14.17 Fossil nuts and seeds from the Clarno Formation of Oligocene age, Washington. (a) Bursericarpum (elephant tree), (b) unidentified Vitaceae (grape), (c) Meliosma (no modern counterpart), and (d) Juglandaceae (walnut). All except the last named are about 0.5 centimeter in diameter. (Specimens collected and photographed by Thomas J. Bones.)

ciate how important this one plant group was to Tertiary life.

During the Tertiary, plant groups migrated extensively in response to climatic changes. In general, these migrations were marked by a withdrawal of tropical and semitropical vegetation toward the equator as the climate of the continents grew cooler and drier. The uplifting of mountain ranges such as the Alpine-Himalayan chain and the American Cordillera profoundly affected these migrations. As seasonal changes grew more pronounced, only those forms of vegetation that could shed their leaves and live through cold seasons in a resting stage were able to survive. In the seas and oceans, plant life also continued to evolve. The diatoms expanded in importance, and have been the basic food source in the marine world ever since the middle Mesozoic. The first known freshwater diatoms are of Cretaceous age.

Figure 14.18 A highly enlarged view of diatomaceous ooze collected from the ocean bottom northwest of Honshu, Japan. (Scripps Institute of Oceanography, photograph by Taro Kanaya.)

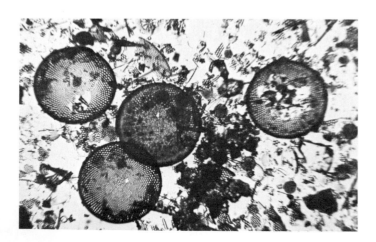

MARINE INVERTEBRATES

The chief marine invertebrates of the Tertiary, arranged roughly in order of decreasing practical importance to paleontologists, are foraminiferal protozoans, pelecypods, gastropods, corals, bryozoans, crustaceans, and sea urchins. Of course, any such arrangement is somewhat arbitrary and is not evident in all places for all subdivisions. Missing from the seas at the dawn of the Tertiary were ammonites, rudistids, *Inoceramus* pelecy-pods, and several types of oysters.

A walk along almost any seashore will indicate how successful the pelecypods and gastropods have been in winning their battle for survival. The shells cast up on the beach or buried in the sediments reveal the amazing variety of shape and size these creatures have attained. Although most of the modern families of pelecypods and gastropods appeared in the Mesozoic, their great expansion occurred during the Tertiary. They are locally so varied and abundant that they serve as a basis for correlating the marine formations.

All the shelled cephalopods except the ancestors of the pearly nautilus had disappeared by the Paleocene, but the shell-less types continued to be well represented by squids, cuttlefish, and the octopus. These forms leave few fossils, and their race histories are poorly known. Squids and their relatives are especially plentiful in the open ocean where they compete directly with the fish.

The importance of the protozoans during the Tertiary is indicated by the number of species that appeared, by the extent to which they contributed their shells and skeletons to the formation of rocks and deep-sea oozes, and by their practical value as guide fossils in correlating oil-bearing rocks. These unicellular, calcareous-shelled organisms have left innumerable remains in all Tertiary epochs. Many forms have been discovered, but the most important guide fossils are the large varieties commonly called nummulites. This group derives its name from the Greek word for "coin," which refers to their flat, disclike shape. These "coin" fossils are common in Egypt and other Mediterranean lands. Some nummulites are more than 2 centimeters across, and they occur locally in such profusion that they make up large masses of rock. Stone containing them went into the construction of the Egyptian pyramids, and they were noticed and recorded by several ancient writers, some of whom thought they were the remains of lentils

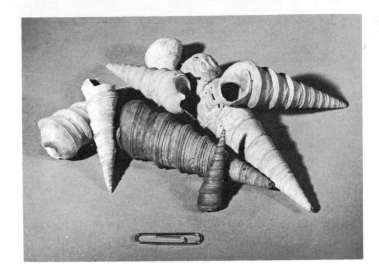

Figure 14.19 *Fossil shells of* Turitella, *a very abundant gastropod of the Tertiary Period. (Smithsonian Institution.)*

Figure 14.20 *Two species of the fossil bryozoan* Metrarabdotos *from the middle Miocene of France (a) and the upper Eocene of Alabama (b). Width of view in each specimen is about 0.5 centimeter. Each opening is the entrance to a boxlike living chamber of a single animal. (Smithsonian Institution.)*

(a)   (b)

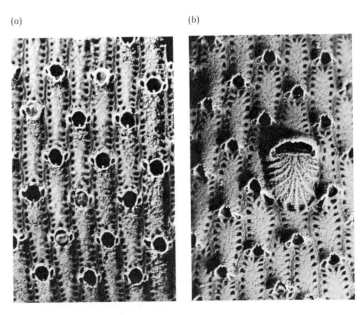

Figure 14.21 Nummulites from the Eocene Giza Limestone of Egypt.

that had been supplied as food to the workmen building the pyramids.

The nummulites lived mainly in warm waters and are highly characteristic of the Eocene and Oligocene of the Tethys Seaway. They swarmed over the coral and algae reefs and contributed their shells to the building of thick limestone formations. They are also found in deposits laid down in the warmer waters of the Western Hemisphere, chiefly adjacent to the Gulf of Mexico and the Caribbean Sea.

Many other protozoans have been found useful as guides for both local and worldwide correlations. They have been especially useful in correlating oil-producing areas of the Gulf Coast, California, Venezuela, the East Indies, and the Near East. As many as thirty zones for each of the Tertiary epochs have been established locally on the basis of protozoans.

No new major groups of invertebrates appeared, and it may even be that Tertiary marine life was somewhat less prolific than it was during the Cretaceous. The relative uplift of the landmasses during the Tertiary stimulated erosion and flooded the offshore waters with heavy loads of mud and silt. Such organisms as sponges, corals, brachiopods, bryozoans, crinoids, and other fixed forms, cannot endure muddy water, and even pelecypods, gastropods, starfish, and sea urchins do not thrive in it. Unfavorable living conditions on continental shelves and the elimination of interior seas may have forced many species into deeper waters, and certainly cut into their living space.

VERTEBRATES

Minor Groups. The Cenozoic is commonly referred to as the *Age of Mammals,* in recognition of the group that contributed the dominant land forms during the period. But in giving due credit to the mammals we should not overlook the fact that birds also progressed during this same period and became adapted to a great variety of environments. Likewise, in the seas the modernized bony fish achieved a supremacy equal to that of the mammals on land. We would therefore be more correct in characterizing the Tertiary as the *Age of Mammals, Birds, and Teleost Fish.* The reptiles had been reduced to relative insignificance at the end of the Mesozoic, and joined the amphibians among the ranks of the dispossessed. The story of Tertiary vertebrates is essentially a tale of the warm-blooded vertebrates. (This statement takes on more significance when we realize that the period in which they thrived was one of relatively cold and changeable climates.)

Mammals. Conditions at the beginning of the Tertiary were ideal for mammals, and they were quick to seize their opportunities. Their ancient reptilian enemies were gone, food was abundant, and living space was almost unlimited. Mammals are characterized by such obvious physical traits as warm blood, hairy covering, efficient reproductive systems, milk glands, strong teeth, and sturdy skeletons. But more important than any of these in the struggle for survival was the gradual enlargement of the mammalian brain, which became an organ capable of storing and retaining impressions that could be used in directing subsequent intelligent action.

Although it is obviously impossible to find out a great deal about the mental processes and behavior of extinct mammals, we do have one important source of information in the form of fossil brain casts. The brain cavity inside a mammalian skull is easily filled with sediment, which may harden to preserve a perfect cast of the brain and its important nerve connections. This cast reveals the size of the brain in relation to the body, the size and position of its various parts, and the all-important foldings in its outer surface.

The study of fossil skulls shows that the mammalian brains gradually increased in size during the Tertiary. If the mass or complexity of the brain is a reliable guide, the stupidest mammal was a mental giant compared with his reptilian predecessors. Even the earliest mammals were relatively "brainy" in the sense that their brains were large in proportion to their bodies. As individual species grew ever larger, their brains also expanded, but at a slower rate. The higher primates, including man, constitute the only known group in which the rate of brain growth kept pace with or exceeded the rate of body growth.

Fossil casts reveal another very important fact about the brain—namely, the phenomenal increase in size and importance of the cerebrum. The *cerebrum,* or forebrain, is divided into two halves called hemispheres, which fill most of the cranial cavity of the mammalian skull. The growth or expansion of the cerebrum resulted mainly from the increase in its outer layer, or *cortex,* commonly known as the *gray matter.*

Figure 14.22   Fossil fish from the Green River Formation near Kemmerer, Wyoming. Literally millions of freshwater fish were buried and preserved in the fine-grained deposits of an extensive lake that covered parts of Wyoming, Utah, and Colorado during the Eocene Epoch. The larger of the two specimens is about 40 centimeters long. (Utah Museum of Natural History.)

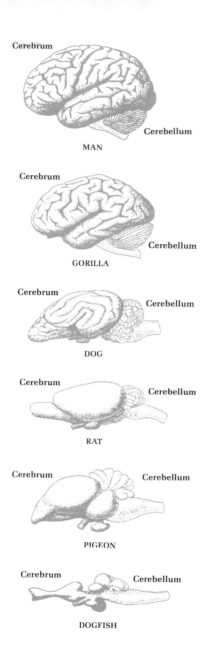

Cerebrum

Cerebellum

MAN

Cerebrum

Cerebellum

GORILLA

Cerebrum

Cerebellum

DOG

Cerebrum

Cerebellum

RAT

Cerebrum

Cerebellum

PIGEON

Cerebrum

Cerebellum

DOGFISH

Figure 14.23   The brain, from fish to man. The diagram shows the increase in size of the cerebrum and the complex foldings that characterize the more advanced forms of life. The illustration is not intended to show true relative sizes. (From Norman L. Munn, "The Evolution of Mind," Scientific American, June 1957, p. 150. Copyright 1957 by Scientific American, Inc. All rights reserved.)

This structure overlaps and dominates the other parts of the brain. We can readily see how the cerebrum has grown in importance by comparing the simple cerebrum of a fish with the progressively more complex cerebrums of an amphibian, a reptile, a bird, a mammal, and man. A comparison of fossil brain cases reveals the same development. The brain of the primitive "dawn horse," *Hyracotherium* (*Eohippus*), the earliest known member of the horse family, from Paleocene rocks of Wyoming, is surprisingly similar to the brain of the modern opossum. Both are primitive and show about the same degree of advancement over reptiles. Fossil evidence also indicates that the portions of the brain having to do with smell were gradually submerged by structures having other functions. The senses of sight and hearing seem to have grown more acute with time.

We shall put off for the time being a discussion of the role of the brain in the evolution of man, but we should mention that intelligent activity is the key to the success of mammals generally. Intelligent activity implies a degree of freedom of action—a mammal's behavior is not stereotyped and is not driven by purely reflex action. Sometimes the animal knows what it is going to do—apparently after having chosen between or among alternatives—before it does it.

Aided by superior brains, mammals are able to compete successfully with animals that are much stronger than they, to escape unfavorable environments, and to survive during periods of danger and stress. Circumstances forced the mammals to "live by their wits" during the Mesozoic. During the Tertiary, as they came to compete with one another and with an unfriendly environment, they became even more intelligent and adaptable.

Mammals of the Paleocene were primitive, unspecialized, and relatively small. Of the fifteen orders of mammals known from Paleocene rocks, only six have been found in the preceding Cretaceous, and only six are alive today. The orders in existence before the Paleocene include the *multituberculates, marsupials, condylarths, pri-*

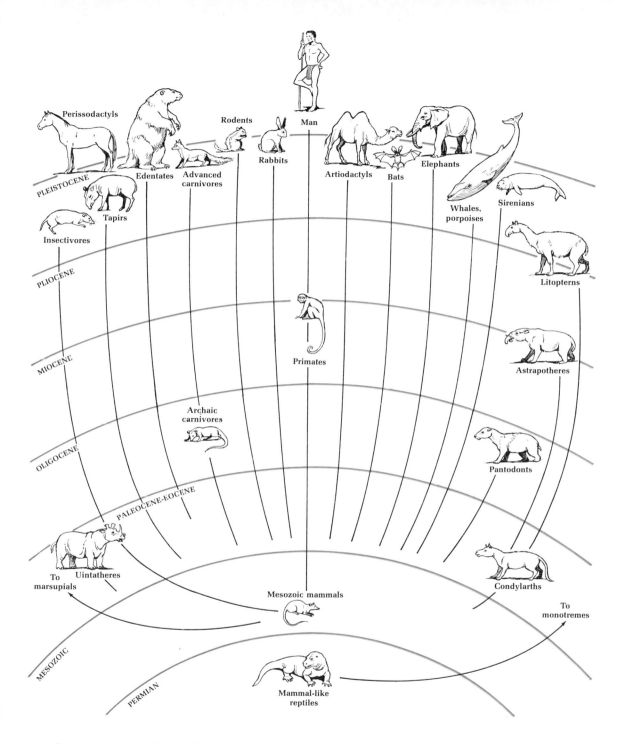

Figure 14.24 Simplified phylogenetic chart of the mammals. This illustration also emphasizes the varied adaptations of the group to different modes of life.

The following labels appear on the chart:

Perissodactyls
Rodents
Man
Edentates
Advanced carnivores
Rabbits
Artiodactyls
Bats
Elephants
PLEISTOCENE
Tapirs
Whales, porpoises
Sirenians
Insectivores
PLIOCENE
Litopterns
MIOCENE
Primates
Astrapotheres
Archaic carnivores
OLIGOCENE
Pantodonts
PALEOCENE-EOCENE
To marsupials
Uintatheres
Mesozoic mammals
Condylarths
To monotremes
MESOZOIC
PERMIAN
Mammal-like reptiles

mates, and *insectivores*. Among the primitive Paleocene forms, we perceive possible ancestors of many succeeding orders, but the differences by which we distinguish living carnivorous types and the various orders of hoofed animals were not yet clearly evident. The Paleocene life of Europe and North America was similar, indicating that a land bridge still connected the two continents. But South America was isolated late in the epoch and gave rise to a number of odd orders that were not able to migrate to other regions.

Nature continued to favor the mammals during the Eocene Epoch. The descendants of the archaic Mesozoic and Paleocene forms lived side by side with the newly evolved and more progressive ancestors of modern families. Ten new orders were added to the fifteen that had carried over from the Paleocene. The multituberculates disappeared at the end of the epoch, after having lived through at least two geologic periods. They may have been pushed aside by the more efficient rodents. A number of clumsy primitive herbivores, such as *Coryphodon* and *Phenacodus*, had teeth that were adapted for eating vegetation but also possessed some carnivorous characteristics such as claws, long tails, and short limbs. Among the important groups that took on recognizable characteristics during the Eocene were the carnivores and the hoofed animals. The early primitive carnivores known as *creodonts* were dull-witted and poorly constructed by comparison with modern forms. Nevertheless, there were doglike, catlike, and hyenalike forms adapted to prey on the contemporary herbivores. For the first time, we find true hoofed animals. There were odd-toed

Figure 14.25   Scene in western Colorado during the Paleocene Epoch. The animal is Barylambda, a large hoofed mammal that has no living descendants. This picture of a well-watered, semitropical landscape is based on study of the sediments and fossil vegetation associated with the vertebrate remains. (Field Museum of Natural History, Chicago.)

Figure 14.26 Middle Eocene life of the Wyoming area. Uintatherium, *a large, six-horned, saber-toothed herbivore, dominates the scene; lower right, the primitive tapir* Helalctes; *lower center right,* Stylinodon, *a gnawing, toothed mammal; center,* Trogosus, *another gnawer; lower left,* Hyrachyus, *a fleet-footed rhinoceros. The tropical vegetation is reconstructed from fossil evidence. (Mural by Jay H. Matternes, U.S. National Museum.)*

forms (*perissodactyls*), including horses, tapirs, and rhinoceroses. The even-toed hoofed animals (*artiodactyls*) spread tremendously during the late Eocene and the ancestors of the deer, pig, and camel appeared for the first time. These last three were small and very much alike. Rabbits were present in a recognizable form, and other small mammals, such as rodents, insectivores, and primates, were gaining in importance. There were bats in the air and whales in the ocean, but the history of these two groups is obscure.

South America was now completely separated from the other continents and supported a number of odd herbivorous groups. The Eocene climate was temperate and equable, and the lands were generally lower than at present. Migration between North America and Europe appears to

have been possible during the early Eocene but became restricted or impossible late in the epoch.

Although no drastic changes mark the close of the Eocene, there was a transitional period during which the archaic mammals were rapidly eliminated and modernized forms began to spread. Thus, the Oligocene faunas are notably different from Eocene types. A few apparently indestructible forms, such as the opossum, moles, and shrews, continued to survive and were joined by other small animals, such as beavers, rats, and mice. The inefficient creodonts dwindled to insignificance, but new families appeared among the artiodactyls and perissodactyls. In addition, horses, deer, tapirs, rhinoceroses, camels, and antelopes, as well as true cats and dogs, were clearly recognizable.

Figure 14.27 *Mammals of the early Oligocene of the South Dakota-Nebraska area. Center and upper right, Brontotherium, a titanothere; lower right, Merycoidodon, a sheeplike grazing animal; lower center, Protapiras, an ancestral tapir; center left, Hyracodon, a three-toed rhinoceros. (Mural by Jay H. Matternes, U.S. National Museum.)*

Among other forms that disappeared during the Oligocene were the gigantic *titanotheres*, typified by *Brontops*, which was as large as an elephant but had weak teeth and a small brain. Another common Oligocene mammal was the *oreodont*, a grazing animal about the size of a modern sheep.

The climates of the northern landmasses were becoming cooler and drier, and the semitropical

Figure 14.28 *Mammals of the early Miocene of Nebraska. Foreground, Parahippus, a three-toed horse; center right, Dinohyus, a giant piglike mammal; near center left, Stenomylus, a small camel; center left, Promerychoerus, an oreodont; under the tree, Diceratherium, a rhinoceros; upper right, Oxydactylus, a long-legged camel; Syndyoceras, a horned artiodactyl, is near horses. (Mural by Jay H. Matternes, U.S. National Museum.)*

forests were retreating southward during the Oligocene. The warmth-loving primates that had been plentiful in North America during the Paleocene and Eocene now disappeared from the continent. There was a marked increase in the number of animals adapted to subsist on grass.

During the Miocene there was a general uplift of the northern continents and major mountain-building along the Tethyan belt. With the uplift came cooler and drier climates and corresponding changes in the plant and animal world. Grasslands expanded, forests retreated, and the environment became even more favorable for animals that could exist by grazing on open plains. It is not surprising that the hoofed animals, able to travel widely over rough ground, should multiply considerably. There were many species of three-toed horses, rhinoceroses, giant pigs, camels, ancestral deer, primitive antelopes, mastodons, and the last survivors of the oreodonts and chalcotheres (extinct relatives of the horse). Also making their appearance were large

pantherlike and tigerlike cats, the first bears and raccoons, and a host of lesser carnivores, such as weasels, wolverines, skunks, and otters.

Uplift had brought Eurasia into contact with North America at the Bering Strait, and Africa was connected with Europe and Asia at several points. However, not all types of mammals found and crossed the available land bridges. South American life continued to evolve isolated from the rest of the world, and Miocene deposits of the southern continent contain abundant fossils of the ground sloths, armadillos, and other bizarre native herbivores. In many ways the Miocene was a high point in mammalian evolution. Conditions were ideal for land life, and wholesale exterminations had not yet begun.

The Pliocene Epoch has been called the autumn of the Cenozoic, for its climate heralded the coming of the ice and cold of the Pleistocene. Pliocene mammals were clearly adapted to cool conditions and seasonal changes. Warmth-loving mammals, such as the primates, which formerly

Figure 14.29 *Mammals of the early Pliocene of the southern High Plains. Center and center left,* Amebelodon, *the shovel-toothed mastodon; lower left,* Teleoceros, *a short-legged rhinoceros; lower right,* Hipparion, *a characteristic three-toed horse. (Mural by Jay H. Matternes, U.S. National Museum.)*

Figure 14.30   One of the richest hills on earth in a famous silver-mining district, Cerro de Potosi, in east-central Bolivia. The deposits are in an intrusion of Tertiary age and have been worked for almost 400 years. (Courtesy of Eugene Callaghan.)

Figure 14.31   The Bingham Canyon Mine near Salt Lake City, Utah. The open-cut workings are descending into a large deposit of low-grade copper ore that originated in association with an igneous intrusion of middle Tertiary age. The surrounding rocks, which contain some of the ore, are of Pennsylvanian and Permian age. (Kennecott Copper Corporation.)

had been widespread, now lived only in restricted tropical areas. The open plains of the northern continents were inhabited by numerous highly specialized forms, many of which grew quite large. A variety of bears, dogs, cats, wolves, antelopes, camels, horses, and mastodons roamed North America and Eurasia. There appears to have been intermittent migration between the northern landmasses and Africa, but South America still remained cut off from the rest of the world until near the end of the epoch. When the land bridge between the two Americas was elevated, a lively exchange of land mammals commenced. The competition was disastrous for the less adaptable South American forms, many of which were exterminated.

The Pliocene Epoch on the whole produced few new types of mammals. The time of rapid expansion and diversification had passed, and an increasingly harsh and inhospitable environment was beginning to eliminate the inefficient.

Economic Products of the Tertiary

Although it was a relatively short period, many important mineral deposits were formed during the Tertiary, including coal, oil, and gas in widely scattered areas of the world and rich deposits of metallic minerals in the Western Hemisphere.

According to a recent statistical survey, 50 percent of the world's oil fields tap Tertiary rocks and are responsible for 38.2 percent of the world's total oil reserves. The Oligocene and Miocene are especially prolific in the rich Middle East fields. Other Tertiary oil and gas fields have been discovered along the Gulf Coast and in California, Venezuela, Colombia, Russia, the East Indies, and the Bass Strait, between Australia and Tasmania. We should also mention here the great Eocene oil-shale deposits in the western United States. Large reserves of coal and lignite occur in Paleocene rocks of Montana and the Dakotas. There are important deposits of Oligocene brown coal in Germany and France.

Although geologists cannot always positively determine when deep-seated ore deposits were formed, most of the metal-bearing ores of western North America and South America are considered to be products of Tertiary activity. Deposits of mercury, gold, silver, lead, zinc, and copper are widespread in and adjacent to the great Rocky Mountain-Andes Cordillera. These deposits are associated mainly with intrusive igneous rocks, and many, out of the hundreds that have been discovered, are bonanzas of the richest type. Similar Tertiary deposits occur around the western margin of the Pacific Ocean and in Japan, the Philippines, the East Indies, and Australia. Almost everywhere these Tertiary deposits have been more or less eroded, and native metals such as gold, platinum, and tin have been released to form placer deposits.

A variety of metallic deposits—gold, silver, lead, zinc, copper, mercury, and other rarer metals—came into being in southern Europe during the Alpine Orogeny. Similar deposits were also formed in the mountain ranges of southern Asia.

Nonmetallic minerals such as clay, diatomaceous earth, gypsum, salt, phosphate rock, and building stone are mined in great quantities from Tertiary rocks at many scattered localities.

A SUMMARY STATEMENT

During the Tertiary Period the continents were elevated, topography was rugged, and rigorous climatic conditions prevailed. The Alpine-Himalayan chain was elevated, and disturbances of various kinds affected the margins of the Pacific Ocean. The East Indies and West Indies also underwent many modifications. Sediments produced during the period chiefly reflect the accelerated erosion of the continents—coarse conglomerates, sandstone, and shale were produced in abundance and fine muds and limestone were rare.

The fringing continental shelves were considerably enlarged, and the rate of marine deposition also appears to have speeded up. All types of reefs expanded, especially the ones attached to Pacific volcanic isles.

Few new major groups of plants and animals appeared during the Tertiary, but many species evolved in response to changed conditions. Mammals were the dominant form of land life. Birds and teleost fish continued to evolve in their respective environments and became highly specialized. Plants migrated widely and there was a general withdrawal of warmth-loving vegetation toward the equator. Grass became the dominant vegetative form in temperate regions. Marine life was somewhat restricted, but all forms that could adapt to life in the open ocean increased.

Many important raw materials came into existence; oil, gas, and coal were formed in connection with Tertiary sedimentation, and many deposits of metallic minerals were formed in association with igneous activity.

FOR ADDITIONAL READING

Osborn, Henry F., *The Age of Mammals in Europe, Asia, and North America*. New York: Macmillan, 1910.

Scott, W. B., *A History of Land Mammals in the Western Hemisphere*. New York: Macmillan, 1937.

Additional references for the material in this chapter can be found at the end of Chapter 11.

THE PLEISTOCENE ICE AGE

Weather and climate are topics of universal interest and importance, and so it is not surprising that the subject of ancient climates should loom large in our reconstructions of the past. We can scarcely think about coal forests, dinosaurs, hairy mammoths, or other forms of prehistoric life without relating them to specific types of climate. The science of *paleoclimatology* deals specifically with what can be discovered about ancient weather and climate.

Of all the climatic events of the past that have been deduced from geologic evidence, the Pleis-

Mount Russell and icefall tributary to Yentna Glacier, from the east. (U.S. National Park Service.)

tocene Ice Age is the best known. Its widespread effects are today plainly in evidence, and snow-fields and glaciers still cover much of the earth. The concept of a great episode of cold and ice has dramatic appeal and has been widely accepted as a well proved event of the prehistoric past.

People who live in the vicinity of glaciers or who enjoy mountain climbing, skiing, and other snow sports, are well aware of the movement and erosive effects of glaciers. Avalanches of snow and rock frequently rain down on the ice, and masses of dirt and rocks of all sizes are picked up and carried along by it. Bodies of persons lost in crevasses have been found at the lower snouts of glaciers, many kilometers from the scene of death. From these and other observations, investigators had already established the rates of movement and studied other habits of glaciers, especially in Switzerland, before the larger ice caps began to be explored.

Glaciated mountains have a special type of topography. Sharp-pointed eminences, such as the famous Matterhorn in the Pennine Alps,

Figure 15.1　A glacier-filled valley in the Swiss Alps. (Swiss National Tourist Office.)

knife-edged ridges, rounded valley heads, or *cirques,* numerous rocky lakes, and large piles of broken rock characterize those areas where ice, rather than running water, has been the chief agent of erosion. Glacial deposition is likewise distinctive. Elongate ridges, undrained depressions, and irregular piles of debris brought down from higher elevations mark the plains and valleys where the glaciers melt away. Long after ice has left the scene, the landscape clearly proclaims its former presence. Nevertheless, it might have taken men some time to appreciate the great erosive power of ice, had they not actually observed glaciers creating their characteristic landforms.

Historical Background

Long before the beginning of modern times, residents had noticed the effects of glaciation in the Alps, far beyond the limits of any living glaciers.

Not until 1821, however, did J. Venetz, a Swiss engineer, present a well-organized argument stating that Alpine glaciers had once been much more extensive than they were then. In 1829, Venetz extended his ideas to include the whole of northern Europe. Until Venetz announced his theory, most investigators had thought that the boulders and chaotic piles of debris scattered over many northern European countries had been deposited by water and floating icebergs, and in the minds of many they constituted evidence for Noah's Flood. In allusion to this supposed origin, the material was called *drift,* a name still retained in some areas.

Until 1837, Venetz's theory encountered more disapproval than approval. In that year Louis Agassiz, a young zoologist, championed Venetz's arguments and suggested that a great Ice Age had once gripped most of northern Europe, as well as the Alps. Agassiz, who was to become the foremost authority on the subject, presented such overwhelming evidence to support his claims, that opposition to the idea gradually ceased. Even Agassiz was amazed at the proof he encountered. The awful magnitude of the Ice Age imparted a solemn and melancholy cast to the thinking of geologists, who realized that even though nothing like Noah's Flood was recorded in the drift, the evidences of catastrophe were no less real. Some prophets of doom even predicted that the world would ultimately end in an endless winter of icy cold.

Once scientists were convinced of the truth of the glaciation theory, they found proof of it on all sides. They discovered that nearly half of

Figure 15.2 *Looking down a Norwegian fjord. This valley was occupied by a glacier that carved the typical U-shaped profile. When the ice disappeared, ocean water flooded the lower part. The stream in the foreground is choked with sand and gravel and is building up its bed with these materials.*

Europe had been submerged by a great sheet of ice that radiated out from the Scandinavian highlands to cover 4.3 million square kilometers. This sheet pushed across the North Sea and joined with local ice masses to cover all of Great Britain except a strip along the southern part of the island. The Alps also supported glaciers, which flowed out over adjacent lowlands but did not join the ice sheet from the north. A considerable portion of northern Asia was under another huge sheet that fanned outward from centers in northwestern Siberia and covered an area approximately equal to that covered by the sheet that originated in Scandinavia. The Siberian sheet eventually coalesced with the Scandinavian sheet in the vicinity of Moscow.

The higher ranges of Asia were covered by ice caps even larger than those of the Alps. Ice streams from the Himalayan ranges descended to within 900 meters of sea level. Even islands in the arctic seas were completely glaciated one or more times.

Agassiz traveled to America in 1846 and found evidences for the former existence of even greater ice sheets. Over 11 million square kilometers of North America showed signs of having once been covered by glacial ice. The area included practically all of Canada and much of the northeastern United States as far south as the Ohio and Missouri rivers. Iceland like Greenland was blanketed by glaciers, and the sea between the islands had been a mass of floating pack ice.

In the Southern Hemisphere the continent of Antarctica is still in the grip of the Ice Age. The great Antarctic ice sheet may have been in existence long before the Pleistocene. At least there is reliable evidence to indicate that it once was thicker, if not more extensive, than it now is. In New Zealand the ice descended below present sea level, and Tasmania was overrun by sizable glaciers. To the north and northeast the higher mountains of Hawaii and New Guinea bore large streams of ice. In South America the glaciers of the Patagonian Andes spread westward to sea level and eastward onto the Argentine pampas.

Figure 15.3 *The Walsh-Logan Glacier in Alaska is a dirt-covered and rapidly melting remnant of the Ice Age. (U.S. Geological Survey.)*

Figure 15.4 *The world of the Pleistocene Ice Age. Major continental and mountain glaciers shown in white and areas of oceanic pack ice in broken patterns. (Adapted from W. L. Stokes and S. Judson, Introduction to Geology: Physical and Historical, Prentice-Hall, Inc., 1968.)*

On Mount Kenya, in eastern Africa, where glaciers still exist, the Pleistocene ice descended 1,600 meters below its present limits.

Estimates indicate that an area of 39 million square kilometers, or 27 percent of the total land surface of the earth, was covered by ice during the last glacial stage.

Multiple Glaciations and Duration of the Pleistocene

As investigators continued to collect evidence of former ice action, it became apparent that the Ice Age was not as simple as they had first supposed. Successive sheets of glacial drift were found piled one upon another, and careful observations revealed that the *glacial stages* had been separated by ice-free, or *interglacial,* stages even longer than the glacial stages themselves. Ice-free periods are indicated by deeply weathered rock, old soils, wind-blown material, and especially fossils of warmth-loving animals. In both Europe and America the Pleistocene has been subdivided into four glacial and three interglacial intervals. There were minor retreats and readvances of the ice within the main stages and the record is complex and varied.

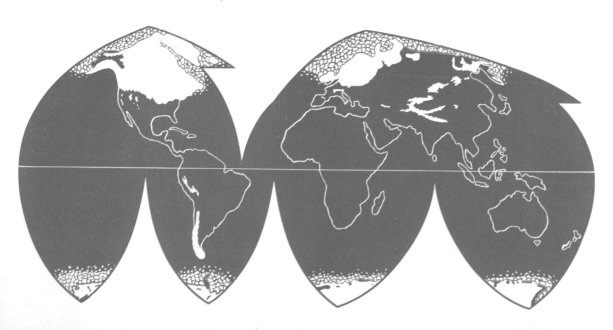

The question of how long the Ice Age lasted is difficult to answer because climatic changes began and ended somewhat gradually and did not affect the whole earth simultaneously in the same way. An inhabitant of the Amazon Valley might say that there had never been an Ice Age; one living in Antarctica might say there had always been one. Three methods for setting limits to the Pleistocene or Ice Age are the following: (1) absolute dating by whatever means are available; (2) the presence or absence of certain key fossils; and (3) climatic changes. Ideally these methods should be used together and should not contradict one another. Obviously, all of these are somewhat arbitrary.

Because the most obvious effects of the Ice Age are in what are now subarctic or temperate lands of the Northern Hemisphere, it is natural that much attention has been given to fossils found in these regions. Those of most use are remains of large vertebrates, and it is generally concluded that deposits of the Pleistocene are identified by the first elephants (in the broad sense), the first modern horses, the first advanced bovids, modern types of camels, and men of the modern species. But fossils of these creatures are rare over vast regions, especially in the tropics, and other key forms are being investigated to supplement them.

Correlating from land to ocean has been difficult. On the reasonable assumption that the Ice Age began with a cooling climate, it is possible to detect the coming of glacial conditions by means of temperature-sensitive organisms, chiefly floating foraminifera. A number of abundant and widely distributed marine fossils have been found to be useful. The foraminifer *Globorotalia truncatulinoides* appears near or at the beginning of the Pleistocene, while its relative *Globoquadrina altispira* became extinct at the close of the Pliocene. The two forms overlap very little if at all. Other species are also being tested as index markers.

It is possible to correlate the onset of general cooling with a number of marine and nonmarine deposits that can be dated by radiometric means. These dates, obtained chiefly by the potassium-argon method, fall within an interval of from 2.5 to 3 million years ago.

The problem of setting an upper age limit to the Pleistocene is also difficult. Are we still in the Ice Age? Looking at the climatic record many observers are convinced that a sudden warming

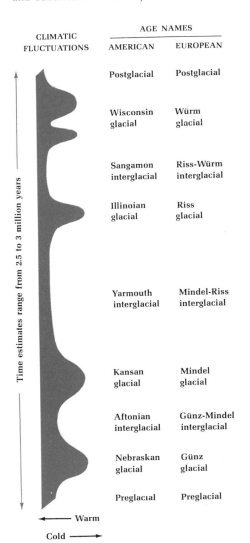

Figure 15.5 Terminology of the subdivisions and correlation of the Pleistocene as commonly accepted in North America and Europe. Because of the intensified study of glacial geology, the terminology and subdivisions are subject to constant revision.

CLIMATIC FLUCTUATIONS	AGE NAMES	
	AMERICAN	EUROPEAN
	Postglacial	Postglacial
	Wisconsin glacial	Würm glacial
	Sangamon interglacial	Riss-Würm interglacial
	Illinoian glacial	Riss glacial
	Yarmouth interglacial	Mindel-Riss interglacial
	Kansan glacial	Mindel glacial
	Aftonian interglacial	Günz-Mindel interglacial
	Nebraskan glacial	Günz glacial
	Preglacial	Preglacial

Time estimates range from 2.5 to 3 million years

← Warm

Cold →

Figure 15.6 *This smoothed and striated rock surface in Alaska indicates that it has been overridden by a glacier.* (U.S. Geological Survey.)

took place from 10,000 to 11,000 years ago. Evidences of this have been detected over wide areas both on land and in the ocean basins. This event is now generally, but not universally, taken as the end of the Ice Age. For the last small segment of time from 10,000 years ago to the present the term *Holocene* is coming into use. This interval is also referred to informally as postglacial or post-Wisconsin. The older term *Recent* is going out of use because it carries a different connotation and its etymology is out of harmony with the other names of the Cenozoic epochs. It is important to realize that carbon 14 dating is possible for the last 50,000 years only. For this reason many dates are available for deciphering the Holocene and late Pleistocene.

Effects of the Ice Age

EROSION BY CONTINENTAL GLACIERS

Areas overrun by continental glaciers are subjected to powerful erosive effects very different from those of running water. When glaciation

begins, the streams of ice are directed and channeled by the preexisting water-worn topography; but as the ice sheets coalesce and thicken, the underlying land surface exerts less and less effect, until the movement of the ice is guided by the slope of its *upper surface,* and not by the original slope or irregularities of the land. Thus, the area that receives the greatest net accumulation of snow eventually becomes the central point of radiation for the entire sheet. A fully formed *continental glacier* may be over 3,000 meters thick, and the ice with its contained rock fragments pushes down on the underlying surface with tremendous force.

Although the movement of the glacier is mainly *within* the ice and there is not a great deal of actual scraping of ice on rock, it is nevertheless true that the prolonged advance of a heavy ice sheet will level and smooth the underlying surface. Local areas of hard rock may be left in relief and softer areas may be more deeply excavated, but the whole topography will usually be considerably smoothed and subdued. Examined in detail, a glaciated surface shows innumerable elongate depressions, grooves, striations, and scratches, which indicate the direction in which

Figure 15.7 Striated glacial cobbles such as these are natural tools that produce the type of surface shown in the preceding illustration. (U.S. Geological Survey.)

Figure 15.8 A diagram showing the effect of a heavy ice mass on the earth's surface. (a) The relation of the land surface to an arbitrary line before the ice has accumulated; (b) the surface depression under the ice and the bulging of nearby areas; (c) the land in the process of recovering its former shape after the ice has melted.

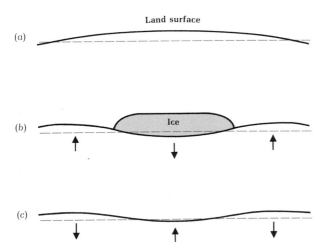

the ice moved. Each successive sheet usually destroys most of the traces left by former sheets, and the best indications of repeated glaciation are preserved in marginal areas where successive sedimentary deposits have accumulated and where the different ice sheets have not quite overlapped.

Favorable locations for study are the east-central part of the United States and northern Europe, where the evidence, though very complex, is far better preserved than in the central glaciated areas. At present, the tundra plains and bare-rock surfaces of northern latitudes are beginning to recover from the effects of the last ice advance. Thousands of lakes, bogs, and ponds that dot these regions are beginning to be drained by streams that are slowly cutting channels among them. Thin soil is forming, and weathering is destroying the smoothed and polished surfaces of the rocks.

DEPRESSION OF THE CRUST BY ICE SHEETS

Observation and theoretical considerations have convinced geologists that the crust of the earth sinks when it accumulates a sufficiently heavy burden of sediment, water, or ice, and that it rises when such loads are removed. The sinking of the overloaded area, they believe, is compensated for at deeper levels of the crust by an outflowing of plastic material. This condition of balance is known as *isostasy* and is of great fundamental importance in shaping the surface and in permitting rocks to rise above sea level.

The concept of isostasy provides a reasonable explanation for the behavior of areas that have supported thick ice caps. In simplest terms, the gradual accumulation of immense volumes of ice eventually depresses the earth's crust until a position of equilibrium is reached. At this point, an ice-covered area that once was above sea level may now be below it. For example, the central part of Greenland is as much as 350 meters below sea level and is covered by ice that is over 3,000 meters thick. Greenland is like an ice-filled bowl rimmed by mountain ranges. Part of the hollowing out undoubtedly can be traced to the scraping and scouring action of the perpetually moving ice, but a considerable amount stems also from the ice's tremendous weight.

After the ice has melted from a depressed area, the crust slowly rises to a new position of equilibrium. The recovery is slow and usually goes on long after the ice has disappeared, which is just what is happening in North America and Europe, where the ice has recently melted. The most obvious example of an ice-depressed area in North America is Hudson Bay. This large epicontinental sea occupies the approximate area once covered by the heaviest ice mass in North America. Old beach lines nearly 300 meters above the present water level on the east side of Hudson Bay show how much upward rebound

has occurred since water flooded the area after the ice had melted. The area is emerging despite a general rise in the sea level, which would, in part, obscure the visible effects of rising land. In areas away from the ice sheets, the rise of sea level assumes its true importance. This will be discussed later in this chapter.

The rise of land around the Baltic Sea is even better known, for the area there is more densely populated, and observations have been carried out over a longer period. Evidence from warped lake levels and from tide-gauge records show that the land is still emerging. Since the ice disappeared, the area has risen at least 76 meters, or an average of about 1 meter per century.

Although we know for certain that the Baltic area is slowly rising, the situation beyond this central area is not so easy to interpret. In northern Holland, for example, where land-sea relations are vitally important, the land has been subsiding at a rate of about 20 centimeters per century. Possibly Holland may once have rested on a bulged area that surrounded the ice and that, with the ice removed, is now subsiding.

The situation around the Great Lakes is likewise not clear. It is known that the lakes are

Figure 15.9 Abandoned beaches on James Bay, Canada. Because this water body connects with the ocean by way of Hudson Bay, the beaches represent either a general fall of sea level or a relative rise of the land. All evidence favors the latter mechanism as the one responsible. (National Air Photo Library, Surveys and Mapping Branch, Department of Energy, Mines, and Resources, Canada.)

being gradually tilted southward because of the more rapid elevation of land to the north, but geologists are not sure whether present movements stem entirely from the effects of unloading the ice sheets. Be this as it may, the lakes are crowding upon their southern shores and creating serious engineering problems.

Geologists do not understand clearly the patterns of atmospheric circulation that prevailed during glacial and ice-free periods, but there are strong signs that wind action in the near proximity of the ice sheets was considerably different from what it is today. The winds not only were unusually strong, but also found quantities of loose sediment to pick up and deposit. Sand and soil were available from emerging ocean and lake beds, from land divested of vegetation by cold or drought, and from other areas that were exposed as the ice retreated, leaving behind finely ground rock waste. Although wind erosion is associated mainly with deserts, the absence of vegetation or other cover ultimately determines the availability of loose surface material. The retreat of an ice sheet probably reflects a drop in precipitation, and while the ice is melting and for some time afterward, there will be periods of dryness characterized by strong wind action. In any event, each of the major periods of glaciation in North America was followed by the deposition of fine material called *loess*.

Loess is composed of relatively small, angular, mineral fragments that lie in blanketlike sheets over irregular topography, strongly suggesting deposition from the air. Loess soils are very fertile and usually support dense human populations. Great sheets lie over such widely scattered areas of the globe as central Europe, the Mississippi Valley, and northern China.

Different from the sheetlike deposits of loess are the wind-formed piles of coarser material we call dunes. Dunes assume many forms, but when properly interpreted they yield information about wind direction, aridity, and climatic fluctuations. Long after the dune surfaces have been covered by vegetation and even after the dunes have been eroded away, they continue to affect drainage in an area. Thus, streams flowing between long, parallel dunes may cut downward and become permanently entrenched, thereby furnishing evidence of former wind directions. Many thousands of square kilometers in the northern Great Plains are drained by numerous parallel, southeastward-flowing streams that geologists believe record the existence of former dunes and the direction in which prevailing winds blew.

In addition to the large quantities of ice that accumulated on the earth's surface during the glacial stages, a less obvious but very large store of frozen water was trapped underground in the soil and rocks. This frozen soil or rock is called *permafrost*. Permafrost forms wherever the temperature remains below 0°C for a period of at least several years. A layer of permafrost from less than 1 meter to about 600 meters thick underlies at least 27 percent of the total land area of the earth. In the Northern Hemisphere about 7.6 million square kilometers are underlain by permafrost.

The present appearance of the northern tundra, spotted with many lakes and bogs, stems mainly from the fact that the frozen substratum prevents surface water from sinking into the ground, and the low temperatures reduce evaporation. Although the edge of the permafrost has been receding rapidly northward during the past hundred years, it is possible to recognize areas it formerly occupied by the distinctive surface features it has left behind. Freezing of the ground tends to split the rocks, and they take on peculiar arrangements and patterns. When geologists encounter such features in unfrozen ground, they assume that permafrost was once present beneath the surface. During the last glacial stage,

Figure 15.10 Peculiar permafrost patterns near Churchill, Manitoba. The Hudson Bay Railroad crosses the area of the photograph. Similar patterns characterize frozen or recently unfrozen ground over wide areas in northern lands. (National Air Photo Library, Surveys and Mapping Branch, Department of Energy, Mines, and Resources, Canada.)

permafrost probably underlay most of Russia and northern Europe, and in the United States extended well beyond the limits of the glacial ice.

LAKES

As the ice sheets melted away, they left in their wake, over the uneven terrain, a profusion of freshwater lakes of all sizes. Other more temporary bodies of water formed where the glaciers obstructed normal river channels. In addition to the lakes caused directly by ice action, others appeared as a result of increased rainfall in regions far removed from the glaciers. A number of examples will illustrate the various factors that tended to create lakes under these different conditions.

The largest known system of ice-dammed lakes occupied extensive areas of central North America south of the ice border. The largest was

Lake Agassiz, which covered about 250,000 square kilometers in Ontario and Manitoba in Canada and North Dakota in the United States. This lake drained southwestward through the Mississippi system until the ice dam that stood against its northern side disappeared and the lake waters discharged into Hudson Bay. The immense, level tract marking the bottom of Lake Agassiz has become part of the fertile wheat belt of North America.

More complex in origin and history are the succession of water bodies whose present remnants constitute the Great Lakes. The history of these lakes commenced in preglacial times, for the areas they occupy were once river valleys that were enlarged and modified by successive glacial events. Omitting complex details, it is sufficient to point out that when the lake basins began to be freed of ice, the lakes drained into the Mississippi system. Later on, as the glacier receded northward, the lakes began draining into the Atlantic by way of the Mohawk and Hudson channels. At a still later stage, the Saint Lawrence outlet was opened.

Supporting evidence for the sequence of events we have just discussed is shown by the fact that fish and other forms of life in the Great Lakes are related to those of the Mississippi drainage, even though the lakes now connect with the Atlantic by way of the Saint Lawrence River. Several whale skeletons have been found in shore deposits of the ancestral Great Lakes. These animals probably swam up the Mississippi

Figure 15.11 *Topography characteristic of the terminal moraines of a continental glacier. Hummocks, drumlins, lakes, ponds, and wooded hills abound in this region of Wisconsin. (U.S. National Park Service.)*

Figure 15.12   Farmland near Larimore, North Dakota. The fertile soil that supports this agricultural area has developed on marginal deposits of the last major advance of the continental glacier. (U.S. Department of Agriculture, photograph by B. C. McLean.)

and Saint Lawrence rivers when these streams were somewhat larger than they are at the present time.

The Hudson River today occupies a great channel obviously cut by a larger, more powerful stream. Because sea level was much lower when the Great Lakes were draining through the Hudson outlet, there is a well-marked river channel running toward the edge of the continental shelf that was submerged as sea level rose. When it became necessary to drive the Catskill Aqueduct under the Hudson River, engineers discovered that the bedrock surface lies at least 210 meters beneath the present river bottom and about 240 meters below sea level.

A spectacular example of the filling and emptying of an ice-dammed lake is furnished by an-

cient *Lake Missoula*, which occupied an area of several thousand square kilometers in western Montana. The lake is believed to have been over 600 meters deep and to have contained more than 2,000 cubic kilometers of water. This vast reservoir was ponded behind an ice dam where the Cordilleran Glacier pushed across Clark Fork and tributary valleys in Montana. The catastrophic breaking of the ice dam during the Pleistocene released tremendous floods of water that rushed across a 38,000-square-kilometer tract of western Washington that is now known as the "channeled scabland." The lake was probably emptied in less than 2 weeks. The erosion and deposition resulting from this flood are so extensive that many geologists have had difficulty believing such a deluge was possible.

LAKES OF THE ARID LANDS

Many lakes occupied basins not carved by, or adjacent to, ice sheets. Throughout the world, and mainly in presently arid or semiarid countries, the wave-cut terraces and abandoned salt-encrusted beds of Pleistocene lakes abound. The basins occupied by most of these lakes were created in pre-Pleistocene epochs of the Tertiary, mainly by earth movements in areas where rainfall, and consequently deposition of sediment, was insufficient to fill them to their rims so that outlets could be cut. As the moisture-producing "belt of westerlies" shifted southward before the ice sheets, the rainfall in these areas increased, the rate of evaporation fell off, and the lakes filled up.

The famous Dead Sea, whose surface now stands nearly 400 meters below sea level, is rimmed by at least fifteen abandoned shorelines and was probably once about 430 meters deeper than at present. In the now arid Sahara Desert there were many large lakes, some covering thousands of square kilometers.

In the Basin-and-Range region of western North America there are 126 closed basins, 98 of which cradled Pleistocene lakes or extensions of existing lakes. The largest and best known of these ancient water bodies are *Lake Bonneville* and *Lake Lahontan*. Lake Bonneville had an area of about 50,000 square kilometers and a maximum depth of about 300 meters. Its shrunken remnants include Great Salt Lake, Utah Lake, and Sevier Lake. The adjacent mountains were glaciated, and geologists have succeeded in correlating the standard glacial stages with the rise and fall of the lake. The site of Mexico City was an extensive lake several times during the Pleistocene, and the sediments and fossils collected from deep borings there provide evidence of many climatic changes. Students of the geology of the Mexico City area are convinced that the periods of glacial nourishment were relatively warm, which challenges the popular notion that glacial growth depends on cold climates.

Figure 15.13 *Ancient beaches of Pleistocene Lake Bonneville in the Terrace Mountains near Great Salt Lake, Utah. Water stood at the highest levels about 20,000 years ago. (Courtesy of Peter B. Stifel.)*

Attempts to correlate glacial deposits with river deposits have not been entirely successful. There is little doubt that streams of all sizes underwent great fluctuations in volume during the Ice Age and that the amount of loose rock material available to be picked up varied with the expansion and contraction of glaciers at their headwaters. At certain stages the rivers filled their beds with sediment; at other stages they cut into and removed it. Thus, a number of steplike terraces might be formed to record the climatic changes that had taken place.

Easier to understand are the ways rivers react to the rise and fall of ocean and lake levels during periods of glaciation. Rivers must adjust throughout their courses to the rise and fall of the level of water bodies they eventually enter. If the level of the receiving body rises, a river will fill its former bed or valley with gravel and sand. If the *base-level* falls, the river will begin to cut downward, leaving steplike terraces to mark its former positions.

The Mississippi River and adjacent parts of the Gulf of Mexico have been intensively studied in connection with glacial effects. The thousands of water wells that have been drilled on the floodplain of the Mississippi and the detailed soundings and submarine maps that have been made of the sea provide excellent samples from which to study the effects of the shifting of sea level in this area.

Geologists who have studied the Gulf are convinced that the fall of sea level at the time of maximum glaciation was about 140 meters. While sea level remained low, the waves were able to cut terraces that are now below sea level and somewhat masked by sediments. During these periods the Mississippi and, of course, other rivers throughout the world followed steeper courses and cut deeply into their floodplains. Because they were flowing with greater energy, the rivers could carry larger fragments and gen-

erally coarser material. River deposits laid down when sea level was low are thus marked by channel cutting and heavy sedimentation. The ancient deeper channels cut by the Mississippi during the ice ages have been mapped very carefully, for they carry large amounts of water that can be tapped by drilling.

When sea level rose, the Mississippi grew more sluggish and dropped its sediment along its course before reaching the ocean. At present, the Mississippi is neither actively depositing nor eroding along its course. This condition may mean that the rise in sea level has slowed down or ceased.

Figure 15.14   Changes in the course of the Mississippi River above Memphis, Tennessee, over a period of about 150 years. (Data from U.S. Army Engineers.)

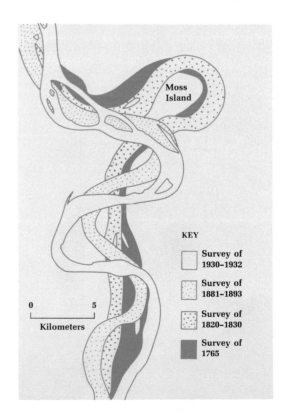

The Oceans during the Ice Age

The Ice Age exerted a profound influence on the larger water bodies of the earth. Today rapid strides are being made in the study of lakes and oceans, and the problems of oceanography are being actively investigated by scientists in many fields. Investigation of the continental shelves, the bottom sediments, the currents, and the chemistry of seawater is proceeding rapidly, but a great deal of work remains to be done.

Geologists assume that the ocean cooled during the individual glacial advances, but it is unlikely that the times of cooling and warming exactly coincided with similar temperature changes on land, for the huge mass of seawater cannot respond to temperature changes as rapidly as do the atmosphere and the land surfaces. The distribution and strength of ocean currents were certainly considerably different, but we know very little about their actual former courses. Temperature of surface waters determines the kinds and, to some extent, the amount of floating shelled organisms and plant life, and this condition in turn is reflected in marked differences in the types of deposits laid down during cool and warm periods.

During extensive periods of glaciation, icebergs become active agents of transportation and deposition, and the material carried by rivers to the oceans is derived partly from glaciers at their headwaters. Temperature variations affect the chemistry of seawater in numerous ways, especially in its capacity to absorb carbon dioxide and to dissolve lime. All these points emphasize the complexity of the problem concerning the reaction of the oceans to changing climates.

THE CONTINENTAL MARGINS

One of the foremost problems confronting geologists today involves the origin and nature of the great sloping continental shelves that surround

Figure 15.15 Extent of the continental shelves, shown in black, about the margins of North and South America. The shallow submerged shelves have a combined area of 5.7 million square kilometers in North America and 2.5 million square kilometers in South America.

the landmasses of the earth. The continental shelves range in width from 16 to 320 kilometers and are essentially submerged portions of the continents. They merge inward with the exposed landscape, but at their outer edge abruptly plunge downward at a much steeper angle toward the depths of the ocean. The outer edges of the shelves lie at depths of between 60 and 180 meters and are remarkably uniform throughout the world. Geologists differ over what may have caused the break in the slope and the relatively smooth surface of the shelves, but the best explanation appears to be that the surface configuration has been modified by the alternate rising and falling of the sea level in response to the freezing and melting of the ice sheets.

Assuming that there were at least four separate glaciations, the shoreline has moved alternately upward and downward across most of the continental shelf eight times, and with each sweep it has smoothed and obliterated existing irregularities, removing high spots and filling low ones. Sediment also has a tendency to shift seaward until it comes to rest below the point where it can be disturbed and moved by waves. The general point marks the break between the continental shelf and the *continental slope.* Soundings show that the shelf is marked in various places by terraces, bars, beaches, river channels, and other features that must have formed in the open air. At other places these features are masked by sediments. Although a great deal of the material of the continental shelves must have been deposited during earlier geologic periods, much of the shaping of the upper surface certainly occurred during the Pleistocene.

DEPOSITS OF THE OPEN OCEAN

Beyond the edges of the continents the effects of glaciation are more difficult to decipher. Wherever glaciers reached the edges of the continents they broke off as icebergs, many of which carried loads of sediment into the ocean. The debris thus deposited ranges in size from huge boulders to the finest clay, and the heterogeneous arrangement of the fragments is distinctive. The chief areas of ice-rafted debris are the North Atlantic, the Arctic Ocean, the northeastern Pacific, and the area surrounding Antarctica.

Beyond the reach of icebergs the chief effects of glaciation were indirect. In the deep oceans,

remote from land, sediment is deposited very slowly, and the materials are extremely fine. Geologists call them *oozes,* in allusion to their soupy consistency. The most common type of ooze consists of the calcareous shells of small floating animals called *Globigerina.* These shells descend slowly from the surface layers of the ocean, and, unless dissolved, form a carpet over the ocean floor. Mixed with the shells, but occurring in pure form when no shells are present, is red clay or ooze that consists of very fine inorganic material. These two constituents, *Globigerina* ooze and red clay, occur in alternating layers from a few centimeters to nearly a meter thick. Geologists assume that the red layers were deposited during the cold periods, when there were fewer living creatures inhabiting the chilled sur-

Figure 15.16 Enlarged model of Globigerina, *one of the most common and abundant types of floating foraminifera. The hairlike extensions are extremely delicate and are not preserved on fossils. Globigerina became common in the Cretaceous and today contributes extensive deposits of its calcareous shells to the ocean bottom. (American Museum of Natural History.)*

Figure 15.17 A deep-sea core taken northwest of Hawaii from a depth of 4,400 meters. The core shows alternations of red clay (dark) and Globigerina-coccolith ooze (light). (Courtesy of Robert J. Hurley, Scripps Institute of Oceanography.)

face water and that *Globigerina*-rich layers mark the warm periods. Cores from the South Atlantic and neighboring areas reveal that an unusual, sudden warming of layers of surface water occurred about 11,000 years ago. As explained in Chapter 6 methods have recently been perfected for correlating oceanic deposits with those of the land. It is too early to predict what will result from the worldwide explorations of ocean bottoms now in progress.

At the present time, sea level appears to be rising throughout the world, and water is encroaching on continents and islands almost everywhere. Geologists attribute this ominous condition mainly to the melting of glacial ice. It is estimated that about 58 million cubic kilometers of ocean water were removed and frozen on land during the Pleistocene Ice Age, an amount roughly equal to 5 percent of all the water of the globe. Of course, most of the last great ice sheets have already melted, and sea level has probably risen about 100 meters since its last low stage, which geologists believe occurred some 40,000 years ago. The exact amount of unmelted ice remaining on the earth is fairly well known, and calculations suggest that if it all melted an additional rise in sea level of from 75 to 90 meters would result. If the ice on Greenland alone suddenly melted, sea level would rise over 7 meters. If distributed uniformly over the entire earth, the Greenland ice would form a layer more than 5 meters thick. The amount of Antarctic ice has been calculated at 29 million cubic kilometers. Measurements made during the International Geophysical Year show that Antarctic ice reaches a maximum thickness of 4,200 meters and has a greater volume than investigators formerly supposed.

The fluctuations in ocean level caused by glaciation are difficult to separate from the effects of actual earth movements, adjustments in the ocean basins, the deposition of sediments, and other obscure factors. If the present trend continues, most of the great cities of the world will eventually be inundated, and present marine installations and port facilities will be rendered entirely useless. In the United States, for example, Long Island, most of New Jersey, the Delaware-Maryland-Virginia peninsula, and most of Florida will be flooded. The submergence of the New Jersey coast has been about 1.5 meters per thousand years for the last 2,600 years, and before that it was about 3 meters per thousand

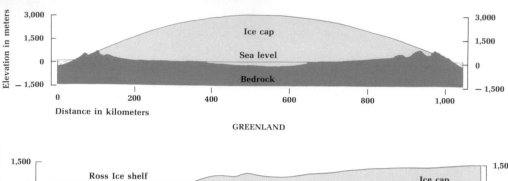

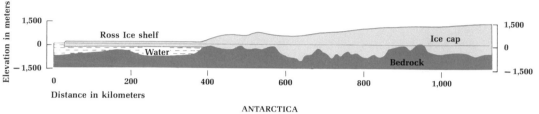

GREENLAND

ANTARCTICA

Figure 15.18   Cross sections showing the Greenland ice cap and part of Antarctica. The vertical scale is greatly exaggerated. The section of Antarctica extends from the IGY (International Geophysical Year) Little America Station to the IGY Byrd Station and does not show the maximum thickness of the ice. The Greenland section cuts the central part of the island and shows nearly the maximum known thickness of ice, about 3,300 meters. The bedrock surface of both areas is entirely or partly below sea level.

years. An independent study of the Connecticut coast shows a submergence of 2.7 meters in the last 3,000 years, and 10 meters in the last 7,000 years.

Life of the Pleistocene

THE GLACIAL ORDEAL

The Pleistocene Ice Age lasted from 2.5 to 3 million years and brought many drastic changes to the lands and oceans of the earth. These alterations affected all forms of life to some degree and set the stage for the appearance of modern man. No radically new or different forms of life, except man, appeared in the Pleistocene, but there were notable evolutionary changes involving the creation of new species and even new genera. The really significant impact of the Ice Age upon life forms came about through *enforced displacement, migration, mixing,* and *isolation,* which resulted as ice sheets expanded and contracted. New living space became available when shallow lands and islands were laid bare of ice and water. Corresponding contractions came as sea level rose in the interglacial stages. Migration routes were opened and closed on a scale not possible during nonglacial times. Land plants were literally displaced thousands of kilometers as the ice sheets grew and melted away. Even the tropics felt the effects of displaced storm tracks and precipitation changes. Not even the depths of the ocean escaped change, as cold, dense water flowed toward the equator and gradually filled the ocean basins.

Each advance of glacial ice totally depopulated millions of square kilometers of the earth's surface. Although most plant and animal species were able to escape extermination by retreating before the ice, their living space was drastically reduced. Put another way, the total food production of glaciated lands was greatly curtailed and fewer individuals, though not necessarily fewer species, were able to exist. Organisms were not forced to adapt to climatic conditions for which they were entirely unsuited; rather, the climatic zones themselves were greatly compressed. Thus, the tundra belt, now perhaps 1,500 kilometers wide in North America, was narrowed to a few kilometers adjacent to the expanded ice fields.

ENFORCED MIGRATIONS OF MAMMALS

It is not unusual to find fossils of cold- and warmth-loving animals alternating with one another in the sediments of glaciated areas. Thus, in central and western Europe the glacial deposits contain woolly rhinoceroses, mammoths, lemmings, reindeer, arctic foxes, and moose, forms now extinct or confined to more northern lands. The interglacial deposits of the same area contain fossils of lions, rhinoceroses, hippopotamuses, and hyenas, animals now characteristic of African climates.

The island of Malta, now isolated in the Mediterranean, has yielded reindeer, arctic foxes, mammoths, bison, horses, and wolves, recording not only much cooler climates but also land connections with the European mainland. Similar less spectacular faunal changes occurred in North America. Reindeer and woolly mammoths reached southern New England, and moose lived in New Jersey during the glacial stages. Fossil sea cows, now found in coastal waters off Florida, ranged as far north as New Jersey, and the tapir and peccary roamed Pennsylvania. Ground-sloth remains have been found as far north as Alaska. Elephants were isolated on the Channel Islands off California.

The history of the musk-ox during the Ice Age is particularly significant. Its bones have been found in Iowa, Nebraska, and Minnesota, recording a time when tundra conditions prevailed in the north-central United States. Today, the musk-ox lives only in the far northern reaches of continental Canada, the Arctic Archipelago,

Figure 15.19 Geologic expedition embarking on a scientific survey in Antarctica. This continent is still in the grip of the Ice Age. (Institute of Polar Studies.)

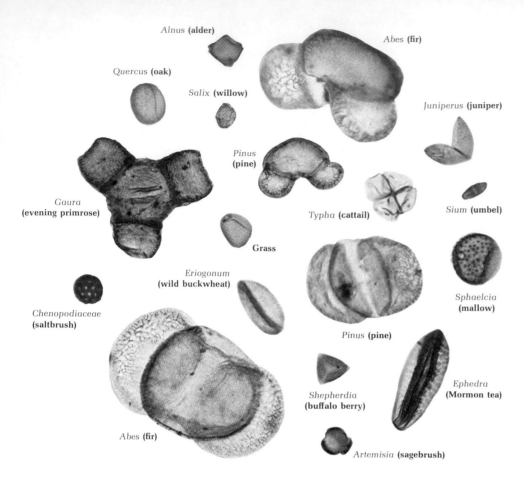

Figure 15.20 Pollen grains from Pleistocene deposits of Las Vegas Valley, Nevada. This assemblage is indicative of semiarid conditions. (Courtesy of P. J. Mehringer, Jr., and the Nevada State Museum.)

and northern Greenland. The animal has evidently moved with its customary environment as the environment has shifted with the ice front across a distance of about 3,200 kilometers.

DISPLACEMENTS
OF PLEISTOCENE VEGETATION

Pleistocene plants belong mostly to still-living species, but there has been a great displacement and mingling of plant groups. The present distribution of plants is difficult to explain without reference to the Ice Age. In Europe the forests of hardwood that characterized the Tertiary were considerably reduced by the advancing Scandinavian and Alpine ice sheets. In North America, where the mountain ranges trend from north to south, the hardwood forests advanced and retreated across open lowlands and were not imprisoned as they were in Europe. As glaciers retreated, the arctic floras followed the margin of the ice and also ascended mountains where favorable cool conditions still prevailed. Thus, once-continuous plant populations became more and more widely separated, one group retreating northward and the other ascending available mountains. Some plants that now grow along the highest slopes of the White Mountains in New

Hampshire and on Labrador have disappeared from the intervening lowlands. On Greenland, species grow that are found also only in the Alps and Himalayas. The great vegetational flux of the Pleistocene is still going on and is of tremendous importance to man.

The study of pollen grains has thrown considerable light on climatic fluctuations and plant migration during the Pleistocene. Pollen is produced in great quantities by many plants and is so resistant to decay that it remains recognizable for extremely long periods. Experts can identify the grains produced by specific plants, and are thus able to reconstruct the general composition of the vegetation of particular times and places. Because plants are very sensitive to environmental factors such as temperature and moisture conditions, the analysis of pollen offers perhaps the best means of discovering what the climates of the past were really like.

Such species as birch, spruce, and fir indicate cold and moist conditions; pine signifies warmth and dryness. Oak, alder, and hemlock suggest warm, moist surroundings; and an absence of tree pollen coupled with an increase of pollen from arctic herbs would indicate a tundra environment. Pollens of grass and drought-resistant shrubs are a sign of dryness. Even the pollens of cultivated plants such as corn yield important clues about the agricultural and food habits of early man.

Figure 15.21 Diagrammatic illustration of a Pleistocene bog deposit with fossil mammoth skeleton. This mode of preservation is common in the glaciated portions of the United States. (New York State Museum and Science Service.)

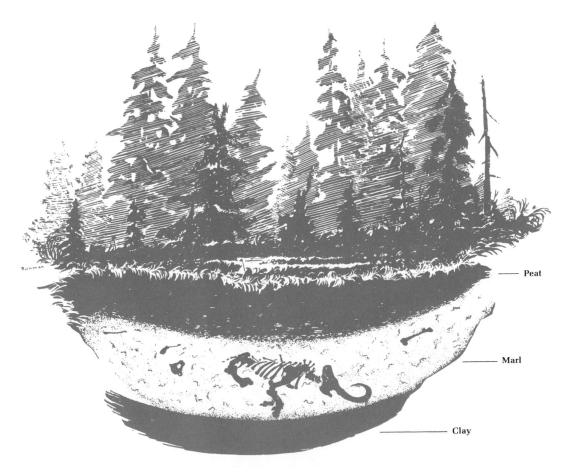

Peat

Marl

Clay

Figure 15.22 Remains of Ice Age elephant and bison as found in Joint Mintor Cave, Devon, England. (Pengelly Cave Studios Trust.)

FATE OF THE GIANT MAMMALS

The Pliocene and Pleistocene were characterized by giant mammals. Almost every group of mammals produced one or several colossal members. These are abundantly preserved on all continents, occasionally in great "graveyards" such as the La Brea tar pits in Los Angeles, California, or the Big Bone Lick in Kentucky. Some of the fossils are so recent that their unpetrified, dried, mummified, frozen, or embalmed remains are still to be found.

Among the most common and well-known giants of the Pleistocene were the mammoths and mastodons. The Imperial Mammoth attained an average height of about 4 meters at the shoulders and had great curving tusks that sometimes were nearly 4 meters long. Mammoth and mastodon bones are surprisingly abundant; over one hundred mastodon skeletons have been recorded from New York State alone. Another famous Pleistocene animal is the saber-toothed cat *Smilodon,* hundreds of which left their bones in the La Brea tar pits. There were true lions much larger than the present king of beasts, and bears more massive than the grizzly. The giant beaver, large as a black bear, could topple the largest trees, and there were other rodents of proportionally large size. Several kinds of bison roamed the American West; one species had a horn spread of nearly 2 meters. Large camels, pigs, and dogs lived in North America, together with the huge ground sloth, heavy as an elephant, which reared clumsily on its hind legs to browse on foliage 6 meters above the ground.

Other continents also had their giants. In Africa there were pigs big as a present-day rhinoceros, sheep that stood 2 meters at the shoulders, giant baboons larger than the gorilla, and ostrich relatives over 4 meters tall.

From South America come fossils of the giant anteater, the glyptodon, and the sloth, together with rodents big as calves. Here also were abundant large, flightless, flesh-eating birds, up to 2.5 meters tall and with 40-centimeter beaks. Even Australia had giant kangaroos and other marsupials.

Most of the giant Pleistocene land animals can be traced to smaller ancestors in preceding epochs of the Tertiary. The gradual increase in size is a tendency that is observed many times in the history of land life and presents no particular problems. It is puzzling, however, to find that most of the giants survived the repeated glacial onslaughts, only to disappear within the last few thousand years. This is particularly true of North America, where glaciation was both extensive and intensive. A recent survey gives the following estimated dates of extermination of some key forms in North America: saber-toothed cat, 14,000 years ago; woolly mammoth, 10,500 years ago; ground sloth, 9,500 years ago; native horse, 8,000 years ago; Columbian mammoth, 7,800 years ago; and mastodon, 6,000 years ago.

(a)

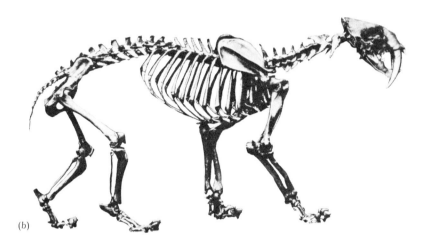

(b)

Figure 15.23   Two typical mammals of the Ice Age.
(a) The large ground sloth; (b) the saber-toothed cat.
Both are found in the famous La Brea tar pits of
California. (Natural History Museum of Los Angeles
County.)

Figure 15.24   The musk-ox, a survivor from the Ice
Age. Scattered remains prove that the musk-ox
migrated widely as the ice sheets advanced and
retreated in the Northern Hemisphere. (Field Museum of Natural History, Chicago.)

Man has been blamed for the destruction of many giant mammals, but such a charge is difficult to prove and can scarcely be true for those animals that disappeared before his arrival. Available evidence suggests that man accomplished by indirect means what he could not do with the primitive weapons then at his disposal. His probable ally was fire which, by accident or design, he applied to forest and prairie with catastrophic results. We know that burning out forests to clear agricultural plots was a common practice among early inhabitants of northern Europe, and certain American Indians also were arsonists when necessary.

The big game that ranges over America today is but a pitiful group of small and medium-sized mammals left over from the Ice Age. Africa today suggests, but does not duplicate, the teeming life that was characteristic of other continents during the Tertiary and Pleistocene. Chapter 19 discusses this subject in greater detail.

Causes of the Ice Ages

No event of prehistoric time has given rise to more speculation than the Pleistocene Ice Age. Theories began to be proposed in the seventeenth century, and new ones are still being formulated; a recent tabulation lists no less than fifty-four "explanations" for ice ages. Newer hypotheses are on the whole more reasonable and at the same time more complicated, because they must account for many facts that were unknown to earlier students. Extensive glaciations are known to have occurred in the Late Precambrian, in the Permian, and in the Pleistocene. We know that these periods of glaciation have been rather short, interrupting longer periods of mild, more equable climate. The one invariable accompaniment of glaciation appears to be mountain building. Intensive mountain building and continental elevation have coincided with all the great glaciations of the past, but there may not be enough well-known cases to prove a cause-and-effect

relationship. The nonglacial periods, nevertheless, have been longer and are marked by widespread shallow seas, low-lying continents, and generally less mountain building. But mountain building alone does not seem to explain the alternating advance and retreat of the glaciers during the ice ages, for there is no proof of topographic changes coinciding with the coming and going of Pleistocene ice sheets.

In general, glacial theories are divided into two groups: (1) those that appeal to conditions entirely outside and independent of the earth (astronomical), and (2) those based on conditions on or within the earth (terrestrial). It may be worth mentioning here that there are three other hypothetical factors in addition to mountain building that could cause an upset in normal climatic patterns: (1) there may be an actual variation in solar energy; (2) a barrier between the sun and the earth's surface, may prevent the usual amount of solar energy from reaching the earth; and (3) something on the earth's surface may interfere with the distribution of heat, causing certain areas to receive more or less heat than usual.

In choosing among these various possibilities, scientists in many different fields have constructed theories that lean heavily on their respective specialties. Not all the theories have been reliable; many are downright ridiculous. A good theory should provide a mechanism or condition to explain the beginning of ice accumulation, must explain the advances and retreats within the broader period of glaciation, and, finally, must describe what factors cause the ice to disappear. Also, as in all good theories, the mechanism should be as simple as possible, appealing to observable or provable forces and avoiding unprovable or hidden ones.

It is, of course, entirely logical to suppose that the fundamental cause of cooling is a decline in the sun's heat. It is true that the sun's output does vary in connection with sunspot cycles, but these cycles are not extreme enough to cause glaciation, and it is doubtful if a decline in the sun's

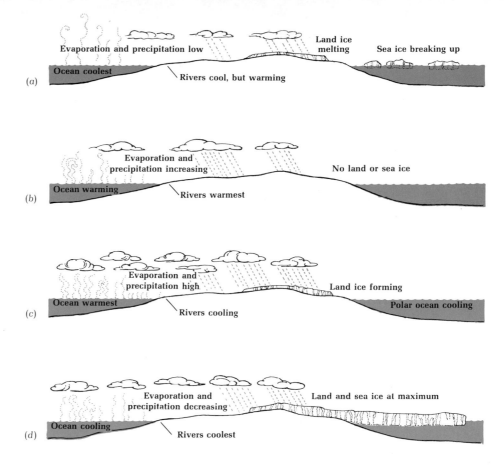

Figure 15.25 *The ocean control theory of glaciation. (a) Glacial retreat; (b) glacial minimum; (c) glacial advance; (d) glacial maximum.*

energy could really cause an ice age, for such a decline would be followed by a decrease in evaporation and eventually by less precipitation. Any appeal to long-term solar changes removes the problem from the realm of observation; sun changes take too long.

Some geologists have advanced the idea that the solar system occasionally passes through clouds of cosmic dust that could cut down the energy received by the earth, and also that fine dust from violent volcanic explosions could have similar effects within the earth's atmosphere. Slightly different from this proposal is the carbon dioxide theory, which is based on the fact that carbon dioxide does not cut down incoming shortwave radiation from the sun but does inter-

fere with the escape from the earth of longer heat waves, thus creating a blanketing or greenhouse, effect, with accompanying greater precipitation and ice formation.

More recent theories have been based on strictly terrestrial factors such as shifts in ocean currents, the effects of precipitation on elevated areas, and isostatic adjustments of land areas. The controlling effect of the ocean has received special attention, for it is a great climatic regulator by which heat and cold are stored and distributed from place to place. The *ocean-control theory* rests on the assumption that cold oceans,

Figure 15.26 *Mount Hood and Bull Run Lake, Oregon. The mountain is one of a series along the Cascade Range that were active volcanoes during the Pleistocene Epoch. It has been deeply eroded by glaciers, a few remnants of which still remain. (U.S. Department of Agriculture.)*

such as result from great ice-forming periods, cannot supply the usual amounts of precipitation to the lands, and that the glaciers are starved by lack of precipitation rather than by accelerated melting. As the oceans regain their heat during ice-free periods, they are able to furnish increasing amounts of precipitation, and the glaciers are reborn if the precipitation falls as snow rather than rain. As the ocean is gradually cooled by ice and cold water, the cycle is repeated.

According to a recently proposed theory, the Ice Age may have been initiated by the uplift of the Isthmus of Panama between North and South America, an event that is known to have occurred near the close of the Pliocene. This incident stopped the westward flow of strong ocean currents and diverted increasing amounts of warm water into northern regions that were sufficiently cold and high and needed only an increasing snowfall to support ice fields. From the fact that the northern ice sheets seem to be under the influence of the Gulf Stream, it is inferred that variations in the strength and position of this great current have caused the alternate spread and retreat of the ice sheets according to the general principle of oceanic control.

Are We Still in the Ice Age?

Only a few decades ago, geologists were confident that the Ice Age was over and past. On the basis of this belief, the name "Pleistocene Epoch" was applied to the Ice Age, and all subsequent time was included under the term "Recent." Now, however, opinions have changed, and most geologists agree that the Ice Age is not over and that we are, therefore, still living in the Pleistocene Epoch. The fact of the matter is that the time that has elapsed since the melting of the last ice sheets in North America and Europe is shorter by far than any of the interglacial periods of the past. In other words, the earth has been just as warm and perhaps just as free from ice several times in the last million years as it is now. There is no good reason to suppose that the glaciers cannot accumulate and push forward again just as they have in the past.

A variety of evidence indicates that the ice sheet melted from the United States about 11,000 years ago and that the Scandinavian ice sheet also disappeared about the same time. Since then, the climate has become generally warmer and drier, but there have been numerous local fluctuations. The most notable event of postglacial time is the so-called *climatic optimum,* a period from about 7,000 to 4,500 years ago when climates were generally warmer than now. The present is a time of very unsettled conditions, with a tendency toward a worldwide rise in temperature and locally increased aridity.

There is a great deal of speculation but no concrete evidence about just where these trends are leading. Some investigators believe that ice caps may begin to re-form in Canada within 10,000 years; others put off the event for 50,000 or 100,000 years. The whole problem is in an interesting state of speculation, and it is certain that considerable information from geology, biology, climatology, and meteorology must be collected and examined before we can tell what to expect in the future. No definite answers can be given on the basis of present information.

A SUMMARY STATEMENT

During the last 2.5 to 3 million years the climate of the earth has been such that glaciers accumulated over at least 39 million square kilometers, or 27 percent of the land area of the globe. Other simultaneous effects include the cooling of the oceans, the formation of freshwater lakes in interior basins, the freezing of extensive tracts of soil and rock, heightened wind action, and a multitude of reactions among the plants and animals of the globe. No portion of the earth, not even the depths of the ocean, escaped the chill influence of the Ice Age.

In spite of an immense amount of study, the duration and subdivisions of the Ice Age are not yet fully known. Estimates of the total length of the Pleistocene range from a few thousand to 3 million years. There is clear proof of at least four major advances of the continental glaciers, but there may have been many minor oscillations, and the history of different areas has varied greatly.

The Ice Age caused great disturbances in the organic world. Plants and animals were forced to migrate with changing climates, and there was an overall reduction in the organic productivity of the lands. There were notable exterminations, especially among the larger mammals, but the greatest extinctions seem to have occurred after the last glaciation was over.

The cause of widespread glaciation is not yet understood, although many theories have been proposed. It seems evident that the Pleistocene Ice Age is not over and that we are now living between two glacial stages during a period of unsettled climatic conditions.

FOR ADDITIONAL READING

Charlesworth, J. K., *The Quaternary Era,* 2 vols. London: Arnold, 1957.

Daly, Reginald A., *The Changing World of the Ice Age.* New Haven: Yale University Press, 1934.

Flint, Richard Foster, *Glacial and Quaternary Geology.* New York: Wiley, 1971.

Nairn, A. E. M., *Problems in Paleoclimatology.* New York: Wiley, 1964.

Rankama, Kalervo, ed., *The Quaternary,* Vol. 1. New York: Wiley, 1964.

Schwarzbach, Martin, *Climates of the Past: An Introduction to Paleoclimatology,* trans. Richard O. Muir. New York: Van Nostrand, 1963.

Stock, Chester, *Rancho La Brea.* Los Angeles: Los Angeles County Museum, 1949.

West, R. G., *Pleistocene Geology and Biology.* New York: Wiley, 1968.

Woodbury, D. O., *The Great White Mantle.* New York: Viking, 1962.

Zeuner, F. E., *The Pleistocene Period, Its Climate, Chronology, and Faunal Succession.* London: The Ray Society, 1959.

THE COMING OF MAN

Although the study of man belongs properly to the field of *anthropology*, it is nevertheless true that early human history is best interpreted in geologic context. The term *paleoanthropology* is used by some authorities to designate the borderline study of ancient prehistoric man. Early human beings lived close to nature, and until they acquired culture and learned to modify their environment they were largely at the mercy of the same forces that influence lower forms of life. They lived directly off the land, close to the

Olduvai Gorge, Tanzania, the site of many important finds of primitive man and associated fossils. (Copyright National Geographic Society.)

plants and animals that provided them with food. Their comings and goings were governed by seasonal changes, by climates, by routes of travel, and by the presence of food sources and shelter. Occasionally, their bones were buried by natural means among the remains of their animal contemporaries.

We must interpret the early history of man by studying his fossil remains, the evidence of associated plants and animals, and the types of sediments in which these fossils occur. We apply the same methods we employed in studying earlier periods and lower forms of life. Only at a relatively late stage, when man began to leave written accounts, do we pass into the *historic period*.

The Descent of Man

Strange as it seems, man finds it difficult to describe himself in precise and scientific terms. In order to think clearly about man and his origin,

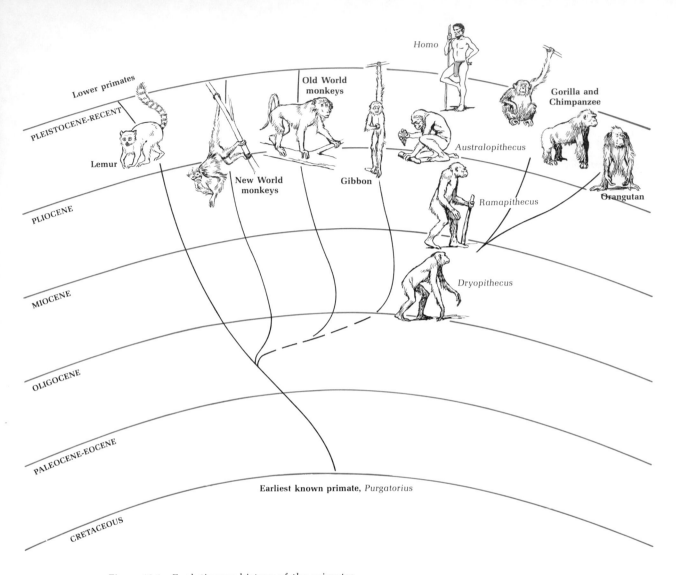

Figure 16.1   Evolutionary history of the primates.

we must be as careful as possible about the way we define him. The broad term "man" is not accurate enough to indicate whether we are thinking exclusively of living types of men or whether we include fossil relatives of living men of our same species, or if we are designating all upright, two-legged, large-brained creatures, regardless of species.

Although man occupies a unique and distinctive position in the biological world, there are many signs that point to his relationship with lower animals. When we apply to ourselves the same logical rules of classification by which we catalog other living things, we see that we are *vertebrates, mammals,* and *primates.* Our internal skeleton of bone, built around a flexible, jointed spinal column, places us unmistakably among the vertebrates. We have a special kinship with the group of vertebrates called mammals, for, like them, we have hair and warm blood and bring forth our young alive and nurse them with milk. Our lower jaw is composed of one pair of bones, our teeth are differentiated into incisors, canines, premolars, and molars, and we have an exceptionally large brain. These details remove all doubt about man's being a mammal.

Among the mammals, man clearly resembles other members of the primate order. This group includes *lemurs, tarsiers, monkeys, apes,* and *men.* All primates possess relatively large brains, five fingers and five toes equipped with nails, *prehensile* (grasping) hands and feet, a very flexible arm, a thumb that can be rotated to oppose the other fingers, an eye socket protected by bone, and a total of thirty-two or thirty-four teeth. This combination of characteristics is so diagnostic that man must undoubtedly be placed among the primates.

Our classification is not complete, however, for we find that man bears a closer resemblance to some primates than to others. We share more physical traits with the monkeys and apes than we do with the lemurs and tarsiers. Thus, the suborder *Anthropoidea* includes monkeys, apes, and man. Continuing our attempt to establish our individuality, we recognize that we are different from both monkeys and apes. Monkeys have tails and relatively small brains, and men are clearly more apelike than monkeylike. In recognition of resemblances between man and apes, both are included in the superfamily *Hominoidea.* Distinguishing man from the apes often narrows down to a hairsplitting exercise on many points, for there are more similarities than there are differences. All apes have relatively longer arms than man. Their feet resemble hands, and some move through the trees by swinging hand over hand from branch to branch. Their great toe is opposable; ours is not. The canines of apes are pointed and interlocked; ours are small and do not rub against one another. Man's pelvis is rounded and

bowl-shaped; the ape's is elongated. These physical differences are more a matter of degree than of kind, but the mental differences are profound. The brains of man and ape resemble each other in a general way, but there is a vast difference in quality. This difference is sufficient, in the opinion of most zoologists, to ensure man a unique position in any zoological classification.

The job of separating man from his living relatives, the apes and monkeys, is therefore not too difficult, and we continue our biological categorizing by placing ourselves in the family *Hominidae,* of which we are the sole living members. We finish the task by assigning to ourselves the generic name *Homo* (Latin, "man") and the specific name *sapiens* (Latin, "wise"). To distinguish ourselves from other varieties or subspecies, we may use the term *Homo sapiens sapiens.*

Like the proverbial skelcton in the closet, a number of extinct man-apes or ape-men have come to light to complicate the classification. Although these relics of the past have time and time again shattered our preconceived schemes of classification, we seek them out and welcome them as the only means whereby we can hope to determine our true lineage.

Fossils not only reveal many stages in the evolution of living primates, but also represent entire branches of the family tree that have disappeared without leaving descendants. In order to accommodate these extinct forms and to place them in proper relation to ourselves, we must frequently look for very fine and subtle anatomical details. The differences between man and gorilla, for example, are comparatively easy to detect, but the distinction between modern man and certain extinct creatures is much more difficult to establish.

With regard to extinct, manlike forms, we may say that, all things considered, we will accept them into the family Hominidae if they appear to have walked erect and had relatively large brains. But many fine points about them must be decided on, and frequent disagreements concerning them are to be expected as studies proceed.

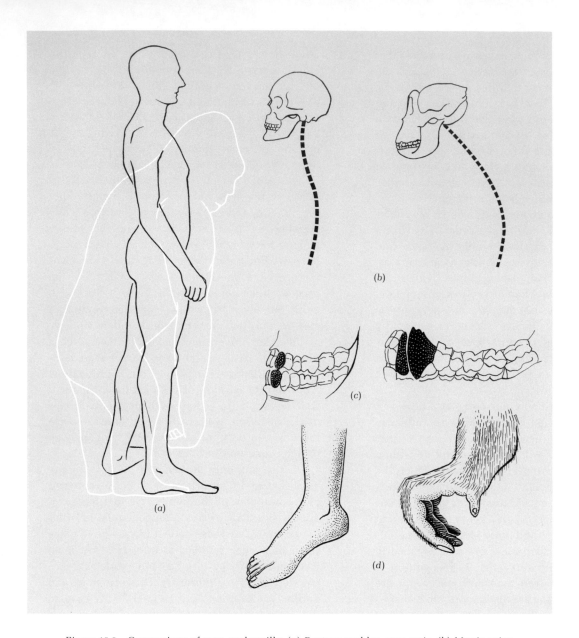

Figure 16.2   Comparison of man and gorilla. (a) Posture and leg–arm ratio. (b) Man's spine vertical and with double curve; ape's inclines forward with single curve. Note man's larger cranium. (c) Man's small canine teeth and ape's large, self-sharpening canines. (d) Man's foot and ape's grasping foot. (Reproduced from M. Bates, Man in Nature, 2nd ed., Prentice-Hall, Inc., 1964.)

(a)   (b)   (c)

In this discussion we will follow the current trend toward simplification by placing all men, fossil and living, within a relatively few species. *Homo erectus* and *Homo sapiens* are accepted as valid by most workers. The positions and proper terminology of *Homo transvaalensis* and *Homo habilis* are still being debated. By using a third term to designate the varieties within these species, we can differentiate them even further. We also avoid becoming involved with some of the problems of nomenclature by referring to fossil forms according to their place of discovery; for example, *Java man, Cro-Magnon man,* and so on.

An Ancient Lineage

If the principle of organic evolution is true, man as a physical being was preceded by a succession of ancestral species that extends back to the beginning of life on earth. We have already reviewed the more ancient stages of this reconstructed ancestry and will now investigate the more recent links within the primate order.

It is significant as evidence for evolution that the members of the primate order appear in the fossil record in the order of their zoological rankings. In other words, the least advanced appear first, the most advanced last. Another unusual feature is the existence today of all essential stages in the primate hierarchy. Thus we may study living specimens of the primitive tree shrews, not greatly advanced over the first placental mammals, and each succeeding type—lemurs, tarsiers, monkeys, apes, and man. Such

Figure 16.3 Modern man differs physically from living and extinct primates chiefly in the large size and complexity of his brain. Shown here are three brain casts: (a) gorilla, which has an average capacity of 500 cubic centimeters; (b) Pithecanthropus, 1,000 cubic centimeters; and (c) modern man, 1,400 cubic centimeters. (American Museum of Natural History.)

comparative living sequences are extremely rare in any group.

The earliest known primate, recently discovered in Montana, has been named *Purgatorius ceratops.* Only small isolated teeth are known, but the assignment to the primate stock seems accurate. *Purgatorius* was contemporary with the last great dinosaurs and its discovery pushes the origin of the primates backward into the Age of Reptiles.

Primates are represented in the Paleocene, earliest epoch of the Cenozoic, by relatively good fossils of creatures such as *Plesiadapis,* an animal that was as small as a squirrel—or as large as a house cat. Remains have been found in France and the western United States. It appears that *Plesiadapis* favored life in the trees, probably subsisted on fruits or insects and has many resemblances to the living lemurs. The primate lineage continues into the succeeding Eocene Epoch and diversifies into a large number of genera such as *Notharctus* and *Smilodectes.* These are slender, tree-living forms with even greater resemblances to present-day lemurs. *Notharctus* must have been relatively abundant; its remains have been discovered in six western states. Although the Eocene record of lower primates is good, none of these specimens can be

Figure 16.4 The Eocene primate Smilodectes. This reconstructed skeleton is based on remains found in southwestern Wyoming. The animal is very similar to certain modern lemurs. (Smithsonian Institution.)

accepted with certainty as a direct ancestor of Old World higher primates.

By the end of the Eocene the primates had virtually deserted North America. One peculiar monkeylike survivor, *Rooneyia,* has been discovered in early Oligocene sediments near the Rio Grande, west of Marfa, Texas. The withdrawal and subsequent absence of primates (except man) in North America may be due to the progressive cooling and drying that characterized the Cenozoic Era.

Incidentally, the fossil record of primates in South America consists of a few scanty remains of monkeys from Miocene rocks of Argentina and Colombia. So far as known, primate evolution produced nothing higher than monkeys in the New World and the search for evidence of the advanced forms shifts to the great landmasses of Africa and Eurasia.

The most significant discoveries of Oligocene primates have been made in the barren Fayum district of Egypt about 100 kilometers southwest of Cairo. Fossil mammals, including relatively abundant primates, are found at several levels ranging in age from 25 to 35 million years. Three groups of primates seem to be represented: a primitive, New World, monkeylike family typified by *Parapicthecus;* a gibbonlike genus, *Aeolopithecus;* and an apelike assemblage including *Aegyptopithecus* and *Propliopithecus.* One authority compares the grade of evolution displayed by the Fayum primates as approximately that seen today in South American monkeys.

Problems of Dating

Before going further with the chronological history of primate evolution, we must digress for a few paragraphs on the subject of dating. As any

Figure 16.5   Skull of Rooneyia, *last known North American primate (except man), from Oligocene rocks near Marfa, Texas. Skull is about 5 centimeters long. (Courtesy of John A. Wilson.)*

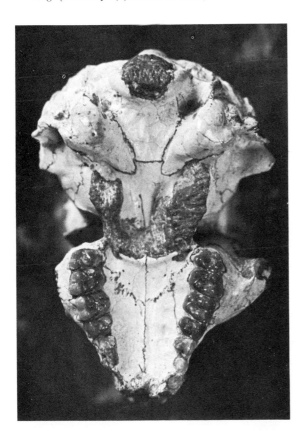

genealogist knows, it is impossible to construct accurate family trees until all the members concerned are assigned correct dates. Thus, John Smith cannot be the father of George Smith if he lived much later than George. However, if John lived before George, he may have been in George's ancestral line. This relationship will remain hypothetical, however, until the investigator can uncover more definite evidence on which to base it. Exactly this same sort of problem confronts the student of fossil remains. Arranging any series of fossil specimens in correct order of succession is a formidable task in itself, and even when completed tells us only that certain relationships are possible and that certain others are not. We should mention here that the correct order of succession can be determined by superposition without establishing the actual age in terms of years.

ABSOLUTE DATING

The discovery of carbon 14 has provided a means of assigning relatively accurate dates to organic remains up to 50,000 years old. This is of great importance, but the method is obviously of no value in dealing with many important events and remains of earlier date. Another method of radiometric dating, based on potassium and argon, is useful in dealing with material over approximately 1 million years old. However, this method is useful in dating fossils only where minerals or rocks containing radioactive potassium can be found in intimate association with the fossils themselves. The gap between the useful limits of the carbon 14 and potassium–argon methods is most troublesome and much research has been conducted to find a suitable means of filling it.

ASSOCIATED REMAINS

Many, if not most, primate remains are found in association with those of other creatures, generally large mammals that lived simultaneously in the same vicinity. The associated bones found

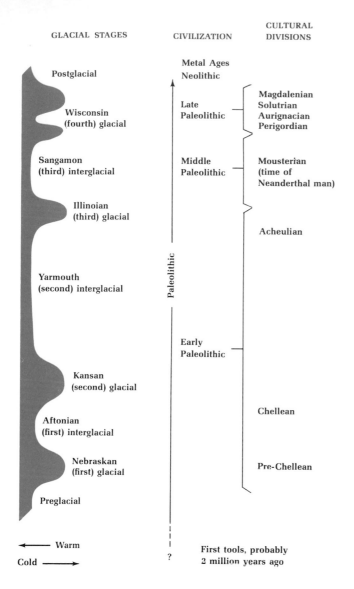

Postglacial

Wisconsin
(fourth) glacial

Sangamon
(third) interglacial

Illinoian
(third) glacial

Yarmouth
(second) interglacial

Kansan
(second) glacial

Aftonian
(first) interglacial

Nebraskan
(first) glacial

Preglacial

Metal Ages
Neolithic

Late
Paleolithic

Middle
Paleolithic

Early
Paleolithic

Paleolithic

Magdalenian
Solutrian
Aurignacian
Perigordian

Mousterian
(time of
Neanderthal man)

Acheulian

Chellean

Pre-Chellean

◄── Warm

Cold ──►

?   First tools, probably
2 million years ago

Figure 16.6 Relation of glacial and interglacial stages to the prehistoric cultural stages of man.

with remains of man may be those of animals hunted or collected for food. These remains, if sufficiently well preserved, serve as excellent references for dating each other. The life-spans, or total period of existence, of many species of

Tertiary and Quaternary mammals are fairly well known and the times of arrival and of extermination of such large and important species as the hairy mammoth, mastodon, cave bear, saber-toothed cat, and others have been established quite closely by superposition, carbon 14 dating, or mutual relations. Any rare bone or tooth found in association with these better-known remains whose date has been established is automatically confirmed as being of the same period.

RELATION TO THE GLACIAL STAGES

Modern man is a product of the Ice Age; his ancestors were well acquainted with cold and storm. The association of human bones and artifacts with glacial deposits proves conclusively that man existed during the Ice Age and that he was profoundly affected by the successive retreats and advances of the glaciers. Geologists usually date human remains and activities in terms of recognized glacial and interglacial stages. No human colony could escape the climatic changes that marked the great Ice Age. Even in the tropics, far from the areas where the ice sheets lay, temperature, sea level, animal life, rivers, and lakes were directly affected by the waxing and waning of the ice. In regions that were free of ice, such as Africa, geologists relate prehistoric events to wet and dry stages that they assume correlate with glacial and interglacial stages elsewhere. This method does not yet yield absolute dates in terms of years before the present, but it does show the correct order of events and cultures. The reader should refer to Figure 16.6, in which a comparison is made between glacial stages and cultural divisions.

TOOLS AND WEAPONS

Early man learned to use a variety of natural materials as tools and weapons, but only the most durable of these have survived the destructive influences of time and weather. Stones shaped for some definite purpose not only indi-

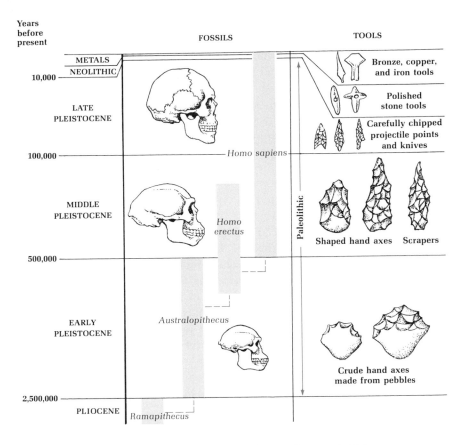

Figure 16.7 Tool-making cultures of Pleistocene man. (Adapted from A. Lee McAlester, The History of Life, Prentice-Hall, Inc., 1968.)

cate the presence of man but also may reveal a great deal about his stages of development and material culture. Stone artifacts are like guide fossils, in that they help geologists assign dates to formations and to other fossils. The study of stone artifacts has become very extensive, and we can mention here only the broadest aspects of the subject.

The long period during which man used stone to fashion his tools is called the Stone Age. We divide the Stone Age into the *Paleolithic,* or *Old Stone Age;* the *Mesolithic,* or *Middle Stone Age;* and the *Neolithic,* or *New Stone Age.* The Neolithic was succeeded by the Age of Metals, during which copper, bronze, and iron successively came into use. The beginnings and endings of

these subdivisions were not simultaneous everywhere, and exact dates are difficult to establish.

The Paleolithic was much longer than the other subdivisions. It began perhaps 2 or 3 million years ago and ended around 10,000 B.C. During the Early Paleolithic, men made tools by pounding one stone on another to produce a crude cutting or scraping edge. These so-called "pebble tools" have been found in Africa and Asia, and date at least to the first interglacial period. The pebble tools gradually evolved into the so-called hand ax, a rather large, tapering, all-purpose tool made to fit the hand. Later, the hand ax was refined and smaller points were made from chips and splinters of flint.

The Middle Paleolithic, known generally as the *Mousterian*, was the time of the Neanderthal man and coincided with the third interglacial stage and the beginning of the fourth, or last, glacial stage. The Neanderthal man was the first to attach flint points to his spears, and he also made a variety of chipped scrapers and cutters.

The last phase of the Paleolithic is distinguished by five cultural stages, all representing advances made by modern-type men. These stages, in order of age, are called *Lower Perigordian, Aurignacian, Upper Perigordian, Solutrean,* and *Magdalenian.* During the Late Paleolithic, man gradually improved his techniques of flaking siliceous rock. The knife appeared by the Lower Perigordian; awls and other objects of bone and antler came into use in the Aurignacian; more improvements during the Solutrean resulted in delicate and beautiful points of many kinds; and in the Magdalenian, awls, needles with eyes, saws, wedges, and harpoons were made from a variety of materials. By the end of the Paleolithic, man had invented all his basic tools.

The Mesolithic was a short interval of transition in cultural development that immediately followed the last glacial advance. During the Neolithic, or most recent part of the Stone Age, man began shaping his stone tools by grinding and polishing rather than by flaking. This period lasted until metals came into use, and its beginning and end were not everywhere simultaneous.

In summary, anthropologists use several methods to date or designate the various stages of human history. Direct absolute dating can be accomplished if suitable geologic materials exist. The *paleontological method* is based primarily on associated fossil animals or plants. The *stratigraphic method* relies on the relationship between successive glacial and interglacial stages and their sediments. If based on tools, the method is *typological.* Of course, the investigator must

Figure 16.8 Dryopithecus (Proconsul) major; the upper jaw with most of the teeth. This specimen from Miocene deposits of eastern Uganda is from 12 to 14 million years old according to radiometric dating. The large canine teeth are diagnostic of an ape grade of evolution. (Wenner-Gren Foundation.)

take all these methods into account and check them against one another to achieve the most reliable results.

History of the Primates Resumed

The history of fossil primates may now be resumed by reviewing what is known about this group in the Miocene Epoch. First to be noted is a troublesome gap of perhaps 10 million years between the latest specimens from the Fayum locality in Egypt and those from later deposits. Although a few specimens are known from the Miocene of Europe, the really abundant and significant contributions come from Africa and southern Asia. A widespread form known as *Pliopithecus,* dated at about 20 million years, seems to represent a line leading to modern gibbons. More important is *Dryopithecus,* first described in 1856, and now known from many specimens found at scattered localities across southern Europe and Asia. Experts have recently concluded that the *Dryopithecus* line probably leads to man as well as to apes (except the gibbon). Another well-known form named *Proconsul*, represented by many specimens from Lake Victoria in eastern Africa, is so close to *Dryopithecus* that some authorities consider them to belong to the same genus. In any event there is ample evidence for a widely distributed group of variable and progressive Miocene primates combining monkeylike, apelike, and even manlike characters.

RAMAPITHECUS—
THE FIRST HOMINID

One of the important distinctions that must be made in the study of primates is that between apes and men. In terms of the fossil record this amounts to distinguishing the earliest creature that is more manlike than apelike, a creature that can confidently be regarded as an ancestor of all members of the hominid family and none of the apes. A satisfactory candidate for this important position exists in the form of *Ramapithecus*, a primate that is known from jaw fragments and teeth discovered in India and Kenya. Specimens of this important link had lain unappreciated in British and Indian museums for many years until properly identified and correlated with more recent discoveries. The remains, though fragmentary, show a short face and dentition that is humanlike in total configuration and individual tooth structure. The canine teeth are small and low, not high and self-sharpening as in the apes. This suggests a side to side chewing motion

Figure 16.9 *Reconstruction of the ancestral ape Proconsul in its native habitat in Africa some 20 million years ago. (By Permission of the Trustees of the British Museum, Natural History.)*

Figure 16.10 *Molar teeth of Ramapithecus show humanlike characteristics. Other skull fragments are known. (Yale Peabody Museum.)*

Figure 16.11 *Australopithecus africanus trans-vaalensis, discovered at Sterkfontein, South Africa. This skull furnished the first certain proof of the small brain size of the australopithecines. Its probable age is in excess of 2 million years. (Wenner-Gren Foundation.)*

rather than up and down chomping. The order of succession of the teeth is also manlike and suggests a long adolescent period for maturing and learning by associating with older, more experienced animals.

There can now be no doubt that the most significant steps in the evolution of man took place in Africa. Africa has contributed many important material clues to the origin of man, in addition to the finds of monkeys and near-apes in the Oligocene of Egypt and the Miocene of Tanganyika (now Tanzania). Fossils of manlike creatures resembling nothing now living or previously known began to turn up in South Africa in 1924. These have been variously termed "ape-men," "man-apes," or "half men," but the name *australopithecines,* or "southern apes," has become well established. The South African fossils have come mainly from limy, hot-spring deposits or from cave fillings being mined for limestone. Over three hundred individual finds are recorded from South African sources.

Additional discoveries of australopithecines have been made in East Africa, thus increasing the number of African sites to at least 9. Additional finds that resemble, and may eventually prove to be, australopithecines have turned up in mainland China, Java, and northern Israel. Although many remains are found isolated in or upon older rocks in such a way that it is difficult to establish their age, many of the newer finds are in stratified deposits where they can be dated

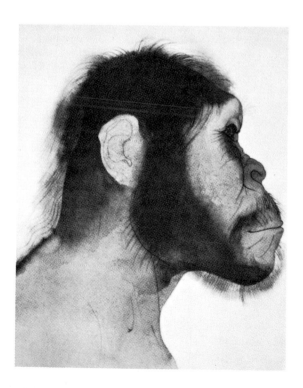

Figure 16.12 *Reconstruction of Australopithecus, a hominid that lived in East Africa during a period of several million years before the appearance of true man. There seems to have been much variation among australopithecines from place to place and with the passage of time. (By permission of the Trustees of the British Museum, Natural History.)*

Figure 16.13 Australopithecus in the environment of southern Africa during early Pleistocene time. (By permission of the Trustees of the British Museum, Natural History.)

The head, however, was not very manlike; the brain was small (average about 500 cubic centimeters), the brow sloping, the heavy lower jaw receding, and the teeth projecting. A covering of hair is assumed by those who have attempted restorations.

The ever-growing number of australopithecine remains reveals a great deal of diversity and variation, and students understandably are taking a very cautious attitude in defining just what should and should not be included in the group. Some subgroups evidently made tools of stone; others did not. Some were exclusively herbivorous; others were carnivorous or omnivorous. If tooth characteristics are any guide, the carnivorous stock appears to have been more successful. All these varieties are to be expected in any adaptable group that had spread widely into a number of environments. We may assume that some freedom of gene flow was possible over wide areas and through long time periods. Conversely, as an inevitable outcome of dispersal and isolation, subspecies and eventually new species arose from the common stock.

The larger of the two australopithecines, Paranthropus, may have evolved into the giant form Gigantopithecus. Several types of uncertain status, perhaps weighing up to 270 kilograms and standing nearly 3 meters tall, have been known since 1935. Uncertainties arise from the fact that only teeth and partial lower jaws are known.

Australopithecines are certainly very manlike, but students do not include them in the genus Homo. The common ancestor has not yet been discovered or identified. Various species, including Australopithecus boisei (formerly Zinjanthropus), Australopithecus robustus and Australopithecus africanus, are representatives of a sterile side branch that became extinct in the middle Pleistocene. More important than the australopithecine remains are those of a more manlike group that many authorities consider to be true men, genus Homo. The first specimens of what was designated Homo habilis were found by Dr. and Mrs. Louis S. B. Leakey at Olduvai Gorge in the same deposits

by associated mammals as late Pliocene to middle Pleistocene in age.

From their relatively abundant remains, a fairly satisfactory picture of what the australopithecines looked like has emerged. In the first place there appears to have been two distinct genera, a smaller one, Australopithicus, and a larger, heavier one, Paranthropus. Australopithicus was a rather small animal, between 27 and 40 kilograms in weight and about 1 meter in height. It was undoubtedly bipedal, all requirements for the erect posture having been acquired late in the Pliocene or early in the Pleistocene.

Figure 16.14 Skulls contrasted. On the right is Australopithecus, on the left a more advanced creature
considered by some to be a man (Homo habilis) and
by others to be another type of australopithecine.
Regardless of classification, the two are known to
have lived contemporaneously at the Olduvai site.
Note the high crest of the skull on the right and the
smooth, more manlike configuration of the one on
the left. The former is also wider at the base,
whereas the latter is wider higher up. The scale is
given in inches (top) and centimeters (bottom). (Photograph by Mary Leakey, copyright National Geographic Society.)

as *Australopithecus boisei*. Many remains are
now known from the Olduvai, Omo, and east
Lake Rudolf localities. The skull is smooth,
with no median ridge, and is widest toward the
top. This creature definitely made tools and
built crude shelters in the open. Contrary to
previous opinions, it is now believed that *Australopithecus* and *Homo* coexisted for several
million years and that one did not give rise to
the other. As is usual with important and critical finds of this nature, there is professional
disagreement as to proper classification and importance. We may confidently expect a long
period of exploration and evaluation in the study
of ancient man in Africa.

ENVIRONMENT OF THE FIRST MEN

At this point a description of the Olduvai Gorge
and Lake Rudolf localities is in order. These rank
among the great fossil repositories of the world,
having yielded thousands of specimens of fish,
reptiles, birds, and mammals. Most important is
the production of a great number and variety of
hominid teeth, bones, and artifacts. Olduvai Gorge
is a 40-kilometer-long drainageway that empties
into the Bilbal Depression in the Great Rift Valley
of northern Tanzania. Although only 90 meters
deep it slices through a succession of abundantly
fossiliferous sediments that contain a record of
over 2 million years of evolution.

The sediments now exposed were laid down
within, and marginal to, a long succession of
shallow lakes that expanded and contracted with
the oscillating climates of the great Ice Age. In-
terspersed are layers of tuff and ash erupted from
nearby volcanoes. These rocks provide minerals
from which radiometric dates have been obtained. The exposed sequence begins with a lava
flow dated at about 2 million years; a bed of
volcanic ash within the overlying fossil-bearing
strata is dated at 1.8 million years old. Fossils
were first found in Olduvai Gorge in 1911. Later,
in 1931, serious research and excavations were

begun by Leakey and have continued under his direction ever since.

Game ranging from fish to elephants has abounded in this area for millions of years and the incentive to hunt and compete in such an animal paradise must be reckoned as important in the early history of mankind.

Other East African localities in addition to Olduvai Gorge are coming into prominence. An international group of American, French, and Kenyan scientists has worked for several years in sediments along the Omo River in remote southwestern Ethiopia. They have discovered many levels with fossil hominids, including both *Homo* and australopithecines. The Omo beds represent almost continuous deposition during an interval from slightly more than 4 million years to about 1.8 million years. The locality is exceptional in the thickness and continuity of its strata and obviously records essential information of greater antiquity than that of Olduvai Gorge.

Another locality on the east side of Lake Rudolf in northern Kenya is being investigated by Richard Leakey, son of Louis Leakey. The extremely fossiliferous beds are similar to those of the Omo area. Among the important discoveries are specimens identified as *Homo,* together with undeniable associated stone artifacts. Although greatest interest centers on hominid remains, all these East African localities are yielding specimens from which the evolution of other mammals can be reconstructed. Chief of these lineages are those of pigs, elephants, antelopes, rhinoceroses, and other plains-living mammals. Research on all these important matters will require an immense amount of labor and the full value will not be known for many years.

Homo Erectus—
Man of the Early Ice Age

Remains of the most famous of all fossils, the so-called Java man, were collected near Trinal, Java, by Eugene Dubois, a Dutch physician, in

Figure 16.15 Mrs. Louis S. B. Leakey directs operations at one of the excavations at Olduvai Gorge. (Courtesy of George H. Hansen.)

Figure 16.16 Richard Leakey examines a skull of Australopithecus on the spot of its discovery, northeast of Lake Rudolf, Kenya. (Copyright National Geographic Society.)

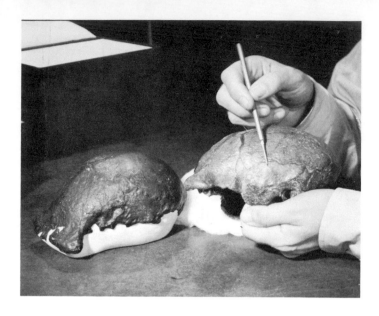

Figure 16.17 · Comparison of Pithecanthropus skulls. A replica of the skull discovered by Dubois in 1891 (left) is less complete than the specimen found by von Koenigswald in 1936. (American Museum of Natural History.)

the period from 1890 to 1893. Originally named *Pithecanthropus erectus* by Dubois, who believed it to be a link between men and apes, this species is now accepted as being entirely manlike and is consequently referred to as *Homo erectus*.

The original discovery in Java consisted of a skullcap, a thighbone, and a number of teeth. In the same deposit, Dubois found fossils of extinct elephants, rhinoceroses, tapirs, pigs, hippopotamuses, antelope, deer, and other mammals. The debate that arose over these finds became so fierce that Dubois, thoroughly discouraged and embittered, locked up his specimens and refused to allow them to be seen again until 1923. He believed that he had discovered a connecting link between apes and man, but he regarded *Pithecanthropus* as being more apelike than manlike.

Since the original discovery, we have learned a great deal more about *Pithecanthropus*. This new information is a monument to the energy and persistence of G. H. R. von Koenigswald, a

Figure 16.18   The banks of the Solo River, Java, where the first discovery of the Java man (Homo erectus) was made in 1890. (Courtesy of George H. Hansen.)

German-born paleontologist. In 1936, working in the Sangiran district of Java, von Koenigswald found a part of a lower jaw with four teeth, and at another site a brain case with the basal parts intact. In 1938 another skull fragment was found, and in 1939 still another with the upper jaw preserved. In addition to these finds, several leg bones that may belong to *Pithecanthropus* came to light. From these remains, anthropologists have been able to reconstruct a satisfactory head of *Pithecanthropus*. The skull is broad in propor-

Figure 16.19 Reconstruction of Homo erectus. (*By permission of the Trustees of the British Museum, Natural History.*)

tion to its length and has a rounded back. The cranial capacity is 700 to 800 cubic centimeters, the brain case is low, the forehead flat, and the eyebrows heavy and protruding. The upper jaw is very large and protrudes forward; the lower jaw is also heavy and lacks a chin. The teeth are large, but the canines project only slightly beyond the level of the other teeth.

There is no evidence that *Pithecanthropus* used fire, and no artifacts can be directly associated with his remains. Nevertheless, one of the skulls shows a wide cleft at the back that looks suspiciously as though it had been made by a stone ax. In the opinion of recent authorities, *Pithecanthropus* was indeed a man and they would prefer to call him *Homo erectus*. He is, of course, quite different from all modern men in brain capacity, in brow and chin formation, and probably in many other ways. Paleontologists generally believe that the Java man lived during the early Pleistocene, but still somewhat later than the australopithecines.

PEKING MAN

Early in the twentieth century, a number of paleontologists became convinced that ancient men may have lived in eastern Asia. They knew that for many centuries the Chinese had been grinding up fossil bones to use in medicines and drugs. A few fossil teeth found in apothecary shops furnished the first clues that ancient men had indeed left remains in China. Some of these so-called "dragon bones" were traced to certain hills near the village of Chou-kou-tien, about 50 kilometers southwest of Peking. Collecting began in this area as early as 1918, and in 1927 full-scale excavations began under the joint efforts of Chinese, American, and Swedish agencies.

The excavations soon revealed that the site was an ancient camping ground where the refuse of human occupancy had accumulated to great depths. Charcoal layers were uncovered along with very primitive stone implements and the burned bones of many animals, including deer, elephants, saber-toothed cats, and giant rodents. Most important were the human bones that eventually were unearthed. Remains of about fifty individuals were recovered—skulls, lower jaws, limb bones, and many teeth belonging to individuals of different ages and of both sexes. Geologists assume that these ancients practiced headhunting and cannibalism because an unusually large number of skulls have been found, basal parts broken open.

The *Peking man,* as he came to be called, received the name *Sinanthropus pekinensis* (Chinese man from Peking). Careful comparisons have revealed that Peking man and Java man were practically identical and that they must have lived at about the same time. Geologists and paleontologists have largely abandoned the name *Sinanthropus* and refer to the Peking and Java men together as *Homo erectus.* In any event, a primitive type of man did inhabit eastern Asia and nearby areas during the early Pleistocene. He was the first known user of fire, perhaps a consequence of his life in more northerly regions.

Unfortunately, many of the remains taken from Chou-kou-tien were lost during World War II, but they had been completely measured, illustrated, and described before the outbreak of hostilities, so all the essential information about them is preserved. The Chinese have reopened the site, and a number of additional specimens have been recovered.

OTHER PITHECANTHROPINES

Among the many discoveries made by Leakey was what he called *Chellean man.* The remains include an excellently preserved braincase with thick walls and massive brows. A number of stone tools typical of the Chellean cultural stage, which has been known since 1846, were also found by Leakey. Some regard Chellean man as a pithecanthropine; others, including Leakey, would consider him somewhat more advanced. A final classification has not yet appeared.

Figure 16.20 *The Rhodesian man, discovered in Northern Rhodesia (now Zambia) in 1921. Some consider this fossil to pertain to Homo erectus, whereas others ascribe it to Homo sapiens. (Wenner-Gren Foundation.)*

Two other African finds are considered to belong to *Homo erectus: Rhodesian man,* from Broken Hill in Zambia, and *Saldanha man,* from Hopefield in South Africa. Both have very heavy browridges, and for this reason have been considered to be pithecanthropines. These creatures could be in the age range from the middle Pleistocene to the early part of the late Pleistocene.

Other pithecanthropine localities are known in Africa. At Ternifine, Algeria, great accumulations of Chellean implements and three human jaws have been found. Although the name *Atlanthropus* has been applied, the remains are considered by experts as definitely *Homo erectus.* From Ambrona Valley in Spain come many tools together with bones of mammoths obviously killed by human hunters. These remains are judged to be pithecanthropine in origin even though no actual skeletal remains have yet come to light.

An important recent discovery of *Homo erectus* has been made on the island of Formosa. Man evidently reached this site from the Chinese mainland during a period of low sea level during the last Ice Age. These scattered remains prove the existence in Europe, Asia, and Africa of a widespread race of humans distinctly unlike those of modern species.

Antecedents of Modern Man

Human remains are extremely rare in the early and middle Pleistocene of Europe, but those that have been found are of paramount importance. European climates were probably too cold for any primates who did not have the intelligence to invent and use special means of obtaining food and protecting themselves against the weather.

Figure 16.21   Skull fragments of Swanscombe man. The large brain and smooth skull leave no doubt that the remains pertain to Homo sapiens. *(Wenner-Gren Foundation.)*

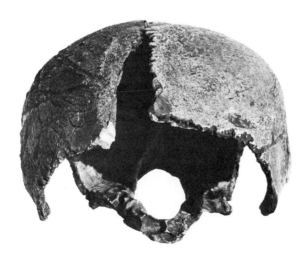

A large, heavy jaw with sixteen teeth intact, discovered near Heidelberg, Germany, in 1907, is a most significant fossil. The age is judged to be late middle Pleistocene, about 500,000 years ago. The absence of other parts of the skull or skeleton makes judgments difficult but *Heidelberg man* is surely somewhere along the route between *Homo erectus* and *Homo sapiens*. He was more advanced than contemporaries in Java and may be the earliest known human of Europe.

Also from Germany is *Steinheim man,* which is represented by the facial regions of a skull found near Stuttgart in 1933. The age is the second interglacial, between 200,000 and 300,000 years ago. Deposits of the same general age at Kent, near London, have yielded another significant individual known as *Swanscombe man.* Discoveries picked up over a period of several years were assembled to constitute the top and back parts of a skull. The brain capacity is estimated at 1,350 cubic centimeters, and many well-formed hand axes are associated with the remains. Swanscombe man was buried with remains of a variety of mammals and the placement in the Ice Age chronology is considered to be very good.

A number of fragmentary skulls and miscellaneous skeletal parts from Java, known collectively as *Solo man,* are considered by some authorities as being intermediate between *Homo erectus* and *Homo sapiens.* Skillfully formed artifacts are also found. Chinese scientists have reported in a preliminary way their discovery of remains intermediate between *Homo erectus* and *Homo sapiens.* Some of these have Neanderthaloid characteristics (see next section), but others resemble *Homo sapiens* and the European forms.

These fragmentary remains from western Europe, China, and Java are about all the described remains we now have of the immediate forerunners of modern man. All that can safely be said is that they differ on one hand from *Homo erectus* in having larger brains and less conspicuous browridges and from Neanderthal man on

Figure 16.22 Reconstruction of Swanscombe man, an inhabitant of the Thames valley during the second interglacial interval. (By permission of the Trustees of the British Museum, Natural History.)

the other hand in having smaller brains and more prominent browridges. It is only fair to state that a number of undescribed finds are in the possession of various students who are being understandably cautious about making definitive pronouncements about their finds, which, after all, may represent that much debated and mysterious "missing link" between man and the lower animals.

Neanderthal Man

Second in interest only to *Homo sapiens sapiens*, or modern man, is the *Neanderthal man, Homo sapiens neanderthalensis*. His skeletal remains are relatively abundant, and we know a great

deal about the implements and weapons he used. Many paintings and reconstructions have been made of him, and these renditions have given rise to the popular notion of a crude, hairy, cave-dwelling race that was somehow pushed aside or superseded in the struggle for existence. Many theories have grown up around Neanderthal man, but we are still far from understanding his true evolutionary position in relation to ourselves. For a long time after his discovery, he was regarded as an ancestor of modern man or as a distinct evolutionary stage through which man had progressed. Later on, fossil specimens were discovered that combine the characteristics of modern man and Neanderthal man, suggesting hybrids that had arisen from a mingling of the two types. Still later, and most revealing of all, remains of modern man at least as old as, and perhaps even older than, Neanderthal man were found, thus proving that the two types had existed simultaneously over long periods and that

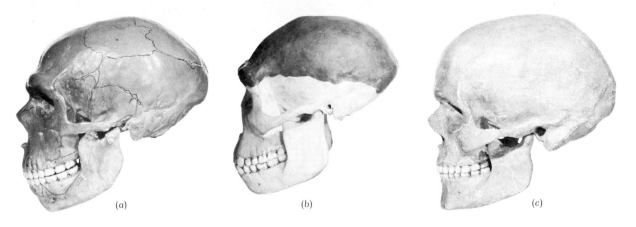

Figure 16.23 *Skull of Neanderthal man (a) compared with* Pithecanthropus *(b) and Cro-Magnon (c). (American Museum of Natural History.)*

Neanderthal man and modern man may have arisen from the same stock.

The first Neanderthal skull was found in 1848 in a cave in the Rock of Gibraltar. This specimen was laid aside and ignored for nearly twenty years until a more complete specimen was found in the Neander Valley near Dusseldorf, Germany. This second specimen was accidentally uncovered in 1856 by workmen clearing out a cave, and a large portion of the skeleton was recovered and preserved. No associated artifacts or other bones were found in the cave. A fierce debate soon arose over the Neander Valley remains. Some investigators held that the bones were the skeleton of an idiot or an individual afflicted with rickets. Others claimed that the skeleton was the remains of a normal being of a type different from modern man, and they gave it the name *Homo neanderthalensis.* The argument was settled in 1886 when additional specimens, particularly two well-preserved skeletons from the Spy Caverns of Belgium, were found mingled with mammal fossils of the glacial fauna. The Gibraltar skull was reexamined and recognized as Neanderthaloid in 1906.

Since 1906, many Neanderthal remains have been discovered. A recent catalog shows that authentic finds have been made in Germany, Belgium, France, Hungary, Italy, Romania, Gibraltar, Channel Islands, Malta, Iran, Russia, Israel, and Rhodesia. Seven almost complete skeletons were found in Shanidar Cave in the northern mountains of Iraq. It is calculated that Neanderthal man lived in this cave for 60,000 years. Many of these finds consist of more than one individual, and accompanying them are

Figure 16.24 *Reconstruction of a Neanderthal family group of about 50,000 years ago. (By permission of the Trustees, British Museum, Natural History.)*

abundant artifacts and animal remains. From these discoveries we learn that Neanderthal man was widespread during the third interglacial interval and that he hunted and killed a variety of animals, including the great cave bear, elk, deer, moose, horse, and the hairy mammoth. He used well-formed, stone-tipped spears and is responsible for the cultural evidences we call *Mousterian*. He also used fire and buried his dead.

Neanderthal man as known in western Europe was rough and brutish in appearance. His skeletal remains reveal very distinct features. He had a squat body with a barrel chest, bowed legs, and flatfeet, characteristics that suggest a slight slouch and a shuffling gait. He had a long, low skull with big, jutting browridges, a flat nose, a protruding, muzzlelike mouth, and a retreating chin; but his brain, as the cranial capacity of 1,200 to 1,600 cubic centimeters indicates, was just as large as, or larger than, modern man's. This description must be modified in connection with remains found outside western Europe. Neanderthal men of Africa and the Middle East show more traits of *Homo sapiens sapiens,* such as less prominent browridges and less rugged skull.

Modern Man

In spite of the many significant discoveries of the past few decades, the precise origin of modern man is still uncertain. It is true that some specimens once thought to be remains of modern man's ancestors have been eliminated from study, their places being taken by better specimens, but a direct line leading to *Homo sapiens sapiens* has eluded the scholars. The abruptness with which modern man replaced Neanderthal man about 32,000 years ago is startling. There is no evidence that the latter gave rise to the former or that there was a transition in culture. So far, the geographical source of modern man has not been located. As usual we need new information to settle this important question.

The first and best-known representative of *Homo sapiens sapiens* is Cro-Magnon man, who takes his name from a French rock-shelter where his remains were first discovered in 1868. Cro-Magnon remains are widespread in western and central Europe, and over a hundred individual specimens have been recovered. The race dates to the waning stages of the last glacial advance, from 10,000 to 37,000 years ago.

The average Cro-Magnon was rather large and sturdily built; many of the men were only slightly under 2 meters tall. There are no traces of Neanderthaloid characteristics in the typical Cro-Magnon skeleton. The forehead and skull vault are high, browridges are small, the nose narrow and prominent, the chin highly developed. His brain capacity averaged 1,700 to 1,800 cubic centimeters, probably more than the average of living men.

Cro-Magnon was progressive and adaptable. During his period of existence he developed at least five distinct successive cultures and became an expert at shaping stone tools and weapons. By 10,000 years ago he had learned to fashion bone and antler and made awls, saws, needles, and delicately fashioned weapons. He apparently did not learn to domesticate animals, but he was a mighty hunter, as bones of his prey and his cave art indicate. Use of fire and clothing enabled him to withstand the elements and to establish more or less permanent settlements.

Man in the New World

The history of primates in the New World is discontinuous and incomplete. North America, lying farther from the equator than any other large continent, became too cold for primates early in the Tertiary, and there are no known fossil remains of the group between the early Oligocene and the late Pleistocene. Even in South America, where we might expect the record to be more abundant, evolution seemingly did not progress past the monkey stage. As far as primates are concerned, the New World was effectively separated from the Old World until late in the Pleistocene, when modern man successfully migrated from one hemisphere to the other. Paleontologists generally agree that America's oldest human inhabitants probably reached North America from Asia via the Bering Strait area, which occasionally formed a dry-land bridge between the two continents.

Strangely, ice was more of a barrier than the sea. From about 17,000 to 4,000 years ago when the Bering bridge was dry land, impassable continental glaciers blocked Alaska. Man had to enter the heart of the continent before or after this period. There is no doubt about his presence in North America after the ice had melted to open a corridor down the Canadian plains or within the Rocky Mountains, but an earlier entry,

Figure 16.25 A site at Dordogne, central France, that was occupied by Cro-Magnon man. (Service Commercial Monuments Historiques.)

Figure 16.26 Discovery sites of important Old World fossil men and premen mentioned in the text.

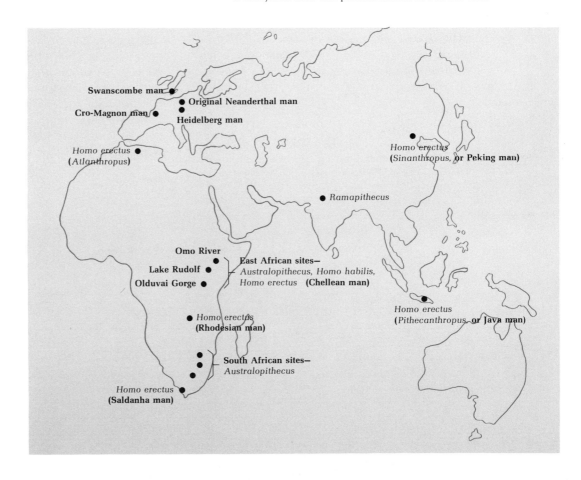

before the glacial barrier appeared, is so far very debatable.

In any event the history of man in the New World is shorter and less eventful than it is in the Old World. In the Americas we deal only with fully developed man, *Homo sapiens sapiens,* and not with his problematical ancestors or relatives. Paleontologists are most interested in the period beginning with the arrival of man and ending about 5,000 years ago. During this interval, ancient Americans lived in association with many large mammals that are now extinct, and pursued a nomadic existence governed by climatic conditions and the distribution of food sources. It seems unrealistic to call these early people Indians, for they may not have been ancestors of the inhabitants Columbus encountered in the fifteenth century. Following the anthropologists, we will call these earliest Americans *paleo-Indians,* or perhaps the *Big Game Hunters.*

The most common traces of ancient man in America, as every schoolboy knows, are stone arrowheads and spear points. But artifacts found on the surface are of little value in helping paleontologists compile a connected history of the people who made them. Only when artifacts are found directly associated with bones and skeletons do they begin to take on meaning. It is far more common to find stone weapons mingled with the remains of animals killed by man than it is to find them with the bones of man himself. We can sum up the situation by saying that stone artifacts of paleo-Indians are fairly common, that stone artifacts and bones of slaughtered animals are occasionally found together, and that artifacts, mammal bones, and human remains have been found together only in a very few places.

We shall mention here only a few of the generally accepted facts about paleo-Indians in America. We know they hunted a variety of now-extinct mammals, because many stone projectile points and bones of these animals have been found mingled together. Among the animals that we know were positively associated with early man in the Americas are several types of extinct bison, the ground sloth, the extinct horse, the mammoth, the camel, the mastodon, and various antelopes. There was a distinct tendency to concentrate on specific animals. At first the mammoth was the chief prey; later, when these were no longer available, the bison came under attack.

America's oldest human beings knew how to use fire, and their stone work is distinctively unlike anything that was done in Europe. They created a characteristic type of projectile point with wide, shallow grooves or channels on one

Figure 16.27 Folsom man hunted the giant bison by killing it with a spear-thrower or driving it over a cliff. (Courtesy of Alfred M. Bailey, Denver Museum of Natural History.)

or both faces. This artifact, best typified by the well-known *Folsom point,* is not the only distinctive American contribution, however, for other types of arrow and spear points, both older and younger, have been discovered.

The first discovery that positively linked ancient man with extinct animals was made at Folsom, New Mexico, in 1926. Here, Folsom points were found with the bones of extinct bison. Actually, these points have been found at various places throughout the United States, but they occur most often in the plains area east of the Rockies. Another important open site is in Lindenmeier, Colorado, where stone implements were found mixed with extinct bison and camels. Other discoveries have been made at Clovis, New Mexico, and Naco, Arizona, both yielding the remains of mammoths. Five localities with the Clovis type of artifacts have been dated at between 11,000 and 11,500 years old.

Cave deposits are naturally associated with mountains, and important cave discoveries have been made in the mountainous western United States. At Sandia Cave, Sandia Mountain, New Mexico, Folsom points and an older type known as the *Sandia point* have been found. Again, no human remains were present; but horse, camel, bison, mammoth, ground sloth, and wolf fossils were found in the Folsom layer, and horse, bison, mastodon, and mammoth remains in the Sandia layer. At Ventana Cave, 40 kilometers south of Phoenix, Arizona, extinct species of wolf, jaguar, ground sloth, horse, and tapir were found in association with crude artifacts. At Danger Cave near Wendover, Utah, five periods of human occupancy have been recorded, the oldest, according to radiocarbon dating, being about 11,000 years old. No human bones or extinct types of animals were found. Gypsum Cave, 25 kilometers east of Las Vegas, Nevada, reveals several levels

of human occupancy, and bones of sloths, camels, and horses are mingled with ancient artifacts of stone and wood.

Of course, paleontologists are more interested in the actual skeletal remains of early man than in his artifacts, but authentic human specimens are extremely rare. We shall mention here only a few of the more widely accepted discoveries. At Vero and Melbourne, Florida, human skeletal remains have been discovered scattered among the bones of fifty types of mammals, twenty-eight of which are extinct. Experts, however, are undecided whether or not the human bones are actually ancient or simply were buried in the older stratum.

An interesting and probably entirely authentic find of human skeletal remains was made in the Valley of Mexico, near the village of Tepexpan, in 1949. A great deal of the skeleton and skull were intact, and bones of mammoths and stone artifacts were found nearby. The exact age of the skeleton has not been positively determined, but authorities generally agree that it is more than 10,000 years old.

The most important North American discovery to date is the so-called *Marmes man* found

Figure 16.28 Discovery site of Marmes man, southeastern Washington. Cave entrance is in shadow on cliff face to left; excavated waste material is in center of photo. (Courtesy of James F. Miller.)

in a cave in southeastern Washington on the floodplain of the Palouse River near its confluence with the Snake River. Here for the first time a long-sought combination of evidence came to light—actual human remains, tools, bones of contemporary food animals, and material suitable for carbon-14 dating are found together. The remains are those of a youth and consist of a skullcap, a piece of a cheekbone, parts of a wrist, and other scraps. Charred bones of elk, antelope, deer, and rabbit suggest some of the food sources; fish and shells of river mussels also occur. Artifacts include a few stone tools and flakes and a finely made bone needle. Radiocarbon dates indicate that this important site is between 13,000 and 14,000 years old. The cave is now under water impounded by filling the lower Monumental Reservoir.

Figure 16.29 Reconstruction of Tepexpan man. Sculptor Leo Steppat of Washington, D.C., gave ancient Tepexpan man a prime-of-life appearance as shown in these progressive photos. Starting with a cast of the skull, he built up the missing front teeth, then added blocks representing tissue thickness as determined on cadavers. Next, he joined the blocks with strips of clay and filled in between the strips, always watching the underlying bone. Modeling of the eyes, nose, and mouth was guided by the bony structure. The rendition of the hair is entirely stylized. (Smithsonian Institution.)

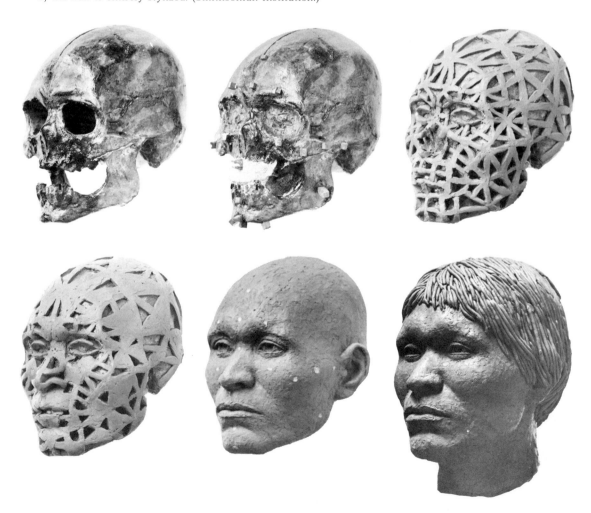

The oldest dated human remains of South America are a lower jaw with teeth found in Guitarrero Cave, northern Peru. Associated charcoal gives a radiocarbon date of about 12,580 years. Fortunately a variety of cultural remains including stone projectile points and artifacts of bone, wood, and fiber are associated. The site evidently was occupied by transient hunters over a long time period. An interesting locality is Fells Cave in Patagonia, where evidence of man is found associated with the ground sloth and guanaco. In Venezuela, man was associated with the glyptodont, horse, and mammoth. Man evidently made a relatively quick traverse of both North and South America once a route was open.

What Makes a Man?

Someone has said that man is a creature made by his enemies. This thought may be extended to include the prehuman stages of evolution also, for if evolution is true, there must have been on earth at all times since life began a lineage capable of producing man. This lineage changed with time and went through countless mutations as it progressed toward humanity. Every favorable genetic change that affected it, every advantage it gained through millions of years of time, must be counted as a contribution to the heritage of the human race. Thus, such characteristics as blood, bone, teeth, jaws, limbs, brain, and sense organs were part of the earlier acquisitions. These are shared with many other vertebrates, however, and are not peculiarly the possession of man. Of more significance are those changes that produced traits unique to man.

A fundamental step was the acquisition of *upright posture.* This came after an *arboreal stage,* during which hands and feet were of roughly equal importance in climbing and grasping. This tree stage had also favored the development of a *prehensile hand* and *flexible arm.* Other important arboreal adaptations are *stereoscopic, sharp-focusing eyes* with *color discrim-*

Figure 16.30 *Footprint of ancient man preserved in hardened sediment near Dilittepe, Turkey. This is one of several dozen footprints left by three persons. Shortly after their formation the prints were covered by a flow of basaltic lava. The age is estimated at about 250,000 years and falls in the latter part of the second glacial interval. Impression is about 28 centimeters long. (Mineral Research and Exploration Institute of Turkey.)*

ination. With these and other mammalian and primate traits already in his possession, the ancestral hominoid came out of the trees to begin an adventurous life of a different sort on the ground.

The descent from the trees may have been forced upon preman by too aggressive tree-living contemporaries, by a dying out of forests, or by the excessive, branch-breaking weight of his own body. In any event, what had been acquired in the trees was all to his advantage on the ground. And not to be overlooked is the *omnivorous diet,* a trait that has certainly allowed man and his ancestors to survive when certain foods were in short supply.

The erect posture was relatively easy to assume because there was no cumbersome tail to be balanced by an inclined or stooping forebody. The hands had become much too useful for feeding and fighting to be used merely for locomotion. One fact that emerges from the study of the australopithecines is that the body, limbs, hands, and feet had already become essentially manlike in the early Pleistocene; the head still was apelike or monkeylike.

Once the upright posture had been perfected so that the skull was more or less balanced on a short neck and a straight spine, the evolution of the brain accelerated. This was a relatively recent development, and probably the most important forward step of all. The growth of the brain was rapid and unprecedented. Table 16.1 shows the sequence of cranial capacities.

Table 16.1 *Cranial Capacity—Ape to Modern Man*

	Cranial Capacity in Cubic Centimeters
Chimpanzee and gorilla	325–650
Australopithecines	450–750
Pithecanthropines	750–1,200
Neanderthal man	1,100–1,550
Modern man	950–2,100

The forepart of the brain expanded and pushed forward so that the cranium began to dominate over the facial portion of the skull. The jaws, teeth, and nose retreated, and even the cheekbones and browridges were subdued. These changes came with food that required less powerful chewing muscles and smaller teeth. The trend was certainly accentuated with the control of fire, the cooking of grains and meat, and the invention of tools with cutting edges, all of which take over tasks previously accomplished by unaided teeth and digestive organs.

The foregoing brief list of biological adaptations and changes would be incomplete without at least mention of certain social and cultural forces that gave man his final status. Speech is certainly a powerful influence, permitting the communication of ideas and transmittal of traditions. Its origin and refinement cannot be traced from fossils, but it is suggested that the development of a chin may signify increasing ability to speak.

Hominoids, including man, tend to be gregarious. The unit is not the individual but the population. The ability to cooperate, to lead and to follow, to act in unison, to share and adjust were essential to survival during the perilous days of numerous hostile contemporaries, food shortages, and rigorous climates.

Several other social traits were certainly significant. Human sexual urges are such that young may be conceived at any season, so there is no distinct time when young are born. In the population group there is a complete gradation from young to old; all live and travel together. One season's offspring are not pushed aside by the next. The young are protected, fed, and taught as long as need be. Parental care must rank high as a trait that favors survival.

It would be foolish to say that man's evolution has ceased, but it seems likely that future significant changes will not be in physical traits. Improvements on the mental and social planes are of more importance to racial survival than mere alterations in the physical body.

A SUMMARY STATEMENT

The appearance of man was a climactic event in the history of the earth. Viewed in the light of his biological relations, man is a *vertebrate*, a *mammal*, and a *primate*; but his superior mental powers place him in a uniquely powerful position with respect to other forms of life. Geology and paleontology contribute to an understanding of man by revealing something about his past and possible origin.

It is certain that erect, two-legged primates (australopithecines and *Homo habilis*) had appeared in Africa by the early Pleistocene. During successive glacial and interglacial stages, more advanced hominids appeared and spread widely through the Old World. *Homo erectus* (the Java ape-man and related forms) inhabited southeast Asia and parts of Africa in the middle Pleistocene. Europe has yielded a number of hominids, including *Heidelberg man, Steinheim man,* and *Swanscombe man,* from the middle Pleistocene. Relationships of these forms are obscure but they are near the ancestral line of *Homo sapiens.*

Homo sapiens neanderthalensis left remains in Europe, Asia, and Africa, and seems to have been well adapted to glacial conditions. *Homo sapiens sapiens,* or modern man, appeared in the second interglacial stage and was contemporary with, rather than subsequent to, Neanderthal man. Cro-Magnon man is the best-known fossil representative of *Homo sapiens sapiens.* Cro-Magnon peoples were progressive and made many of the fundamental discoveries and inventions that existed when history dawned.

Man colonized the New World at a relatively late date. In the Americas he lived with and hunted a variety of gigantic mammals that are now extinct.

FOR ADDITIONAL READING

Buettner-Jannsch, J., *Origins of Man: Physical Anthropology.* New York: Wiley, 1966.

Clark, W. E. L., *History of the Primates.* London: British Museum (Natural History), 1950.

———. *The Fossil Evidence for Human Evolution.* Chicago: University of Chicago Press, 1957.

Cornwall, I. W., *The World of Ancient Man.* New York: Day, 1964.

Dobzhansky, Theodosius, *Mankind Emerging, the Evolution of the Human Species.* New Haven: Yale University Press, 1962.

Greene, J. C., *The Death of Adam.* Ames, Iowa: Iowa State University Press, 1959.

Howell, F. Clark, and François Bourliere, *African Ecology and Human Evolution.* Chicago: Aldine, 1963.

Howells, W. W., *Early Man.* New York: Time Inc., 1967.

Leakey, Louis S. B., *et al., Olduvai Gorge, 1951–61,* Vol. 1, *A Preliminary Report on the Geology and Fauna.* New York: Cambridge University Press, 1965.

Oakley, K., *Frameworks for Dating Fossil Man.* Chicago: Aldine, 1964.

Wormington, H. M., *Ancient Man in North America.* Denver: Denver Museum of Natural History, 1957.

SURVIVAL BY EVOLUTION

17

The word "evolution" has appeared a great many times in foregoing chapters with the assumption that the student has an understanding of what the term implies. It is worth emphasizing that evolution in its very general sense means the opening, unrolling, unfolding, development, or emergence of anything from a lower, simpler, or worse to a higher, more complex, or better state and that organic evolution specifically designates the derivation of all animals and plants from preexisting ancestral species.

In this chapter we shall take a fairly close look at evolution as a concept and at those factors that influence the evolutionary process.

Darwinian Evolution

The most widely accepted explanation of how organic evolution operates is that of Charles Darwin. His theory has been briefly epitomized in such terms as "a struggle for existence," "descent with modification," and "survival of the fittest." All these phrases reveal something of the theory, but the essence is *natural selection*, meaning that living things without exception are exposed to natural environmental forces that eliminate some and preserve others according to their fitness or ability to survive.

Charles Darwin's great work *On the Origin of Species by Means of Natural Selection, or the Preservation of Favoured Races in the Struggle for Life,* published in 1859, is a classic of science and is said to be the most influential nontheological book ever written. Natural selection has been accepted as the best explanation of what has been learned about the interrelationships not only among present life forms but also among all life forms of the past. A brief review of what is known about prehistoric plants and animals has been outlined in preceding chapters of this book. It is now appropriate to consider whether these facts fit the Darwinian theory and whether or not we may go beyond the original theory to other worthwhile concepts.

As we saw in Chapter 1, a hypothesis is merely a proposition or principle put forth without firm evidence from observation or experiment, whereas a theory is a firm explanation grounded on further observation, facts, and experiments. Darwin had already developed the concept of natural selection from a hypothesis to a theory before he wrote his book, but whether or not Darwin's theory of evolution by natural selection is a fact depends on the tests one wants to apply. To those who demand eyewitness or documentary substantiation it must remain a theory—such proof is impossible. Those who are satisfied with abundant circumstantial evidence consider the theory to be a proved fact.

Darwin's theory is immensely successful because of the many ways it can be tested. Investigators have been collecting data for over a century to refute or support it. The search eventually became world wide and extends backward in time by way of fossils into the remote past. Every aspect of life science has been involved; every group of organisms, no matter how obscure or how prominent, has been under investigation. Even man is not immune from scrutiny, and this, as everyone knows, is where the theory ran into its greatest obstacles.

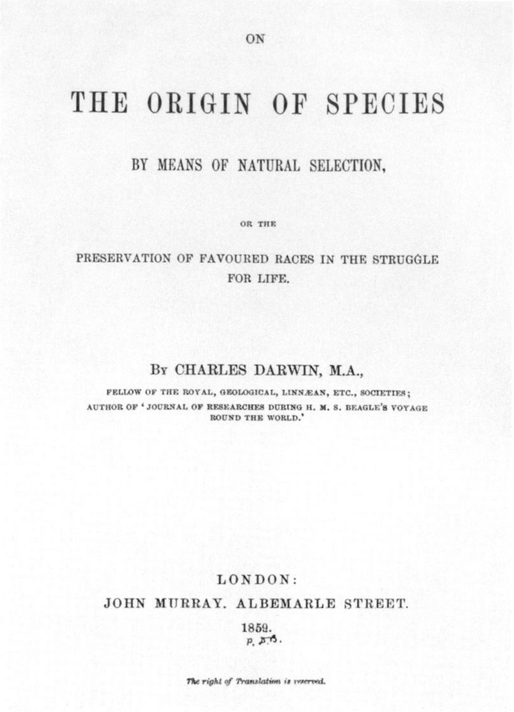

BASIS OF DARWIN'S THEORY

Darwin derived his theory of natural selection from three great facts of nature: overproduction, variation, and constancy of species. It is a matter

ON

THE ORIGIN OF SPECIES

BY MEANS OF NATURAL SELECTION,

OR THE

PRESERVATION OF FAVOURED RACES IN THE STRUGGLE FOR LIFE.

By CHARLES DARWIN, M.A.,

FELLOW OF THE ROYAL, GEOLOGICAL, LINNÆAN, ETC., SOCIETIES;
AUTHOR OF 'JOURNAL OF RESEARCHES DURING H. M. S. BEAGLE'S VOYAGE ROUND THE WORLD.'

LONDON:
JOHN MURRAY, ALBEMARLE STREET.
1859.
p. 176.

The right of Translation is reserved.

Figure 17.1 Title page of the first edition of The Origin of Species. (*Rare Book Division, The New York Public Library, Astor, Lenox, and Tilden Foundations.*)

of everyday observations that normal reproductive processes result in many more potential individuals (spores, eggs, seeds, and embryos) than can possibly become adults—there is, in a word, a great overproduction and tendency toward overpopulation. A single corn plant can produce 18 million pollen grains; the fern *Marattia* can produce 30 million spores in one season. A single fish can spawn as many as 120 million eggs, and a female frog lays about 20,000 eggs every year. Darwin observed that the descendants of a single pair of elephants, if all survived to their natural age limits, would total 19 million individuals in only 750 years. This immense potential overtaxing of environmental resources at all levels becomes an actuality as growth continues. Struggle and death are difficult to reconcile with any concept except that of natural selection. There is, beyond question, a struggle for existence, and this is a problem for evolutionists as well as for anti-evolutionists.

It is also well known that individual organisms vary—each is different in some way or in some degree from all others. Living things differ from their parents and from their offspring, and there are appreciable differences among offspring of the same parents. True, "life begets life," but never exactly. This being so, the struggle for existence that arises from overproduction is not among equals. Some individuals have an advantage over their fellows in being larger, stronger, or more intelligent. Those best suited for survival, everything else being equal, will live to carry on the race. This is the heart of Darwin's theory.

In spite of great overproduction and intensive competition at all levels the result has not been a chaotic proliferation of types of living things. Organisms can still be sorted out into distinct groupings (taxa; singular, taxon) according to their presumed natural relationships. Darwin accepted these evidences of relationship in terms of literal descent but was never exactly clear on how relationships are maintained, how differences and similarities are transmitted from one generation to another, and how one generation might influence the next.

Darwin fully realized and often deplored this lack of understanding of heredity and variation. The work of Gregor Mendel, pioneer student of genetics, was unknown when Darwin was writing The Origin of Species. Mendel's work was published obscurely in 1865 and was not "rediscovered" and verified until 1900. The mechanism of inheritance as understood today was not even guessed at by Darwin in spite of the fact that it was a cornerstone of his theory. Until about the middle of the nineteenth century, many biologists held that the male sperm contains a minute replica of the adult; in the case of man this is the homunculus, or "little man." The process of development is merely the enlargement of the individual in the sperm.

In his book Variation of Animals and Plants under Domestication (1868), Darwin developed the theory of pangenesis which takes a different view of reproduction. Darwin theorized that the cells of the body throw off minute particles, or gemmules, and that these were gathered from all parts of the body into the sexual cells. Thus, all characteristics came from the body tissue, not from previous germ cells, and are passed on to descendants without much possibility of change. This concept provides no obvious answer to the question of why offspring appear to show varying degrees of resemblance to their parents or even possess entirely new characteristics. Certainly this theory was of little help in explaining mutations.

We now know that germ cells come from germ cells, not from body cells, and that all permanent transmittable changes must take place in these cells. The idea that acquired characteristics can be inherited is no longer tenable.

Because of its inability to explain how inheritance was accomplished, the theory of natural selection was criticized as being essentially negative. It could account for the success and survival of some forms and the extinction of others; it could explain the "survival" but not the "arrival" of the fittest. Such criticisms have largely disappeared because modern genetics has contributed a mass of evidence on the nature of genes and chromosomes that shows with great exactness how inheritance is possible. It is now known that genetic mutation, spontaneous changes in chromosome structure and number, and genetic recombinations are capable of bringing into existence new and novel organisms. This information would have been most welcome to Darwin.

DARWIN AND THE FOSSIL RECORD

More pertinent to the subject of prehistoric life are anti-evolution arguments based on the fossil record. It is pointed out that many essential links are missing. In fact, in the view of some critics, it is not just occasional links but entire chains that are missing. On this point Darwin said: "The crust of the earth must not be looked at as a well-filled museum, but as a poor collection made at hazard."

How a person reacts to this situation undoubtedly depends on his mental frame of reference. On one extreme are those who regard the fossil record as so incomplete and imperfect as to be meaningless and unreliable. These critics naturally tend to emphasize areas where the evidence is weakest. Thus birds, bats, insects, and flowering plants are singled out as having essentially no fossil background. They take this to mean that these organisms actually had no developmental history, no ancestors, no gradual evolution, no continual lineage, and no connecting links. Therefore, in their view, evolution loses by default and special creation must be admitted. Critics of the geologic record say little with regard to the family histories of groups that leave more abundant remains such as elephants, horses, rhinoceroses, and oysters. But even here, those who are looking for gaps and missing links will always find them.

Those who are inclined to view the fossil record in a more optimistic way will emphasize its completeness rather than its incompleteness. Past successes encourage them to believe that links are not missing because they never existed but rather because they were not preserved or have not yet been found. Evolutionists are inclined to explain the failure of the record by appealing to everyday observations that prove that of the millions of individual living things that exist today only a few will leave their remains in a permanent repository. Zoologists know that such delicate organisms as birds, bats, and insects only rarely leave intact remains.

Has Darwin's case been strengthened or weakened by fossils discovered since his time? The question in this form is absurd and the answer is obvious. Since Darwin's day so many important links have been found that it would

be impossible to list them all. Ancestral plants and animals have been discovered in the Precambrian, a period that once was thought to be lifeless; early chordates have been found in the Ordovician; an advanced amphibian (*Ichthyostega*) clearly linked to previous fish has been discovered. Several good links between amphibians and reptiles have come to light, and an almost perfect transition form from reptile to bird was found (*Archaeopteryx*). So many repto-mammals or mammal-like reptiles are now known that it is difficult to designate any one form as a link. Darwin didn't know about *Hyracotherium*, the ancestral horse, and only a few dinosaurs had been named. Even in South America, where Darwin first saw the bones of large vertebrates in the rocks, the record has increased at least tenfold and has provided an excellent record of South American mammals.

Most important of all are the finds that bear on the origin of man. Darwin had not examined a single specimen of prehistoric man; only two were then known and but one of these, Neanderthal man, had been described. There are now literally thousands of significant specimens of prehistoric men and lower primates. These specimens include perfect skulls and complete skeletons. Although it is not yet possible to arrange these in orderly "family trees," they do prove the existence of many species that are important steps in the evolutionary panorama.

One fact emerges: no matter what specific forms may yet be revealed by the searchings of paleontologists *the fossil record must always improve.* It would be foolish to assert that links now missing will never be found, and those who try to belittle the contribution that fossils make to an understanding of evolution will face ever-increasing difficulties.

Although a century of investigation has revealed a great number of weaknesses in Darwin's theory, the central idea that natural selection is the guiding principle in evolution remains as powerful as ever. Critics of natural selection as such insist that the principle is incapable of producing the intricate and far-reaching effects attributed to it. They claim that most variations are minute and not of life-and-death significance, that variations have little or nothing to do with fitness, and that most individuals die as "unfit" before they are tested by nature in any meaningful way. Other critics contend that no conceivable series of small variations could bring into existence some of the complicated or intricate characteristics of living things. Still another argument is that no amount of natural selection can transform one species or kind into another.

The limitations of this book do not allow extended discussion of these objections. The effects of small variations and the meaning of apparently useless characteristics are being intensively studied with the result that many previous arguments against their importance are losing their strength; investigators are becoming very cautious in evaluating variations of any type. Seemingly unimportant characteristics have proved to be significant when fully understood. What appears to be trivial, useless, or ineffectual has proved in case after case to be significant in itself or perhaps linked with something else that is significant.

Although it is true that nature may not effectively test every organism to determine its fitness or unfitness, enough are tested to produce a statistically significant result. Everyone knows that most mutations are harmful or deadly and that some may be classed as essentially neutral and hence of no significance. Although all this is true it is really no argument against natural selection. It overlooks the fact that there is tremendous overproduction and waste and that nature in an impersonal way seems not to care about the failures. One successful mutation preserved is worth a thousand failures. Even the loss of what appears to be a success is not important; it will probably be repeated in the same population before long. The mathematics of man is apparently not yet able to deal with the complicated interplay of factors that decide which living things are fit and which unfit.

What the Struggle Is All About

In thinking about the struggle for survival among living things, we must consider the instinct for self-preservation. Manifestations of self-preservation range from the simplest motions of withdrawal or avoidance in the amoeba to the reasoned long-range planning of man. The ancient and deep-seated reaction for self-preservation is explained by the simple fact that organisms without it have always been systematically eliminated—natural selection at its very best!

No less important than self-preservation is preservation of the race or type. Self-preservation is prerequisite to preservation of the race, and preservation of the race ensures self-preservation—one without the other is pointless.

Any reaction aimed at individual or racial survival requires energy. Even a resting spore or seed is expending energy in just being alive. The struggle for existence is readily translatable into the search for energy and the attempt to avoid becoming energy for something else. The search for and preservation of energy sources taken in connection with the reproductive drive goes far in explaining the varied adaptations, forms, functions, and reactions of living things. A basic question regarding any organism, living or extinct, is how it stands in the prevailing food chain—what it ate and what ate it. If this question can be answered most other characteristics can be understood.

FOOD CHAINS, WEBS, OR PYRAMIDS

When photosynthesis became a basic food-making process, life entered a phase of expansion and development potentially capable of continuing as long as sun and earth maintain their present relationships. Green plants were assured a

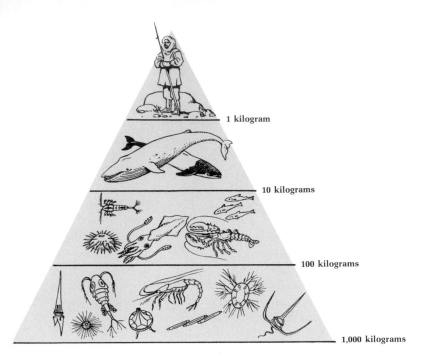

1 kilogram

10 kilograms

100 kilograms

1,000 kilograms

Figure 17.2 A simplified food pyramid of Antarctic life. The basic food supply is floating plant plankton. This is eaten by animal plankton, which is in turn consumed by whales and other marine vertebrates. Man occupies the summit of the pyramid when he captures whales or fish. The numbers indicate that for a gain of 1 kilogram of human flesh, from consumption of whales and other marine invertebrates, approximately 1,000 kilograms of plant plankton must have been produced.

permanent existence and their stored food supplies made possible the existence of animals. Because photosynthesis depends on light, it cannot be continuous, and so if plants used up their energy as fast as they captured it, they would die when deprived of light. The alternation of day and night and the recurrence of cold or dry seasons dictate that plants must be able to store food. Stores of simple starches, oils, and sugars synthesized during favorable periods are necessary for uninterrupted existence. Such stores invite attack from other organisms that either lack the power of photosynthesis or have lost it as

they achieved more efficient shortcuts in the energy struggle.

Thus the first animals were dependent on plants. This association may have begun when mutations allowed the passive extraction of a small store of food from one cell to another by a simple process such as osmosis. Once under way the carnivore–herbivore relationship became increasingly complex until the present-day situation was achieved. The arrangement now appears to be in a state of equilibrium, with benefits accruing both to plants and to the animals that feed on them. Animal respiration supplies carbon dioxide, which is essential to photosynthesis. Plants, if living alone, could remove all the carbon dioxide from the atmosphere within a few thousand years. With the passage of time plants and animals have achieved the most important self-sustaining and balanced ecological system of all time.

Under the universal law that every organism must obtain energy or food in competition with others, a struggle for existence became inevitable. The constant search for ways and means of obtaining energy led not only to diverse ways by which animals could utilize plants but also to ways in which animals could prey upon other animals. This resulted in what are called *food chains, webs,* or *pyramids.* As the struggle intensified, each new food source invited attack by ever-present hungry predators. This led to better protective devices or to more efficient methods of storage and utilization. The sharper tooth was counteracted by the thicker shell or tougher hide, the swift attack by the still swifter means of escape. Every challenge brought forth a suitable response as the struggle to survive drove each successive generation towards greater efficiency, economy, and intelligence in searching for or protecting its energy quotient. In a world of finite

resources peopled by creatures capable of overproducing their kind, no other means of existence is possible or even conceivable.

BUILDING THE FOOD PYRAMID THROUGH TIME

Today food pyramids are more or less static arrangements for the environments in which they are found. From the viewpoint of historical geology they are not static but have come into being in a developmental way. Obviously they have arisen by gradual evolutionary steps, with the older, more primitive forms occupying the lower levels and progressively younger ones at higher

Figure 17.3 *Building food pyramids through time. Each successively larger pyramid embodies all surviving life in the pyramid below.*

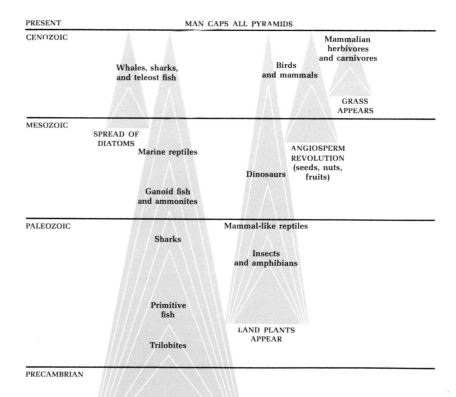

positions. This must be so in view of the fact that no organism, plant or animal, can exist for long without a dependable food source, and no evolutionary line can begin unless it is preceded by some animal or plant on which it can feed.

We scarcely need mention that the first extensive fossil record was left by the algae and that these simple photosynthetic plants still support the great mass of aquatic life. The excess food stores accumulated by algae and bacteria undoubtedly made possible the emergence of a horde of invertebrate animals including filter feeders and mud-grubbers. The food-getting mechanisms of early animals, as shown by fossils, were relatively passive and certainly nonaggressive. No actual teeth are known from the entire 100 million years of the Cambrian Period; even the ubiquitous trilobite had no jaws. Defensive or food-protecting devices were correspondingly weak. Most food pyramids were low, consisting of one or two stages and a great deal of unused or surplus energy must have wasted away in the ocean.

As time passed, active predators appeared, chiefly vertebrates with sharp teeth, strong jaws, acute sense organs, and intelligent reactions. The food pyramid took on additional steps and less stored energy was safe from predators. Finally, the oceanic food pyramid was completed, and a lengthy period of competition began to determine which animals might from time to time occupy the summit of the food pyramids.

For literally billions of years there appears to have been no land life. Plants came first and were thriving when animals emerged. Arthropods such as scorpions and centipedes were probably the first to come ashore and these may have been important food sources for the vertebrates that followed. Thus the basic pyramid of land life, plants–herbivores–carnivores, was established. Every advance in plant life was followed by corresponding advances in the animal world. Any new device for generating or storing food was followed by adaptations for taking it away. Eventually the struggle became worldwide, and successively more complex forms, culminating with man, occupied the summit of most food pyramids.

Adaptability and Adaptation

Each species must carry on the struggle to survive in its own way—that is what makes it something different and apart from every other species. Each species is adapted in many unique ways for the life it leads and for the time and place of its existence. The fact that organisms are fitted to live in their individual environments is appreciated by even the unschooled savage, but he sees no particular problem in the situation. His myths explain how plants and animals came into existence with special characteristics that "fit" them for existence in a world already prepared to receive them. The scientific study of just how organisms are fitted to exist in their environments is an important branch of biology, but it covers only the superficial and restricted aspects of a much broader subject—not only the problem of how organisms became adapted in the first place, but also how they manage to readjust continually to changing conditions. The term "fitness" implies only that an organism has the necessary characteristics to exist under a certain set of conditions; the term "adaptation" has a broader meaning, implying an ability to change or adjust to new situations. In other words, the facts of adaptation focus attention on the great problems of evolutionary change.

The discovery that neither the physical world nor its inhabitants has remained static is one of the chief contributions of historical geology. Without this discovery, the problems of adaptability and evolution might never have been fully appreciated. Living things have apparently not only appropriated new environments as these have become available, but they have also become better adapted to these environments after they have entered them. The exploitation of new or additional environments by various groups of organisms has been a fundamental reaction of life.

Some types of organisms have been much more adaptable than others. Compare, for example, the sedentary, simple sponges and the mobile, complex arthropods. The sponge has remained essentially unchanged throughout geologic time; it has never left the water, has only one method of feeding, and has no means of locomotion. The arthropods, regarded as the dominant phylum insofar as numbers is concerned, have diversified greatly with time, have achieved many methods of locomotion, and have very diversified food habits. If we were to attempt to show the evolution of these two groups graphically, the sponges could be represented by a simple stem with few parallel branches, the arthropods by a much thicker stem with many spreading branches going in different directions. Drawings of *family* or *phylogenetic "trees"* are commonly used to express the various ways in which groups have diversified. The branching figures thus constructed are not to be taken as true in all respects, but they do serve to illustrate in a graphic way the basic features of diversification.

SIMPLE LINEAGES

When biology was in its infancy, species were considered to be unchanged and unchangeable, and it was customary to portray the history of any group as a single straight line or lineage, without branches or bends of any kind. This idea has been called the *"totem pole"* or *"ladder"* concept. It is true only for those portions of race histories that are indeed without branches or deviations, and is untrue for any known group over a long period of time.

DIVERGENCE

Whenever a species becomes isolated into two or more groups, there appears to be an inevitable tendency for the groups to acquire distinct differences. The longer the separation continues, the more unlike the descendants become and the more they depart from the common ancestral form. This process is called *divergence* and is visualized by the branching of a tree into more or less equal limbs. At the time of the initial splitting of the group the differences may be very minute, but later they may be extreme. An example among the mammals is the derivation of the bears from doglike ancestors. In the Miocene the distinctions between bears and dogs were few, and the name "bear dogs" is appropriately used. With time, the two groups became very distinct: divergence had created a new family of animals. It is supposed that dogs are also distantly related to seals through some remote, but still undiscovered, ancestor. The seals have become adapted to sea life, the dogs to land life.

CONVERGENCE

Convergence is the opposite of divergence and is visualized as the near approach (but never actual joining) of two separate lineages. In other words, it is the tendency of two unrelated groups to produce similar forms. Convergence illustrates the power of the environment in shaping living things to suit particular ends. One of the most striking examples of convergence among vertebrates is the attainment of similar body shape by fish and water-living reptiles and mammals. This particular convergence has occurred many times and has affected animals now alive as well as others that have long been extinct. The ultimate product is a long, spindle-shaped, streamlined body driven by a symmetrical tail that provides the propulsive force for swimming. There are usually two pairs of fins or flippers for balancing and steering, and a triangular dorsal fin rising from the back. Size does not seem to be a factor, for the distinguishing features may be repeated on all levels, from the whale, largest of all vertebrates, to tiny tropical fish, which are the smallest bony animals known.

Specific examples illustrating the convergence

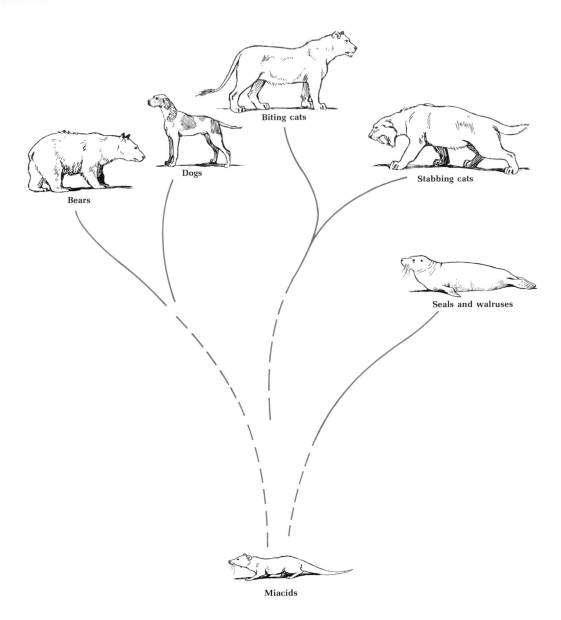

Figure 17.4   Divergences among the carnivorous mammals.

toward an ideal aquatic form are living sharks representing fish, extinct ichthyosaurs (literally fish-lizards) representing reptiles, and porpoises representing mammals. Incomplete attainment of this form is illustrated by seals and walruses, and by the ungainly sea cows (Sirenia), which are considered to be relatives of the elephant!

Convergences of form to meet the requirements of life in the air are fully as striking as those displayed by aquatic forms. The ability to fly has been achieved by three groups of vertebrates—the birds, the bats, and the extinct reptilian pterodactyls. In each group the forelimbs have taken over the function of wings and the body shape is strikingly similar. The hind limbs are small and relatively weak and are adapted mainly for perching or grasping. The body is shortened and compact, and the rib cage is large to accommodate the powerful heart and lungs. The tail is generally short and stubby and may be modified into a rudder or braking device. There is a pronounced tendency to lighten the head; all unnecessary bone and tissue is eliminated from the skull, and even the teeth may be missing. The trend has been to lighten the body and to compress it into a boxlike form between the wings.

There is evidently an ideal aerial form just as there is an ideal aquatic form. The insects approach this form within the possibilities of their fundamental patterns, and aircraft engineers incorporate the general shape in their models and finished designs. In living things this form must be a mechanically efficient flying machine that will allow its possessor both to capture food and to escape from enemies. The physical properties of air dictate what the ideal shape must be, and any animal having this shape has been molded to it by a series of modifications in structure.

Among the many uncanny resemblances brought about by convergence among now extinct animals is the saber-tooth adaptation that was achieved independently by a South American marsupial carnivore in the Pliocene, by a true carnivore during the Oligocene, and by a primi-

tive creodont in the Eocene. In all these forms the adaptation involves not only the huge, saber-like, canine teeth but also other portions of the head. The canine teeth in each instance are greatly elongated and the lower jaw has a downward-projecting flange, analogous to a scabbard or sheath, that protected the protruding teeth from being broken when the mouth was closed. The adaptive value of the flange structure is obvious, for an animal's very life depended on keeping the giant stabbing teeth sharp and unbroken. These teeth were the animal's main mechanism for food gathering and for defending itself against its enemies.

Even the highly distinctive dinosaurs show marked convergent tendencies. The common Jurassic carnivorous dinosaur *Allosaurus* had a remarkably perfect mimic in the form of a contemporary called *Ceratosaurus*. Although the remains of *Ceratosaurus* are scarce, it is evident that the two forms had reached almost identical proportions and bodily structure. So close is the resemblance that the experts have expressed doubts that the two are really distinct species. But before a final decision can be made, additional specimens of *Ceratosaurus* must be discovered.

The dog is a typical carnivorous animal. Strong, tough, quick-witted, with sharp teeth and claws, able to run swiftly and to drag down his prey by brute strength, he has succeeded in holding his own in many environments. This general description of the dog family would fit a number of other animals that evolved at various times and places from totally different parent stocks. Some of these types are known only from fossils, but a few are still in existence. Most remarkable of dog mimics is the Australian *thylacine,* the marsupial "wolf," or "tiger." It is thinner than a true wolf, has a long and slender muzzle, medium-sized ears, and a long tail. Details of the skull are remarkably doglike, even to the arrangement of the teeth. It utters a coughing bark to add to its doglike character. As the name implies, the animal is a marsupial. The young are carried in

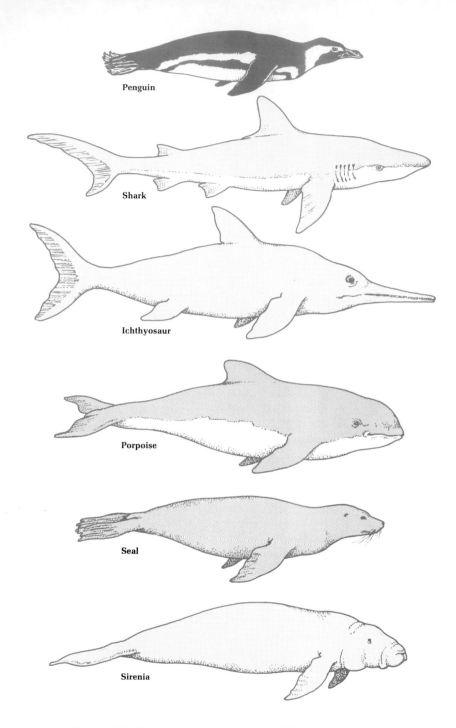

Penguin

Shark

Ichthyosaur

Porpoise

Seal

Sirenia

Figure 17.5   Convergence among aquatic life.

a pouch for several months. In competition with the true dog, the marsupial form proves to be weaker, and is almost surely doomed to early extinction.

SMALL- AND LARGE-SCALE CONVERGENCE

What we have already said about convergence shows how overall similarity may be attained by animals that were originally quite different. There are other manifestations of convergence that also should be mentioned. One of these is illustrated by the close similarity of specific organs in different animals. An outstanding case is the vertebrate eye and the eye of certain cephalopods. The eye of an octopus possesses all the essential features of the vertebrate eye, including a crystalline lens, retina, and eyelid. There can be little doubt that the eyes of mollusks and vertebrates developed separately, perhaps even from totally eyeless ancestors. The conclusion seems justified that the ultimate perfected eye could be patterned only along very restricted lines, guided by the properties of light and the possible methods by which it is received and perceived by living matter. The differences between man and octopus are extreme, yet in one organ at least they are similar. (Incidentally, eyes, or light-perceptive organs of some kind, have appeared independently in eight different groups of animals.)

Another remarkable convergence of specific organs, in this case feet and legs, is shown by horses and the fossil *litopterns* of South America. The legs of both groups are specialized for running, and in both instances toes were gradually reduced and the foot bones were elongated. The South American *protoreotheres* and the horses are the only strictly one-toed animals known. Because the habits, environment, and food sources of both forms are similar, it is small wonder that the end results were also nearly identical.

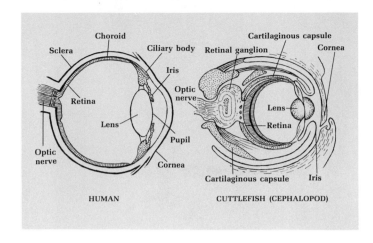

Figure 17.6 *The human eye compared with the eye of a cuttlefish.*

Even more significant and impressive are the similarities of entire assemblages of animals, in which not only individuals but entire populations living at one time or in one particular place may be closely matched by products of other times and places. The best-known instance of this large-scale convergence involves the *marsupial* mammals of Australia and the true, or *placental,* mammals elsewhere. Nature has produced among the Australian marsupials a large number of opposites, or look-alikes, that match typical placental forms in the rest of the world. The Tasmanian "wolf" we have already mentioned. There are marsupial "mice" and "rats" complete with long curving tails, pointed noses, and beady-eyed expressions, and even a marsupial "mole" with powerful digging claws and a shape adapted for subterranean burrowing. The hopping bandicoots with their long ears and hind legs resemble the rabbit, although not too closely. Squirrels are mimicked by the tree-living marsupials. There are even marsupial "flying squirrels" that glide from tree to tree exactly like their counterparts in other parts of the world.

PLACENTALS

Wolf

Ground hog

Rabbit

Flying squirrel

Mole

Mouse

MARSUPIALS

Tasmanian wolf

Wombat

Rabbit-eared bandicoot

Sugar glider

Mole

Mouse

Figure 17.7   Some typical marsupials and the placental, higher, mammals to which they are distantly related. An example of large-scale convergence.

CONVERGENCE IN PLANTS

Evidence showing convergent evolution abounds in the plant world, but the steps by which the observed results were attained are not always clearly demonstrated by fossil remains. The almost universal tendency of vascular plants to take on a treelike form is perhaps one of the most obvious examples. The necessity of exposing their food-making organs (leaves) to the light results in competition that drives plants higher and higher into the air. The ones that succeed in overtopping their neighbors have a decided advantage, and those species that achieve the proper combination of treelike characteristics become dominant. In the teeming forest no species can escape the necessity of fighting for light—the rule is plainly "compete or perish." A tropical forest is composed of hundreds of different types of trees, many of them members of families that in the temperate zone produce only shrubs or herbaceous plants. Thus, bamboo, a member of the grass family, in the competitive atmosphere

of the jungle becomes a giant struggling for its share of light. On the Galapagos Islands even the sunflower and cactus have become treelike. The generalized tree form is possessed by literally hundreds of kinds of plants and must be impressed on them by the nature of the environment in which they grow. Again, nature seems to have favored one form above other forms, and plants have responded to nature's bidding, each species in its own way and each within the framework of whatever it has had in the way of capabilities.

PARALLEL DESCENT

Closely related to, and in some instances not easily distinguishable from, convergence is *parallel descent*. Parallel descent is the tendency of two originally divergent lineages to evolve along essentially similar paths. A good example is furnished by the mastodons and mammoths. Although both of these animals are elephants in the general sense, the fossil record shows that elephants sprang from the mastodon line in the

Figure 17.8   Convergence among plants.

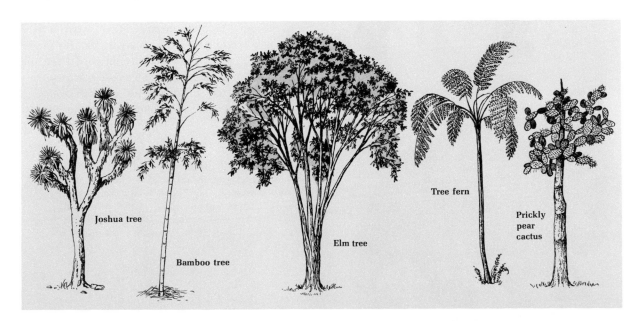

Joshua tree

Bamboo tree

Elm tree

Tree fern

Prickly pear cactus

(a)   (b)

Figure 17.9 *The close resemblance between the New World porcupine (a) and the Old World porcupine (b) is considered by some zoologists to be the result of parallel evolution. The common generalized rodent ancestor gave rise to two lines, both of which eventually acquired porcupine characteristics. (From E. D. Hanson,* Animal Diversity, *2nd ed., Prentice-Hall, Inc., 1964.)*

Miocene and evolved rapidly into many genera, including the extinct mammoths and the elephants that are familiar sights to us today.

During the Pleistocene the mammoths and mastodons both attained large size, developed long, heavy tusks in the upper jaw, had the same peculiar type of growth and replacement of the molars, and acquired a heavy coat of hair. Thus the two lineages had the same ancestor, but originally diverged, only to later evolve along similar lines.

Adaptive Radiation

All the patterns of evolutionary change mentioned in the foregoing pages are found in the comprehensive process known as *adaptive radiation* or *adaptive branching*. This may be visualized in a crude and imperfect way by an imaginary view looking directly downward into a large, spreading tree in such a way that the major branches are seen to radiate outward from the central trunk. Examples of convergence, divergence, and parallel evolution appear over and over in the maze of branches, but the general

(a)

(b)

Figure 17.10 *The mastodon (a) and mammoth (b) are culminating end products of lengthy parallel evolution. Originally unlike in many ways, they eventually became similar in size, tusks, and teeth, and in a variety of adaptations for life in cold climates. (New York State Museum and Science Service.)*

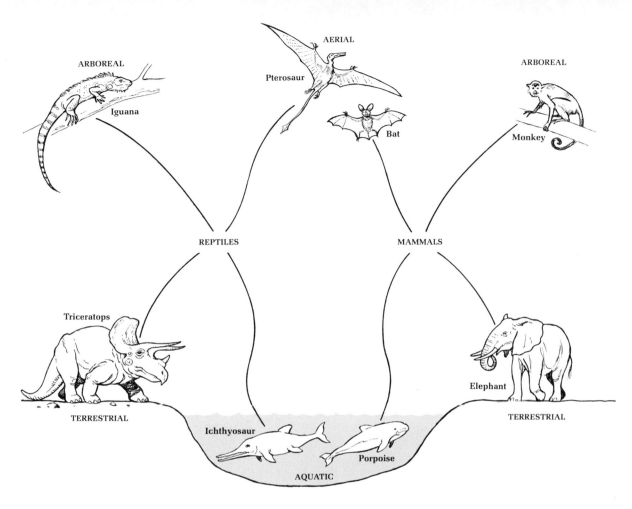

ARBOREAL

Iguana

AERIAL

Pterosaur

Bat

ARBOREAL

Monkey

REPTILES

MAMMALS

Triceratops

Elephant

TERRESTRIAL

Ichthyosaur

Porpoise

AQUATIC

TERRESTRIAL

Figure 17.11   Radial adaptation of reptiles and mammals.

tendency is an outward spreading into unoccupied living space.

This analogy is good in another respect, for it is the nature of vegetative branches to divide and subdivide but not to reunite. Life forms, once separated, are likewise prevented by genetic differences from merging or interbreeding.

Viewed in the broadest possible way, the spread of life through time and into all available environments represents adaptive radiation on a vast scale. The drive to enter all possible envi-

ronments is evident on every level of existence and in practically all groups, so that the principle may be illustrated in connection with any well-known groups of animals. The classes of vertebrates will serve our purposes best, because the animals concerned are well known and the fossil record fortunately happens to be quite good.

Considering the lands and waters of the earth as a complex mosaic of potential habitats or niches available for use by living things, it is possible to assess the success of any basic stock by noting the extent to which it has managed to move into the available environments.

Fish, the lowest class of vertebrates, are represented by about 30,000 living species and 4,000 extinct ones. There are numerous kinds of fish in fresh, marine, and brackish water almost everywhere. Fish live at all depths in the ocean and can tolerate a wide range of temperature. Individuals range in size from midgets less than a centimeter long to the 18-meter-long basking shark. Means of getting food are varied and there are herbivorous and carnivorous varieties. The radiation of the fish has been extensive and can be said to include even a branch that adapted to land life.

Amphibians descended from fish and are a link between aquatic and land life. They are the least numerous and perhaps the least successful of the vertebrates. It is puzzling that there are no marine forms. They do not thrive in dry or cold situations. All are rather weak and have poorly developed limbs and limited ways of capturing food. The radial adaptation of amphibians is not impressive, but their descendants succeeded well where they failed.

Reptiles now occupy a rather restricted position in the world of life but their past success during the Age of Reptiles was truly spectacular. During a span of 175 million years they were dominant on land and in the air and water. Reptilian modes of locomotion and food getting are diverse and effective. Certain dinosaurs were the largest of all land animals. The greatest reptilian weakness appears to be lack of tolerance for low temperature—the cooler lands and waters are closed to them.

Birds have been called "glorified reptiles,"

thus calling attention to adaptive advances over their ancestors. Their trademark is the feather, one of the most important bodily structures to appear among living things and the one that makes flight a way of life. Birds are mostly small but there are and have been a few large flightless varieties. Birds have succeeded in practically all environments including the inhospitable polar regions. They are particularly abundant in forests, on islands, and in the vicinity of water bodies. Their food habits are diverse but insect eating is particularly notable. Because they are warm-blooded they are able to go beyond their reptilian ancestors into many cooler environments.

Mammals, like birds, arose from reptilian ancestors during the Mesozoic Era. They are without doubt the most efficient and successful living things, even if we leave man out of the picture. Their many admirable traits, including temperature-regulating devices, live birth, milk production, diverse diets, effective locomotion, and intelligent behavior patterns, have carried them into practically every environment of the earth. The Age of Mammals has already spanned 60 million years and is obviously not over.

It is apparent that the successive classes of vertebrates were able to populate increasingly more and more of the available living space: fish in the waters; amphibians in warm water and on adjacent land; reptiles on dry land as well as in warm water; birds and mammals on cooler land and in cooler water as well as in environments previously pioneered by lower vertebrates. Adaptive radiation or branching is obviously a measure of the evolutionary powers of the bone-bearing animals.

A SUMMARY STATEMENT

Organic evolution is the gradual progressive change of living things through time. Implicit is the idea that all life has descended from one or a few simple beginnings. The most widely accepted theory as to how present complex organisms have been derived from simpler ancestors was proposed by Charles Darwin and is known as

Darwinism. Such terms as "struggle for existence," "survival of the fittest," and "descent with modification" tell something about this theory, but the central feature is natural selection. All living things are subject to tests that eliminate some and preserve others, so that over long time periods and many generations those best adapted will remain while others will be eliminated. The fossil record consists mainly of species that have been eliminated, with a few types that are ancestors of living species.

Because nothing can exist without an adequate energy supply, it is instructive to examine the fossil record and observe that the basic sources of the food chain such as algae came first, followed by successively more efficient dependent or predatory forms. With time, the search for food has led to development of more efficient, intelligent creatures. Man now sits at the summit of most food pyramids.

Among the patterns of survival that have been followed time after time by living things are: *divergence,* the development of different modes of life by closely related animals; *convergence,* the development of similar modes of life by distantly related animals; *parallel evolution,* the maintenance of a similar sequence of changes by related organisms; and, *adaptive radiation* (or adaptive branching), the attainment of adaptations by a single group that enable it to exist in as many situations as its basic genetic possibilities and prevailing opportunities permit. Thus mammals have radiated into arboreal, aerial, aquatic, underground, and plains-dwelling types.

FOR ADDITIONAL READING

Barnett, Lincoln, and the Editors of *Life, The Wonders of Life on Earth.* New York: Time Inc., 1960.

Dobzhansky, Theodosius, *Genetics of the Evolutionary Process.* New York: Columbia University Press, 1970.

Dodson, Edward O., *Textbook of Evolution.* Philadelphia: Saunders, 1952.

Eiseley, Loren, *Darwin's Century.* Garden City, N.Y.: Doubleday, 1961.

Jepsen, G. L., *et al.,* eds., *Genetics, Paleontology and Evolution.* Princeton, N.J.: Princeton University Press, 1949.

Mayr, Ernst, *Populations, Species, and Evolution.* Cambridge, Mass.: Belknap Press of Harvard University Press, 1970.

Moore, Ruth, and the Editors of *Life, Evolution.* New York: Time Inc., 1964.

Simpson, G. G., *The Major Features of Evolution.* New York: Columbia University Press, 1953.

Stebbins, G. Ledyard, *Processes of Organic Evolution.* Englewood Cliffs, N.J.: Prentice-Hall, 1966.

Wallace, B., and A. M. Srb, *Adaptation,* 2nd ed. Englewood Cliffs, N.J.: Prentice-Hall, 1964.

Watson, D. M. S., *Paleontology and Modern Biology.* New Haven: Yale University Press, 1951.

MIGRATION, ISOLATION, AND SPECIES FORMATION

18

It is not at all easy to discover the significant connections between the present world and the geologic past. This is chiefly because we know so much about the present and so little about the past. The past may be likened to a hazy view of a distant forest, a picture painted in broad strokes, or a lengthy reel of motion-picture film. The present, by comparison, is similar to a close-up view of the leaves of a single tree, a painting filled with minute detail, or the last few frames of a motion picture.

Migrating caribou in the interior of Alaska. (Charlie Ott, from National Audubon Society.)

Although we are confident that we are dealing with cause and effect and that the present world and its inhabitants are natural products of past events and evolutionary processes, we have difficulty in discovering and sorting out the past events that have had the most significant consequences. In spite of these difficulties, however, we must give due attention to all lines of evidence in seeking answers to the question of how the present world came to be. We must draw from modern biology as well as from paleontology, from living things as well as from fossils.

In this chapter we attempt to correlate changes in the physical world with changes in the biological world. In other words, we deal with the problem of how the origin and dispersal of species relates to the long-term changes that are the subject of historical geology. Some of our examples are from existing situations and some are from the reconstructed past. Many of the facts and principles presented earlier will be applied and substantiated in this discussion.

The Challenge
of Physical Change

Living things are found almost everywhere and their modes of life and ultimately their very existence depend on the environments provided by nature. There are at least 2 million species of plants and animals on earth, each of which occupies a unique niche in the natural world for which it is peculiarly adapted. To the untrained or time-bound mind each species was apparently created to fit its particular niche, and neither the species nor the niche have changed since the time of creation. This naïve idea has been rudely shattered by the findings of biology and historical geology, which indicate clearly that environments as well as organisms are continually changing.

IMPORTANCE OF CLIMATE

The greatest single factor that determines where plants and animals can live and thrive is climate. We include here the "climate" of the seas and oceans as well as that of the emergent lands. Climate includes a diversity of influences such as cold, moisture, winds, atmospheric transparency, and particularly the distribution of these on a seasonal or yearly basis. As a general rule climate is regarded as being a long term, stable factor, but from the geologic viewpoint this is not so. Climates do change in a slow, inexorable way and this is more important to an understanding of the past than the mere knowledge that conditions were thus and so at a particular time.

Figure 18.1 *The Valley of Ten Thousand Smokes, Alaska, showing the devastation caused by a layer of volcanic ash that gushed out of numerous vents in the valley floor at the time of the eruption of Mount Katmai on June 6, 1912. (U.S. National Park Service.)*

Figure 18.2 Inhospitable Antarctic terrain offers little opportunity for colonization by any type of life. (Official photograph, U.S. Navy.)

Figure 18.3 The effects of a rise or fall of sea level on living things.

Why do climates change? To some scientists climatic change is strictly a geologic matter, to others it is not. In other words, some regard all changes of climate as being due to terrestrial, or earthbound, influences; others seek extraterrestrial or astronomical causes. Both viewpoints have much to support them, but additional information will be needed before a final decision can be made. There is a branch of geology, known as *paleoclimatology*, that deals with ancient climates. For purposes of the present discussion we shall accept climate and climatic change as being of great importance in the present and past distribution of living things and shall turn our attention to matters that are unquestionably in the geologic realm.

SEA-LEVEL CHANGES

Changes in sea level must head any list of strictly geologic processes that have affected the nature and distribution of living things. Because the

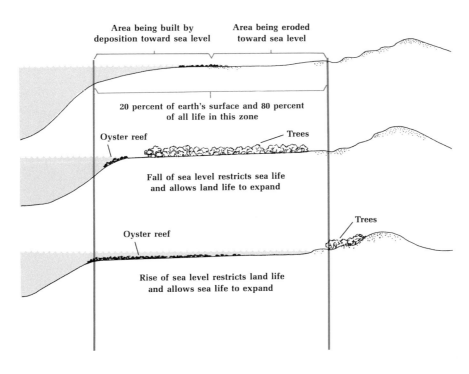

advance or retreat of a shoreline is a slow process, it is not usually appreciated or emphasized in the short-term observations of biologists; only through knowledge of the prehistoric past does it assume true importance. It is worth pointing out here that the level of the sea tends to lie on or near the vast sloping areas known as the continental shelves and their continuation inward in the form of emerged coastal plains. Through weathering and erosion by running water the level of the exposed lands is being brought slowly to sea level, while the sea bottom is being built upward by deposition toward the same level. If this tendency were not counteracted in various

Figure 18.4 Mountain building and speciation. When land is completely flat (a), a given species can wander where it will. The formation of a new mountain range (b) will have little effect at first. The animals cannot cross the barrier, but the species remain the same on both sides for some time. After a long period of separation, the animals on either side of the mountains become separate species (c). After the mountain range has eroded (d), the animals can again intermingle, but they are no longer able to interbreed, because they are now two separate species.

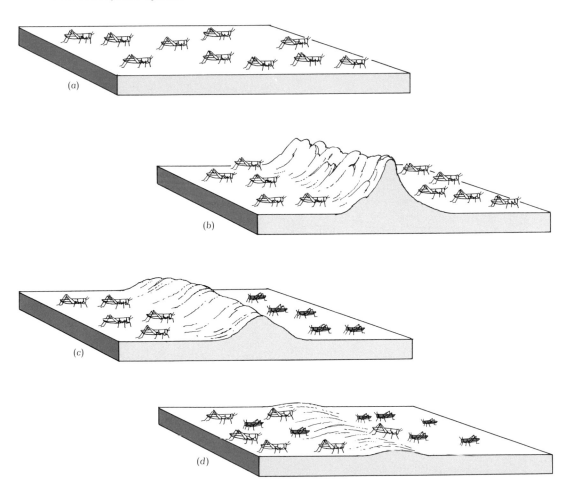

ways it is possible that most of the land areas would eventually come to be within a few feet of sea level.

The point of the foregoing paragraph is that a very wide and significant strip of territory is constantly being shaped and affected by any change in sea level. As sea level becomes lower, regression of the water lays bare a wide tract of former sea bottom; as sea level rises the water returns and a wide belt of lowland is flooded. Effects on life forms are obviously very drastic. If continued long enough, a lowering of sea level might literally crowd shallow-water sea life over the edge of the continental shelf; conversely a rising sea might push the world's land life into inhospitable higher regions and perhaps out of existence.

Changes of sea level have other important effects. As seas retreat, former shallow banks become islands, and former separate islands may be connected with one another and with adjacent continents. On the other hand, as sea level rises some islands may be completely submerged, while other large elevated tracts may become separate islands or archipelagoes. Any geologic mechanism that tends to be cyclic results in alternating fusion and breaking of land connections on a grand scale. Such changes are particularly significant insofar as migrations and the formation of new species are concerned.

ICE AGES

Nothing known in nature is more thoroughly devastating to life than a continental glacier. An advancing ice sheet wipes out all life in its path and, by cooling the surrounding land or water, alters the habitats of life on a wholesale basis. It is through repeated glacial stages that sea level is raised and lowered in a cyclic manner as described in Chapter 15. Cooling of deep ocean water is another result of glaciation that greatly effects bottom-dwelling organisms.

Large glaciers do not actually crush the life out of plants and animals. Their cooling effects

go before them to create situations which inhabitants cannot tolerate. If an avenue happens to be open, a threatened organism may move away and survive in its customary climatic zones even though these may be much compressed and distorted. In many instances, however, the only escape routes lead to destruction in the sea or on mountain peaks.

DIASTROPHISM

Diastrophism, meaning large-scale earth movements including mountain building, has been a constant force in the history of the earth. The creation of an uplift, no matter if it be a small volcanic cone or an extensive mountain chain, means the production of new habitats for life.

Later, as a mountain range is eroded, it provides less and less environmental diversity. The cooler and usually wetter environments disappear as the peaks are lowered or destroyed. Areas of rugged topography, cliffs, and talus piles steadily decrease as steeply sloping ground fades into flatter terrain. Eventually the area consists of low rolling hills that are only slightly above the surrounding lowlands. Such must have been the history of the ancient Appalachian chain. Over the geologic ages, habitats for life have been created and diversified by upbuilding forces and destroyed or simplified by leveling effects of sedimentation and erosion.

Islands may be thought of as emergent lands and also as submarine elevations analogous to mountains on land. They support different types of life according to their configurations both below and above sea level. A knoll or seamount thousands of meters below the surface of the sea will support an entirely different assemblage of life than one that reaches near the surface. The most prolific and varied zone of all is just below sea level where reefs flourish. Above sea level inhabitants of an island are governed by much the same factors that affect organisms that exist on continental mountains.

Consider now the effects of the appearance

Figure 18.5   Scene in the Andes Mountains of Bolivia. This range was elevated early in the Tertiary Period and is so high and long that few species of plants or animals have been able to cross it. (Courtesy of Eugene Callaghan.)

Figure 18.6   The Blue Ridge, Virginia, showing the low, moderately eroded profile of the Appalachian Mountains. Old ranges, such as this, present no effective barrier to migration or dispersal of plants and animals. (U.S. National Park Service.)

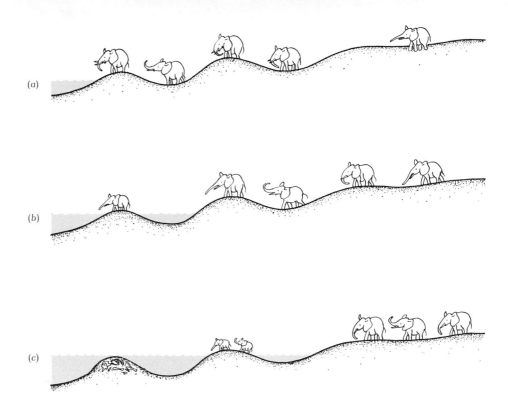

(a)

(b)

(c)

Figure 18.7   Sea-level changes and speciation. In (a), the level of the sea is low enough to permit the free intermingling of all members of a particular species. As sea level begins to rise (b), one group of animals is isolated and a new species is formed. All other animals can still interbreed. A second rise of sea level (c) means extinction for the new group formed in stage (b), but causes the formation of a second new species by isolating still another group of the original animals.

and subsequent destruction of a mountain uplift on the life of adjacent lower tracts. Any strong uplift will obviously disrupt the free exchange of living things that formerly took place and it may be geologic ages before communication will again be possible. An analogous situation comes with the emergence of a submarine ridge. It will separate formerly continuous populations of fish and other marine life for long periods; for example, the Isthmus of Panama arose between North and South America and has been in existence for about 2.5 million years. During this time of separation many evolutionary changes have affected the life on either side.

Not to be overlooked in any discussion of land and sea barriers is their modifying effect on currents of air or water: for example, the towering Sierra Nevada has created a desert in its rain shadow to the east.

LAKES AND RIVERS

A very important environment is that of freshwater lakes and streams. Lakes are especially sensitive to geologic change; they appear, expand, and disappear over relatively short periods geologically speaking. Thus during wet stages of the Pleistocene Ice Age many large interior basins were filled with water. This was the origin of Lake Bonneville and Lake Lahontan in the Great

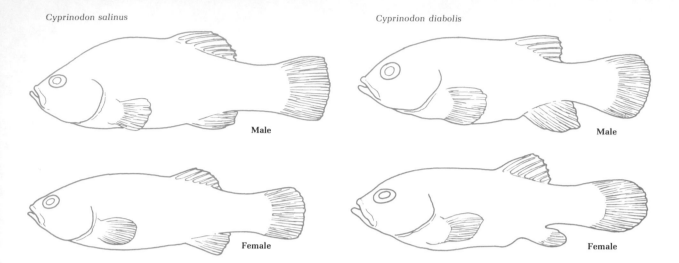

Cyprinodon salinus   *Cyprinodon diabolis*

Male   Male

Female   Female

Figure 18.8   *Speciation by isolation. The species on the left is found in Salt Creek, Death Valley, California; the one on the right in a single, small spring in Ash Meadows, Nye County, Nevada. Formerly, during more humid phases of the Pleistocene, the localities were connected by a river system at which time the total population of Cyprinodon would have been one species. With isolation, which may have lasted 20,000 years, two species have emerged. Note relative sizes of males and females and differences in fin outlines and placement.*

Basin and Lake Agassiz in the east-central part of North America. The filling of these basins was of course fatal to land life within them, and the subsequent drying up of the lakes was no less of a catastrophe to fish and other water-dwelling creatures that had come to occupy them.

Rivers, especially long ones, are important migration routes for fish and for streamside vegetation. At the same time rivers and their canyons are barriers to many forms of life. Thus the Grand Canyon of the Colorado separates closely related yet different species of the flying squirrel. Before the canyon became an effective barrier these species were probably one.

Migration and Dispersal

When confronted with an unfavorable alteration in its environment such as described in foregoing paragraphs, an organism has three alternatives: (1) it may leave the area of disturbance; (2) it may remain and adapt itself; or (3) it may neither adapt nor migrate and therefore perish.

The most common response of plants and animals to environmental change has been to move away from the disturbance. This is not to be understood as sudden, precipitous flight. Escape is usually slow and involves all the adaptations that organisms have developed for migration and dispersal. The word "migration" has come to be widely used to designate the slow spread of organisms over the earth even though the usage is not quite accurate. In its most technical sense migration means either a seasonal or a permanent movement of a large number of animals from one place to another. Some degree of instinctive or purposeful action is implied. In contrast, dispersal means scattering or spreading in different directions; no instinctive or conscious goals seem to be involved. Birds *migrate* to warmer climates as winter approaches; dandelion seeds are *dispersed* by the wind.

It is obvious that both migration and dispersal

Figure 18.9   *The Kaibab and Abert squirrels inhabit opposite sides of the Grand Canyon and seldom, if ever, interbreed. Even though they probably originally were members of the same population, they have become visibly different. The Kaibab squirrel has darker underparts and a whiter tail than does the Abert form.*

Figure 18.10   *Migrating waterfowl illustrate the ability of organisms to move and spread from place to place. (Allan D. Cruickshank, from National Audubon Society.)*

have served to spread plants and animals over the lands and in the waters of the earth. But not all migrations succeed, and very few of the eggs, spores, seeds, or larvae that are dispersed haphazardly from their parents will even take hold in a new environment—and few that do take hold will survive.

The process of distribution is opportunistic; it involves prodigious waste and is achieved by fantastically complex and varied means, all of which prove its essential importance in the history of life. From the first appearance of abundant fossils, dispersal has been at work. In succession, the earth has been inhabited by characteristic forms of life; hence, the Age of Trilobites, the Age of Fish, the Age of Reptiles, and so on.

How did the successive floras and faunas of the past reach the places where they are found? Answers differ from group to group. Large land-living mammals definitely require actual dry-land connections over which they can walk; smaller mammals, together with reptiles, birds, amphibians, and freshwater invertebrates, can be dispersed occasionally over water by flying, swimming, or floating.

Because they lack locomotive powers, land plants must be dispersed as seeds or spores. The varied devices by which seeds are dispersed are known to everyone and need not be discussed here. Slow and haphazard as these methods are, it is still obvious that plant species have spread widely, some even halfway around the globe. The fact that plants produce prodigious quantities of seeds at great expense to their vital processes proves that dispersal is not only important but absolutely essential to their survival as species.

MIGRATION BETWEEN CONTINENTS

There are three favored explanations as to how large land-living creatures can have reached opposite sides of wide and presently impassable bodies of water. The first depends on continental drift. The idea that the large landmasses were once in very close proximity if not in actual contact is now accepted as a fact by most geologists. It provides a ready explanation for the fact that essentially the same plants and animals, and even some identical fossil species, are found in Antarctica, South Africa, and South America. The discovery of a large fossil reptile (*Lystrosaurus*) in Triassic rocks of Antarctica is taken as proof of continental drift because it is inconceivable that this large land animal could cross

Figure 18.11 Dr. Edwin H. Colbert collecting the remains of the Lystrosaurus *fauna, Coalsack Bluff, Queen Alexandra Range, Antarctica. (Institute of Polar Studies, Ohio State University, photograph by David H. Elliot.)*

the hundreds of kilometers of open water that now separate its fossil remains. Continental drift explains the migration of land life during those periods of time when the landmasses were in contact, in other words up until about the middle Mesozoic. After this time other explanations must be called upon.

After the middle Mesozoic, land bridges may have been the means by which animals passed from one continent to the other. Examples include presently existing connections such as the Isthmus of Panama or shallowly submerged ones such as the Bering Strait. Needless to say many land bridges have been "invented" in the past to explain certain puzzling facts of animal distribution. Most of these have been demolished now that the sea bottoms are fairly well explored and the validity of continental drift established.

Still a third explanation is that migration may take place on or across islands that have been completely destroyed or submerged. The fact that fossils of shallow-water organisms have been dredged from the summits of submerged Pacific seamounts that are now over 1,200 meters below sea level strengthens this idea. Incidentally, it is estimated that there are at least 10,000 submerged seamounts in the Pacific. If all or even some of these were emergent, it is obvious that many plant and animal species could migrate rather easily over wide areas.

As it turns out, each of these explanations does not contradict or exclude the others. Each will explain certain situations when the facts are known.

DISPERSAL OF OCEANIC LIFE

The problem of dispersal of ocean-dwelling life is of course somewhat different. Several mechanisms by which dominantly aquatic animals or plants may be dispersed over water are: (1) by clinging or attaching themselves to floating logs, natural rafts, or pieces of pumice; (2) by being borne by currents; (3) by clinging to birds, fish, or aquatic mammals; and (4) by being wafted by strong winds. The chief problem is how floating marine organisms can survive long intervals of travel over stretches of deep water before finding shallow shores with favorable temperatures. The larvae of corals may float from 1 to 30 days. In an average Pacific Ocean current flowing with a velocity of 4 kilometers per hour, the coral might travel a maximum of close to 2,900 kilometers during this time. The common American oyster exists in a floating stage for about 2 weeks, and could be carried over 950 kilometers by an average current. In view of the many uncertainties that make it difficult to determine the possible effectiveness of these several methods of dispersal, especially during periods in the long-distant past, we can learn relatively little from the geologic record regarding specific migrations of the lower forms of life.

Isolation and Species Formation

Isolation is a term that has grown increasingly more important in biological thinking since it was used by Darwin and his contemporaries in the 1850s. There are different types of isolation, but as used in connection with living things, the word usually means "the separation of groups of individual organisms so that they cannot interbreed." The separation may be geographic, such as that imposed by an impassable body of water. Such isolation, caused by external factors, is called *extrinsic isolation*. Isolation that arises from differences within an organism, such as outright physical, genetic, or physiological incompatability, is called *intrinsic isolation*. If either extrinsic or intrinsic isolation results in the splitting of a formerly homogeneous interbreeding population into two populations that cannot interbreed, we may say that *speciation* has taken place. The organic world is replete with examples of populations in all stages of becoming isolated and differentiated. Although we may observe the process in operation, we have not, in the time available since detailed study began, seen and recorded the appearance of a new species through modification of an ancestral species.

Ignoring the many important studies that have been made of isolation in the modern world, we shall assume on the basis of ample evidence that the isolation of segregated populations provides the most effective and common opportunity for the species-making process to operate. Furthermore, it is apparent from actual observation as well as from statistical analysis that the effects of genetic mutations, good or bad, depend on the size of the group of organisms in which they occur. If the population is too small, there will be too few mutations to have any effect, and inbreeding may soon destroy the entire group. If the group is too large, a favorable mutation may be swamped, and may never be able to prove its superiority.

Theoretical calculations suggest that populations of between 250 and 25,000 individuals provide ideal conditions for species formation. Groups of this size provide a sufficient supply of mutations, and yet are small enough to limit strong competition that would eliminate a favorable new adaptation. Within this population range we may get a variety of mutations both good and bad, but it is neither the number of mutations nor the number of failures that counts. A few significant successes are all that is required to give decisive advantage to the group out of which they arise.

Geologic changes make isolation inevitable. Groups of all sizes are cut off from their contemporaries by alterations in land and water bodies. Darwin understood the importance of isolation and segregation and he sensed that there must be a relationship between geologic change and organic change. It was the genetic mechanism or link that managed to elude his grasp.

Fossils reveal almost nothing of the delicate shades of difference that accompany isolation. The all-important differences in plumage of birds, for example, would be lost in fossils. But geology does furnish examples of many past physical changes, and occasionally it gives very specific information as to when and where migration and isolation have taken place. The presence of the same or related fossil species on two separate landmasses or in deposits of two distinct bodies of water is sufficient proof that migration of some sort has taken place. In many cases it is impossible to determine which of two occupied areas may have been the original site; indeed, if a group seems to have appeared in two areas simultaneously, it may be that it migrated from yet a third, unsuspected locality.

In trying to prove the existence of actual dry-land connections between landmasses (which also serve simultaneously as barriers to sea life), we should best direct our attention to animals that have no other means of dispersal than walking. Animals that fly, swim, or float may lend corroborative evidence, but such evidence is usually subject to more than one interpretation. The larger land mammals offer the best material for study because they are distinctive; they have relatively short geologic histories, they must migrate across actual dry-land connections, and they are sensitive to climatic changes and to other geographic factors that are relatively easy to isolate and to understand.

ISLANDS AND THEIR LIFE

With the exception of large continental islands such as the British Isles, the vast majority of islands are too small to reveal much in the way of geologic history. Most islands are of volcanic origin or have been built up by corals and other organisms in relatively recent time. Such environments are obviously unfavorable for the preservation of extensive fossil records or large thicknesses of sedimentary rock. On the other hand islands furnish some of the clearest evidences of evolutionary processes, such as dispersal, colonization, adaptation, competition, speciation, and extinction. The flora and fauna of islands show many novel and unusual organic productions such as giant reptiles, flightless and almost defenseless birds, dwarf mammals, flightless insects, species on the verge of extinction, and others just newly arrived.

The study of island life has been highly instructive and yields its most meaningful results when carried on in proper historical context. Only by taking into account the time factor and the total past history can the life of islands be fully understood.

Although an island in the usual sense is defined as an area of land surrounded by water the concept may be extended to any isolated situation. Thus a high peak is islandlike; so is an oasis surrounded by sand, a grove of trees in the prairie, a meadow in a forest, a swamp, a glacier, a submerged reef or volcano. In fact, in thinking about living things, almost any situation where a species may be surrounded and more or less isolated may be referred to as insular. It is worthwhile and instructive to study insular situations as they now exist, but it is much more meaningful to seek answers to other questions concerning such matters as how the islands originated, how the inhabitants got where they are, where they came from, how long they have been isolated, what was the order in which they arrived, what the outside disturbances have been, and what the ultimate fate of the entire system might be.

The Galapagos Islands. From the viewpoint of evolutionary theory, the desolate Galapagos Islands, situated on the equator nearly 1,000 kilometers west of South America, are without equal. After visiting them over 100 years ago, Charles Darwin wrote that here ". . . we seem to be brought somewhat near to the great fact . . . that mystery of mysteries . . . the first appearance of new beings on this earth." The observations Darwin made in the Galapagos profoundly influenced his theory of the origin of species. Later

students have added to his original findings and have verified the fact that 37 percent of the coastal fish and 40 percent of the plants are indigenous to, and must have evolved within the confines of, these islands and their waters.

The islands are of volcanic origin and show no evidence of having been connected with the mainland of South America. Whatever land life they support must have been carried over the intervening sea, and few indeed are the animals that have succeeded in establishing themselves. There are two genera of native mammals, five of reptiles, six of songbirds, and five of other land birds, not including the many forms recently introduced by man that now run wild and have entirely altered the original situation over most of the islands.

In the absence of competition from mammals, the reptiles, represented by lizards and tortoises, grew huge; in fact, the tortoises originally found there are the largest and oldest living representatives of their kind. The birds known as *Darwin's finches* include fourteen species; all show signs of having descended from a common ancestor. They have diversified into ground-living and tree-living species with significantly different food habits and structural modifications. The three common species of the ground finches are large, medium, and small, and feed on large, medium, and small seeds, respectively. Apparently, when the ancestral form appeared, there were no competitors, and its descendants were able to diversify unchecked and to utilize practically all available food sources.

The origin of the Galapagos flora and fauna is best explained on the theory that the ancestral forms were supplied by haphazard means from other areas. The islands are all of relatively recent volcanic origin and were of necessity barren of life where they emerged. Waifs and castaways propelled by winds and currents began to arrive; a few survived and a great natural experiment was under way. The finches were probably among the earliest arrivals. They found a variety

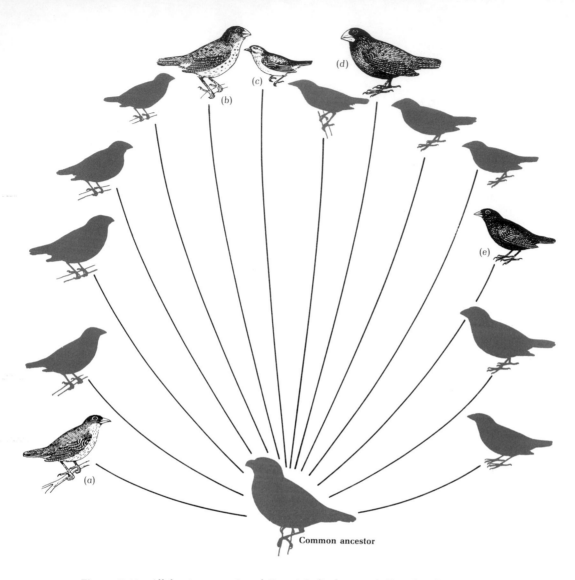

Common ancestor

Figure 18.12 *All fourteen species of Darwin's finches are believed to have evolved from a single ancestral type. Five different varieties are illustrated here:* (a) *a woodpeckerlike finch;* (b) *a vegetarian;* (c) *a single species of warbler finch;* (d) *the largest of the ground finches;* (e) *a cactus-eating finch. The first three are tree finches, the last two, ground finches.*

of food sources and no competitors. Vegetation and insect life differed from island to island, and the finches, through evolutionary modifications, became adapted to these different food sources. The descendants of the common ancestor now

resemble as many as half a dozen different families of birds on the mainland.

Strangely, many Galapagos animals are related not to those of South America but to those found on islands of the West Indies. This is ex-

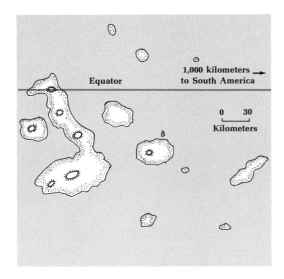

Figure 18.13 *The Galapagos archipelago is a natural laboratory of evolution. Built up by volcanic action from the bed of the Pacific Ocean, these islands have never been connected with the mainland of South America, which is 1,000 kilometers away. Ancestors of the present inhabitants arrived over the intervening water at various times during the past and diversified and adapted according to opportunities they encountered on the different islands.*

all are the Galapagos penguins, whose ancestors drifted up the cool Humboldt Current from distant Antarctic shores. Seals are also found. The effects of westward drifting ocean currents and winds are obvious. Only one migrant from the opposite direction, a land mollusk from Polynesia, is known.

The Hawaiian Islands. The Hawaiian Islands are isolated from all other large land areas by wide sea barriers and have never been connected with any continent. They have arisen gradually from the floor of the Pacific Ocean by volcanic action. It is now established that there is a gradual decrease in age from west to east. Kauai, on the west, is about 5.5 million years old, and Hawaii, on the east, is still building, with no known surface rocks over 700,000 years old. All native plants and animals of the Hawaiian chain must have arisen from waifs, castaways, and accidental travelers from over the ocean. Only one mammal, a small bat, is native to the islands, but many other mammals have been introduced by man. There are 7 native reptile species, 70 species of birds, 1,064 species of land snails, over 3,772 insect species, 1,064 seed plants, and 168 ferns

plained by the fact, established on other evidence, that the Isthmus of Panama is of relatively recent origin, and strong ocean currents from the Atlantic could once pass between North and South America from one ocean to the other. The great tortoises that give the Galapagos Islands its name have their nearest relatives among fossil forms of South America. Flamingoes, with relatives in the West Indies, are found in the Galapagos and nowhere else in the Pacific. Oddest of

Figure 18.14 *Giant tortoise (Testudo elephantopus porteri) as photographed in its native habitat on Indefatigable Island in the Galapagos. A rare species that may yet escape extermination. (UNESCO/ Herzog.)*

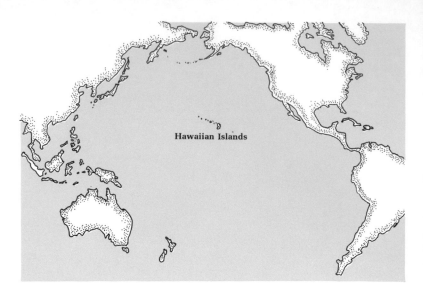

Figure 18.15 The Hawaiian Islands are located in the mid-Pacific many hundreds of kilometers from any of the major continents.

and pteridophytes, all found nowhere else. Most amazing are the fruit flies of which more than 250 species are known. There are 21 species of these flies on the island of Hawaii alone, and because none of these are found elsewhere it must be true that 750,000 years is sufficient for clear-cut speciation of fruit flies. It is assumed that the pioneer species came to Hawaii from Maui, only 48 kilometers away, and that a passage from one island to another must be an exceedingly rare event. As a matter of fact, only about 3 percent of all Hawaiian fruit fly species are found on more than one island.

Two other groups have made significant contributions to our knowledge of evolution—the land and freshwater snails and a songbird family, *Drepanididae*, or honeycreepers. The honeycreepers live everywhere from high mountains to the seashore and have evolved many striking differences in plumage and food habits, the latter being very clearly shown by shape and size of beak. The snails are so variable that different species are often found in adjacent narrow canyons or on different sides of a lava flow.

The importation of new species into the Hawaiian Islands promises to prove fatal to many of the indigenous species. At least half of the native birds are in danger and less than 25 percent of the total area of the islands is now covered by the types of vegetation that were there when Captain Cook arrived in 1778. The long freedom from competition and enemies was at first a distinct advantage for the original pioneers but their failure to acquire protective reactions and adaptations has now become the chief reason for the extermination of their descendants.

The Dodo. We must also mention the famous dodo bird that lived and developed in isolation on the island of Mauritius in the Indian Ocean. The island was discovered in 1598, and by 1684 the dodo had been exterminated by man and the creatures he introduced—the rats and pigs. Other birds, now also extinct, inhabited the island. The dodo typifies in popular fancy a great deal of what scientists have learned about the reactions of life to isolation and subsequent competition.

Krakatoa. One of the most instructive natural experiments in the dispersal of organisms was initiated by the devastation of the island of Krakatoa by a tremendous volcanic explosion in 1883. The island, located between Java and Sumatra in the East Indies, was completely sterilized of all life and left covered by a layer of pulverized rock and ash up to 60 meters thick. A naturalist visited the island 3 months after the eruption, and the only visible form of life he encountered was a single spider. Three years later, a number of algae, 11 ferns, and 16 flowering plants had appeared. Thirteen years after the eruption there were 50 flowering plants, and in 1905 there were 137 varieties and the island was beginning to support heavy vegetation. By 1930 the whole island was covered by young forest growth and there were plenty of birds and reptiles. Although it is only 40 kilometers to the nearest land, these figures illustrate that life cannot be dispersed over water immediately.

It is instructive to speculate about what would have happened on Krakatoa if it had appeared in past geologic ages before the flowering plants had evolved or at an even earlier time when algae were the highest forms of life. It is obvious that in these hypothetical situations the colonization would have had to await the development of higher forms of life and that such islands would probably have been largely barren of life, even until the forces of erosion destroyed them.

It is likewise interesting to speculate on what happened to the seeds and spores of the plants that were growing on Krakatoa when the island exploded. The dust from the eruption rained down over an area of thousands of square kilometers, and floating masses of volcanic ash spotted the surrounding ocean for months afterward. The finer particles of ash and dust reached the higher levels of the atmosphere where they circled the earth for over a year. Spores and seeds comparable in size to dust fragments may have landed at far distant places, thus carrying on another phase of dispersal.

Continental islands display the same geologic structure and characteristics of the larger continental masses of which they are fundamentally a part. Their life has been derived mainly from the nearby land, but they nevertheless reveal many instances of species formation. Practically all the continental islands were repeatedly attached to, and separated from, their respective continents by the rise and fall of sea level during the Ice Age. Ireland joined England and both were connected with continental Europe during periods of low water. Ceylon and India were joined, and the Persian Gulf was a marshy plain. Large tracts in the East Indies were laid bare, and Sumatra, Borneo, and Java became attached to the Asiatic mainland. New Guinea and Australia were connected by a broad lowland. We have already mentioned many of the effects of the Ice Age on continental islands, so we need not elaborate on any of the details here.

Great Britain. Geologists have learned a great deal about the Pleistocene history of the British Isles by studying the distribution of its plants and animals in relation to those of continental Europe. Students of the subject generally agree that the last glacial advance covered all Ireland and nearly all of Britain and forced most forms of land life to retreat to the Continent. With glacial sea level at least 90 meters lower than it now is, there was a land connection between Ireland and England and between England and the Continent. A vast, swampy lowland, which is now submerged to form the English Channel, extended from Denmark to France. Through this region flowed a large river, of which the present Rhine and Thames were the chief tributaries.

As the glaciers melted, sea level began to rise and Britain became open to recolonization from the mainland. Cold-tolerant plants and animals —among them elephants, lions, hyenas, bears, rhinoceroses, man, and many smaller mammals—took up early residence. Some of these newcomers also reached Ireland, but the rising waters closed behind them and Ireland was severed from England before England was separated from the mainland. This is the explanation biogeographers offer for the old question of why there are no snakes in Ireland. By the time the climate became warm enough for snakes, the route to Ireland had been closed. England continued to receive not only snakes but a variety of other organisms, including oak, elm, and alder.

Figure 18.16   *Great Britain and nearby parts of continental Europe, showing the wide tracts that lie less than 180 meters below sea level. During the Ice Age there was free communication between the islands and the Continent, and plants and animals migrated in both directions. A large river, of which the Rhine and the Thames were tributaries, drained into the North Sea. The last separation of Great Britain from the mainland took place about 7,000 years ago. Present lands are shown in light blue, chief rivers and water over 180 meters deep in dark blue, and shallowly submerged land in white.*

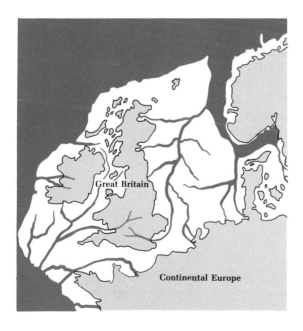

Man himself crossed dry-shod. A substantial part of the story is preserved in the Dogger Bank, a now submerged peat bog lying between England and Denmark that has been relatively undisturbed by marine erosion. This area is famous for its fishing, and the nets of fisherman have brought up not only masses of drowned peat, or *moorlog*, but also bones of extinct mammals and, strangely enough, a man-made harpoon.

According to carbon 14 dating of the peat beds, the last ice melted from Great Britain about 10,000 years ago, and the islands were finally separated from the Continent about 7,000 years ago. Since then, no plants, animals, or men have been able to gain entry into Britain by land.

Indonesia. The numerous islands constituting the western part of Indonesia rise from the shallow seas of the Sunda Shelf. Included are hundreds of small islands and large ones such as Sumatra, Borneo, and Java. The interesting fact insofar as our present topic of past and present island connections is concerned is that the entire Sunda Shelf would become continuous land with a slight drop in the level of the sea. This tract would constitute a continuation of the Asian mainland and would almost connect with the Philippines to the north.

The island of Java has yielded seventy-five species of fossil mammals of late Cenozoic age including the famous ape-man, *Pithecanthropus* (*Homo erectus*). All of these mammals are related to forms from the Asiatic mainland. The Java rhinoceros is the only large mammal remaining from the extensive fauna of the not-too-distant past. Pleistocene mammalian remains are also found less abundantly on Sumatra and Borneo.

Lowering of the sea level by only about 100 meters would not bring the islands to the east and south into connection with the Sunda Shelf, for the water is about 1,000 meters deep between Borneo and Celebes, and even deeper between Celebes and New Guinea. It is therefore surprising to find elephant and giant tortoise remains on Celebes. A possible migration route from

China by way of Formosa and the Philippines has been suggested. The life of Celebes has affinities with Australia; about 25 percent of the inhabitants resemble those of that continent. It is evident however that relatively few plants and animals succeeded in crossing all the deep-water barriers between Australia and Asia, or vice versa. The break in the migration route is known as *Wallace's line*. It must have been in existence for millions of years, for the fossil life shows the same great differences as do the living forms.

INTERCONTINENTAL EXCHANGE

North and South America. Perhaps the most striking and clear-cut case of isolation followed by intercontinental migration and competition involves North and South America. These continents are connected at present by the narrow, low-lying Isthmus of Panama, which has been alternately submerged and elevated a number of times in the geologic past. A study of the earliest mammalian forms inhabiting North and South America indicates that the two landmasses were joined at the beginning of the Cenozoic, about 65 million years ago. At that time, the ancestors of the South American armadillos and tree sloths entered the continent, together with certain primitive marsupials. Following this early connection, the land bridge sank beneath the ocean.

Over the millions of years during which North and South America were separated, their respective plants and animals evolved into distinct species, genera, families, and even orders. Especially significant were the curious, hoofed herbivores of South America, which are now entirely extinct. These creatures were preyed on by old-fashioned marsupial carnivores, because no advanced, or placental, flesh-eating forms were present. During the period of isolation a few small mammals, including monkeys and rodents, entered South America, probably by island-hopping on floating vegetation. While North America was receiving periodic waves of highly competitive migrants from Asia, the South

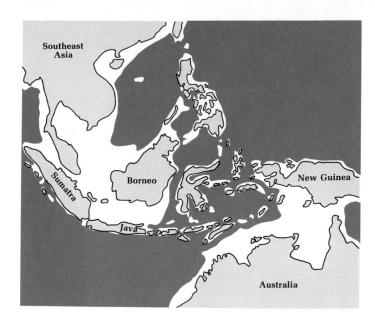

Figure 18.17 Southeastern Asia, the East Indies, and Australia, showing the area that lies within approximately 180 meters of sea level. During the stages of maximum glacial accumulation, these tracts were laid bare and became routes of travel between what are now islands. This explains the presence of the rhinoceros in Java and of many large fossil mammals on other islands. New Guinea and Australia were connected and their plants and animals are similar. The deep water between Australia and the large islands to the northwest has effectively halted the passage of most plants and animals. Present land areas are shown in light blue, water over 180 meters deep in dark blue, and shallowly submerged areas in white.

American forms led a placid, sheltered existence free from the stimulation (or, more realistically perhaps, the harassment) of sharp competition.

A slow adjustment of the earth's crust again brought the Panamanian land bridge above sea level in late Pliocene time. In very short order there was a wholesale exchange of land mammals. Competition between the long-separated groups commenced in an arena of continental size. Most seriously affected were the South American hoofed herbivores, which not only were subjected to the competition of northern

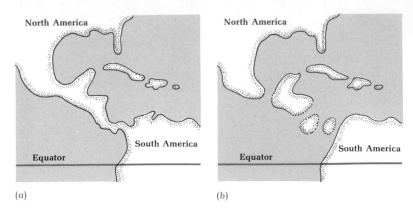

(a)   (b)

Figure 18.18   The Panama land bridge as it is today
(a) and the same region as it may have appeared
during the general submergence of the middle Tertiary (b). When the isthmus is above water, it forms
a migration route for land animals and is a barrier
to sea animals. When the area is submerged, it becomes a barrier to land life but opens a route for
the exchange of Atlantic and Pacific sea life.

herbivores such as horses, tapirs, peccaries, camels, and deer, but also were preyed on by wolves and members of the cat family, likewise new arrivals. The native South American carnivores and herbivores made only a short and ineffectual stand before they were exterminated by the northern invaders. Perhaps in many cases the competitors never met in actual combat, for by gathering their food more efficiently, a successful group could starve less efficient forms to death at a distance.

A number of South American forms invaded North America and for a time gave its fauna a distinctive aspect. These immigrants included several kinds of ground sloths, glyptodonts, large armadillos, and capybaras (large rodents related to guinea pigs). For a while these creatures ranged widely and even became common. Ground sloth remains have been found as far north as Alaska, and the recency of some of their fossils discovered in the United States and Argentina indicates that they were exterminated only within the last few thousand years. All that now remain in the Northern Hemisphere of these South American forms are the armadillo and porcupine. Most of the larger mammals of South America are descendants of northern stock. The llama, a member of the camel family, has found refuge there, although its relatives have become extinct in their original North American homeland. The puma prowls from Alaska to Tierra del Fuego and is one of the widest ranging of all mammals. A relative of the Virginia and mule

deer inhabits the northeastern part of South America.

The Panama land bridge not only influenced the migration of land animals but also had a simultaneous effect on life in the adjacent coastal waters. When the isthmus was submerged and closed to land animals, the straits opened to sea life, and when the isthmus emerged, it became a barrier to sea creatures. We can understand the sea life on opposite sides of the isthmus only in the light of geologic history. Surprisingly, most of the fish species on opposite sides of the isthmus are very similar. Detailed study, however, reveals that the species found on either side occur in almost identical pairs. A given species on the Atlantic side usually has a related and almost identical twin species on the Pacific side. Also, the nearest relative of an Atlantic species is not found in a nearby part of the Atlantic side but in the Pacific. There are over one hundred twin species of fish in the area. The degree of difference between the two members of a pair varies somewhat, but in the opinion of experts who have studied the subject, it is doubtful that any species of offshore fishes are actually identical on the two sides of the isthmus. The same is true of crabs and other invertebrates.

It is quite obvious that the marine fauna was split into two colonies by the uplift of the land bridge during the Pliocene. This event occurred perhaps 2.5 million years ago, which is therefore the approximate length of time required for species formation among fish under these particular conditions. Although isolation is involved, the two populations were both very large, and evolu-

Figure 18.19 The Bering Strait between North America and Asia, showing the shallow, submerged shelf that would become land with a drop in sea level of only 45 meters. During the Ice Age and at various times during the Tertiary Period, this land bridge was open to migrations between the Americas and Eurasia. In many ways it is the most important land bridge on earth. Present land areas are shown in light blue, water over 45 meters deep in dark blue, and water less than 45 meters deep in white.

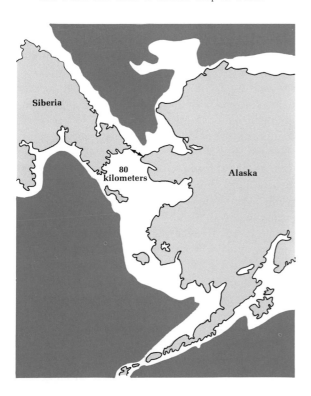

tion had not proceeded with the speed it could have attained in smaller groups. Many unanswered questions about land bridges still confront geologists, but obviously they must be examined in the light of both their underwater effects and their dry-land effects.

Eurasia and the Americas. Less spectacular, but more important, are the migrations that passed over the Bering Strait between Alaska and Siberia. Here a relative lowering of sea level by about 45 meters would today provide a dry-land crossing, and there is evidence that changes of this magnitude occurred many times in the past to connect the Old World and the New. If the land were to rise again and remain above water for any appreciable length of time, the animals of all the great continental masses could potentially be brought into contact, thus creating a "world continent" of land life. Careful study of the available evidence indicates a strong intermingling between North American and Eurasian stocks in the late Eocene and early Oligocene, in the late Miocene, and again in the late Pliocene and Pleistocene. Between these times, as at present, the exchange of forms was at a low ebb. We must not assume that there was a general two-way rush of animals across the Bering bridge every time the land rose. The area is situated in a relatively cold latitude, and many warmth-loving forms would not have utilized it. Early in the Eocene there was a great deal of similarity between the mammals of the two landmasses, and we assume that migration at this time was heavy. It now appears likely that northeastern North America was in contact with northwestern Europe, so that two possible migration routes were available. Migrations after the Eocene were more selective. During the Miocene the migrants included dogs, members of the elephant family, rhinoceroses, and deer. Eurasian stay-at-homes included hyenas, gazelles, pigs, hippopotamuses, and giraffes. Animals that did not leave the New World included the raccoon, ground sloth, pronghorn antelope, and peccary. During the

Pleistocene the coldness of the Bering area seems to have definitely discouraged the migration of warm-climate animals. At this time North America received mammoths, musk-oxen, bison, mountain goats, mountain sheep, and moose from Eurasia, while in the other direction camels, beaver, and marmot padded their way to more comfortable climes.

Australia. Man's sudden introduction of foreign plants and animals into Australia provides a striking historic example of the effects of isolation and subsequent competition. Australia was separated throughout the Tertiary Period from the other great landmasses, and only occasionally have land mammals been able to reach it. Ancestors of the famous platypus, or "duckbill," may have arrived as long ago as the Triassic or Jurassic. Later on, during the Late Cretaceous or early in the Tertiary, the marsupials arrived, and still later, by a process of island-hopping, came rats, mice, the native dog, or dingo, and man. The arrival of these newcomers, together with bats, is not difficult to understand, especially in view of the close—indeed, almost oddly cliquish—company kept by man, dogs, and rodents.

When white men colonized Australia, they upset irreparably the balance of nature and set in motion a biological reaction that must ultimately prove fatal to most of the original inhabitants of the island continent. Among the animals innocently but ignorantly introduced were the fox, brown rat, and rabbit. Domesticated cats and dogs also took to the brush and became, to all intents and purposes, wild beasts. The rabbit is responsible for the almost complete disappearance of the rat kangaroo, rabbit bandicoot, and other native marsupials. Of more immediate interest are the harmful effects of rabbits on the sheep industry; this has touched off an endless campaign of shooting, poisoning, fumigating, fence building, and biological warfare, but it appears it has been all to little avail.

From India came two kinds of deer and the water buffalo, now running wild by the thousands in Arnheim Land. The American grey squirrel is also spreading widely near Melbourne.

About a dozen kinds of foreign birds have become acclimatized, the most detrimental to human welfare being the house sparrow and the starling. Nor should we overlook the plants that have been introduced. The cactus has taken over millions of acres and apparently finds Australia well suited to it's needs. In this case man was able to combat the cactus menace by introducing a small insect that eats it, thus keeping it in check.

Strategy of Change

Our illustrations of dispersal, isolation, and speciation have been drawn from the most obvious cases known to biologists and paleontologists. The reactions of large land animals or plants to expanding or contracting living space are relatively easy to understand, for man has reacted in similar ways in his conquest of the earth. In general we have emphasized what happens when additional territory becomes available to a species. If the new environment is empty of life, the food chain must be built up from the base, starting with the formation of soil, as on the devastated island of Krakatoa. If food is adequate but is not exactly the customary fare of the newcomer, he may have to adjust to the new diet, as did Darwin's finches. If the native food is of a kind that is already well suited to his needs, the invader may spread with wildfire speed, as did the rabbit in Australia. If the food resources are already being used by inefficient or noncompetitive forms, these may be pushed aside with relative ease, as were the inhabitants of South America. But if the territory is held by hardy and efficient species, the invader may be repulsed, or at most will make but little inroads, as the South American fauna did in North America.

In all these instances we have assumed that life has been able to reach and at least make an initial attempt to occupy newly available territory. The record proves life has indeed reached and colonized all but the most hostile portions of the earth. It is also clear that the process takes time and that it is partly an accidental matter as to what arrives and when it arrives. It is plain to see that highly specialized land animals may never reach out-of-the-way areas such as remote islands. But more significant is the fact that the arrangement of continents and oceans has been such that all major forms of life have in time reached all large areas for which they are adapted or could possibly adapt.

We are tempted to refer to the reactions of organisms in protecting themselves and preserving their individual species by using such human terms as competition, struggle, victory, and defeat. Call the process what we will; reduce it to cold science or elevate it to poetic levels; the fact remains that *life as we know it is capable of certain reactions that have enabled it to survive and proliferate despite a wide variety of radical changes in its surroundings.*

The following facts seem to be important to the discussion at hand:

1. The physical world is in a state of dynamic change. The changes have at times exceeded the tolerances of individual species and have frequently destroyed all life in limited areas. But never have physical changes killed all the protoplasm on earth or totally exterminated the life of a major environment.

2. Most physical changes have been of a gradual nature. The growth of mountains, spread of glaciers, submergence or emergence of land, and other widespread effects have come about over time periods equal to thousands or millions of generations of the organisms that were affected.

3. Surviving life forms have a few standard reactions for survival. Probably most important is the production of an excess of offspring and a means of dispersing them so that some will escape local conditions. Dispersal thus allows a species to slowly follow a shifting favorable environment or to escape an unfavorable one. The production of an excess of offspring is wasteful, but a balance has been achieved by most forms between overproduction and underproduction.

4. The genetic material of life as we know it is capable of being altered so as to bring about changes in *form, function,* and *behavior.* These mutations may be severe enough to prevent the development of the potential individual, or they may be so mild as to have no influence on its future. There are, however, on the average enough variations in a population to provide individuals capable of existing in environments slightly different from those in which their predecessors lived. *The process of genetic change seems geared to the process of physical change—* neither too fast nor too slow, but with enough margin to ensure that the poorly adapted will invariably be reduced in number, pushed into out-of-the-way places, or exterminated, while the well adapted will survive and thrive and reach out into new environments.

5. Genetic change may be induced by various means, but in nature practically all of it results from *radiation* coming from unstable materials in the earth, or from *cosmic rays* from space. The genetic material is held together by bonds that can be broken by types of penetrating natural radiation that have pervaded the environment on earth since life began. Again, the reaction seems related to the intensity of the cause—too much radiation would totally disrupt the genetic material, too little radiation would freeze the pattern into permanent forms. The result in any event is a waste of individuals, but, again, sufficient new forms are produced to meet the shifts of the environment.

Ignoring any explanation calling for supernatural design or purpose, we begin to see how the reactions of life as we know it have produced evolution and achieved progress. In this inquiry experience tells us that the principle of uniformity should not be ignored and that Darwinian evolution may have been at work even in such distant times past as the remote prefossil eras.

From what we know about the success and

failure of present-day organisms, we may assume that early accidental combinations of matter may have produced a variety of living or near-living molecules, including some not now in existence. Among those which could not survive and were consequently eliminated we might include the following:

1. Forms that produced no new individuals, or at most too few to escape local conditions.

2. Forms that had no means of dispersal from local conditions and were subject to destruction by localized natural upsets.

3. Forms with genetic material too readily broken down by prevailing destructive influences such as heat, radiation, or chemical reaction—in other words, too much mutation.

4. Forms with genetic material too firmly con-structed to be affected by prevailing destructive influences—in other words, not enough mutation.

Eliminating these unsuccessful forms, we are left with life as we know it, which is able to change with the environment and in time even to alter that environment, to some degree at least.

In spite of the many safeguards and tolerances achieved, as it were, by the experience of its ancestors, no organism (including man) is able to look ahead and foresee all possible emergencies. Life has not been prepared for every type and degree of physical change; the wayside of evolution is strewn with the remains of those which a nonmalicious but nevertheless very exacting and inexorable environment has tried and found wanting. The subject of extermination will be treated in the next chapter.

A SUMMARY STATEMENT

Individual species have spread outward from their places of origin according to prevailing opportunities and conditions. Geologic processes have been constantly operating to create and destroy land areas and bodies of water, and earth movements are chiefly responsible for opening up migration routes between the major land areas and ocean basins. Under the influence of changes in their environment, plants and animals have been forced to migrate, have become adapted to new conditions, or have been exterminated. Destructive and constructive topographic effects have had great influence in bringing about evolutionary advances. On occasions where relatively small populations of organisms have been isolated by geographic or geologic changes, the formation of new species has been favored.

A few well-known instances of geologic control of biological distribution and evolution are the colonization of the Galapagos Islands and the Hawaiian Islands, the formation of the Panama land bridges between North and South America, the opening and closing of the Bering Strait between North America and Eurasia, and the Pleistocene migrations into, and away from, the British Isles. The many radical changes brought about by man in Australia indicate how invasion and competition may have operated in the past in the absence of man.

FOR ADDITIONAL READING

Barnett, Lincoln, and the Editors of *Life, The Wonders of Life on Earth*. New York: Time Inc., 1960.

Bowman, Robert I., ed., *The Galapagos: Proceedings of the Symposia of the Galapagos International Scientific Project*. Los Angeles: University of California Press, 1966.

Cain, A. J., *Animal Species and Their Evolution.* New York: Harper, 1960.

Darlington, C. D., *The Evolution of Genetic Systems,* 2nd ed. New York: Cambridge University Press, 1958.

Hopkins, D. M., ed., *The Bering Land Bridge.* Stanford, Calif.: Stanford University Press, 1967.

Mayr, Ernst, *Animal Species and Evolution.* Cambridge, Mass.: Harvard University Press, 1963.

West, R. G., *Pleistocene Geology and Biology.* New York: Wiley, 1968.

19 EXTINCTION AND REPLACEMENT

Charles Darwin's great work *The Origin of Species* gives a rational scientific explanation of how the earth came to be populated by thousands of diverse organic species. The central theme is epitomized in the popular phrase, "survival of the fittest," which implies that as some organisms survive, others must perish. Another great book remains to be written as a companion piece to Darwin's *Origin;* it might bear the title *The Disappearance of Species.* Darwin knew that many forms of life had disappeared, but the true mag-

The great cave bear was contemporary with man in the caves of prehistoric Europe. There can be little doubt that man had much to do with the extermination of this creature. However, many large Pleistocene mammals seem to have disappeared without the intervention of man. (Field Museum of Natural History, Chicago.)

nitude of extinct life is only now beginning to be appreciated as geologists uncover more and more evidence of the past.

The Scope of Extinctions

The problem of extermination began to receive serious attention when in 1801 Baron Georges Cuvier announced his discovery of twenty-three species of animals no longer on earth. He believed that they had been eliminated by divinely ordered catastrophes. Before the earth had been thoroughly explored, many people believed that no forms of life were really extinct, but still lived in remote jungles, on undiscovered islands, or deep in the ocean. Thomas Jefferson, on the basis of a large fossil claw (of a ground sloth) discovered in a cave in Virginia, believed that a huge lion still existed in the western forests. It is said that he hoped Lewis and Clark would find the beast alive. Other thinkers appealed to

Noah's Flood as the agent that had wiped out all extinct animals. But as exploration and study of living and fossil organisms progressed, all these explanations were found wanting, and the fact emerged that the fate of the vast majority of created species has been outright extinction.

It has been calculated that 98 percent of all living families of vertebrates have descended from 12 percent of those alive in the Mesozoic Era. Death is commonplace enough, but when it is known to have removed entire species, genera, families, orders, classes, and even phyla, it becomes a factor of major importance in the understanding of life.

The realization that organisms have been eliminated on a large scale shook the theological thinking of the Western world almost as much as the idea of evolution. The concept seems to reflect adversely on the wisdom of the Creator and the perfection of His works: Must organisms be created only to be destroyed? Must cruel, ruthless competition be the means of eliminating the weak, helpless, and less fit? Philosophers of religion are still struggling with these questions.

The dying out of species is one of the problems of biology that paleontology is best qualified to study and solve. In this chapter we will briefly consider what has been observed in historic time, what the prehistoric unrecorded influence of man may have been, what has been learned about extermination in the remote past, and what may be inferred from so-called "living fossils" and other observable phenomena.

Man the Exterminator

A recent study concluded that man has been directly or indirectly responsible for the disappearance or near disappearance of more than 450 species of animals. Since the year 1600 more than 125 distinct species of birds and mammals are known to have disappeared, plus nearly 100 subspecies, races, or varieties. Among the large birds and mammals known to have been destroyed by

man are the following (the date indicates the death of the last known specimen): Steller's sea cow (1768), great auk (1844), dodo (1861), Labrador duck (1875), auroch (wild ox) of Europe (1878), wild horse, *Equus quagga* (1878), sea mink (early 1900s), passenger pigeon (1914), Wake Island rail (1945). Dozens of less well known species, especially birds have gone the same route. One authority has concluded that if the current rate of killing goes on, most of the remaining 4,062 species of mammals not beneficial to man will be gone in about 30 years. A list of endangered species would be too lengthy for available space. Notable are the following: the Asiatic lion, glorified for centuries in Eastern art and literature, and now down to about 300 individuals; the beautiful Arabian oryx, driven to extinction by being hunted from motorized vehicles; the Indian tiger, hunted for its pelt; the blue whale, prized commercially, with only 2,000 estimated remaining; the Tasmanian "wolf," a marsupial, only a few of which have been seen in the past decade; the Java rhinoceros, whose horn is prized as a medicine, only a few dozen of which remain.

Also in a precarious condition from human interventions and depredations are the African lion, African rhinoceros, cheetah, gorilla, roan antelope, wild dog, koala bear, snow leopard, panda, American bald eagle, American alligator, musk-ox, and the giant sea turtle.

Island faunas have suffered the most. Of the approximately 225 large animals known to have become extinct in the past 4,000 years at least 166 were island dwellers. These forms had developed no reaction to protect them against the sudden appearance of man and his domestic animals such as dogs, pigs, and goats.

There are a few bright spots. The American bison was reduced to a few dozen individuals before coming under adequate protection. Similarly the Alpine ibex has recovered from less than 50 individuals to at least 6,000. The trumpeter swan once thought to be entirely extinct is now 2,000 strong. Other American species on the way back are the sea otter and whooping

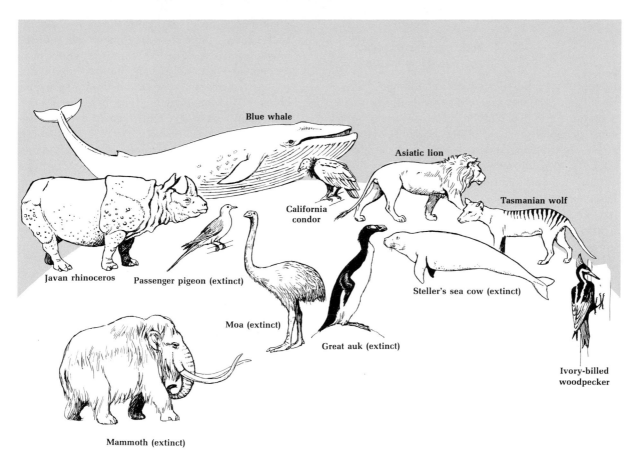

Figure 19.1 *Some animal species that have become extinct in recent history and others that are presently in danger of extinction.*

crane. Alert action may save others but merely to put a few individuals in zoos will only delay the ultimate dismal outcome. Any meddling with the so-called balance of nature is certain to be fatal to some species.

This is not the place to expand on historical events, but a few general comments may be applicable to prehistoric situations. Man has brought about exterminations in a number of ways that have parallels in the natural world. Actual outright killing for food or sport is the most obvious—overpredation is the general term.

Elimination or restriction of living space by the removal of forests, draining of swamps, or destruction of grasslands has been devastating to large areas. Introduction of dangerous competitors, pests, and enemies, as with the rabbit in Australia, is especially deadly to island species. Alterations of the environment by pollutants and wastes is proving detrimental in many unexpected ways. Although many of man's harmful activities were perfectly innocent or ignorantly unintentional, the ultimate effect is still the same—other species have suffered.

It now seems fairly certain that humans have been capable of creating and using death-dealing weapons for almost 2 million years. At first came such elemental acts as killing a tortoise with a stone or knocking down a bird with a stick. Clubs, spears, throwing sticks, bolas, and other devices came in due time, and nothing was safe from man's depredations. The use of fire in driving and killing game no doubt played a large part in the spread of man and the death of contemporary animals. Organized hunting and a taste for cooked meat are naturally associated.

We are especially suspicious of man as the agent which brought about the death of many large mammals between 5,000 and 15,000 years ago. In North America the list of the dead includes the extinct peccary, the imperial mammoth, the woolly mammoth, the ground sloth, the glyptodont, the giant bison, the dire wolf, the native horse, the Columbian mammoth, and the mastodon. Many of these are known to have been associated with the ancient human inhabitants of the Americas. As mentioned in a previous chapter, it seems puzzling that these large mammals could have lived through four successive ice advances, only to die out within the last few thousand years. The circumstantial evidence against man is strong, for his entry into this hemisphere coincides with the beginning of the exterminations just enumerated.

Animal against Animal— North and South America in the Early Pleistocene

The idea of evolution seems to have taken root in Charles Darwin's mind during his explorations in South America, perhaps while he contemplated the bones of extinct animals in Argentine

Figure 19.2   The dodo (Dodo ineptus). (a) The skeleton; (b) the reconstruction. This clumsy, stupid, and flightless bird was exterminated in its home island of Mauritius within historic time but enough original material was saved to give the reconstructions shown here. (a, Smithsonian Institution; b, by permission of the Trustees of the British Museum, Natural History.)

(a)　　　　　　　(b)

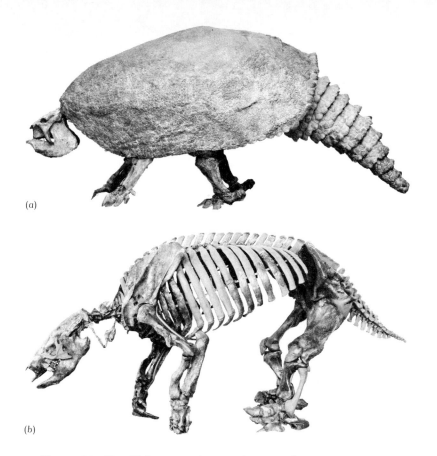

(a)

(b)

Figure 19.3 Two Pleistocene edentates from South America. (a) The armor-plated skeleton of a glyptodon from Argentina; (b) the skeleton of the giant ground sloth Scelidotherium from Bolivia. Relatives of both these animals migrated to North America and thrived for a short period in the Pleistocene. (a, Field Museum of Natural History, Chicago; b, Smithsonian Institution.)

soil. Later studies have revealed a drama of death and extermination in South America that is illustrative of Darwin's classical concept of direct competition for food and living space—a clear example of the survival of the fittest.

North and South America apparently were in contact during the Cretaceous and early Paleocene, and both had similar primitive mammals. By the late Paleocene the Panamanian bridge was broken and the South American fauna was left

in isolation to diversify into many unique and unusual forms. Disregarding certain rodents and monkeys who migrated in the middle Tertiary, no other large, land-living animals reached South America until the Panamanian land bridge reappeared at the beginning of Pleistocene time. A literal wave of foreign invaders swept into South America. The list includes bears, weasels, cats, pigs, deer, horses, rabbits, dogs, camels, mastodons, and tapirs. At the same time several South American groups traveled northward, the most successful being certain rodents including the porcupine and various edentates such as the ground sloths and armadillos.

The picture of superior North American animals entering South America brings to mind a well-equipped and organized army sweeping into a land of primitive, unprotected aborigines. No less than four orders of South American mammals disappeared almost immediately. South American forms were insignificant in North America. This extinction could scarely have been due to climatic or geomorphic changes in South America; it appears to be due to direct competition among species and the ascendency of the more efficient organisms.

Major Geologic Exterminations

As the paleontologist sees it there are two chief classes of exterminations: those for which he has a satisfactory explanation and those for which he has none. Those exterminations just described appear to have been due to the depredations of man or the sudden onslaught of stronger animals upon weaker ones. In the more distant past there have been even greater exterminations for which no obvious causes have been found. These affected many types of life over the entire globe. They brought total destruction to many important families of plants and animals and left many others in a greatly weakened condition. Notable extinctions characterized Late Cambrian, Late Ordovician, Late Devonian, Late Permian, and Late Cretaceous time. So marked were the last two of these critical periods that they serve as major punctuation marks in the history of the earth. Because these critical episodes had worldwide effects we look for correspondingly widespread causes or agents. Explanations range from outright catastrophism to extremely subtle and insidious reactions. The following have been suggested: (1) the impact of comets or asteroids with resultant heat, blast, shock, and tidal waves; (2) bursts of lethal radiation from space such as might accompany supernovae, solar flares, or geomagnetic polarity reversals; (3) widespread simultaneous mountain building with resultant climatic changes; (4) chemical changes in the ocean, such as an increase or decrease in oxygen

or salt content; (5) worldwide transgressions of the sea caused either by ordinary erosional and depositional effects or by strong movements of continents or submarine mountains; (6) atmospheric effects, such as an increase or decrease of oxygen content arising from greater or lesser photosynthetic activity (the spread of angiosperms on land or diatoms in the ocean may have had such effects); (7) outbreaks of volcanism with widespread dispersal of ash and dust that may have altered the weather and climate of the entire earth.

There may be elements of truth in any of these explanations. They all have catastrophic overtones, but drastic consequences suggest drastic effects and we are dealing with death on a large scale in the two great exterminations that serve to divide the major fossil-bearing eras in an unmistakable way.

CLOSE OF THE MESOZOIC— DEATH OF THE GREAT REPTILES

The most popular of all extinct animals are the dinosaurs; their very name has come to suggest something dead and out-of-date. Their disappearance took place 60 million years ago. Man obviously had nothing to do with it, and we cannot reasonably apply the concept of displacement by superior forms of life as seen in the extermination of the South American mammals. There were no bigger and better dinosaurs, and the mammals seem too small and ineffectual to be of significance in disposing of the reptile giants.

Eight orders of reptiles were in existence in the Cretaceous; when the Cenozoic opened there were only four. Among the groups that disappeared were the flying reptiles, or *pterosaurs,* the long-necked marine *plesiosaurs,* and two orders of dinosaurs. In the main, the animals that vanished were large and relatively specialized for certain modes of life. Those groups that survived, including turtles, crocodiles, snakes, lizards, and

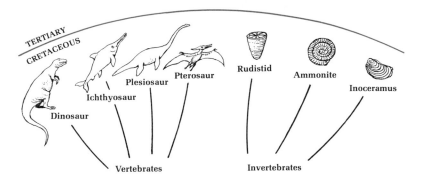

Figure 19.4 *Animals exterminated at or near the close of the Mesozoic Era.*

rhyncocephalians, are all relatively small and perhaps less specialized. Chief interest has centered on the problem of the disappearance of the dinosaurs. This spectacular group had dominated the earth for at least 140 million years and was apparently holding its own up to the time it disappeared. The dinosaurs had already weathered many severe changes that seem, at least from our viewpoint in the present, to be fully as critical as the ones that finished their careers. They had passed through a tremendous revolution in the plant world during the middle of the Cretaceous, which saw almost all the older vegetative types replaced by the flowering plants. In North America, especially, they had successfully lived through the extensive marine inundations that had eliminated probably at least 40 percent of their former living space, restricting them to areas that were certainly not their most favored environments. Even the great sauropods had survived the destruction of large areas of their low-lying habitat, and were still in existence at the very end of the Cretaceous.

Many theories have been proposed to explain the extinction of the dinosaurs. One attributes their decline to the rise of egg-eating mammals or even egg-eating reptiles. Although it is probably true that dinosaurs did not instinctively pro-

tect their nests from potential danger, it seems unlikely that dinosaur eggs could have gone unappreciated by predators for over 140 million years, only to become suddenly desirable at the close of the Cretaceous. Moreover, the egg-laying crocodiles and turtles seem to have escaped this menace.

Extremes of temperature have naturally been invoked as potential death-dealing influences. Like all reptiles, the dinosaurs avoided cold areas, and a general worldwide lowering of temperatures could have restricted vital activities and perhaps eventually caused death. The inferred effects of heat are not so obvious. Proponents of the "too hot" theory argue that higher temperatures, unfavorable to the survival of sperms, induced a fatal decline in the birth rate.

Another theory links the extinction of dinosaurs to the rise of modern types of plants. Proponents of this idea argue that the dinosaurs originated with and became adjusted to such plants as conifers, cycads, and ginkgoes, which release relatively little oxgen, and that they could not adapt to the more vigorous flowering plants that came into prominence during the Cretaceous. The contention is that the dinosaurs were literally overoxidized, or burned out, by the more rapid life processes and greater food requirements impressed on them by an oxygen-rich environment.

Cosmic or astronomical events have also been

Figure 19.5   Triceratops, a common dinosaur of the latest Cretaceous. This and contemporary dinosaurs disappeared at the end of the Mesozoic. (By permission of the Trustees of the British Museum, Natural History.)

suggested as responsible for dinosaur extinction. The charged particles known as *cosmic rays*, which reach the earth from outer space and which are known to be responsible for abrupt genetic changes, may have suddenly increased to the point where harmful mutations multiplied and destroyed the race. Such effects may have accompanied reversals of the earth's magnetic field as discussed in Chapter 6. Even more fantastic is the theory that the impact effects of a tremendous meteor from space may have killed all dinosaurs simultaneously over the entire earth. Heat and shock waves, it is suggested, might wreak more destruction among large animals that could find no shelter, than among smaller animals.

Lastly, there are theories that attribute extinction to physiological changes that occurred within the dinosaurs themselves. One authority contends that dinosaurs may have suffered from overly active pituitary glands, which caused them to grow larger and larger and to develop an excess of horny or bony useless excrescences. With time these irregularities placed an excessive drain on vital energies, and the dinosaurs succumbed.

Although the greatest interest and aura of mystery attaches to the dinosaurs, we must not forget that other great reptilian orders died out in the Cretaceous Period. Ichthyosaurs and plesiosaurs disappeared from the seas and pterosaurs from the air. Among the invertebrates there were equally impressive exterminations, specifically the ammonites and belemnites, certain lines of oysterlike pelecypods, and the huge rudistids. Whatever affected the dinosaurs seems also to have affected these other forms of life, and the most acceptable theory ought to explain as many of the disappearances as possible. It must be emphasized that the exterminations were neither instantaneous nor simultaneous. Most of the dinosaur genera had already gone before the end of the Cretaceous, and the same is true of the pterosaurs. The final extinction of the plesiosaurs does not coincide with that of the ichthyosaurs, and not all ammonites died together.

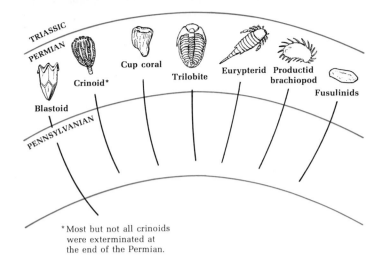

*Most but not all crinoids were exterminated at the end of the Permian.

Figure 19.6 Animals exterminated at or near the close of the Paleozoic Era.

CLOSE OF THE PALEOZOIC— TIME OF THE GREAT DYING

One hundred and twenty-five million years before the extinction of the giant reptiles there was an even greater extermination. At or near the close of the Permian Period, nearly half of the known families of animals of the entire earth disappeared. This included 75 percent of the amphibian families and 80 percent of the reptile families. The most drastic changes, however, were among the invertebrates that lived in the sea.

The list of marine animals exterminated includes not only groups that were apparently almost at the end of their careers, but also others that were numerous and worldwide in distribution. Among the ancient lines that were finally exterminated were the trilobites and eurypterids. The trilobites had been steadily declining from their high point in the Late Cambrian and Ordovician Periods, and during Permian time were represented by only one family with a relatively few species. The elimination of the lingering trilobite family would merit no special comment if it had not coincided with other, more striking, exterminations. The eurypterids, like the trilobites, were nearing their racial end and apparently needed only a final nudge to push them over the brink.

Other groups were at or near the peak of their development when they were exterminated. Most notable of these are the productid brachiopods, a prolific group represented by many species in the Devonian, Mississippian, Pennsylvanian, and Permian. None has been found in the Triassic. Remains of Permian productids have been found on all continents and Greenland; the group was

apparently adapted to a variety of marine conditions. They disappeared completely near the close of the Permian, leaving no descendants whatsoever. Another eminently successful late Paleozoic group was the protozoan subfamily, the fusulinids. These small shell-bearing *Foraminifera* were very abundant in late Paleozoic seas, and at least five hundred species are known from the Mississippian, Pennsylvanian, and Permian Periods. Fusulinid remains have been found in rocks of all continents; they were apparently successful in a variety of environments. No survivors or descendants have been discovered in Triassic beds however.

The animals that survived the critical late Paleozoic Era may be divided into two groups: those that lingered on in a much weakened condition and shortly disappeared, and those that revived and increased. Three groups that survived only to vanish in the Triassic were the orthocone (straight-shell) cephalopods, the conodonts, and the conularids. The straight-shell cephalopods reached their high point long before, during the Ordovician, and were rapidly tapering off at the time of the critical period. The

Figure 19.7   Marine arthropod, Eurypterus lacustris, *from the Silurian of New York. Eurypterids became extinct at the close of the Paleozoic. (Buffalo Museum of Science.)*

Figure 19.8   Productid brachiopod from the Permian Period. All species of this type of shelled marine invertebrate disappeared at or near the close of the Paleozoic Era. Total width about 4 centimeters.

conularids never were very important, and their extermination appears to have been merely a matter of time.

Most amazing is the history of the stony, or lime-secreting, corals during this period. A great variety of forms, including thousands of species, are found in middle and late Paleozoic rocks. They showed signs of decline during the Pennsylvanian and Permian, but the tendency became a landslide near the close of the Paleozoic. There must have been a few forms that survived into the Early Triassic, but these holdouts were so rare that few coral fossils have been discovered in Early Triassic deposits. By the Middle Triassic a few reef-building forms had appeared, and from then to the present the group has diversified and spread throughout the warmer seas.

The extermination near the end of the Paleozoic presents some strange contrasts with later critical periods. Marine invertebrates show the greatest effects; relatively few land animals were exterminated; and there is little evidence that more efficient life forms actively displaced weaker ones. Geologists have a strong suspicion that climatic change was the critical factor. The earth had just passed through a great ice age, and climatic extremes prevailed that influenced both lands and water bodies. A tentative theory is that the elevation of the continental masses during the late Paleozoic interfered with the equable distribution of heat, that great glaciers eventually formed and that the overall temperature of the ocean was considerably reduced. Marine animals, being more closely adjusted to temperature conditions than to any other factor of their environment, were forced to adapt to the increasing cold. Those that could not adapt became extinct.

Displacement and Replacement

We have deliberately emphasized the most abrupt, widespread and spectacular exterminations of the past. These seem to reflect severe regional or worldwide stresses of one kind or another. Such exterminations tend to overshadow the more gradual, selective exterminations, which one is tempted to call ordinary or routine, that have taken place throughout geologic history. Thus the archaeocyathids died out in the Cambrian, the receptaculitids in the Devonian, the graptolites and ostracoderms in the Mississippian, spiriferoid brachiopods, stegocephalian amphibians and conodont animals in the Triassic, and creodonts (archaic carnivorous mammals) in the Pliocene.

In all these instances the evidence is strong that the victims were gradually displaced by more efficient competitors or were driven slowly to extermination by enemies against which they had no defense. It takes no great stretch of the imagination to believe that the jawless and toothless ostracoderms were eliminated by other fish with these structures; or that the sprawling, weak-backed amphibians must have given way to agile, better-constructed reptiles; or that the defenseless, graptolites were a favorite food of fish; or that one reef-builder was replaced by another that could construct a stronger, more coherent reef.

Such prosaic, everyday matters as food getting, reproduction, and protective mechanisms must have weighed heavily in the struggle to survive. However, under ordinary conditions, each contest appears to have been a long one and the outcome long delayed.

It seems strange that the principle of uniformitarianism has been largely disregarded as a part of most hypotheses having to do with prehistoric extermination. This may be due to our lack of observation of true "natural" examples of extermination. As we mentioned before, man has been more or less involved with all known exterminations within historic time, and has also hastened the end of some species and retarded the final demise of others.

If we could imagine a world without either man or extraordinary natural crises and try to reconstruct what would then determine the success or failure of organisms, it would probably be obvious that simple, everyday factors are more important than rare and spectacular ones. The adaptations that characterize successful forms seem to center around the commonplace matter of obtaining suitable food. Food (or energy) capture is basic, but the subject is complex and difficult to analyze, even when observed in living organisms. Also, obtaining food is only one aspect of the energy problem: digestion, storage, utilization, and conversion to energy are equally important, and none of these is very obvious from fossil remains. The most successful phyla of animals are those that have developed the most varied means of gathering food. Members of the chordates, echinoderms, and arthropods, for example, have become mud-grubbers, filter-feeders, scavengers, and predators, according to their opportunities. Other groups such as the sponges, brachiopods, and bryozoans rely on only one chief method of capturing food. It goes without saying that those groups that utilize many methods of food gathering embrace numerous species; those whose food-gathering capacity is limited to one method include but a few groups of organisms.

Those species that can live on a variety of foods are better equipped to survive than those that can exist on only one type of food or particular prey. Among the world's successful omnivores are men, pigs, bears, and crows, all of which have enjoyed more than average success in the struggle to survive. The koala bear can eat only eucalyptus leaves, and the anteater only ants or termites. Obviously, if these food species should dwindle or disappear, the animals they support could not survive. This is not to say that animals with specialized diets are unsuccessful; they are merely more vulnerable to extinction.

Relative efficiency in gathering food seems to explain many of the successes and exterminations of the geologic past. The trilobites were abundant and varied, and entered many marine environments at a time when they apparently had no enemies, and but few competitors for the available food supplies. Included were herbivo-

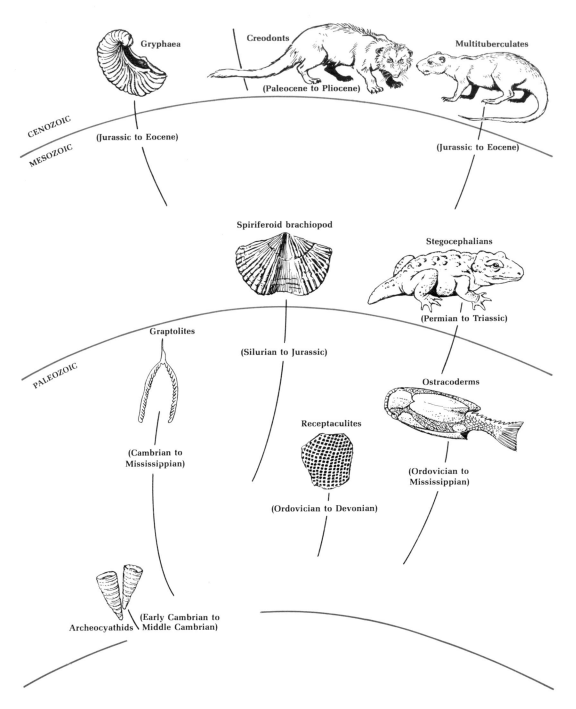

Figure 19.9 *Geologic ranges of some selected species, from origin to extermination.*

rous, carnivorous, and probably omnivorous types.

Soon came the primitive, jawless and toothless ostracoderms as potential competitors of the trilobites, but their racial life was short and they were soon displaced by their jawed and toothed descendants. The spread of active, intelligent, predaceous fish coincides with the decline of the defenseless, thin-shelled trilobites and the soft, floating graptolites. About this time, thick- or spiny-shelled brachiopods became common and sharks with shell-crushing teeth appeared, as if to feed on them. When the prey declined, so did the sharks. A little later, in the Mesozoic, we find fossils of the great marine reptiles, the ichthyosaurs and plesiosaurs, with shells of cephalopods in their stomach cavities. When the ammonites declined, so did the big reptiles.

Turning to land animals, we note the importance of improved methods of locomotion. Those creatures with better feet and legs seemed invariably to gain the ascendancy. The clumsy, weak-backed, sprawling amphibians gave way to the reptiles with stronger feet, limbs, and backbones.

The bipedal dinosaurs seem to dominate their quadrupedal associates. The front limbs of the bipedal dinosaurs, unlike those of their quadrupedal counterparts, were useful in gathering plant food, in tearing flesh, and in fighting—three advantages that obviously helped to stay for a time the doom of these monsters.

As food supplies diminished at the time of the Cretaceous–Tertiary transition, the huge dinosaurs were at an obvious disadvantage. Whatever else may be said about them, we know that dinosaurs were exceptionally large, a characteristic

Figure 19.10 Specimens of Ischadites, an Ordovician organism of unknown relationships. Individuals are shown in the original matrix and as freed specimens. (Smithsonian Institution.)

Figure 19.11 Coiled graptolite, Cryptograptus, from Silurian rocks of Cornwallis Island, northern Canada. This specimen has been freed from the matrix by acid solution and shows the original three-dimensional form. Specimen about 2 centimeters across. (Courtesy of R. Thorsteinsson.)

that always places an animal in a precarious position as far as nourishment is concerned. We may also note that changes in the *nature* of the vegetation may have been important. The dinosaurs began their racial histories amid ferns, cycads, and evergreens of various types. As angiosperms took over, there came quantities of small, compact nuts, fruits, and seeds. These were suitable for small creatures such as mammals, but were not sufficient fare for satisfying either the needs or the appetites of the towering, bulky dinosaurs.

There is no doubt that the success or failure of many nonaggressive forms is determined by their relative efficiency in such matters as food-getting, reproduction, and protection. The geologic histories of such great groups as the pelecypods and the brachiopods may illustrate this fact. The brachiopods were once the dominant shellfish in the sea. Over 20,000 extinct species have been identified. Today there are only 250 species, existing in out-of-the-way places. The pelecypods, on the other hand, were neither as plentiful nor as advanced as the brachiopods during the Paleozoic. They are now represented by 35,000 species and appear to have increased steadily, while the brachiopods have declined. Both forms live in similar environments and seek essentially the same food. The brachiopods are mainly stationary, incapable of any extensive movement. The pelecypods, with few exceptions, can swim, float, crawl, or burrow. The brachiopods are mainly passive feeders, depending on favorable currents to deliver their food. The pelecypods are capable of generating powerful currents of their own to gather food, and also can actively seek it in various places. Granting that there may be other unsuspected factors, these two differences may easily explain the gradual decline and near extermination of the brachiopods, and the appearance of correspondingly new varieties of pelecypods. The competition between these creatures is worldwide and has been going on steadily throughout all post-Cambrian time, accelerating during the so-called "revolutions."

Although admittedly oversimplified, this case illustrates the process by which a group may gradually be eliminated over vast periods of time, independent of catastrophes or of physical changes of a geologic nature. The reason for the success of one group and the failure of another may be as outwardly obvious as the possession of a sharper claw or a more rapid reaction; it may also be very subtle—perhaps a single small genetic mutation, a new digestive enzyme, a quickened nervous reaction, or a bit of protective coloration.

It is no coincidence that not only the large land and sea reptiles, but also the larger cephalopods (ammonites) and pelecypods (inocerami and rudistids) became extinct near the close of the Mesozoic. Perhaps smaller creatures survived at this and other times because their food requirements were more easily met during times of general shortage.

The mammals seem to show with unmistakable clarity the effects of competition for food. In the widespread, humid forests of the early Tertiary, there were many small forest dwellers and browsers. These evolved into larger herbivores with heavy teeth to take advantage of the abundant foliage of trees and shrubs. As forests withdrew or thinned out, grass became a dominant plant form, and grazing animals replaced the browsers. Among mammalian carnivores we see a constant decline in the less brainy, poorly constructed, blunt-toothed creodonts, and their replacement by intelligent, more agile, and sharper-toothed species. Teeth, the structures which evolved specifically to cope with food procurement, show increasing specialization, diversity, and improvement. It seems pointless to deny that the story of extinction and replacement of Tertiary mammals can be read in brains, teeth, and feet.

Finally, by a devious route came man, able and willing to exploit all energy sources and to

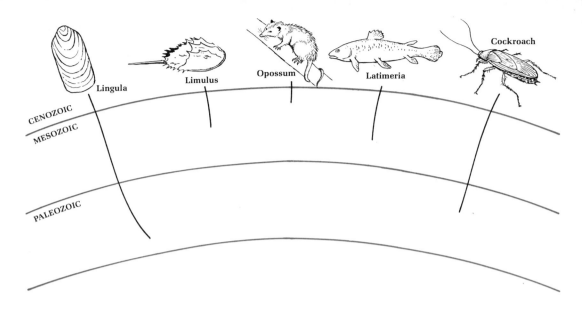

CENOZOIC

MESOZOIC

PALEOZOIC

Lingula   Limulus   Opossum   Latimeria   Cockroach

Figure 19.12   Geologic ranges of several of the so-called "living fossils," species that have shown an amazing ability to survive over vast periods of geologic time.

command the summit of every food pyramid. In the process of achieving domination he exterminated many forms of life, domesticated and preserved others, and even gained the power of life and death over his own species.

Postscript—
Some Notable Survivors

Here and there among the roll of living things are names made famous by the long family histories of their possessors. We marvel at these veterans of the struggle for existence and seek to understand why they have survived while their contemporaries long ago vanished. The secrets of their success may be difficult to uncover or to comprehend, for there seems to be no common characteristic that accounts for their survival.

Aside from simple organisms that reproduce by fission and are, as we have said, in a sense immortal, the most ancient known genus alive is the brachiopod *Lingula*. This creature is a small, tongue-shaped, shell-bearing animal that now

inhabits warm, shallow, offshore waters in many places throughout the world. *Lingula* appeared first in the Silurian and has persisted almost without change through a life history of innumerable generations spread over at least 350 million years. This organism can live for long periods in partly fresh water or in water so foul that it would kill most mollusks; it can even exist for hours on open flats exposed between tides, with no apparent injury resulting. Another an-

Figure 19.13   Model of Latimeria, a coelacanth fish widely publicized as a "living fossil." Model is about 1.2 meters long. (Courtesy of Hiroshi Ozaki.)

Figure 19.14 Tuatera, the last living member of the ancient reptilian order Rhynchocephalia. This small, lizardlike reptile survives only on several small islands off the coast of New Zealand. (From the film Message from a Dinosaur, copyright 1965 by Encyclopaedia Britannica Educational Corporation, Chicago.)

Figure 19.15 The durable opossum, a venerable survivor from earlier ages. (U.S. Department of Agriculture.)

cient ocean-dweller is *Neopilina galatheae,* a mollusk the size of a silver dollar. Its fossil relatives have been discovered in rocks estimated to be some 350 million years old, where they are associated with trilobites.

The most widely known of all "living fossils" is the coelacanth fish *Latimeria.* This creature is truly a relic from the days of the dinosaurs. Its discovery in 1939 was literally the biological find of the century, for up to that time investigators believed that it had disappeared during the Cretaceous Period, at least 60 million years ago. *Latimeria* provides a most valuable link between ordinary fish and land vertebrates, for some of its close relatives made the important step from water to land during the Devonian Period. Although a dozen or more specimens have now been taken, it is a very rare animal and lives only in a restricted area in the western region of the Indian Ocean.

On land, one of the most venerable of all four-legged creatures is the famous *Tuatera,* a lizardlike reptile 30 centimeters or more in length that has come down from the Triassic Period practically unchanged. This animal is the last

living survivor of its order, which was represented in Mesozoic time by larger and more numerous representatives. The few remaining members inhabit some small offshore islands near New Zealand, a sanctuary of sorts where they are protected by law.

Among mammals, the opossum boasts the oldest known racial history. The most ancient fossil specimen comes from Cretaceous rocks, and the group underwent relatively few changes through succeeding epochs of the Tertiary. The opossum is a small animal with primitive five-toed feet, teeth capable of dealing with a variety of foods, a prehensile tail, and a small brain. It is highly prolific, and is not easily killed. There can be no doubt that its conservative habits and uncomplicated mode of life have favored its survival, and it is instructive to compare it with many larger and brainier forms it has outlived during its 60-million-year existence.

A SUMMARY STATEMENT

Paleontological research first suggested and later overthrew the doctrine of universal catastrophes. The evidence strongly suggests now that local mass destructions are of little influence in shaping the overall pattern of evolution. Man has been directly or indirectly involved with most exterminations of the past few thousand years and must be regarded as having totally upset the balance of nature that existed before he came. Study of the prehistoric past indicates that ultimate survival depends on the slow acquisition of adaptations that enable their possessors to capture more food, to move efficiently from place to place, to reproduce sufficient young, and to protect themselves better than their contemporaries. The implication is very clear that it is dangerous for a group to become too perfectly adapted. Change in the physical world is inevitable, and those organisms with the power to change with the environment, or to shift to new conditions if necessary, are most successful over long periods of time. Spectacular temporary success seems to be gained only at the expense of specialization and ultimate extermination. Finally, the beneficial effects of competition are clearly indicated. Struggle with the blind forces of nature and with other living things eliminates the less fit and strengthens the survivors.

FOR ADDITIONAL READING

Burton, Maurice, *Living Fossils*. New York: Thames and Hudson, 1955.

Colbert, E. H., *Evolution of the Vertebrates*. New York: Wiley, 1969.

Martin, P. S., and H. E. Wright, *Pleistocene Extinctions: The Search for a Cause*. New Haven: Yale University Press, 1967.

Newell, Norman D., "Crises in the History of Life." *Scientific American,* Vol. 200, 1963, 76–92.

Simpson, G. G., *The Major Features of Evolution*. New York: Columbia University Press, 1953.

Vinzenz, Ziswiler, *Extinct and Vanishing Animals*. New York: Springer-Verlag, 1967.

SOME SELECTED FAMILY HISTORIES

Here and there among the many organisms that have lived on earth, a few stand out because they are interesting in themselves and important to man. Some of these have long since died out and are known only by their fossils; others belong to lineages that have passed their prime and are declining toward extinction; still others are in their racial youth and probably will continue to develop for many years.

A good fossil record is essential to a satisfactory understanding of any group, and for our present purposes we will emphasize those with

The horse in rock art of the western United States. The family history of the horse, especially the domestic horse (Equus caballus), is thoroughly interwoven with the story of man. (Courtesy of R. E. Borstadt.)

relatively well documented past histories. As a rule, the most complete and understandable fossil records are left by plants and animals that were relatively numerous, possessed diagnostic hard parts such as bones, shells, or woody tissue, and lived in localities where their remains could be buried and fossilized rather easily. Organisms that inhabited floodplains or the nearshore areas of lakes and seas have left more abundant records than those of the open oceans or of elevated continental areas where erosion is a dominant process.

A number of race histories are summarized in the following pages to illustrate what we can infer when fossils are sufficiently abundant. Groups selected include the trilobites, the cephalopods, the dinosaurs, the horses, the oysters, and the redwoods. Each of these groups has an extensive fossil record and a notable past history; otherwise, they have little in common, for their modes of existence, habits, and environments differ considerably. Nevertheless, each group succeeded in its own way in attaining a prominent place in the history of the past.

Some Selected Family Histories 487

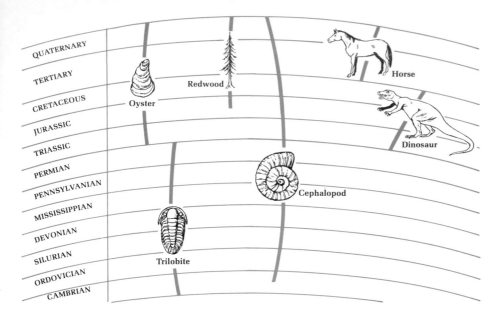

QUATERNARY
TERTIARY
CRETACEOUS
JURASSIC
TRIASSIC
PERMIAN
PENNSYLVANIAN
MISSISSIPPIAN
DEVONIAN
SILURIAN
ORDOVICIAN
CAMBRIAN

Oyster
Redwood
Horse
Dinosaur
Cephalopod
Trilobite

Figure 20.1 Geologic ranges of several important groups with unusually complete fossil records.

The Trilobites

Perhaps the most widely known of all extinct invertebrates are the trilobites, which we have already mentioned briefly in connection with the Paleozoic periods. They have left an extensive fossil record on all continents, and their remains have been collected by amateurs as well as by professional paleontologists for many years. They have often been called "fossil butterflies," which indicates that they are recognizable representatives of the great phylum of segmented animals known as the arthropods. Although trilobites are only remotely related to present-day animals, a rough comparison can be made between them and the common wood louse, or pill bug, and the king crab. So far as we know, trilobites left no close relatives or direct descendants.

The trilobites were adapted for life in the shallow oceans, where they swam, floated, crawled, or burrowed, seeking food and protection. They lacked jaws and biting mouth parts

and were incapable of capturing large, active prey. Instead, they were probably content to feed on fine organic particles obtained from the water or ooze of the sea bottoms. Well-preserved specimens reveal that the stomach was located in the head, underneath a bulging protuberance known as the *glabella*. Some experts, half seriously, have suggested that the pressure of the stomach may have hindered the development of a brain. The mouth, which was provided with a sort of hinged cover, opened backward, directly beneath the stomach. The legs, which could push material forward into the mouth, evidently helped the animal gather food. Most trilobites had well-developed eyes, and we assume that these specimens lived within the zone of light penetration and were not buried in the bottom sediments. A few were blind, however, and may have lived in deeper water or within the mud. In any event, practically all trilobites had a rounded, shovel-shaped front margin, well adapted for pushing through or over soft sediments.

The typical trilobite fossil does not duplicate the complete animal, although it does give a fairly accurate idea of the shape of the body. The softer parts of the animal, which are rarely preserved, were covered by the hard, chitinous armor, or *carapace*, which is preserved in the typical fossil. A few specimens have been found that display insectlike antennae and numerous fringed "legs" that protrude from the carapace. The trilobite skeleton consisted of a number of pieces that might or might not become separated before fossilization. The head, in most forms, consisted of three pieces, which could be split apart along definite lines during moulting. The middle part, or *thorax*, was made up of as many as two dozen separate segments that came apart easily; the tail, or *pygidium*, constituted still another separate piece. Trilobite remains usually consist of a mixture of all these parts, appropriately termed "trilobite hash"; the perfect specimen is a rare exception. Because the creature may have cast off its shell periodically like the crab or lobster, it is possible that one individual left a number of fossils. But so diagnostic are the cast-off pieces that an expert can usually recognize a species simply from its head or tail.

The ultimate origin of the trilobites is a paleontological mystery. The first positive fossils appear with their distinguishing characteristics well developed at the very beginning of the Cambrian Period, suggesting a long period of previous evolution. Students of the situation believe that trilobites were derived from a segmented, thin-skinned, wormlike ancestor sometime in the Late Precambrian. The segments in the head region of this hypothetical creature gradually fused (a process not uncommon among arthropods), and when fossils of the trilobites first appeared, the head had already been formed by the fusion of six segments. No more segments were added to the head, but the tail region underwent extensive fusion and enlargement after the Early Cambrian. Some of the oldest known trilobites were relatively long and narrow, with as many as twenty body segments and a small tail. With the passage of time, successive thoracic, or body, segments were fused to the tail, so that it became correspondingly larger. In some later trilobites, the tail makes up about one-third of the total length and

Figure 20.2 *The Devonian trilobite* Phacops, *with the large eyes and simplified outline typical of late Paleozoic forms. (Field Museum of Natural History, Chicago.)*

Figure 20.3 *Fossil "hash" made up chiefly of disarticulated trilobites. Two different types of tails and other pieces are visible. An expert is able to identify and date correctly specimens of this type. Section is about 6 centimeters across.*

Figure 20.4 Specimens of the Ordovician trilobite Flexicalymene. The curled-up condition is shown by most fossils of this genus. (Smithsonian Institution.)

bears traces of as many as sixteen segments, but the thorax is correspondingly shortened and has only nine segments.

As the size of the tail increased, some forms developed the ability to "enroll," or curl up, like a pill bug. In these species, the head and tail are exactly the same shape and fit closely together. The ability to roll up may have served a protective function, but we know nothing about the nature of the trilobites' enemies unless they were the fish that were taking possession of the sea during the time the trilobites were declining.

Other evolutionary trends are apparent. Some families became smooth and relatively flat and seem to have become adapted to withstand the

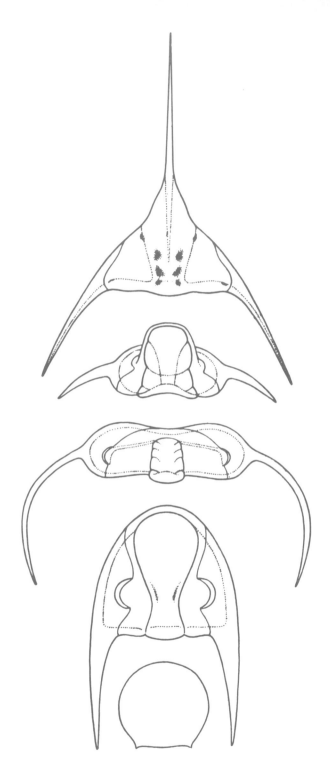

Figure 20.5 Variety in trilobite heads. The different shapes probably correlate with varied modes of life. (Reproduced from Treatise on Invertebrate Paleontology, Part 0-1, University of Kansas Press.)

force of the waves upon the reefs where they lived. Another development was an increase in spines and "ornaments." These appeared chiefly in the Ordovician and Silurian forms. Such outgrowths may have protected their owners or may have helped them to float near the surface; the spines inhibited the tendency to sink.

The history of the trilobites through the entire Paleozoic Era includes an earlier phase of success and dominance during the Cambrian and Ordovician Periods, followed by a prolonged decline to extinction in the Permian. The total period of existence was at least 350 million years. A recent compilation shows a total of ninety-three families in the Cambrian, sixty-five in the Ordovician, sixteen in the Silurian, fourteen in the Devonian, four in the Mississippian, four in the Pennsylvanian, and two in the Permian. About 5,000 species have been described, and new forms are being discovered yearly on all continents, including Antarctica.

The extermination of the trilobites in the Late Permian appears to have been complete and final. The niche they once occupied has been taken over by other creatures that live in much the same way and on similar food but do so more efficiently. There can be little doubt that the decline of the trilobites was partly the result of competition from more efficient forms. The final push into extinction may have come through drastic physical changes, but the fate of the trilobites had already been sealed long before the final blow fell.

The Cephalopods

Man's excursions beneath the surface of the ocean have brought him into contact with many fantastic animals, including a number of forms belonging to the cephalopoda, or "head footed" mollusks. The group includes the semilegendary Kraaken, or *giant squid,* immortalized in Jules Verne's *Twenty Thousand Leagues Under the Sea.* Here is included the "devil fish," or *octopus,* whose writhing arms, keen eyes, and intelligent behavior betoken a high degree of development. Not to be overlooked is the *pearly nautilus,* whose shell has long been one of the supreme prizes of the shell collector. The squid, octopus, and pearly nautilus are in reality the lingering remnants of one of nature's most venerable families. The group has produced through the geologic ages a succession of important wide-ranging forms that rank high in the esteem of collectors and are of foremost importance in the practical field of correlating sedimentary deposits around the earth.

The ancestors of the cephalopods first appear as fossils in the Upper Cambrian rocks with small, cone-shaped shells about 1 centimeter long. The shells are divided into compartments by thin partitions, so that the interior is distinctly chambered. A circular opening through the center of each successive partition is also characteristic. Apparently the animal lived in the last and largest compartment, with fleshy, tubelike extensions of its body running backward into the chambers it had previously occupied. Thus the earliest fossils show the basic chambered shell that was to undergo some very amazing changes through succeeding geologic periods.

Although the device of building successively larger chambers was obviously an effective way to ensure protection and permit growth, the resulting composite structure was by no means always perfectly adapted to the needs of its owner. A relatively long shell, with the living animal occupying the larger end, and a succession of hollow chambers trailing behind, was mechanically cumbersome. If the vacant spaces became filled with fluid, the dead weight would be difficult to drag about, and if they became full of gas, they would tend to rise and tip the animal on its head. A number of ingenious solutions to the problem of weight distribution gradually evolved. The species that retained the original straight form developed the ability to deposit additional shell material in the vacated chambers. These deposits were laid down mainly on

Figure 20.6 *Cephalopod shells. Left, the large coiled shell of Parkinsonia, cut to show the internal chambers, some filled with sediment, others partly filled by crystals. Center, the shell of the living pearly nautilus, showing the color pattern. Right, the fossil shell of Placenticeras, a Cretaceous ammonite from Wyoming. In the foreground is a broken shell of the pearly nautilus that shows the chambers.*

the bottom of the chambers and presumably were just heavy enough to keep the shell in a horizontal position. The straight-shelled forms predominated during the Ordovician and Silurian and reached a length of nearly 5 meters and a diameter of about 20 centimeters, which made them the giants of their time. They declined thereafter and became extinct in the Middle Triassic.

Another adaptation permitted the shell to swing upward, but the outer edges of the living chamber were pinched together so that the animal was in no danger of being dislodged from its shell. This tendency produced a number of short, stubby forms with a restricted front opening, like a modified keyhole, just large enough to permit the tentacles to be extended. This group became extinct in the late Paleozoic.

The most successful and probably only mechanically sound solution to the shell problem was achieved in a number of individual lines and involved coiling the shell into a compact spiral.

The various stages of coiling can be traced in fossils of early Paleozoic age. Some show merely a slight curvature, others a partial coil or one complete coil, and still others several coils, which may or may not touch each other. Finally, in those forms that achieved the logical culmination of the coiling tendency, the outer coils overlap and hide the inner ones. The nautilus shows this so-called *involute stage* of coiling, in which only the last coil is visible. In its final, perfected stage, the center of gravity of the shell is located near, and directly below, the center of buoyancy, thus following a principle familiar to all boat builders. The chambers of the living nautilus, and presumably of extinct forms that were similarly shaped, are filled with gas that permits the animal to rise or sink with only a slight expenditure of energy.

The history of shell building among the cephalopods also involves a series of structural modifications of internal parts. The edges of the platelike partitions that subdivide the interior became progressively more crinkled with the passage of time. The junction between the outer shell and inner partition is called the *suture*, and the line it follows is the *suture pattern*. This pattern becomes visible when the internal cavities are filled with mineral or rock and the outer shell peels off. The edges of the cross-partitions, or *septa*, that are thus revealed can be traced or photographed. In its simplest form, the suture line was either straight or gently curved. In the modern nautilus, which has remained relatively simple in this respect, the suture line follows this pattern. We should mention that the septa mark successive positions of the back portion of the body and that a new septum must be built whenever the animal grows in size and pulls itself forward within the shell.

In the nautilus and its relatives, the suture never passed beyond the simple curved pattern. In the ammonites (now extinct) the suture pattern underwent further modifications with the passage of time. The term goniatitic is used to describe sutures that take the form of sharp, angular bends. This type of suture pattern is found in

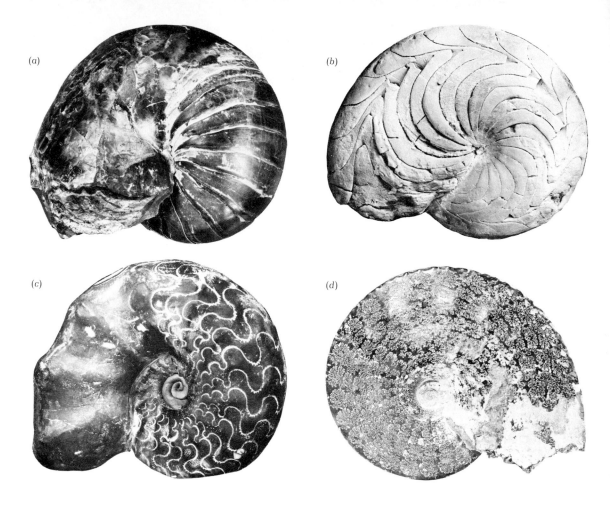

various ammonite groups ranging from the Devonian Period to the Triassic. During the Mississippian a number of families began to develop minute, toothlike crenulations along the edge of the portions of the suture line that curve away from the opening in the shell. This type of structure is called *ceratitic,* and it persisted well into the Triassic. Once the crenulation commenced, it spread rapidly and appeared independently in many groups. The intricate splitting and subdividing eventually involved the whole suture line, which became extremely complex. The ammonite type of suture pattern ranges from the Middle Permian to the close of the Cretaceous.

Figure 20.7 *Cephalopod specimens illustrating the various types of suture patterns. Shell material has been removed to expose the sutures and the rock material filling the spaces between the partitions. (a) Eutrephoceras, from the Cretaceous of Wyoming, showing the relatively straight nautiloid suture pattern. (b) Imitoceras, from the Mississippian of Indiana, showing the goniatitic type pattern, with sharp angular bends. (c) Ceratites, from the Triassic of Germany, with the ceratitic suture line in which small crenulations are beginning to appear on the larger curves. (d) Placenticeras, from the Cretaceous of South Dakota, with the complexly developed ammonitic suture patterns. Specimens range from about 7 to 17 centimeters in diameter but are shown the same size for comparative purposes. (a and c, courtesy of Julian Maack; b, courtesy of G. Arthur Cooper, Smithsonian Institution; d, Ward's Natural Science Establishment, Inc.)*

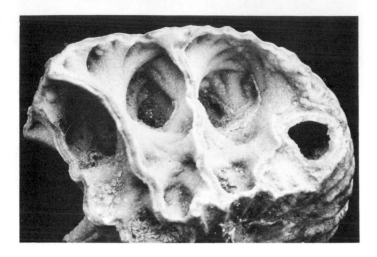

Figure 20.8 Interior of a naturally broken cephalopod shell. The branching folds of the septa are shown to have a buttresslike configuration that suggests that the folding serves to strengthen the shell against external pressure. Specimen is about 4 centimeters across.

Figure 20.9 A specimen of Parapuzosia bradyi, the largest ammonite fossil ever discovered in North America. The huge shell was taken from the Cody Shale of Late Cretaceous age, Big Horn, Wyoming. (Graybull Rockologists.)

Many theories have been proposed to explain the possible meaning or value of the increased complexity of the suture pattern. Because it must have been reflected in the soft tissue of the animal, it obviously provided a means of keeping the body firmly fixed in its living chamber. The strengthening of the shell against water pressure was also advantageous, for the cross-bracing and buttressing effect of the partitions prevented the shell from being crushed in deep water. All in all, it seems that the end result of these adaptations was a relatively light and extremely strong shell adapted to a mobile mode of life and to active, predaceous feeding habits.

Another large group of cephalopods, the *Coleoidea*, underwent a long and ultimately successful struggle tending toward the complete disposal of the shell. At first the animal lived inside its shell like an ordinary cephalopod, but later it more or less outgrew and even perhaps engulfed the shell, which became progressively less important. The shell became lighter and smaller, and present-day squids and cuttlefish emerged as the successful end-products. In these groups the shell has taken over the function of a stiffening-

Figure 20.10 Didymoceras, *an ammonite of unusual shape and complexity, from the Late Cretaceous of Colorado. (U.S. Geological Survey.)*

rod similar to the backbone of the vertebrate. Shell-less forms are not usually fossilized, but it seems safe to conclude that the octopus (of which there are a few fossils) is the culmination of the tendency among the Coleoidea to rid themselves of the shell.

The history of the cephalopods is notable for its ups and downs, for its extremes of success, and for the many close brushes it has had with extinction. The nautiloids reached a high point in the Ordovician, after which they gradually declined until none remain today but the nautilus. The goniatitic types dominated cephalopod life in the Mississippian and were followed by the ceratites and the ammonites, each becoming a dominant group in its time.

The recorded history of the ammonites is instructive. Ten families existed at the beginning of the Triassic, and these had evolved into twenty-nine families by its close. The end of the Triassic appeared to have been a critical time, and only

two of the twenty-nine families survived. These successful forms repopulated the oceans of the world and by the close of the Jurassic had increased to twenty-two families. Again, there was a weeding-out process, and only eight families survived into the Cretaceous Period. These families were in turn very prolific and successful, but were completely exterminated at the close of the period, and the entire race of ammonites vanished. The last surge of shelled cephalopods came during the early Tertiary, when the ancestors of the pearly nautilus expanded into a number of genera, all of which achieved worldwide distribution. But all except the pearly nautilus vanished during the middle part of the period.

The history of the cephalopods seems to revolve largely around their struggle to lighten or escape from the ancestral molluskan shell. Animals such as oysters, clams, and snails appear to have found the shell an absolute necessity, but there is little doubt that it interferes with an active life. The shell-bearing cephalopods were able to achieve mobility and also buoyancy while retaining the advantages of protection and strength that the shell provides. The squid and the octopus ultimately became essentially shell-less mollusks, but they more than made up for the dubious advantages of a shell by combining speed, camouflage, advanced sense organs, and keen intelligence.

The Redwoods

The redwoods are the tallest, largest, and most famous trees in the world. They rank high on any list of natural wonders and in many ways are the most impressive of all living things. At present, two redwood species exist. The most abundant is *Sequoia sempervirens*, the Coast Redwood, which grows only within the summer fog belt of California and southwestern Oregon. The Sierra Redwood (*Sequoia gigantea*) grows natively only in California on the western slope of the Sierra Nevada between elevations of about 1,500 and

Figure 20.11   Giant sequoias. (U.S. National Park Service.)

2,500 meters. The *Metascquoia,* a relative of the redwoods, lives only in the Red Basin of China's Szechwan Province. The Coast Redwood flourishes in regions of high precipitation, summer fogs, and mild temperature. The Sierra Redwood favors cooler temperatures and snow. The Chinese form appears to thrive in areas of relatively high temperature and precipitation.

The redwoods are famous for their great size and long life-span. The General Sherman tree, "largest living thing in the world," weighs an estimated 5,600 metric tons and has a volume of 1,500 cubic meters. The tallest known redwood is about 112 meters high and over 13 meters in circumference at the base. The rate of growth is not unusually high; the great size results from the tree's long life-span. There are thousands of Sierra Redwoods between 2,000 and 3,000 years old. Certainly some of them are more than 3,000 years old, and a few are possibly as much as 4,000. The oldest accurately dated tree is over 3,230 years old.

The present greatly restricted range of the redwoods gives no hint of their flourishing past. As representatives of the cone-bearing trees, they have had an unusually long and notable history, extending over at least 140 million years of time. The earliest known redwoods lived in the Juras-

Figure 20.12 A petrified redwood in Petrified Forest Park near Santa Rosa, California. The log is about 24 meters long and nearly 4 meters in diameter. A large, live oak is growing from the fossil trunk. (Courtesy of Jeanette Hawthorne, Petrified Forest of California.)

sic Period and were contemporary with the dinosaurs. Their great period of expansion was during the early and middle Tertiary, when vast forests flourished in many parts of the world.

Petrified redwood forests have been found in Texas, Pennsylvania, Colorado, Wyoming, Washington, California, Canada, Greenland, Alaska, and on St. Lawrence Island in the Bering Sea. Fossils also occur in many European countries, including Great Britain, and in Russian Siberia, Japan, and other Asiatic lands. The fossils are highly distinctive; logs are common and the foliage and cones are locally abundant. The most common fossils are very much like the living Coast Redwood and the Chinese Metasequoia, but no fossil record of the Sierra Redwood has been found.

The discovery of the living Metasequoia in 1944 was a dramatic event in the history of botany. A single enormous tree near the village of Mo-tao-chi in central China proved to be unlike any living tree previously identified, but the twigs and cones were found to be identical with fossil specimens from Japan and elsewhere. It was as if a fossil had come to life, for investigators had assumed that the genus had been extinct for 20 million years. Over a thousand living trees of the species are now known.

The success of the redwoods, both individually and collectively, is due to a number of significant adaptations that were probably achieved early in their career. They resist three of the great natural enemies of the forests: insects, fungi, and fire. Both the living tree and the deadwood are distasteful to insects, even to those notorious wood-destroyers, the termites. Whereas most types of wood are susceptible to destruction by fungi and bacteria, the redwoods

Figure 20.13 Fossil leaves of Metasequoia occidentalis, from sandstones of middle Tertiary age, Stevens County, Washington. The leaves are about 1 centimeter long (U.S. Geological Survey.)

are resistant to these agents even when dead. A redwood log entangled in the roots of a 340-year-old hemlock tree was still in sound condition when it was found. Experts believe that the redwoods' great resistance to living enemies stems from the presence of *tannin*, a strong astringent, in the wood.

Redwoods contain neither pitch nor resin, and the bark has an asbestoslike resistance to fire. It is more common for fire to destroy the central heartwood of the large redwoods than it is for it to harm the outside. Many of the older trees show evidence of fire damage, but it apparently has not been a common cause of death. Aside from the inexorable changes of climate and erosion to which all things are subject, the redwoods are vulnerable chiefly to the competition for living space that must inevitably accompany the expansion of the individual trees.

Through a series of adaptations, nature endowed the redwoods with an amazingly long life-span. To enjoy this seemingly supreme endowment, an organism must obey the universal requirement that life be accompanied by growth. Only by the addition of new wood and foliage can a tree survive. These conditions can have but one possible outcome: the individual will eventually either run out of living space or reach the limit of its mechanical strength, or of its food, air, or energy supply, and will then die of starvation—a reminder that even the best of adaptations has its restrictions.

The Dinosaurs

The dinosaurs are the best-known and most spectacular of all prehistoric animals. The average schoolboy knows more about them than he does about most living animals, and the names of many of these strange creatures have become common household words. Dinosaurs have been popularized in many ways. Literally millions of diminutive models have been distributed as toys, ornaments, and curiosities, and life-size mecha-

nized replicas have acted the parts of monsters at many world fairs and in numerous movies. Writers of science fiction have peopled their lost worlds and prehistoric scenes with dinosaurs or dinosaurlike forms ever since the remains of the huge beasts were first discovered. Dinosaurs have enlivened countless animated and still cartoons, providing both comic and awe-inspiring atmosphere. Dinosaur skeletons are the chief attraction in natural-history museums, where they attract more viewers than any other type of exhibit. Small wonder that dinosaurs have attained a position in popular fancy unsurpassed by any other extinct group.

Despite our tendency to sensationalize the dinosaurs or to present them in a comic or ridiculous light, the fact remains that they reveal many sobering and significant things about the meaning and destiny of life. Dinosaur fossils have been found on all major continents. They were almost as cosmopolitan in their time as man is at present. They have left their bones and footprints in many different kinds of sediment as testimony that they could survive in many types of environment. By a conservative estimate, about five hundred dinosaur species have been described, and it seems probable that this figure represents only a minor fraction of the total. Their remains are abundant only at local spots or in certain formations, but they probably existed in many other places where they left no record at all. They were a dominant form of life for three periods of geologic time, embracing a total span of 150 million years. By any standard they were eminently successful, and are well worth the attention of students of life on earth.

Dinosaurs were reptiles. Like crocodiles, turtles, snakes, and lizards, they must have been egg-layers, cold-blooded, and without hair or feathers. Certainly the skull structure is unmistakably reptilian, as are the teeth, ribs, legs, and tail; but all these parts were modified in various ways to meet the specific opportunities of the Mesozoic environment. Although not all dinosaurs were giants, they were on the average much

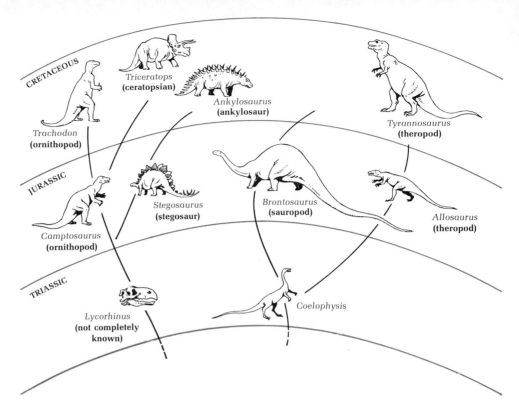

CRETACEOUS

Triceratops
(ceratopsian)

Ankylosaurus
(ankylosaur)

Tyrannosaurus
(theropod)

Trachodon
(ornithopod)

JURASSIC

Stegosaurus
(stegosaur)

Brontosaurus
(sauropod)

Allosaurus
(theropod)

Camptosaurus
(ornithopod)

TRIASSIC

Lycorhinus
(not completely known)

Coelophysis

Figure 20.14   *Simplified phylogenetic chart of the dinosaurs.*

larger than any group of land animals before or since. They did not appear on the scene fully developed, or of maximum size, but evolved into giants only with the passage of millions of years. It is worth mentioning that large reptiles in general possess the power of almost unlimited growth. In other words, they never cease growing entirely, and thus achieve a measure of eternal youth. Unlike mammals, which grow rapidly to an adult stage and then degenerate and die, the reptile grows slowly and is usually killed by accident or disease. This important difference between mammals and reptiles may explain in part why dinosaurs grew to be so huge. (Of course, many living reptiles find being small a definite advantage.) The opportunities that permitted dinosaurs to grow so large have not been repeated since, and probably never will be.

The earliest dinosaurs and their immediate ancestors were relatively small, lightly built,

lizardlike animals with what seems to have been an impelling tendency to get up on their two hind legs. Many living lizards can run on their two hind legs, and the ability seems natural in speedy, four-legged, long-tailed forms.

Dinosaurs may not have been the first bipeds, but they were the first to succeed in establishing the two-legged stance on a permanent and successful basis. This accomplishment was of great importance, for it freed the forelimbs from the tasks of body support and locomotion and opened up many possibilities that are closed to four-legged animals. Of course, the back legs and especially the hip region of bipeds must be mechanically strong and solidly built to assume the entire burden of carrying the body. It is natural,

Figure 20.15 *Typical of many restorations of dinosaurs is this model of the Jurassic carnivore* Allosaurus. (*Courtesy of Hiroshi Ozaki.*)

therefore, that the hip region assumed great importance, and furnishes the fundamental basis for classifying dinosaurs into their two groups, or orders. Avoiding technical details, all dinosaurs are easily divided into two orders: *Ornithischia,* with a birdlike pelvis, and *Saurischia,* with a reptilelike pelvis.

The first unmistakable saurischian dinosaurs were small and lightly built, but their basic skeletal structure could be modified to carry increasingly great weight. The forelimbs could assume a variety of special functions and at first were not much smaller than the hind limbs. The teeth were sharp and of uniform size and shape, suggesting that the animals were probably carnivorous. One of the best-known early dinosaurs that stands near the base of the family tree of the saurischians is *Coelophysis,* a large number of whose skeletons have been discovered in the Upper Triassic Chinle Formation in northern New Mexico. An abundance of dinosaur tracks

in these and other Triassic rocks proves that *Coelophysis* was not without numerous contemporary forms. From what we know about the tracks and skeletons, it is apparent that practically all Triassic dinosaurs represented modifications of the *Coelophysis* body type. Some were small and of delicate construction with birdlike feet. Others were more sturdily constructed and left blunt-toed, heavy-footed impressions. Perhaps the heavier ones were already feeling the necessity of reverting to a four-footed pose. In the Triassic rocks of the eastern United States, over two dozen kinds of dinosaur tracks have been found. These remains were among the first fossils to be discovered in American soil, and because dinosaurs were then unknown, it is only natural that these tracks at first were attributed to gigantic birds, which were supposed to have roamed the area.

During the Jurassic Period, dinosaurs became unquestioned masters of the lands. Although there was considerable variety in the surface of the earth, the Jurassic landscape was typified by shallow seas, low-lying swamps, and vast alluvial plains. The climate was mild and without extremes of glaciation. Dinosaurs continued to evolve along lines initiated in the Triassic, but in many types there was a distinct tendency toward gigantism. The dinosaurs seemed almost to have achieved a self-contained world of their own. They were no longer the prey of other reptiles, for there were none able to do more than merely annoy them. They were their own chief competitors. The herbivorous dinosaurs competed with other herbivorous dinosaurs, and the carnivores contended chiefly with one another. They were so large that they could no longer protect themselves by hiding; as a matter of fact, such a precaution was almost unnecessary. They towered above the trees, stalked the river banks, and waded in the marshes, each form well adapted and successful in its particular niche. Competition among the many Triassic forms had evidently eliminated many incompetent types, and the species that populated the Jurassic scene belonged mainly to a few standardized genera that made up a fairly well-balanced fauna.

The Jurassic was a high point in dinosaur evolution, and the *sauropods* were the most characteristic forms. The sauropods were long-tailed, long-necked, four-footed dinosaurs, exceedingly large and lizardlike in shape. They had evolved from relatively large long-necked herbivorous forms in the Triassic and must have reverted to the four-legged pose chiefly because of the tremendous weight of their massive bodies. There were several types of sauropods that were similar in outward shape but differed in many minor details of anatomy, indicating a diverse ancestry.

The sauropods were clearly adapted to live in marshes, rivers, and lakes. No animal before or since has been better fitted to exploit the rich food resources of these environments. Safe from predatory animals that dared not venture into water-covered areas, they browsed on the abundant vegetation, growing larger and larger with the passage of time until they became the greatest animals ever to walk the earth.

The sauropod was essentially a colossal food-gathering machine. The long neck must have evolved to permit the animal to graze easily over extensive areas. The head and teeth seemed to serve no protective or defensive purposes but

Figure 20.16 *Petrified bones of the foot and lower leg of a large dinosaur uncovered by removing the matrix of hard conglomeratic sandstone. Dinosaur National Monument, Utah. (U.S. National Park Service.)*

Figure 20.17 *Restoration of the Early Cretaceous dinosaur Deinonychus. Specimen was recently discovered and described by John H. Ostrom, Yale University. (Courtesy of John H. Ostrom.)*

Figure 20.18 Skull of the "bonehead" dinosaur, Troödon. The brain is covered by 15 centimeters of solid bone. It is thought that these animals butted their heads together as a way of combat. (Smithsonian Institution.)

were well adapted for underwater feeding. Perhaps the only limits to the size of these animals were correlated with their ability to gather enough food to keep the body supplied with energy. It is not difficult to understand why sauropod bones are among the most common remains in Jurassic river deposits. They were too large to be eaten or rapidly weathered away and so were buried and fossilized in great numbers. In the dinosaur beds of Tanzania, Africa, German scientists discovered the limb bones of a number of sauropods standing vertically in sandstone that must have once been a beach or quicksand deposit. The upper part of the skeletons had been destroyed before fossilization occurred.

On the banks of Jurassic streams lurked a number of carnivores. Again, the outward appearance of the various species was similar, but internal details of anatomy varied. The typical carnivore was a biped with powerful hind legs and small, armlike forelimbs. His tail was long and slender and when walking he held it off the ground. His head was massive and his mouth contained many sharp, pointed teeth. The pattern was typified by *Allosaurus* of the western United States, but similar forms existed in Europe, Asia, and Africa. The principle of parallel evolution appears to be well illustrated here.

Among the minor Jurassic dinosaurs were armored or plated varieties typified by *Stegosaurus* of the western United States. Here again the hind legs were long and the front ones were awkwardly placed, suggesting descent from a bipedal ancestor. Most unusual were the large, bony plates along the neck and back and a pair of clublike spikes at the end of the tail. Considering his small brain, weak teeth, and the burden of bony plates, it is somewhat of a mystery how *Stegosaurus* managed to survive as long as he did.

Cretaceous dinosaurs were similar in many ways to their Jurassic predecessors, but the individual species were quite new. The period was a time of more varied landscape and of great changes in the plant world. Among the dinosaur population, the greatest change appeared among the semiamphibious forms. There were relatively fewer sauropods and many more *trachodonts,* or *duckbills,* so called from the shape of the front of the skull. A possible reason for this shift was the progressive appearance of tougher types of woody vegetation in the Cretaceous, which the

duckbills, with as many as 2,000 solidly packed teeth, could successfully chew, but which the sauropods, with their weak, peglike teeth were unable to.

Representative of the carnivores were *Tyrannosaurus, Gorgosaurus,* and other bipeds. These creatures generally resembled *Allosaurus* and earlier Triassic forms, but there were a few general changes that seemed to affect the entire line. Most notable was a trend toward shortening of the forelimbs and a loss of fingers and toes. The shortening of the forelimbs, which had begun in the Triassic, came close to ending in *Gorgosaurus,* whose small arm and two-fingered hand must have been useless for defense or food gathering.

As we might expect, the armored dinosaurs of the Cretaceous went far beyond their Jurassic counterparts in the extent to which they became encased and plated with bone. The fantastic *Ankylosaurus* became literally a walking fortress. He could present a solid and impregnable array

Figure 20.19 A variety of dinosaur skulls. (a) Euparkeria, a small, ancestral form; (b) Tyrannosaurus, the great carnivore; (c) Diplodocus, a long-necked sauropod; (d) Stegosaurus, the plated dinosaur. The adaptations to varied types of food are obvious.

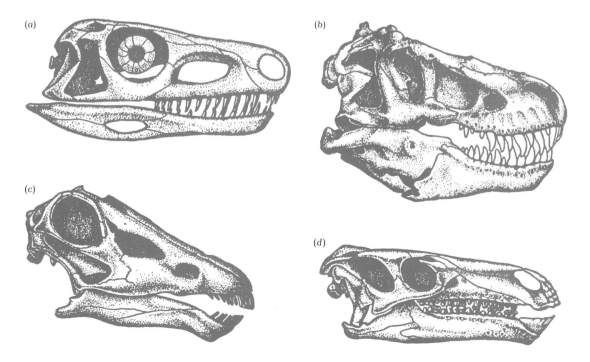

Figure 20.20 Dinosaur tracks of Early Cretaceous age from the Paluxy River near Glen Rose, Texas. (American Museum of Natural History.)

of armor to his foes by merely squatting down and drawing in his limbs and head. The entire upper surface of his body was covered by bony plates several centimeters thick, and along his flanks was a row of sharp, protruding spikes.

Not all Cretaceous dinosaurs were forced to flee before the giant carnivores or to rely on unwieldy armor for protection. One group, at least, was obviously willing and able to take a stand before *Tyrannosaurus* and his kind. This group, aptly called *Ceratopsians*, after their heavy horn-bearing skulls, is absent from Jurassic

rocks, but became quite common during the Cretaceous. The earliest known forms were small dinosaurs 2 meters or so long that lived in central Asia during Early Cretaceous time. Only one small, simple horn projected from the nose, and at the back of the head rose a short collar or frill. By successive elaborations on this simple pattern there arose larger and larger forms, approaching 15 metric tons in weight and sporting a pair of horns above the eyes, in addition to the one on the nose. The short collar became a huge, expanded frill of bone, rimmed with hornlike spikes and competely covering the neck. The best-known specimen of this group was *Triceratops*, which is frequently portrayed in combat with his contemporary, *Tyrannosaurus rex*. Breaks and scars in fossilized dinosaur bones indicate that fierce battles often did take place.

The reasons for the success of the dinosaurs, as far as such reasons can be reconstructed, seem to hinge chiefly on their great size. Their huge bulk removed them from petty competition with all other groups and imparted to them a certain degree of immunity, even from natural forces such as floods and storms. They were atop the very summit of the food pyramid of the time. They could easily take as food any land plant or animal they so desired. Only those things too small or obscure to be worth the taking were excluded from the dinosaurs' diet, which may explain why all known Jurassic mammals and birds were small and insignificant.

As long as food was abundant and climates were right, dinosaurs ruled the world. That they were eminently successful for three geologic periods entitles them to a prominent place in the history of life on earth.

The Oyster

The seas contain very few animals outwardly as unattractive as the oyster. Misshapen, drab, practically motionless, and devoid of expression, the oyster presents little to stir the imagination

Figure 20.21   Fossil oysters. Above, the tiny shell of Ostrea strigilecula, *from the Jurassic of Wyoming. Below,* Ostrea titan, *a 30 centimeter-long giant from the Miocene of California.*

or aesthetic sensibilities. Nevertheless, in its prosaic way, the oyster contributes more to the welfare of man than any other invertebrate of the sea, and its geologic history is long and informative. Kilometer after kilometer of oyster banks and reefs fringe the warmer borders of the continents, furnishing food and raw materials for man. The approximate annual world production of oyster meat approaches 54 million kilograms, and in several countries, especially the United States, France, the Netherlands, Japan, and Australia, oyster-fishing is an important industry. Oyster shells are dredged by the thousands of tons from shallow banks and used for construction material and other purposes. A number of mollusks popularly called oysters are not really members of the genus *Ostrea;* among these are "pearl oysters," genus *Melagrina.* More than 1,000 species of oysters, including fossil and living species, are known.

Oyster shells consist of two parts, or *valves,* held together by muscles and an elastic ligament.

The upper valve is usually thin and flat and fits over the lower valve like a lid. The lower valve may be massive and heavy, and because it cannot be moved the oyster is completely immobile. When many oysters cluster together in a *bed,* they may press together and deform one another so that the individual shells are not always the same shape. The *mantle,* a blanket of tissue that surrounds the oyster's soft body, takes calcium carbonate from sea water and secretes it to form the shell. The shell-building operation progresses rapidly in both cold and warm water.

Oysters feed by straining minute organic particles, such as algae and bacteria, from seawater. Powerful circulating currents are created by the beating of small hairlike structures on the gills. An oyster may pump as much as 38 liters of water an hour, but its feeding habits are difficult

Figure 20.22 *An oyster reef exposed at low tide on the Texas coast. (U.S. Department of the Interior, Fish and Wildlife Service.)*

to predict. It may become very temperamental and go without feeding for long periods, even though food may be abundant. Unfortunately for man, the oyster is able to take in and retain poisonous substances and harmful bacteria that may cause disease and death. Many oyster beds have been polluted and rendered useless by industrial wastes. New oysters require years to reach commercial size. They may spawn every year, but with the common American oyster (*Crassostrea virginica*) a successful *spatfall,* or new crop, is achieved only once in three or four years.

Individual oysters function alternately as males or females. A single female-functioning oyster may produce up to 125 million eggs a year. Those young oysters that survive the first few

days must find a suitable place to grow and develop. The most favorable sites are the old shells of dead oysters. Successive generations, settling on the shells of their dead predecessors, build up immense reefs. Commercial fishermen, aware of the oyster's habit, strew shells in favorable places to attract the larvae, and in this way cultivate a valuable crop.

Although the origin of the oyster is obscure, one authority claims to have found a suitable ancestral form in the Permian. The first reeflike accumulations appeared in the Jurassic, and progress thereafter was rapid. By Cretaceous time oysters had established themselves in most warm, shallow seas, and were building great reefs consisting of countless individual shells. Their habit of building upward on their own abandoned shells enabled them to live for generations in the same spot and to keep themselves above the muddy sediment that collected around the reefs. These vertical accumulations of fossil shells, when exposed by erosion, are called *tepee buttes* in the arid regions of the American West. Present-day reefs may stand on foundations of semifossil Pleistocene accumulations. Still older reefs are found in shallow-water formations millions of years old. Oysters are the chief contributors to the so-called "kitchen middens," huge piles of shells heaped up along the seashore by the many primitive tribes that practiced oysterfishing on a large scale in prehistoric times.

The tough and adaptable oyster is beset by numerous enemies, which not only seek it as food, but find protection and living space on its reefs and shells. It is plagued by internal parasites that live in its tissues and by boring sponges that riddle its shell with holes and drive it to deposit more and more shell material. Perhaps its most implacable foe is the starfish, which clamps itself over the oyster's shell and slowly but irresistibly forces it open and then digests the oyster inside. Starfish of all ages prey on oysters; young starfish live on young oysters and older starfish on older oysters. Gastropods also drill holes through the shell, and the oyster is helpless against these attacks. Another deadly enemy of the oyster is the crab, which is capable not only of opening the shell but of crushing it in its powerful claws.

A surprising variety of harmless creatures settle on the oyster shells and reefs merely to find a solid footing or a place of refuge. These squatters in turn attract other animals, and eventually a whole community of organisms based on the oyster and its shell is established. The nature of the oyster community varies from place to place, and the whole subject is a classic field of research. One oyster bank was found to contain almost three hundred other animal species including twelve sponges, fourteen coelenterates, eight flatworms, four nemerteans, ninety mollusks, forty-two annelids, two sipunculids,

Figure 20.23   Tepee buttes east of Pueblo, Colorado. These hills up to 18 meters high are composed mainly of oyster shells. The limy accumulations are left in relief with the removal of the softer surrounding shale. The ability of oysters to grow on the shells of former generations permits the accumulation of these moundlike forms. (U.S. Geological Survey.)

seventy-six arthropods, twenty bryozoans, five echinoderms, and nineteen chordates.

The oyster has succeeded in maintaining and enlarging its domain by a number of fundamentally important adaptations. Most basic of all is its ability to exploit the abundant food resources immediately above the ocean floor. The powerful current action it is able to generate is probably the chief factor here. The oyster converts and stores its surplus food and is able to produce prodigious numbers of offspring. These plentiful replacements have enabled it to succeed in spite of the many enemies, parasites, and predators that have come to seek it as a food source in all stages of its existence. Last but not least, the oyster can exist in a variety of environments, especially on muddy bottoms where most animals would be suffocated by sediments. This ability correlates with the oyster's habit of settling and building on the shells of its dead predecessors, which makes possible the growth of reefs. As a more or less passive creature almost without sense organs and defensive mechanisms, the oyster has succeeded in spite of overpopulation, competition, disease, intensive crowding, and the depredations of man. As long as the seas remain, the oyster will probably be one of its chief inhabitants.

The Horse

What is a horse? This may seem a pointless and unnecessary question, for almost everyone has a mental image of what a horse looks like. We must, however, be rather specific if we are to get a clear picture of the living horse in relation to his cousins and to his ancestors as well. Horses belong to the order *Perissodactyla,* which includes all odd-toed hoofed animals. The horse family (*Equidae*) includes all living horses, donkeys, zebras, and onagers, as well as all their ancestors back as far as we have been able to trace them. The genus *Equus* includes all the living members of the family as well as their immediate ancestors and relatives from the Pleistocene Epoch. The horses, donkeys, zebras, and onagers are usually considered as separate subgenera of the genus *Equus.* The species *Equus caballus* includes only the domestic horse and its wild relatives that interbreed and produce fertile offspring. We need not list here all the numerous breeds, races, and types.

The history of *Equus caballus* has become inextricably interwoven with the story of man. Most horses in the world today are either domesticated or are descended from domesticated stock, which indicates the extent of man's influence in the past, and suggests that the horse has no future independent of humanity. Through the ages, the association between men and horses has taken various forms. At first, horses were evidently merely a source of food. They were possibly driven over rocky cliffs or into natural traps of various kinds and then killed for their meat and skins. There is some evidence that they may have been captured and kept alive as a convenient food source. This possibility, together with the natural curiosity and docility of certain individual horses, may have encouraged its domestication and led to its becoming the slave of man. As a bearer of burdens, a mount, a puller of the plow, and a companion in warfare, the horse has been indispensable to man's conquest of the earth.

THE HORSE BEFORE MAN

During the early Cenozoic, in the Paleocene Epoch, primitive members of the horse family appeared in western North America. This area may have been the true place of origin because the earliest known species, *Hyracotherium* cf. *H. angustidens,* is small and closely resembles ancestral rhinoceroses and tapirs from the same formations. Such resemblance would indicate that the perissodactyl families had just commenced to acquire their individuality and were not far from a common ancestor that probably could not be assigned to one family in preference to another. Perhaps such an ancestor will be forthcoming. In any event, by the Eocene Epoch primitive horses inhabited Europe as well as America.

A specimen discovered in Eocene sediments in the environs of London was for a long time the earliest fossil horse. But no one recognized it as a horse and the specimen was given the name *Hyracotherium* in allusion to its supposed resemblance to the hyrax (coney), a small, rodentlike, ungulate mammal still living in Asia Minor. Later, primitive horses were found in equivalent strata in America and were given the name *Eohippus,* while *Hyracotherium* reposed unrecognized in a museum. Eventually, it became known that the two animals are one and the same. The European name has priority and should rightfully apply to all specimens, but to preserve the much used and popular American

term the designation, *Hyracotherium* (*Eohippus*), is permitted.

Paleocene and Eocene horses ranged between 25 and 50 centimeters high at the shoulders. All of these earlier varieties had simple, low-crowned teeth, three-toed hind feet, and four-toed front feet. There seemed to be relatively little evolutionary advancement during the Eocene. It was evidently advantageous to remain small in the semitropical forests where *Hyracotherium*, browsed on the soft vegetation and sought protection from his carnivorous contemporaries.

Changes in the physical world altered the environment and created challenging new opportunities for the horse family. America was severed from Europe and became the center of horse evolution. The continents gradually became higher and cooler, the semitropical vegetation withdrew southward, and grass for the first time became a dominant plant form. These tendencies began to be evident in the Oligocene, at which time the most common horse species was *Mesohippus*. This animal was slightly larger than *Hyracotherium*, standing about 60 centimeters high, and was distinctly horselike in appearance. The outer toe of the hind foot had disappeared,

Figure 20.24 Not recognized as a horse, this specimen, discovered near London in 1856 or 1857, was named Pliolophus vulpiceps. Later, when better connecting links were discovered it became an essential link in the horse line under the name Hyracotherium.

Figure 20.25 A fanciful sketch by Thomas Henry Huxley showing a hypothetical "Eohomo" riding the early horse Eohippus. Needless to say, the association has absolutely no basis in fact.

Figure 20.26 Reconstruction of Mesohippus, the three-toed Oligocene horse. (Field Museum of Natural History, Chicago.)

Figure 20.27 Semiwild horses rounded up at a waterhole on a range in the western United States. (U.S. Department of Agriculture.)

and both feet were functionally three-toed. The cheek teeth were still low-crowned and best adapted for browsing, but the premolars and molars had become very similar in form and were well suited for crushing and grinding rougher vegetation.

With the spread of grass, the scene of horse evolution shifted from the forests to the plains. Conditions favored speedy animals capable of foraging over wide areas and possessing strong teeth for crushing and grinding harsh, gritty herbage. The horse family now entered a period of great expansion and diversification. The environment permitted various lines of development, although many are difficult to trace and correlate. Of major importance during the Miocene was the emergence of the land bridge between North America and Asia, which permitted the horse to spread rapidly over all parts of the Old World. Some conservative American types preferred to remain in the woodlands and were browsers until the end of their careers.

The most significant adaptation of this middle period of development centers around *Merychippus*, a key form that spread widely and became the most successful grazing animal of the time. The chewing teeth were flattened and squared off and had developed permanently open roots that permitted long-continued growth. This single tooth adaptation may have been as crucial to the survival of the horse family as any other it ever achieved. The enamel of the tooth began to be enfolded deeply, and the folds were filled with cement, creating a pattern of ridges well suited for breaking down rough plant material. The feet still retained three toes, but the central one was much larger than the others—the lateral ones were small and practically useless. *Merychippus* stood about 1 meter high and looked a great deal like a small, modern pony.

Once the transition to the open plains and a grass diet had been achieved, the future course of horse evolution was fairly well determined. By the end of the Miocene there were at least six distinct lines of grazing horses. Most had three toes and generally were growing larger and developing more efficient teeth. The Miocene seems to have been a high point in mammalian evolution, and the horse family constituted one of the most abundant groups of the time. The succeeding Pliocene Epoch was a time of extermination for all but a few of the horse genera. The one-toed condition was apparently perfected by only one form, *Pliohippus*, which was evidently the ancestor of *Equus*, the genus living today. The final loss of all external traces of the side toes was merely the logical end of a long series of changes and was neither sudden nor particularly important at this time. There was considerable variation in tooth patterns, however.

In the late Pliocene, *Pliohippus* gave rise to the modern genus, *Equus*; the transition was a matter of refining details, for no major changes seemed possible. *Equus* appears to have migrated to Europe soon after it arose in North America. Through the million or so years of its existence, the genus *Equus* has left its bones in Pleistocene deposits almost everywhere. Among the unexplained events of the late Pleistocene is the abrupt disappearance of horses in the New World. Horses were thriving in both North and South America and were contemporary with man until

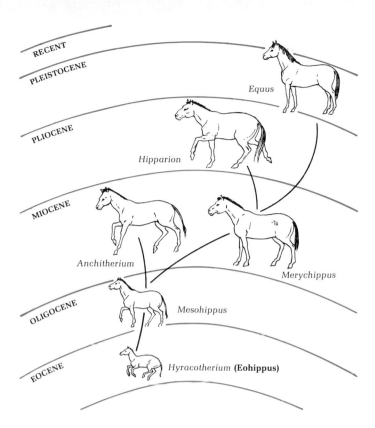

Figure 20.28 *The evolution of the horse. (Adapted from A. Lee McAlester,* The History of Life, *Prentice-Hall, Inc., 1968.)*

only a few thousand years ago. They then disappeared and were not reintroduced until transported across the Atlantic in the wooden ships of early explorers from Europe.

The lengthy transition from *Hyracotherium* to *Equus*, said by one investigator to have taken 15 million horse generations, brought the following changes: (1) increase in size; (2) lengthening of legs and feet; (3) loss of side toes; (4) increased complexity of premolar and molar teeth, flattening and widening of the incisors, and loss of the canines; (5) lengthening and deepening of the face and lower jaws, mainly to accommodate the longer teeth; (6) increase in size and complexity of the brain; and (7) straightening of the back.

It is interesting to consider whether or not the horse has reached an evolutionary dead end. Clearly the limbs, feet, and teeth have attained a high degree of specialization. A foot with only one toe cannot be reshaped to function like a five-fingered hand, and teeth highly adapted to grazing cannot cope efficiently with other types of food. The arrangement of muscle and bone that permits great speed and endurance is apparently incapable of much improvement or change. A longer leg would necessitate a longer neck and face, and these might add more weight to the animal, which would in turn require him to take more food, and could reduce his speed. The whole organization may have reached a state of equilibrium of parts, so that a change in one character would require adjustments in many others. As a forager over grassy plains and as a speedy match for the ever-present carnivores, the horse has had no equal. As long as the natural challenge existed, the horse evolved; now that the environment that produced it has practically vanished, the horse has likewise declined. The question of whether or not further evolutionary changes are possible may never be answered.

Many students of evolution have been impressed by the seemingly straightforward way in which the descendants of *Hyracotherium* progressed toward the modern *Equus*. It appears to some of them almost as if the end were clearly in view from the beginning. They note that there were few unsuccessful side branches, no detours down blind alleys, and little effort lost through the acquisition of useless or harmful adaptations. Belief in undeviating evolution toward a seemingly predestined goal is called *orthogenesis,* and the horse has long been regarded by many authorities as an outstanding example of this process. When the record is examined in great detail, however, the apparently simple history shows many complications, and the family tree bears many side branches that were unable to survive.

A SUMMARY STATEMENT

In spite of its many obvious imperfections, the fossil record permits us to reconstruct the general history of life and even to know certain groups in considerable detail. The best-known groups are, of course, those with an abundant fossil record. We have summarized the histories of the trilobites, cephalopods, oysters, dinosaurs, horses, and redwoods as examples of what can be pieced together from adequate fossil remains and study of living descendants or relatives. This varied group illustrates the general truth that the diversity of the physical world permits a corresponding diversity of life forms to succeed, each organism in its own way, in its own realm, and in its own time.

FOR ADDITIONAL READING

Colbert, E. H., *The Dinosaur Book.* New York: McGraw-Hill, 1951.
————, *Evolution of the Vertebrates,* 2nd ed. New York: Wiley, 1969.
Holton, Nicholas, III, *Dinosaurs.* New York: Pyramid, 1963.
Simpson, G. G., *Horses.* New York: Oxford University Press, 1951.
Stirton, R. A., *Time, Life and Man—The Fossil Record.* New York: Wiley, 1959.

Additional references for the material in this chapter can be found at the end of Chapter 5.

TABLES OF MEASURE

1. Selected Metric Measures with English Equivalents

LENGTH

1 micron (μ) = 0.001 millimeter (mm) = 0.00004 inch (in.)
1 mm = 0.1 centimeter (cm) = 0.03937 in.
1,000 mm = 100 cm = 1 meter (m) = 39.37 in. = 3.2808 feet (ft)
1 m = 0.001 kilometer (km) = 1.0936 yard (yd)
1,000 m = 1 km = 0.62137 mile (mi)
1 in. = 2.54 cm
12 in. = 1 ft = 0.3048 m
63,360 in. = 5,280 ft = 1 mi = 1.60935 km

AREA

$1 \text{ mm}^2 = 0.00155 \text{ in.}^2$
$1 \text{ m}^2 = 10.764 \text{ ft}^2$
$1 \text{ km}^2 = 0.3861 \text{ mi}^2$
$1 \text{ in.}^2 = 6.452 \text{ cm}^2$
$1 \text{ ft}^2 = 0.09290 \text{ m}^2$
$1 \text{ mi}^2 = 2.58 \text{ km}^2$

VOLUME

$1 \text{ mm}^3 = 0.000061 \text{ in.}^3$
$1 \text{ cm}^3 \text{ (cc)} = 0.061 \text{ in.}^3$
$1 \text{ m}^3 = 35.315 \text{ ft}^3$
$1 \text{ km}^3 = 0.239911 \text{ mi}^3$
$1 \text{ in.}^3 = 16.387 \text{ cm}^3$
$1 \text{ ft}^3 = 0.02832 \text{ m}^3$
$1 \text{ mi}^3 = 4.1682 \text{ km}^3$

WEIGHT

1 gram (g) = 0.001 kilogram (kg) = 0.035 ounce (oz) = 0.002205 pound (lb)
1,000 g = 1 kg = 2.205 lb
1 lb = 16 oz = 453.6 g = 0.4536 kg
1 oz = 28.35 g
1 short ton = 2,000 lb = 0.907 metric ton
1 metric ton = 1,000 kg = 2,205 lb

2. Temperature Conversion Formulas

0° Kelvin (K) = −273.16° Centigrade, or Celsius (C) = −459.69° Fahrenheit (F)

F = 9/5 C + 32
C = 5/9 (F − 32)
K = C + 273.16

GLOSSARY OF SELECTED TERMS

Many of the following terms are quoted from "Glossary of Selected Geologic Terms" by William Lee Stokes and David J. Varnes. Permission to quote has been kindly granted by the Colorado Scientific Society.

ABIOGENESIS The theory that living things can originate from nonliving matter, or, more commonly, the evolution of nonliving organic molecules. Compare *biogenesis*.

ABSOLUTE DATE A date expressed in terms of years and related to the present by a reliable system of reckoning.

ACIDIC A term applied to oxides that react with water to form acids in the chemical sense. Also loosely applied to igneous rocks with 65 percent or more silica.

ADAPTATION Any specific characteristic of an organism that is of value to its survival in a given situation. The term may also apply to the process by which such beneficial characteristics are achieved or accumulated.

ADAPTIVE RADIATION The complex evolutionary changes by which any basic stock of organisms is able to enter and exploit a number of environments.

ADENINE An amino acid that links with sugar-phosphate chains to form DNA and RNA.

AEOLIAN Pertaining to wind. Designates rocks and soils whose constituents have been carried and laid down by atmospheric currents. It is also applied to erosive and other geologic effects accomplished by wind.

AGE A period of earth history of unspecified duration characterized by a dominant or important life form, for example, the Age of Fishes. Also, a time when a particular event occurred, such as the Ice Age. Age may also refer to the position of anything in the geologic time scale; if possible, it may be expressed in years.

ALGAE Any member of a numerous group of simple plants usually classified as a subdivision of the *Thallophyta*. Algae contain chlorophyll and are capable of photosynthesis. They may appear blue-green, green, red, or brown and are classified accordingly. Forms capable of secreting calcium carbonate are important rock builders.

AMINO ACID Any of a large group of nitrogenous organic compounds that serve as structural units in the proteins and are essential to all forms of life.

AMMONITE Any member of an extinct group of marine mollusks, whose fossil remains are especially important as index fossils in the Permian Period and the Mesozoic Era. Typically, they possess a coiled, many-chambered shell with complex crenulations along the edges of the septa between the chambers.

ANAEROBIC Pertaining to, or operating in, the absence of free oxygen.

ANGIOSPERM Any member of the advanced group of plants that carries its seeds in a closed ovary and has floral reproductive structures.

ANGULAR UNCONFORMITY An unconformity or break between two series of rock layers such that rocks of the lower series meet rocks of the upper series at an angle; in other words, the two series are not parallel.

ANORTHOSITE A variety of igneous rock formed at depth and composed of 90 percent or more of the feldspar mineral anorthite.

ANTHRACITE Metamorphosed bituminous coal, or hard coal having about 95 to 98 percent carbon.

ANTHROPOLOGY The study of man, especially his physical nature and the ways in which he has modified or been modified by external forces.

ANTICLINE In simplest form, an elongate fold in rocks in which the sides slope downward and away from the crest.

ARAGONITE Calcium carbonate ($CaCO_3$) with crystals in the orthorhombic system (three unequal axes at right angles to each other). As a constituent of some shells it is less stable than calcite.

ARCHAEO- (archeo-) Combining form meaning "ancient."

ARCHAEOCYATHID Any of a group of extinct spongelike animals restricted to the Early Cambrian.

ARTHROPOD Any member of the great animal phylum that is characterized by jointed appendages, a bilaterally symmetrical body, and usually an external chitinous skeleton.

ARTIFACT Anything of a material nature produced by human skill.

ARTIODACTYL Any of the even-toed hoofed mammals.

ASEXUAL REPRODUCTION Reproduction by one individual independently of others.

ASH In geology, the finest rock material from volcanic explosions.

ASTHENOSPHERE A world-circling zone of soft, hot, plastic material extending roughly 100 to 400 kilometers below the earth's surface. This mobile layer separates the lithosphere above from the mesosphere below and on it the brittle lithospheric plates shift slowly in a complex interlocking pattern.

ATOLL A roughly circular, elliptical, or horseshoe-shaped island or ring of islands of reef origin, composed of coral, algal rock, and sand and rimming a lagoon in which there are no islands of noncoral origin.

AUTOTROPH An organism that can create its constituents from carbon dioxide, nitrogen, water, and external energy. Green photosynthetic plants are the best examples.

BASALT A word of ancient and unknown origin that describes a group of related igneous rocks that comprise the most common and widely distributed of all lavas. As generally used, the term includes the majority of fine-grained, dark, heavy volcanic rocks.

BASE LEVEL The level, actual or potential, toward which erosion constantly works to lower the land. Sea level is the general base level, but there also may be local and temporary base levels such as lakes.

BATHOLITH A very large mass of intrusive rock, generally composed of granite or granodiorite, which in most cases cuts across the invaded rocks and shows no direct evidence of having a floor of older solid rock.

BASIC Widely, but loosely, used to describe rocks with a relatively low content of silica and a correspondingly high content of minerals rich in iron, lime, or magnesia, such as amphibole, pyroxene, and olivine. The content of silica in so-called basic rocks is on the order of 45 to 52 percent. Because basic, as applied to rocks, has no direct relation to *base* in the chemical sense, it is being replaced by the term *subsilicic*.

BASIN A depression in the land surface. In geology, it is an area in which the rocks dip toward a central spot. Basins tend to be accentuated by continued downsinking and thus receive thicker deposits of sediment than surrounding areas. An example is the Michigan Basin.

BED Bed, and layer refer to any tabular body of rock lying in a position essentially parallel to the surface or surfaces on or against which it was formed, whether these be a surface of weathering and erosion, planes of stratification, or inclined fractures.

BEDDING PLANE The surface that separates any layer of stratified rock from an overlying or underlying layer.

BEDROCK The more or less solid undisturbed rock in place either at the surface or beneath superficial deposits of gravel, sand, or soil.

BELEMNITE A general name for any ancient squidlike cephalopod with a pointed, cylindrical, internal skeleton of solid calcium carbonate.

BENTONITE A rock composed of clay minerals and derived from the alteration of volcanic tuff or ash. The color range of fresh material is from white to light green, or light blue. On exposure the color may darken to yellow, red, or brown.

BIOGENESIS The origin of life from preexisting life, or, as now commonly used, the origin of life without regard to process.

BIOSPHERE All earth life considered together, usually as an interdependent system.

BIOTA The plants (flora) and animals (fauna) of a particular time and place.

BIOTITE Dark or "black" mica ranging in color from dark brown to green. A rock-forming mineral useful in age determination by the potassium-argon method.

BITUMINOUS COAL A compact, brittle coal of a gray-black to velvet-black color. It burns with a yellow flame and gives off a strong bituminous odor. Generally, there are no traces of organic structures visible to the eye.

BIVALVE An invertebrate animal whose shell is divided into two equal or subequal parts, or valves. Brachiopods, pelecypods, and ostracods are examples.

BLASTOID Any of a large group of extinct marine echinoderms that possess a stem and a budlike head, or theca. They range from the Ordovician to the Permian.

BORDERLAND An actual or hypothetical landmass occupying a position on or near the edge of a continent and supplying sediment to a geosyncline or site of deposition on the continent.

BRACHIOPOD A type of shelled marine invertebrate now comparatively rare but abundant in earlier periods of earth history. Brachiopods are common fossils in rocks of Paleozoic age. They have a bivalve shell and are symmetrical with regard to a plane passing through the beak and the middle of the front margin.

BRECCIA A rock consisting of consolidated angular rock fragments larger than sand grains. It is similar to conglomerate, except that most of the fragments are angular, with sharp edges and unworn corners.

BRYOZOANS Aquatic invertebrate animals that individually average less than 1 millimeter in length but that construct large colonial structures that have been preserved as fossils in rocks of all ages from Late Cambrian upward. Only lime-secreting varieties are common as fossils, and bryozoan limestone or marl is widespread. In older textbooks the bryo-

zoans are combined with the brachiopods to constitute the phylum *Molluscoidea;* most recent students treat the bryozoans as a distinct phylum.

CALCITE A common, rock-forming, carbonate mineral whose chemical formula is $CaCO_3$. It has a hardness of 3 and a specific gravity of 2.7. A rock with much calcite is said to be calcareous.

CARBONACEOUS Containing carbon. In geology, containing either coal in well-defined beds or small disseminated particles of carbon mingled with inorganic constituents.

CARBONATE Any compound formed when carbon dioxide contained in water combines with the oxides of calcium, magnesium, potassium, sodium, and iron. Among the common carbonates are dolomite, siderite, and calcite.

CARBON 14 DATING A method of obtaining approximate age limits of carbon-bearing materials based on the radioactive disintegration of carbon 14, which enters living materials along with other nonradioactive isotopes of carbon. Because the half-life is only 5,730 years, the method is not very useful in dating materials over 40,000 years old.

CARBONIZATION The process of converting a substance into a residue of carbon by removing other ingredients, as in the charring of wood, the natural formation of anthracite, and the fossilization of leaves and other plant organs.

CAST A natural or artificial reproduction of an object showing its outward shape. Natural casts are common as fossils.

CATASTROPHISM The belief that the past history of the earth and of living things has been interrupted or greatly influenced by natural catastrophes occurring on a worldwide or very extensive scale.

CEMENT The material that binds the particles of a consolidated sedimentary rock together. Various substances may act as cement, the most common being silica, calcium carbonate, and various iron oxides.

CEMENTATION The process by which sediments are consolidated into hard rocks through the deposition of cement in the pore spaces.

CEPHALOPOD Any member of a large class of mollusks whose head and mouth are circled with muscular tentacles. Water is drawn in and expelled through a siphon. The eyes are well developed and the whole animal is highly organized for rapid, intelligent action. Many fossil remains have been discovered in rocks of Cambrian age and younger. Modern representatives include the squid, octopus, and pearly nautilus.

CHALK A soft, white, fine-grained variety of limestone that is especially characteristic of rocks of Cretaceous age in Britain and northwestern Europe.

CHERT A very dense siliceous rock usually found associated with limestone, either in the form of nodular or concretionary masses or as distinct beds.

CHITIN A horny organic substance, chemical composition $C_{32}H_{54}N_4O_{21}$. It is present in the

skeletons and protective coverings of most arthropods and in some sponges, coelenterates, and worms.

CHLOROPHYLL The pigment of green plants that is involved in photosynthesis.

CLASS A group of organisms next in rank below the phylum and next above the order in ordinary biologic classifications.

CLASTIC ROCKS Include those deposits that are made up of fragments of preexisting rocks or of the solid products that are formed during the chemical weathering of such older rocks.

COAL A general name for combustible, solid, black, or brownish-black carbonaceous materials formed through the partial decomposition of vegetable debris. Its formation is distinctly traceable through a series of gradational steps starting with peat and passing through lignite, bituminous coal, and anthracite to a final theoretical limit of nearly pure carbon. It is usually distinctly stratified and is found in association with ordinary sedimentary rocks such as shale and sandstone and, more rarely, limestone.

COELENTERATE A member of the animal phylum Coelenterata, characterized by a hollow body cavity, radial symmetry, and stinging cells; includes jellyfish, corals, and sea anemones.

COLONIAL ANIMAL An individual that lives in close association with others of the same species. Usually it cannot exist as a separate individual.

COMPACTION The act or process of becoming compact. The term is usually applied in geology to the process whereby loose sediments are changed to rocks.

CONCRETION A spherical or irregular concentration or localized mass of mineral matter formed in a matrix of different composition. Many concretions contain fossils.

CONFORMITY The mutual relationships between sedimentary beds laid down in orderly sequence with little or no evidence of time lapses and, specifically, without any evidence that the lower beds were folded, tilted, or eroded before the higher beds were laid down.

CONGLOMERATE The consolidated equivalent of gravel. The constituent rock and mineral fragments may be of varied composition and range widely in size. The matrix of finer material between the larger fragments may be sand, silt, or any of the common natural cementing materials such as calcium carbonate, silica, clay, or iron oxide. The rock fragments are rounded and smoothed from transportation by water or from wave action.

CONODONT Any small, toothlike fossil of phosphatic composition. Conodonts are found in rocks ranging in age from Ordovician to Triassic. Their origin is uncertain. They have been variously assigned to vertebrates and several invertebrate phyla.

CONSOLIDATION In geology, any or all of the processes whereby loose, soft, or liquid earth materials become firm and coherent. Any action that increases the solidity, firmness, and hardness of earth materials is important in consolidation.

CONTACT The surface, in many cases irregular, that constitutes the junction of two bodies of rock.

CONTINENTAL DEPOSITS Deposits laid down on land or in bodies of water not connected with the ocean. The term is applicable whether the landmass is a true continent or only an island. The continental environment embraces fluviatile, lacustrine, glacial, and aeolian conditions.

CONTINENTAL DRIFT The process, considered by some to be theoretical and by others to be a fact, whereby one or more large landmasses split apart and drifted laterally to form the present-day continents.

CONTINENTAL GLACIER A large ice sheet that completely covers a large section of a continent, covering mountains and plains under an unbroken expanse of ice.

CONTINENTAL SHELF The gently sloping belt of shallowly submerged land that fringes the continents. It may be broad or narrow. The slope is roughly 1 in 540 (1 meter of vertical drop per 540 meters of horizontal distance) and the break of slope into deeper water is generally at a depth of about 180 meters. The geology of the continental shelves is similar to the geology of the adjacent emergent land.

CONVERGENCE In terms of living things, convergence is the gradual process by which two or more originally unlike organisms become more and more similar in form, function, or reactions.

CORAL A general name for any of a large group of marine invertebrate organisms that belong to the phylum Coelenterata, which are common in modern seas and have left an abundant fossil record in all periods later than the Cambrian. The term "coral" is commonly applied to the calcareous skeletal remains. As found in the fossil state, coral consists almost exclusively of calcium carbonate.

CORAL REEF A reef, usually very large, and made up chiefly of fragments of corals, coral sands, and the solid limestone resulting from their consolidation.

CORE A cylindrical piece of material cut and brought to the surface by special types of rock-cutting bits during the process of drilling. Also, the innermost part of the earth.

CORRELATION The process of determining the position or time of occurrence of one geologic phenomenon in relation to others. Usually, and in the narrowest sense, it means determining the equivalence of geologic formations in separated areas through a comparison and study of fossil remains or lithologic peculiarities. In a wider sense, it applies to the cause-and-effect relationships of all geologic events in time and space and to the arrangement of these phenomena in a logical and complete chronological system, such as the geologic time scale.

COSMOLOGY The science that deals with the universe, its parts, and the laws governing its operation.

COSMOPOLITAN As applied to fossil organisms, the term "cosmopolitan" implies a widespread geographic distribution.

CREODONT One of the groups of early, primitive, carnivorous mammals included in the suborder Creodonta. They flourished early in the Tertiary Period.

CRINOID An exclusively marine invertebrate animal belonging to the phylum Echinodermata. Fossil crinoids are found in Late Cambrian and younger rocks. Typically, they are attached by a jointed stem, and their shape suggests a lilylike plant, hence the name "sea lily," by which they are commonly known. Crinoids were especially abundant in Devonian and Mississippian time, declined at the end of the Paleozoic Era, and achieved a secondary maximum in the middle of the Mesozoic Era. About 650 species are still in existence.

CROSSOPTERYGIAN A type of fish considered to be ancestral to land vertebrates and characterized especially by a stout, muscular fin with a bony axis.

CROSS SECTION A geologic diagram or actual field exposure showing the geologic formations and structures transected by a given plane. Cross-section diagrams are commonly used in conjunction with geologic maps and contribute to an understanding of the subsurface geology. The formations, faults, veins, and so forth, are shown by conventional symbols or colors, and the scale is adapted to the size of the features present. Unless otherwise noted, cross sections are drawn in a vertical plane.

CRUST In a general sense, the crust of the earth is the crystalline shell that encloses the weaker, less well known, central part of the earth. The term "crust" is frequently used to mean the outermost part of the earth, in which relatively low velocities of earthquake waves prevail, above the first major discontinuity, the so-called Mohorovičić discontinuity.

CYCAD A plant of the class Gymnospermae having a short, pithy trunk, a thin, woody covering marked by numerous leaf-base scars, and palmlike leaves and cones.

CYCLOTHEM A succession of beds deposited during a single sedimentary cycle of the type that prevailed during the Pennsylvanian Period. The orderly repetition of a sequence of various kinds of strata in a series of cyclothems reflects a similar repetition of conditions of deposition over fairly wide areas of shallow sea and adjacent low-lying land areas.

CYSTOID Any member of a class of echinoderms with a boxlike or cystlike body constructed of numerous plates that may be arranged regularly or irregularly. The plates may be perforated by many pores, and the creature may have a stem and short food-gathering appendages. The known geologic range is from the Ordovician to the Devonian.

CYTOPLASM The contents of a cell, excluding the nucleus.

DECOMPOSITION The breaking down of minerals and rocks of the earth's crust by chemical activity. Complex compounds usually are broken into simpler ones that are more stable under existing conditions. Decomposition is usually attended by mechanical disintegration to produce the comprehensive changes called weathering.

DENDROCHRONOLOGY The science of dating and correlating that involves matching growth rings of trees or other vegetation.

DEPOSIT Anything laid down. A natural accumulation of mineral matter in the form of solidified rock, unconsolidated material, useful ores, or organic materials such as coal and oil.

DERIVED A term applied to fossils or rock fragments that have been removed by erosion or by some other process from their original sites and redeposited in later formations.

DETRITUS Any material worn or broken from rocks by mechanical means. The composition and dimensions are extremely variable. The deposits produced by accumulation of detritus constitute the detrital sediments.

DIAGENESIS Taken collectively, the processes that take place as loose sedimentary material hardens to rock. Does not include metamorphic effects.

DIASTEM A minor or obscure break in sedimentary rocks that involves only a very minor time loss. A short interval.

DIASTROPHISM The process or processes that deform the earth's crust.

DIATOM A microscopic aquatic plant (one of the algae) that secretes a siliceous skeleton, or test.

DIKE A sheetlike body of igneous rock that fills a fissure in older rocks, which it entered while in a molten condition. Dikes occur in all types of material—igneous, metamorphic, and sedimentary; if in sedimentary rocks or bedded volcanic rocks, dikes "cut" the formations or transect the beds at an angle.

DINOSAUR Any of a large number of extinct reptiles, usually of large size and belonging to either the order Saurischia or the order Ornithischia. The dinosaurs were confined to the Mesozoic Era and were characterized by diapsid skull structure, three bones uniting to form the hip joint, and other peculiar structural features.

DISCONFORMITY A break in the orderly sequence of stratified rocks above and below which the beds are parallel. The break is usually indicated by erosion channels with sand or conglomerate, which indicate a lapse of time or absence of part of the rock sequence.

DISINTEGRATION The reduction of rock to small pieces mainly by mechanical means. Most writers restrict the term to the effects of purely physical agents such as changes of temperature, frost, the rifting effects of roots, undermining, and abrasion produced by wind, water, and ice. Others include chemical actions, in which case the term is practically synonymous with "weathering."

DISPERSAL The spread of a species from its point of origin into other territory where its existence is possible.

DISTILLATION The process of creating fossils by eliminating the liquid or gaseous constituents from an organic substance so that only a carbonaceous residue remains.

DOLERITE (1) Loosely, any dark igneous rock whose constituents cannot be easily determined megascopically. (2) Any coarse basalt. (3) Any rock of the composition of basalt regardless of grain size. (4) A rock of the diorite or gabbro clan with uniform medium to small grains.

DOLOMITE A mineral that is composed of the carbonate of calcium and magnesium $CaMg(CO_3)_2$. The term also designates a rock composed chiefly of the mineral dolomite.

DOME An upfold in which the strata dip downward in all directions from a central point or area. It is the reverse of a basin.

DRILL HOLE An artificial hole cut or drilled in the earth to explore for valuable minerals or to secure scientific data.

EARTH (1) The solid matter of the globe as contrasted with water and air. (2) The loose or softer material composing part of the surface of the globe as distinguished from the firm rock. The word is rather indefinite in this sense, meaning about the same as, but not technically synonymous with, the term "soil."

ECHINODERM Any member of the phylum Echinodermata. The chief characteristics are radial symmetry and a spiny skin. Common living examples are the starfish, sand dollar, and sea urchin. Extinct forms include blastoids and cystoids.

ECHINOID Any of a number of marine invertebrate animals belonging to the class Echinoidea of the phylum Echinodermata. Recent forms are variously known as sea urchins, sand dollars, and sea porcupines. They are abundant in present-day seas and have left a fossil record extending back to the Ordovician Period. Certain forms are useful guide fossils for some formations of Mesozoic and Cenozoic age.

ECOLOGY The study of the relations between organisms and environment. *Paleoecology* is the same study applied to past conditions.

EDENTATE Any member of the mammalian order Edentata, a group characterized chiefly by degenerate teeth. Living examples include the armadillo and the sloth.

ELECTRIC LOG A record of the electrical responses of the geologic materials encountered in a drill hole. Electric logs are useful in locating changes in composition and in making local correlations.

EPICONTINENTAL Resting on a continent, as an epicontinental sea.

EPOCH A unit of geologic time; subdivision of a period. Some geologists restrict the term to the equivalent of a rock series, such as the Eocene Epoch or Series of the Tertiary Period or System.

ERA One of the major divisions of geologic time, including one or more periods. The eras usually recognized are the Archaeozoic, Proterozoic, Paleozoic, Mesozoic, and Cenozoic.

EROSION The wearing away and removal of materials of the earth's crust by natural means. As usually employed, the term includes weathering, solution, corrosion, and transportation. The agents that accomplish the transportation and cause most of the wear are running water, waves, moving ice, and wind currents. Most writers include under the term all the mechanical and chemical agents of weathering that loosen rock fragments before they are acted on by the transporting agents; a few authorities prefer to include only the destructive effects of the transporting agents.

EROSIONAL UNCONFORMITY A break in the continuity of deposition of a rock series that is made manifest by evidences of erosion. The strata above and below the break may be parallel, with no evidence of folding of the lower beds during the lapse in sedimentation.

ERRATIC A rock fragment, usually large, that has been transported from a distant source, especially by the action of glacial ice.

EUCARYOTE A type of living cell having a nucleus enclosed within a nuclear membrane and with well-defined chromosomes and plastids. Eucaryotic cells reproduce by miotic division in which genetic material is segregated among successive cells or descendants.

EURYPTERID An extinct type of arthropod with pincherlike claws and thirteen abdominal segments.

EVOLUTION In the broadest sense, the progressive change of any entity from simple to complex, lower to higher, or worse to better. See *organic evolution.*

EXOSKELETON A hard, outer skeleton or protective covering to which muscles are attached. The integument of the arthropod is a typical exoskeleton.

EXPOSURE An unobscured outcrop of either solid rock or unconsolidated superficial material. In one sense or another, the term embraces all earth materials appearing at the surface that are not hidden by vegetation, water, or the works of man.

EXTRUSIVE ROCK A rock that has solidified from material poured or thrown out upon the earth's surface by volcanic action.

FACIES In general, facies designates the aspect or appearance of a mass of earth material different in one or several respects from surrounding material. The features on which facies are named and recognized are usually selected more or less arbitrarily and may be lithologic (lithofacies) or biologic (biofacies).

FACIES CHANGES Lateral or vertical changes in the lithologic or paleontological characteristics of contemporaneous deposits. Because facies relationships are usually complex, the exact features selected for mapping or discussion should be clearly designated.

FAULT A break in materials of the earth's crust along which there has been movement parallel with the surface along which the break occurs. A fault occurs when rocks are strained past the breaking point and yield along a crack or series of cracks so that corresponding points on the two sides are distinctly offset. One side may rise or sink or move laterally with respect to the other side.

FAUNA The aggregation of animal species characteristic of a certain locality, region, or environment. The animals found fossilized in certain geologic formations or occurring in specified time intervals of the past may be referred to as fossil faunas.

FAUNAL SUCCESSION The observed sequence of life forms through past ages. The total aspect of life at any one period is different from that of preceding and succeeding periods. Faunal succession implies but does not prove evolution.

FISSION TRACKS Minute tubes formed in certain minerals, such as mica, by fragments of radioactive elements that have undergone fission in place.

FLINT A dense, hard, siliceous rock composed of very finely crystalline and amorphous silica.

FLOODPLAIN A strip of relatively smooth land bordering a stream, built of sediment carried by the stream and dropped in the slack water beyond the influence of the swiftest current. It is called a living floodplain if it is overflowed in times of high water but a fossil floodplain if it is beyond the reach of the highest flood.

FLORA The assemblage of plants of a given geologic formation, environment, region, or time interval.

FLUVIAL Pertaining to streams or stream action.

FORAMINIFERA An important order of one-celled animals (protozoa) that have left an extensive fossil record in rocks of Ordovician and younger age. They are almost all marine and have durable shells, or tests, capable of fossilization. Being small, their remains are readily recovered from well cores and cuttings and have become very important in correlating oil-bearing rocks. Thousands of fossil species have been discovered and they are especially useful as guide fossils in rocks of late Paleozoic, Cretaceous, and Tertiary age.

FORMATION The fundamental unit in the local classification of rocks. The larger units, groups, and series may be regarded as assemblages of formations and the smaller units as subdivisions of formations. The discrimination of sedimentary formations is based on the local sequence of rocks, lines of separation being drawn at points in the stratigraphic column where lithologic characters change or where there are significant breaks in the continuity of sedimentation or other evidences of important geologic events.

FOSSIL Originally, any rock, mineral, or other object dug out of the earth. Now restricted to any evidence of the existence or nature of an organism that lived in ancient times and that has been preserved in materials of the earth's crust by natural means. The term is not restricted to petrified remains—that is, those of a stony nature—and includes besides actual remains such indirect evidences as tracks and trails. Fossils are, with few exceptions, prehistoric, but no age limit in terms of years can be set. Fossils are useful in studying the evolution of present life forms and in determining the relative ages of rock strata. The term also is frequently, but loosely, used in connection with ancient inorganic objects and markings, such as fossil ripple marks or rain prints.

FOSSIL ASSEMBLAGE All the fossil organisms that can be found or collected from a single bed or formation and that are assumed to have lived at the same time.

FUSULINID Any of an important group of extinct, marine, one-celled animals (class Sarcodina, phylum Protozoa) that have left an extensive fossil record from late Paleozoic time. Owing to their small size, they are easily recovered from well cuttings and are of great value in correlating oil-bearing rocks.

GALAXY A portion of space in which stars, dust, gas, and matter in general are concentrated.

GASTROPOD Any member of a large and important class of mollusks that typically possess a coiled, single-chambered shell. Marine, freshwater, and terrestrial forms exist, and the group has left fossil representatives in Cambrian and all younger rocks. The gastropods are extremely numerous at present and have been important throughout the Cenozoic Era. Snails are the best-known representatives.

GEOCHEMISTRY The chemistry of the earth.

GEOCHRONOLOGY The study and classification of time in relation to the history of the earth.

GEOLOGIC AGE The time of existence of a fossil organism or the occurrence or duration of a particular event as stated in terms of the conventional geologic time scale. Any event not datable in terms of years is usually assigned a relative geologic age.

GEOLOGIC COLUMN A diagram showing the subdivisions of part or all of geologic time or the rock formations of a particular locality.

GEOLOGIC MAP A map on which geologic information is plotted. The distribution of the formations is shown by means of symbols, patterns, or colors. The surficial deposits may or may not be mapped separately. Folds, faults, mineral deposits, and so on, are indicated by appropriate symbols.

GEOLOGIC TIME The segment of time that elapsed before written history began. Although no precise limits can be set, the term implies extremely long duration or remoteness in the past.

GEOLOGY The science that treats of the origin, composition, structure, and history of the earth, especially as revealed by the rocks, and of processes by which changes in the rocks are brought about. Included is the study of the origin and evolution of living organisms, especially in prehistoric times. There are many subdivisions of the science, of which the following are important: historical geology, physical geology, economic geology, structural geology, mineralogy, mining geology, physiography, geomorphology, petrography,

petrology, vulcanology, stratigraphic geology, and paleontology.

GEOPHYSICS Broadly, the physics of the earth, including the fields of meteorology, hydrology, oceanography, seismology, vulcanology, magnetism, and geodesy. In the more popular and practical sense, the term implies the application of electrical, thermal, magnetic, gravimetric, and seismic methods to the search for petroleum, metals, and underground supplies of water.

GEOSYNCLINE Literally, a great, elongate downfold in the earth's crust. In general, the surface dimensions must be measured in terms of scores of kilometers and the thickness of accumulated rocks must be on the order of 10,000 to 15,000 meters. A typical geosyncline comes into being through long-continued, gradual subsidence with simultaneous filling by shallow-water sediments. Geosynclines usually originate between or adjacent to the more solid shield or platform areas of the globe. They may become, with suitable structural evolution, the sites of large-scale deformation, and many major mountain systems are formed of compressed geosynclinal sediments.

GENE A hereditary determiner, located in a chromosome.

GENUS A group of organisms next in rank above the species and next below the family in taxonomic rank.

GLACIAL DRIFT As used today, the term "glacial drift" embraces all rock material in transport by glacial ice, all deposits made by glacial ice, and all deposits predominantly of glacial origin laid down in the sea or in bodies of glacial meltwater, whether rafted in icebergs or transported in the water itself. It includes till and scattered rock fragments.

GNEISS A banded metamorphic rock with alternating layers of unlike minerals. Usually, equigranular minerals alternate with tabular minerals.

GONDWANA A hypothetical continent formed by the union of South America, Africa, Australia, India, and Antarctica. This landmass is thought to have broken into its present fragments during the Mesozoic Era.

GRANITE A true granite is a visibly granular, crystalline rock of predominantly interlocking texture, composed essentially of alkalic feldspars and quartz.

GRANITIC Having the general character of granite, especially the structure of interlocking crystals. The mineral composition may or may not be the same as true granite.

GRAPTOLITE Any of a large number of extinct marine invertebrates that occur as fossils from late in the Cambrian Period to the Mississippian Period. Their zoological affinities are obscure, but they have been recently assigned to the phylum Protochordata and are thus distantly related to vertebrates. They are especially useful as guide fossils in Ordovician rocks.

GRAVEL Loose, or unconsolidated, coarse granular material larger than sand grains, resulting from erosion of rock by natural agencies. The lower size limit is usually set at 2 millimeters.

GROUP A unit of stratigraphic classification. A local or provincial subdivision of a system based on lithologic features. A group is usually less than a standard series and contains two or more formations.

GUIDE FOSSIL Any fossil that has actual, potential, or supposed value in identifying the age of the rocks in which it is found. Also called index fossil.

GUYOT A flat-topped submarine mountain whose summit is supposed to have been exposed to wave action and planed away until it reached the ocean level.

HALF-LIFE The time required for the radioactive decay of one half of the initial number of atoms in a specimen of radioactive material.

HALITE Common rock salt (NaCl). Occurs in massive, granular, or compact form, and, if crystallized, in cubes. The taste is distinctive.

HIATUS A break or gap in the geologic record, as when rocks of a particular age are missing. The hiatus of an unconformity refers to the time interval not represented by rocks or to rocks missing by comparison with other areas.

HISTORICAL GEOLOGY The study of the history and development of the earth, including the life forms that have inhabited it, and the sum of that knowledge. Historical geology encompasses what astronomy and geophysics can tell of the earth's origin, the paleontologic evidence of the nature of ancient life and its development through geologic time, and the relations developed by stratigraphy, structural, and other branches of geology that place the events of earth history in a sequential order.

HORIZON A surface of contact or an imaginary plane without actual thickness that marks a certain level in stratified rocks.

HYDROCARBON Any of a considerable number of chemical compounds consisting only of hydrogen and carbon.

ICHTHYOSAUR Literally "fish lizard." Any of the extinct, aquatic, fishlike reptiles belonging to the order Ichthyosauria.

IGNEOUS Pertaining to, or having the nature of, fire. As used in geology to distinguish one of the three great classes of rocks, the name is a misnomer, for there is actually no fire involved; it should be interpreted to mean high temperatures.

IGNEOUS ROCKS Rocks formed by solidification of hot mobile rock material (magma), including those formed and cooled at great depths (plutonic rocks) that are crystalline throughout and those that have poured out on the earth's surface in the liquid state or have been blown as fragments into the air (volcanic rocks).

INSECTIVORE A member of the Insectivoria, an order of placental animals. Living examples include the shrews and hedgehog.

INTRUSIVE ROCK A rock that has solidified from a mass of molten material within the earth's crust but did not reach the surface.

INVERTEBRATE An animal without a backbone; pertaining to such an animal or animals.

ISLAND ARC A group of islands having an arclike pattern. Most island arcs lie near the continental masses, but inasmuch as they rise from the deep ocean floors, they are not a part of the continents proper.

ISOLATION A term used in biology to designate any process or condition whereby a group of individuals is cut off and separated for a considerable length of time from other areas or groups. The situation need not arise from actual geographical factors. Animals may become isolated as a result of food preferences or because of purely psychological reactions.

ISOSTASY An inferred condition of balance within the earth's crust by which large high landmasses are sustained not by the strength of the crust but by floating on a heavier, plastic interior similar to the way icebergs float in water.

ISOTOPE A variety of a chemical element that differs in atomic weight but is otherwise very similar to other isotopes of the same element.

KEY BED A well-defined and easily recognizable bed that serves to facilitate correlation in geologic work. The term is also applied to the horizon or bed on which elevations are taken or to which elevations are finally reduced in making a structure contour map. The term is used interchangeably with "key horizon."

LABYRINTHODONT Pertaining to a peculiar tooth structure characterized by deep infolding of the enamel. This type of tooth is possessed by extinct amphibians and related ancestral fish.

LACUSTRINE Pertaining to lakes.

LAND BRIDGE A land area, usually narrow and subject to submergence, that connects landmasses and serves as a route of dispersal for land plants and animals.

LAND FORM A term applied by physiographers to each of the multitudinous features that taken together make up the surface of the earth. It includes all broad features such as plains, plateaus, and mountains, and also all the minor features, such as hills, valleys, slopes, canyons, arroyos, and alluvial fans. Most of these features are the products of erosion, but the term also includes all forms that result from sedimentation and from movements within the crust of the earth.

LAURASIA A hypothetical land mass composed of Asia, North America, and other minor landmasses of the Northern Hemisphere.

LAVA A general name for molten rock poured out on the surface of the earth by volcanoes and for the same material that has cooled and solidified as solid rock.

LIMESTONE Strictly defined, limestone is a bedded sedimentary deposit consisting chiefly of calcium carbonate ($CaCo_3$) which yields lime when burned.

LITHIFICATION The process or processes by which unconsolidated rock-forming materials are converted into coherent solid rock.

LITHOLOGY The study of stones or rocks, especially those of sedimentary origin. Also, the description of the total physical characteristics of specified samples or formations.

LITHOSPHERE The solid outer shell of the earth from 20 to 50 kilometers thick. It includes the upper part of the mantle and the crust. It is divided into six major and many minor slablike plates that are shifting about on the plastic asthenosphere in a complex interlocking pattern.

LITTORAL Pertaining to the nearshore environment.

LIVING FOSSIL A term applied to any organism with a long geologic history, usually one that has outlived the forms with which it was once associated.

LOG A record of the earth materials passed through in digging or drilling a test pit or well. It may contain, in addition, notes regarding geologic structure, water conditions, casing used, and so on. Special types of logs are electric, caliper, radioactivity, sample, and so forth.

MAFIC Pertaining to dark-colored igneous rocks that are mostly relatively rich in magnesium and iron.

MAGMA Hot mobile rock material generated within the earth, from which igneous rock results by cooling and crystallization. It is usually a pasty or liquid material, or a mush of crystals together with a noteworthy amount of liquid phase having the composition of silicate melt.

MAMMAL A vertebrate animal characterized by warm blood, a covering of hair, live birth (there are two egg-laying exceptions), and the ability to suckle its young.

MANTLE In the geophysical sense, the part of the earth between the surface and the core, excluding the part above the Mohorovičić discontinuity. The term is occasionally used without this exclusion (see *crust*). In a more general sense, mantle refers to the loose material at or near the surface, above bedrock. In zoology, the mantle is the membrane lining the respiratory cavity of mollusks or brachiopods. It also secretes the shell substance.

MARL Sedimentary material that is mostly soft and clayey, containing shells or other calcareous matter.

MARSUPIAL Any of the group of mammals that lack a placenta and have an abdominal pouch in which the immature young remain for some time after birth. Examples are the kangaroo and opossum.

MEGA- Combining form meaning "great" or "large" (*mega*fossil).

MEMBER A subdivision of a geologic formation that is identified by lithologic characteristics such as color, hardness, composition, and similar features and that has considerable geographic extent. Members may be given formal names.

MESOSPHERE The solid interior of the earth below the hot and plastic asthenosphere. Also the layer of the atmosphere above the stratosphere.

META- Combining form meaning "changed" or "altered."

METAMORPHIC ROCKS One of the three great groups of rocks. Metamorphic rocks are formed from igneous or sedimentary rocks through alterations produced by pressure, heat, or the infiltration of other materials at depths below the surface zones of weathering and cementation. Rocks that have undergone only slight changes are not usually considered metamorphic; for practical purposes, the term is best applied to rocks in which transformation has been almost complete or at least has produced characteristics that are more prominent than those of the original rock.

METAZOAN Any animal in which more than one kind of cell makes up the tissues or organs.

METEORITE A mass of mineral or rock matter coming to the earth from space.

MICRO- Combining form meaning "small."

MICROFOSSIL Any fossil too small to be studied without magnification. Includes single organisms, fragments of complete organisms, or colonies of many organisms.

MICROPALEONTOLOGY The branch of paleontology dealing with fossils so small that they require magnification for identification and study.

MINERAL DEPOSIT A local accumulation or concentration of mineral substances or of a single mineral, either metallic or nonmetallic, which is of actual or potential economic value.

MOBILE BELT A belt or tract of the earth's crust, usually long and relatively narrow, that displays evidence of greater geologic activity such as geosynclines, folds, faults, and volcanic activity. Contrasts with stable blocks.

MOHOROVIČIĆ DISCONTINUITY A level of major change in the interior of the earth. It is found just beneath the crust at depths ranging from 5 to 50 kilometers.

MOLD An impression of the exterior or interior of an object from which it is possible to obtain a cast or reproduction of its outward shape.

MOLLUSK Any member of the numerous group of animals constituting the phylum Mollusca. In general, they are soft-bodied and are protected by a calcareous shell of their own making. There are marine, freshwater, and terrestrial forms, and the range of the phylum is from the Early Cambrian to present.

MOSASAUR A large extinct marine lizard commonly found in Upper Cretaceous rocks.

MULTITUBERCULATE An extinct mammal of the order Multituberculata, which existed in the Late Mesozoic and Early Tertiary. Their teeth are characterized by numerous cusps, and their habits and appearance were evidently somewhat like those of modern rodents.

MUTATION An inherited change stemming from modification of the hereditary material in the reproductive cells. The change may be slight or great.

NANNOFOSSIL A very small microfossil, usually less than 100 microns in diameter.

NATURAL SELECTION The complex process whereby organisms are eliminated or preserved according to their fitness or adaptation to their surroundings, especially to changes in the environment.

NAUTILOID Any of a large group of marine invertebrate organisms constituting a division of the class Cephalopoda of the phylum Mollusca. Typically, they have a straight, curved, or coiled, many-chambered shell. The edges of the septa between chambers have a straight or curved pattern and are not acutely angular or crenulated as in the ammonoids. They range from the Cambrian to the present with a maximum development in the Silurian.

NEBULA A cloud of gases or dust in space. Sometimes the term is applied to distant clusters of stars because of their hazy appearance.

NONCONFORMITY A type of unconformity in which an older, eroded sequence of rocks meets a younger, overlying sequence at an angle. Tilting and erosion of the lower sequence before deposition of the higher beds is implied. Some geologists use nonconformity only for cases where the older rock is of plutonic origin; both usages are evidently correct.

NOVA Literally a "new" star but more accurately one that increases suddenly in size and brilliance. After expending tremendous energy for a short while, it fades to obscurity.

NUCLEIC ACID An organic acid characteristic of the nucleus of living cells; DNA (deoxyribonucleic acid) and RNA (ribonucleic acid) are best known examples.

NUMMULITE Any of a large group of foraminiferal protozoans having coinlike shells. They are common in the Early Tertiary rocks of the warmer regions of the earth.

OFFLAP The arrangement of nonconformable sedimentary units in a depositional basin whereby the shoreward edge of each succeedingly younger unit is farther offshore than the unit on which it lies.

OLIVINE One of a number of important iron-rich or magnesium-rich rock-forming minerals. Olivine is common in oceanic basalts.

ONTOGENY The sequential changes or events in the development of an individual organism.

OOZE Ooze in a sedimentary sense is any soupy deposit covering the bottom of any water body. Specifically, the term relates to more or less calcareous or siliceous deposits that cover extensive areas of the deep ocean bottom. The marine oozes contain in greater or less quantities the shells of small organisms whose presence in quantities of 25 percent or more leads to differentiation into varieties based on their presence. Thus, there are the *Globigerina*, pteropod, radiolarian, and diatom oozes. The percentage of the shells of these organisms may range from 0 to nearly 100.

Other constituents of the oozes are minerals of a wide range and various other kinds of organic matter.

ORDER The group of organisms next in rank below a class and above a family in biologic classification.

ORGANIC Pertaining to, or derived from, life or from an organism. Chemically, an organic compound is one in which hydrogen or nitrogen is directly united with carbon.

ORGANIC EVOLUTION The theory that all living things have descended from one or a few simple beginnings. Includes the progressive changes in all lineages.

OROGENY The process by which great elongate chains and ranges of mountains are formed. Although the process or processes are not well understood, many orogenic movements appear to start with the downwarping of a large trough in the earth's crust, which is filled with sediments. The trough and its included sediments are then mashed, and the width of the belt is greatly shortened by folding and faulting. Igneous activity generally accompanies or follows deformation, and many of the largest bodies of intrusive igneous rocks lie within orogenic belts. The episode of deformation by which a specific system of mountains comes into being may be called an orogeny. The word thus seems to signify not only a process but also an event.

ORTHOGENESIS Evolution or development along definite lines as the result of a supposed directing influence.

OSTEOLOGY The study of bone as such and the ways it is organized into skeletons of bony animals.

OSTRACOD Any of a great number of small aquatic invertebrates of the phylum Arthropoda (class Crustacea, subclass Ostracoda). Typically, their bodies are small, segmented, and encased in a bivalved, horny, or calcareous shell. There are both marine and nonmarine species, and they range from the Ordovician to the present. They are valuable guide fossils for many marine formations and are of special importance in correlating nonmarine continental rocks.

OSTRACODERM Any of a large group of extinct fishlike jawless chordates with the head region encased in bony plates or scales.

OUTCROP A part of a body of rock that appears bare and exposed at the surface of the ground. In a more general sense, the term also applies to areas where the rock formation occurs directly beneath the soil, even though it is not exposed.

OVERTHRUST FAULT In a general sense, any reverse fault with low dip; more specifically, a low-angle fault on which the mass above has demonstrably moved or been pushed over a relatively stable mass below the fault. Usually a reverse fault having a dip of less than 20 degrees and a displacement measured in kilometers.

OZONE A molecule consisting of three atoms of oxygen. Ozone is very reactive and is highly

concentrated in a layer about 25 kilometers above the earth's surface.

PALEO- Combining form denoting the attribute of great age or remoteness in the past.

PALEOECOLOGY The study of ancient ecology or the relations of fossils to their environment.

PALEOGEOGRAPHIC MAP A map that shows the reconstructed geographic features of some specified period in the ancient past.

PALEOGEOGRAPHY The study of ancient geography.

PALEOMAGNETISM Magnetism imparted to susceptible minerals and rocks in the distant past and preserved to the present time.

PALEOLITHIC Pertaining to the earliest stage in use of stone by mankind, the Old Stone Age.

PALEONTOLOGY A study of the plant and animal life of past periods. It is based on the fossil remains found in the earth.

PANGAEA A hypothetical supercontinent composed of all the major landmasses of the globe.

PARACONFORMITY An obscure or uncertain unconformity above and below which the beds are parallel and there is little physical evidence of a long lapse in deposition.

PERIOD The fundamental unit of the standard geologic time scale, the time during which a standard system of rocks was formed.

PERMAFROST Permanently frozen ground, or more correctly, ground that remains below freezing temperatures for two or more years.

PETRIFACTION The process of petrifying, or changing into stone; conversion of organic matter, including shells, bones, and the like, into stone or a substance of stony hardness. Petrifaction is produced by the infiltration of water containing dissolved mineral matter such as calcium carbonate, silica, and so on, which replaces the organic material particle by particle, sometimes with original structure retained.

PETRIFY To convert organic material such as wood or bone into stone.

PETROLEUM A complex mixture of naturally occuring hydrocarbons.

PHANEROZOIC The part of past time and the rock record in which "visible", mostly metazoan life has existed.

PHOTOGEOLOGY The study of geology from photographs, usually those taken from aircraft.

PHOTOSYNTHESIS The synthesis of carbohydrates by green plants in the presence of sunlight.

PHYLOGENY The history of any race or group.

PHYLUM A large group of plants or animals; a major division of a kingdom. Usually divided into subphyla or classes.

PHYTOSAUR An extinct reptile somewhat like a crocodile in appearance that is included in the order Thecodonta. Phytosaurs are characteristic of the Late Triassic.

PILLOW LAVA Lava in its fresh or hardened form that shows pillowlike or baglike forms. The structure results when molten rock encounters water.

PLACENTAL Pertaining to, or possessing, the embryonic organ known as the placenta, which attaches the embryo to the uterine wall. Hence, any of the advanced mammals in which the young go through a considerable period of growth within the mother and are born in a comparatively advanced stage of development.

PLANETESIMAL Literally, "little planet," meaning a small body in space that behaves like a small planet in following an orbit around the sun. Planetesimals may be somewhat hypothetical entities but are required in certain theories of earth origin.

PLANKTON Floating organisms of seas or lakes.

PLATE TECTONICS The concept that the crust and outer mantle of the earth are divided into large plates or tabular blocks that are capable of mutual reactions that produce earthquake belts, mountain ranges, and other major features.

PLESIOSAUR Any of the extinct marine reptiles characterized by a long, flexible neck and flattened body which make up the order Plesiosauria.

PLUTONIC Pertaining to igneous rocks occurring or solidifying deep in the earth.

POLAR WANDERING The apparent movement of the magnetic poles during past geologic time in relation to the present positions.

PORPHYRY An igneous rock containing a considerable proportion, say 25 percent or more by volume, of large crystals or phenocrysts set in a finer groundmass of small crystals or glass, or both.

POSITIVE AREA A relatively large tract or segment of the earth's crust that has tended to rise over fairly long periods with respect to adjacent areas.

PRECAMBRIAN Pertaining to or designating all rocks formed prior to the Cambrian Period.

PREHISTORIC TIME The interval of time preceding the historic period or the invention of writing.

PRIMATE Any member of the placental mammal order Primates, characterized by large brains, prehensile hands, and five digits on hands and feet.

PROCARYOTE An organism consisting of cells that lack nuclear walls, well-defined chromosomes, or organelles. They cannot reproduce by miotic cell division.

PRODUCTID Any of a great number of extinct, spine-bearing brachiopods that were common in the late Paleozoic.

PSEUDOFOSSIL An object of inorganic but natural origin that might resemble or be mistaken for a fossil.

PTERODACTYL Any member of the reptilian order Pterosauria, or flying reptiles.

RADIOACTIVE DISINTEGRATION The change an element or isotope undergoes during radioactivity.

RADIOACTIVE ISOTOPE A variety of any element that has a different atomic weight from other isotopes of the same element and is radioactive. Carbon 14 is a radioactive isotope of carbon.

RADIOACTIVITY The continuous emission of energy in the form of particles or waves from the nucleus of an atom. Radioactive substances such as radium occur naturally in the earth. Other elements may be made artificially radioactive by bombardment.

RADIOGENIC Formed by, or resulting from, radioactive processes, such as radiogenic heat.

RADIOMETRIC Pertaining to measurements based on radioactive processes.

RED BEDS Red sedimentary rocks of any age.

REEF An aggregation of organisms with hard parts that live or have lived at or near the surface of a body of water, usually marine, and that build up a mound or ridgelike elevation. Reefs are considered to be sedimentary accumulations.

REGRESSION The large-scale withdrawal of seawater from a land surface. Also the succession of sedimentary deposits that results.

RELATIVE AGE The age of a given geologic feature, form, or structure stated in terms of comparison with its immediate surroundings, and not in terms of years or centuries.

RELATIVE DATING The placement of an object or event in its proper chronological order in relation to other things or events without reference to its actual age in terms of years.

REPLACEMENT The process whereby the substance of a rock, mineral, ore, or organic fragment is slowly removed by solution and is replaced by material of a different composition.

REWORKED A fragment of rock or a fossil removed by natural means from its place of origin and deposited in recognizable form in a younger deposit is said to be reworked.

RIPPLE MARK The undulating surface sculpture of ridges and troughs produced in noncoherent granular materials such as loose sand by the wind, by currents of water, and by agitation of water in wave action.

ROCK In the popular and also in an engineering sense, the term "rock" refers to any hard, solid matter derived from the earth. In the strict geologic sense, a rock is any naturally formed aggregate or mass of mineral matter, whether or not coherent, constituting an essential and appreciable part of the earth's crust. A few rocks are made up of a single mineral, as a very pure limestone. Two or more minerals usually are mixed together to form a rock.

SALT (1) Any of a class of compounds derived from acids by replacement of part or all of the acid hydrogen by a metal or metal-like radical; $NaHSO$ and Na_2SO_4 are sodium salts of sulphuric acid (H_2SO_4). (2) Halite, common salt, sodium chloride $(NaCl)$.

SALT DOME A more or less circular uplift of sedimentary rocks caused by the pushing up of a body of salt or gypsum. The salt usually occurs as a central core or plug and the surrounding sediments are pushed up at sharp angles.

SANDSTONE A consolidated rock composed of sand grains cemented together. The size range and composition of the constituents are the same as for sand, and the particles may be rounded or angular. Although sandstones may vary widely in composition, they are usually made up of quartz, and if the term is used without qualification, a siliceous composition is implied.

SCHIST A crystalline metamorphic rock that has closely spaced foliation and tends to split readily into thin flakes or slabs.

SEAMOUNT A discrete, isolated submarine peak. Seamounts are usually volcanic and flat-topped, many are capped with reef deposits formed when they were near sea level.

SEDIMENT In the singular the word is usually applied to material in suspension in water or recently deposited from suspension. In the plural the word is applied to all kinds of deposits from the waters of streams, lakes, or seas, and in a more general sense to deposits of wind and ice. Such deposits that have been consolidated are generally called *sedimentary rocks*.

SEDIMENTARY ROCK Sedimentary rocks are composed of sediment: mechanical, chemical, or organic. They are formed through the agency of water, wind, glacial ice, or organisms and are deposited at the surface of the earth at ordinary temperatures. The materials from which they are made must originally have come from the disintegration and decomposition of older rocks, chiefly igneous.

SEDIMENTATION Strictly, the act or process of depositing sediment from suspension in water. Broadly, all the processes whereby particles of rock material are accumulated to form sedimentary deposits. Sedimentation, as commonly used, involves not only aqueous but also glacial, aeolian, and organic agents.

SEISMOGRAPH A sensitive apparatus that records the various earth motions created by an earthquake.

SHALE A general term for lithified muds, clays, and silts, that are fissile and break along planes parallel to the original bedding. A typical shale is so fine-grained that it appears homogeneous to the unaided eye, is easily scratched, and has a smooth feel. The lamination, or fissility, is usually best displayed after weathering.

SHELL (1) The crust of the earth or some other continuous layer beneath the crust. (2) A thin layer of hard rock. (3) The hard outer covering of an organism or the petrified remains of the covering.

SHIELD The Precambrian nuclear mass of a continent, around which, and to some extent on which, the younger sedimentary rocks have been deposited. The term was originally applied to the shield-shaped Precambrian area of Canada but is now used for the primitive areas of other continents, regardless of shape.

SIAL A term derived from *silica* and *aluminum* that is applied to the whole assemblage of relatively lightweight, high-standing, continental rocks including the granites, granodiorites, and quartz diorites. The contrasting term is *sima*.

SILICA Silicon dioxide (SiO_2). Silica forms the

natural crystalline minerals quartz, cristobalite, and tridymite, and the noncrystalline mineral opal, which carries from 2 to 13 percent water. Quartz is the most abundant mineral in the visible portions of the earth's crust and occurs in a great variety of igneous, metamorphic, and sedimentary rocks and as the filling in veins.

SILICATES A group that contains the most important and numerous of the rock-forming minerals. Silicates are combinations of silicon and oxygen with the metallic or basic metals.

SILICIC Pertaining to, or derived from, silica or silicon; specifically, designating compounds of silicon, as silicic acid.

SILICIFICATION The process of combining with, or being impregnated with, silica. A common method of fossilization.

SILL A tabular body of igneous rock that has been injected while molten between layers of sedimentary or igneous rocks or along the foliation planes of metamorphic rocks. Sills are relatively more extensive laterally than they are thick.

SIMA A term derived from silica and magnesium that designates a layer of dense, heavy rock that is worldwide under continents and oceans. It may resemble basalt or peridotite.

SKELETON The hard structure that constitutes the framework supporting the soft parts of any organism. It may be internal as in the vertebrates, or external as in the invertebrates.

SPECIES A group of plants or animals that normally interbreed, produce fertile offspring, and resemble each other in structure, habits, and functions.

SPIRIFER A general name for brachiopods with wide, pointed, or winged shells.

SPONTANEOUS GENERATION The appearance of life or living things from dead material without the intervention of outside or supernatural forces.

STAGE The time-stratigraphic unit next in rank below a series. It is the fundamental working unit in local time-stratigraphic correlation and therefore is employed most commonly to relate any of the various types of minor stratigraphic units in one geologic section or area to the rock column of another nearby section or area with respect to time of origin.

STEGOCEPHALIAN Any large extinct amphibian with a broad, flat, bone-covered skull.

STRATIFICATION The characteristic structural feature of sedimentary rocks produced by the deposition of sediments in beds, layers, strata, laminae, lenses, wedges, and other essentially tabular units. Stratification stems from many causes—differences of texture, hardness, cohesion or cementation, color, mineralogical or lithological composition, and internal structure.

STRATIGRAPHER One who studies, or who has expert knowledge of, stratigraphy.

STRATIGRAPHY The branch of geology that deals with the definition and interpretation of the stratified rocks, the conditions of their formation, their character, arrangements, sequence, age, distribution, and especially their correlation, by the use of fossils and other means. The term is applied both to the sum of the characteristics listed above and to the study of these characteristics.

STRATUM (plural, STRATA) A single layer of homogeneous or gradational lithology deposited parallel to the original dip of the formation. It is separated from adjacent strata or cross strata by surfaces of erosion, nondeposition, or abrupt changes in character. Stratum is not synonymous with the terms "bed" or "lamination" but includes both. "Bed" and "lamination" carry definite connotations of thickness.

STROMATOLITE A laminated deposit built up by various associations of simple organisms, primarily blue-green algae. The layers are mostly calcite but some inorganic constituents may be bound into the structure. Moundlike to fingerlike forms are characteristic.

STRUCTURAL GEOLOGY The study of the architecture of the earth insofar as it is determined by earth movements. "Tectonics" and "tectonic geology" are terms that are synonymous with structural geology. The movements that affect solid rock result from forces within the earth and cause folds, joints, faults, and cleavage. The movement of magma, because it is often intimately associated with the displacement of solid rocks, is also a subject that lies within the domain of structural geology.

SUBSIDENCE A sinking of a large area of the earth's crust.

SUBSURFACE Pertaining to, formed, or occurring beneath, the surface of the earth.

SUPERPOSITION The natural order in which rocks are accumulated in beds one above the other. The law of superposition means that any undisturbed sequence of sedimentary rocks will have the oldest beds at the base and the youngest at the top.

SYNCLINE An elongate, troughlike downfold in which the sides dip downward and inward toward the axis or did so in the early stages of formation.

SYSTEM A fundamental division or unit of rocks. It is of worldwide application and consists of the rocks formed during a period, as the Cambrian System.

TAXON A group of organisms making up any specific taxonomic category such as species, genus, family, and so on.

TAXONOMY The science of classification, especially the classification of living things.

TECTONIC Pertaining to rock structures formed by earth movements, especially those movements that are widespread.

TETHYS A dominating seaway that lay between the northern landmass (Laurasia) and the southern landmass (Gondwana) during Paleozoic and Mesozoic time. The Alpine-Himalayan ranges reveal the deposits of this seaway.

THERAPSID Any member of the extinct reptilian order Therapsida. The various species had many mammal-like characteristics, and there were both herbivorous and carnivorous forms of varied size.

THRUST FAULT A fault in which the upper, or hanging, wall appears to have moved upward at a relatively low angle.

TILL Unstratified and unconsolidated debris deposited directly by glacial ice.

TILLITE Indurated till. The term is reserved for pre-Pleistocene tills that have been indurated or consolidated by processes acting after deposition.

TIME-ROCK UNIT Also called a time-stratigraphic unit, this term designates a mass of rock defined on the basis of arbitrary time limits and not physical characteristics. The Cambrian System is a time-rock unit including all rocks deposited during the Cambrian Period.

TONGUE A subdivision of a formation that passes in one direction into a thicker body of similar type of rock and dies out in the other direction.

TRANSGRESSION The gradual large-scale encroachment of water across a land surface. As a result the water-laid deposits overlap progressively landward.

TRILOBITE A general name for an important group of extinct marine animals (phylum Arthropoda, class Crustacea) whose remains are found in rocks of Paleozoic age. They have a compressed trilobate body with numerous segments in the thoracic region. Some forms are especially valuable guide fossils for the Cambrian Period.

TRITIUM A radioactive isotope of hydrogen that is produced in the upper atmosphere as a result of cosmic-ray activity. It has a half-life of 12.5 years and can be used in tracing and dating water masses in the ocean or underground.

TYPE LOCALITY The place from which the name of a geologic formation is taken or from which the type specimen of an organism comes.

UNCONFORMITY A surface that separates one set of rocks from another younger and bedded set and that represents a period of nondeposition, weathering, or erosion, either subaerial or subaqueous, prior to the deposition of the younger set. Although most commonly applied to an interruption in the continuity of a sequence of sedimentary rocks, the term is also used for breaks in a sequence of layered volcanic, or pyroclastic, rocks, if erosion or weathering is evident, and for the break between eroded igneous or metamorphic rocks and younger bedded rocks.

UNIFORMITARIANISM The belief or principle that the past history of the earth and its inhabitants is best interpreted in terms of what is known about the present. Uniformitarianism explains the past by appealing to known laws and principles acting in a gradual, uniform way through past ages.

UPWARP A broad area uplifted by internal forces.

VALVE The one or several pieces that make up the shell of an invertebrate animal.

VARVES The regular layers or alternations of material in sedimentary deposits that are caused by annual seasonal influences. Each varve represents the deposition during a year and consists ordinarily of a lower part deposited in summer and an upper, fine-grained part deposited in the winter. Varves of siltlike and claylike material occur abundantly in glacial-lake sediments and are believed to have been recognized in certain marine shales and in slates. The counting of varves and correlation of sequences have been applied toward establishing both the absolute and relative ages of Pleistocene glacial deposits. Varved "clays" of glacial origin commonly consist largely of very finely divided quartz, feldspar, and micaceous minerals rather than mostly true clay minerals.

VARIETY A subdivision of a species.

VASCULAR PLANT Any plant with tissues specifically adapted to conduct liquids or gases.

VERTEBRATE An animal with a backbone.

VOLCANIC ASH The unconsolidated, fine-grained material thrown out in volcanic eruptions. It consists of minute fragments of glass and other rock material, and in color and general appearance may resemble organic ashes. The term is generally restricted to deposits consisting mainly of fragments less than 4 millimeters in size. Very fine volcanic ash composed of particles less than 0.05 millimeter may be called *volcanic dust*. The indurated equivalent of volcanic ash is *tuff*.

VOLCANISM The phenomena related to, or resulting from, the action or actions of a volcano.

ZONE A subdivision of stratified rock based primarily on fossil content. It may be named after the fossil or fossils it contains. Strictly speaking, the zone of paleontological stratigraphy is based not on one but on two or more designated fossils. No fixed thickness or lithology is implied by the term "zone."

INDEX

Fish (*cont.*):
as growth measure, 18, *18;* teleost, 323, 354, *355;* Tertiary, 354, *355*
Fission tracks: dating by, 28–30, *30*
Floodplains, 76
Fluorine analysis: dating by, 31
Folding, 79–80, *80*
Folsom man, *416,* 417
Food-gathering: and species extermination, 321–22, 479–82
Food pyramids, 427–30
Foraminifera (*see also* Sarcodina), 112, 286–87, 353–54
Footprints (*see* Tracks)
Formations: angular unconformities, 62; bedding surfaces (planes), 59–60; columnar section, 129, *129;* conglomerate, 67–68; contacts, 62–63; correlation of, 134, 148; cross sections, 129, *130;* definition, 127; economic importance, 146–48; exposures, 63; and facies, 129; folding, 79–80; geologic, 127–29, *130;* in Grand Canyon, *128;* interpretation of, 127–29; mapping, 128–29; maps of, 132–33; meaning, 68–70; naming, 127–29; and paleogeographic maps, 133; and paleontologic zone, 131–32, *132;* law of superposition, 70–72; strata, 59; study of, *59;* and time-rock units, 131; unconformities, 61–62
Fossilization, 86–91, 487
Fossil record: beginning of, 225; criticisms of, 426; and Darwinism, 425–26
Fossils (*see also* Species; Paleontology; and zoological groups): anatomy in identification, 105–7; assemblages, 143; biometrics, 99; biotas, 102; burrows, 91; carbonization of, 90; casts, 87–88; classification, 110–116; collection of, *144;* collections, *103;* colors of, 90; conditions creating, 85–91; coprolites, 92; correlation by, 104–141; deep-sea, *146,* 146; definition, 84, 102, 140, 143; early interpretations of, 95–97; and evolution, 109–10; faunal succession, 93; faunas, 102; floras, 102; frozen, 90–91; guide (index), 142–43, *143;* homotaxis of, 135; imperfections, 90; impressions, 90; methods of formation, 87–91; molds, 87, *87;* molecular, 225; oldest, 225, 226; and organic evolution, 109–10; origin, 85–87; permineralized specimens, 88;

Fossils (*cont.*):
petrifaction, 88–90; petrifying agents, 90; phylogenies, 108–9; Precambrian, 225–32; preparation, 95; principle of faunal succession, 102–4; pseudofossils, 92–93, *92;* racial histories, 487–512; reconstruction of, 104–7; replacement, 88; study of, 93–95; trace fossils, 91; tracks, 91; trails, 91; zones, 131–32, *132*
Fragments: derived (included), 74, *74*
Franciscan Chert, *80*
Frogs, 323
Frozen animals, as fossils, 90–91
Fungi, 112
Fusulinids, 283, 286–87, *286,* 477

Galapagos Islands, 455–57
Galaxy, 151–55
Gamma rays, 22
Gamow, George, 152
Gastroliths (stomach stones), 92
Gastropoda, 114
Gastropods (*see also* Invertebrates): late Paleozoic 284; Tertiary, 353, *353*
Genera, 111
Genes (*see* Genetic change; RNA)
Genetic change, 464–66
Geologic eras: source of names, 121
Geologic maps, 132–33
Geologic processes: diastrophism, 38–43; erosion, 43–45; sedimentation, 45–47; volcanic action, 47–50
Geologic time scale, 122–23
Geology: applications, 8, 11; contributions, 8; and dispersal of organisms, 442–66; and evolution, 428–39; and "fossil controversy," 95–96; historical, 1, 6, 120; methods, 6–7; in oil exploration, 148
Geophysics, 138, 169–75
Geosynclines: Appalachian, 194; characteristics, 193–96; concept of, 193–96; evolution of, 193–96; Franklyn, 245; in mountain building, 195; Rocky Mountain, 244–45; Tasmanian, 245; Tethys, 241, 264, 298; types, 193; Uralian, 241–42
Gibraltar, Rock of, *300*
Ginkgo, 279, *351*
Glaciation (*see also* Pleistocene Epoch, Tillite): in Alps, 365–67, *366;* continental, 367–70; effects, 370–82; glacial cobbles, *371;* High Sierras, *312;* in Paleozoic, 373, 376; relation to prehistoric man, *400;* in South America, 373; stages, 400; theories of, 388–90; tillite, 373; topography produced by, 365–66; varves, 18–20

Glaciers (*see also* Glaciation), 70, *365,* 366
Global tectonics (*see* Plate tectonics)
Globigerina, *380*
Glossopteris flora, *281,* 281–82
Going-to-the-Sun Mountain, Montana, *191*
Gorgosaurus (*see also* Dinosaurs), *7, 503*
Gorilla: compared to man, *396*
Gondwana, 176, *264,* 264
Gneiss, *196, 203*
Grand Canyon: formations of, *59, 128, 129;* lava flows in, *82;* Precambrian, *200;* role in species formation, 450, *450;* strata, *59, 128*
Grand Teton Range, Wyoming, *337*
Granite, 175, 312
Graphaea, *318*
Graptolites: *Cryptograptus, 481;* extinction, 479; facies, 244; life habits, 258; Ordovician 252, *252;* Silurian, *481*
Graptolithina, 115
Grass, 350–52
Gravel, *65,* 74, *74*
Great Barrier Reef, 85
Great Britain: in Mesozoic, 301; migrations to, 459–60; in early Paleozoic, 238; in late Paleozoic, 265; Paleozoic mountain building in, 241; as part of continent, 459–60; in Pleistocene, 459–60; time scale, 125–26
Great Sphinx, 5
Great Spiral Galaxy, *154*
Greenland: basalt, 306; glaciation in, 367, 371–72, 381; ice cap, 382; in Mesozoic, 306; oldest rocks, 197; rock age, 197
Green River Formation, *344*
Grenville Orogeny, 198
Ground sloths, *387*
Growth rings, 16–18, *16, 17*
Guadalupe Mountains, 271, *274*
Guiana Shield, 200
Guide fossils: early Paleozoic, 250; late Paleozoic, 285–87; Mesozoic, 319; Tertiary, 353–54; use in correlation, 142, *143*
Gulf of California, 183
Gulf of Mexico, 378
Gulf Stream, 390
Gunflint Iron Formation: fossils in, *211,* 227, *227*
Guitarrero Cave, 419
Guyots, 307–8, *308*
Gymnospermae, 112
Gymnosperms: late Paleozoic, 279; Mesozoic, 313–14
Gypsum Cave, Nevada, 417

Half life, 23, *24*
Halite (*see* Salt deposits)
d'Halloy, J. J. d'Omalius, 125

Halysites, *255,* 283
Harding Sandstone, 256
Hawaiian Islands, 49, 457–58
Heat: as condition of life, 213, 215; of earth's interior, 174; in formation of the earth, 171; range in universe, 215
Heidelberg man, 411
Helium, 152–53, 155–56
Herculaneum, 48
Hercynian Revolution (Orogeny), 264, 265–66, 294
Hercynian interval, 236
Hermit Shale, 277
Herodotus, 50
Hesperornis, 331–32
Hexactinellida, 113
High Plains, 347
Himalayas, 300, 367
Homestake Mine, South Dakota, *208*
Historical geologist, 6–10
Historical geology: assertions of, 8; definition of, 3; description of, 6–8; methods of, 7–8; scope of, 120
History: as science, 6
Holocene Epoch, 126
Holothuroidea, 115
Hominidae, 395
Hominoidea, 395
Homo erectus (Pithecanthropus and Java ape-man): in Africa, 410; appearance, *409;* characteristics, 407–11, *409;* in China, 409–10; in Formosa, 411, in Java, 407–9; in Spain, 410
Homo habilis, 406
Homo neanderthalensis (Neanderthal man), 412–14
Homo sapiens, 414–19
Homo sapiens sapiens, 395, 414–20
Homotaxis: of fossils, 135
Hoover Dam, 47
Hornea, 277
Horse, 508–12
Horseshoe Falls (Niagara), 44
Horsetail, 279
Hot spring, 219
Hoyle, Fred, 152
Hudson Bay, 372, *372*
Hudsonian Orogeny, 198
Hudson River, 303, 376
Humbolt, Alexander von, 125
Hutton, James, 51, 52–53, *53*
Hydrocarbons, 217
Hydrogen, 152–53, 155, 188
Hydrosphere (*see* Oceans; Water)
Hydrozoa, 113
Hypothesis, 4
Hyracotherium (*see also* Horse; Eohippus), 356

Icarosaurus, *330,* 330
Ice ages (*see also* Glaciation, Pleistocene Epoch): and continental islands, 459–61;

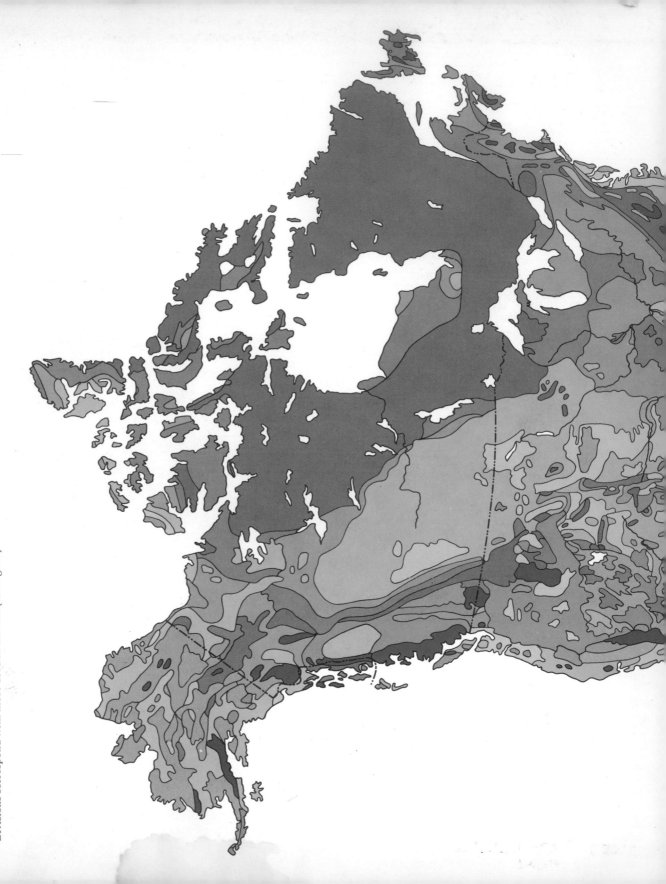

GEOLOGIC MAP OF NORTH AMERICA

Divisions correspond with those used in the text (see legend)